T0297231

Oxford Texts in Applied and Engineering Mathematics

OXFORD TEXTS IN APPLIED AND
ENGINEERING MATHEMATICS

Titles marked with an asterisk (*) appeared in the Oxford Applied Mathematics and Computing Science Series, which has been folded into, and is continued by, the current series.

L. RAMDAS RAM-MOHAN

Worcester Polytechnic Institute
Worcester, Massachusetts

Finite Element
and Boundary Element Applications
in Quantum Mechanics

OXFORD

UNIVERSITY PRESS

*This book has been printed digitally and produced in a standard specification
in order to ensure its continuing availability*

OXFORD
UNIVERSITY PRESS

Great Clarendon Street, Oxford OX2 6DP

Oxford University Press is a department of the University of Oxford.
It furthers the University's objective of excellence in research, scholarship,
and education by publishing worldwide in

Oxford New York

Auckland Cape Town Dar es Salaam Hong Kong Karachi
Kuala Lumpur Madrid Melbourne Mexico City Nairobi
New Delhi Shanghai Taipei Toronto
With offices in
Argentina Austria Brazil Chile Czech Republic France Greece
Guatemala Hungary Italy Japan South Korea Poland Portugal
Singapore Switzerland Thailand Turkey Ukraine Vietnam

Oxford is a registered trade mark of Oxford University Press
in the UK and in certain other countries

Published in the United States
by Oxford University Press Inc., New York

ISBN 978-0-19-852522-6

To

Sita, Arun and Sumati

and

my family

Preface

The interplay between physics, mathematics, and computational issues is particularly well represented in the application of the finite
element method (FEM) and the boundary element method (BEM)
to physical problems. This book provides a physical introduction to
the FEM through the discretization of the action integral for the solution of Schrödinger's equation. Being a variational approach, the
FEM provides the means for systematically improving accuracy in
calculations in a natural manner. Throughout this book the action
integral and variational principles are shown to provide the means
for adaptively improving results in the FEM. Both the FEM and the
BEM allow one to transcend geometrical constraints associated with
traditional treatments based only on analytically solvable problems.

The FEM applied to quantum heterostructures has allowed the
development of the novel paradigm of *wavefunction engineering*. This
has permitted the optimized design of heterostructures for optoelectronic applications – the design of the quantum interband cascade
laser is described in detail. Nonseparable potentials do not pose any
difficulty in the FEM, as exemplified by calculations for the energy
levels and wavefunctions for the hydrogen atom in extremely large
magnetic fields. The FEM can be used for the solution of nonlinear Schrödinger equations such as the Ginzburg–Landau equation
in superconductivity. The freedom to employ the Dirichlet, Neumann, or Cauchy types of boundary conditions (BCs) in the FEM
is shown to be an important advantage. A detailed treatment of the
Schrödinger–Poisson self-consistency problem in heterostructures is
given, and also a chapter is devoted to calculating electronic energy
levels in planar heterostructures with an in-plane magnetic field, the
so-called Voigt geometry.

In Chapter 8, I have presented the concept of wavefunction engineering of quantum semiconductor devices that emerges through the
application of the FEM to band structure calculations.

Absorbing elements, which I have labeled as "stealth" elements,
permit the modeling of scattering in open regions while discretizing

finite regions only around the scattering center. Local details of "near-field" wavefunctions and evanescent modes in the scattering region are obtainable using these methods. The study of the time evolution of systems described by Schrödinger's equation is steadily coming to the fore, and the computational issues in this area are discussed.

The reader may find a somewhat greater emphasis on the treatment of quantum semiconductor structures. I found it useful to employ examples from this area in order to provide a framework for the discussion of complex BCs in realistic, practical applications; other applications in molecular, nuclear, and gravitational physics, as well as other subdisciplines of physics, could also be envisaged, and these too can be very fruitful areas that can be explored using the FEM.

The action integral is the integral over space and time of the Lagrangian density. In the time-independent case, I have taken the liberty of calling just the spatial integral over the Lagrangian density the "action." This allows me to refer to variational principles as the principle of stationary action throughout the book.

The Green's function approach to the solution of differential equations leads to the boundary integral method. It is shown in considerable detail that the discretization of this boundary integral leads to the BEM. The method is applied to the calculation of electric field enhancements at metallic surfaces due to surface plasmon excitations, to quantum mechanical scattering, and to quantum waveguides. Part V of this book, containing the three chapters on the BEM, together with the appendices of Part VI, constitute a complete introduction to this powerful method of computation with its own special strengths that I have tried to bring out in the presentation.

A chapter introducing some of the recent methods in the solution of large matrix equations and in matrix diagonalization of large matrices is included. The treatment presented here is a physicist's approach to these issues. The appendices on the theory of the Gauss quadrature method, generalized functions, and the Green's function method for solving inhomogeneous equations are self-contained introductions to these topics, and are appropriate supporting material for the text.

The material presented in this book is aimed at senior undergraduates, graduate students, and those interested in computational methods and their applications to physical problems. The wide range of applications of the FEM and BEM with their concrete examples presented here should be of interest to researchers specializing in many areas of physics. This is not a book with cook-book recipes for computer programs – other resources exist for that purpose; it is directed towards giving the physical basis for the two computational approaches presented here together with physical examples.

I have successfully engaged undergraduate students, including promising Freshmen, in exposing them to variational methods and the principle of least action, and the Green's function method using chapters from this book. After much agonizing, I decided to retain the "atomic" system of units that is still widely used in quantum mechanics.

I wish to thank my colleagues who have influenced the contents of this book in various ways: Adriaan Walther read through the manuscript and his comments have enhanced the clarity of the presentation. A. K. Rajagopal provided comments on the material presented here, especially on the chapter on time evolution in quantum mechanics. The chapters on the BEM were looked over thoroughly by John Albrecht and Peter Knipp. The calculations on surface plasmons stem from work done with Peter Wolff. I wish to acknowledge a long and fruitful collaboration with Raymond Goloskie, who passed away in 1997, on the theoretical and numerical issues on the BEM presented in this book. Interactions with collaborators including Don Dossa, Jacek Furdyna, Hong Luo, Jerry Meyer, Anant Ramdas, Sergio Rodriguez, Janine Shertzer, Igor Vurgaftman, Peter Wolff, and with students listed in my articles referenced in the following have enriched this book substantially. The final touches to the manuscript were completed while I was at the Air Force Research Laboratory (AFRL) at the Wright–Patterson Air Force Base, and I wish to thank the AFRL for its support and hospitality. This manuscript has taken shape over the past eight years. It could not have been completed without the active support and help of my family, and I thank them.

L. R. Ram-Mohan
lrram@wpi.edu

Contents

Contents

Contents

PART VI   APPENDICES

Part I

Introduction to the FEM

1

Introduction

The dramatic advances in computational methods over the past several decades have provided a remarkable realization of approximate numerical solutions to physical problems which have been considered to be analytically intractable. In particular, the finite element method (FEM) represents a stratagem of proven success for solving boundary value problems in quantum mechanics. This method originates in the principle of stationary action and other variational principles, which are central unifying principles in all areas of physics. In fact, essentially all the differential equations of interest in physics can be derived by the calculus of variations from a corresponding action integral. With its basis in variational principles, the FEM can be expected to hold an increasing appeal among physicists. This book provides an introduction to the FEM through applications in quantum mechanics.

In parallel with the variational method we consider the Galerkin method of weighted residuals for the solution of differential equations. The two formulations are related to each other when the differential equation is derivable from a variational principle. It should be noted that the Galerkin method is applicable even if the latter does not hold. We introduce the Galerkin method in Chapter 5.

Problems in potential theory and in scattering theory are often cast into the form of an integral equation by the use of Green's functions and the well-known Green's theorem of vector analysis. The integral equation is a self-consistency condition that has to be satisfied by the solution. The volume integrals in n dimensions ($n = 2$ or 3) can be reduced to surface integrals in $n - 1$ dimensions. This reduction in the dimensionality of the problem is a major factor in making the boundary element method (BEM) an attractive alternative to the FEM. The BEM is presented in Part V of this book.

This chapter is devoted to three seemingly disparate themes. We provide an elementary introduction to quantum mechanics and summarize its postulates. This is merely a brief introduction – the reader

is referred to standard texts on quantum mechanics which provide
a full treatment of the subject. Next, we introduce the variational
method and show that the principle of stationary action applied to
the Lagrangian for a given system allows us to derive the differential
equation obeyed by the physical observable in the problem. This is
illustrated by constructing the Lagrangians from which we derive the
Poisson equation and Schrödinger's equation. Thirdly, a short introduction is given to the finite element approach for the evaluation of
the spatial dependence of the action integral. This is done in terms
of as-yet-unknown parameters which are the field amplitudes at discrete points. The principle of stationary action is now implemented
through the variations of the parameters. In traditional variational
methods the trial functions are defined globally over the entire domain, thereby drastically reducing their flexibility. Instead, in the
FEM the domain is divided into small regions, called *elements*, and
a local polynomial representation is used for the fields within each
element.

The three topics are brought together in the following chapters
through the application of the FEM to quantum mechanical problems. We conclude this chapter with a brief historical note on the
development of the FEM.[†]

1.1 Basic concepts of quantum mechanics

1.1.1 Schrödinger's equation

By the beginning of the twentieth century, it became obvious that
a number of physical observations in the emission of radiation from
objects in thermal equilibrium, in atomic spectroscopy, and in the
specific heats of solids, could not be understood in terms of classical electromagnetic theory or mechanics. In order to explain the
novel observations it was necessary to introduce new rules for the
behavior of particles and of radiation. These new rules are essential in providing the correct description of physical processes in the
sub-microscopic, atomic domain. They are also consistent with the
classical theory as the scale of energy in a physical system becomes
macroscopic.[1,2] The quantum theory of radiation and matter has
successfully explained all the basic issues of atomic and molecular

---

[†]The reader will find it useful to go over the first three chapters and then
revisit this chapter to develop an overview of the material presented.

structure.[‡] Radiation of energy is *postulated* to occur in quanta of energy $E = h\nu = \hbar\omega$. Here $\omega = 2\pi\nu$ is the angular frequency of the radiation, ν being the frequency, and h is called Planck's constant, with $\hbar = h/2\pi$. The value of $\hbar$ is

$$\hbar = 1.0546 \times 10^{-34} \, \text{J s} = 6.582 \times 10^{-16} \, \text{eV s}. \tag{1.1}$$

The quantum of radiation is called a photon, and its momentum is defined by $p = E/c$, where $c = 2.9979 \times 10^8$ m/s is the velocity of light. Equivalently, we have $p = h\nu/c = h/\lambda$, where λ is its wavelength. The quantization of light radiation leads to a natural explanation of the black-body radiation throughout its entire spectral range. The photo-electric effect, according to Einstein, is understood in terms of the energy needed by the electron to escape its binding to the metal. This energy is called the electron's work-function ϕ. The photon must deliver this energy $h\nu = \phi$ before the electron can exit the metal.

It was proposed by de Broglie that all particles are quanta and have the same wave–particle duality as that displayed by radiation. Depending on the measurement process, we can observe a quantum's wave-like or particle-like behavior. An electron of energy $E = p^2/2m$ has a momentum p; it would therefore have a wavelength, the de Broglie wavelength, of $\lambda = h/p$.

The quantum hypothesis of radiation provided a clear appreciation of the light scattering process called the Compton effect in which radiation is scattered by electrons. The experimental angular distributions of scattered electrons and photons are obtained when the process is treated as a collision between point particles which obeys the conservation of energy and of momentum as in point-particle collisions.

The fact that a particle has wave-like properties suggests that we describe a particle by a *wavefunction* $\psi(\mathbf{r}, t)$ which represents the probability amplitude, with $|\psi(\mathbf{r}, t)|^2$ giving the probability density of observing the particle in a volume element d^3r located at $\mathbf{r}$. The probability of observing the particle somewhere in all of space is unity:

$$\int |\psi(\mathbf{r}, t)|^2 \, d^3r = 1. \tag{1.2}$$

[‡]This understanding in terms of quantum mechanical considerations now extends to relativistic quantum electrodynamics and also to condensed matter theory through the development of quantum statistical mechanics.

The wavefunction $\psi(\mathbf{r}, t)$ is complex in general. We can "derive" Schrödinger's wave equation for a free particle by considering the Fourier expansion of the probability amplitude. We have[§]

$$\psi(\mathbf{r}, t) = \int_{-\infty}^{\infty} \frac{d^3k\, dw}{(2\pi)^4} A(\mathbf{k}, w) e^{i\mathbf{k}\cdot\mathbf{r} - i\omega t}, \qquad (1.3)$$

where $A(\mathbf{k}, w)$ is the Fourier transform of $\psi(\mathbf{r}, t)$. Let us construct the quantity

$$\Lambda = \left[i\hbar \frac{\partial}{\partial t} \psi(\mathbf{r}, t) + \frac{\hbar^2}{2m} \nabla^2 \psi(\mathbf{r}, t) \right],$$

using equation (1.3). Taking the derivatives under the integral sign we obtain

$$\Lambda = \int_{-\infty}^{\infty} \frac{d^3k\, dw}{(2\pi)^4} A(\mathbf{k}, w) \left(\hbar\omega - \frac{\hbar^2 k^2}{2m} \right) e^{i\mathbf{k}\cdot\mathbf{r} - i\omega t}. \qquad (1.4)$$

Letting $E = \hbar\omega$ and $p = \hbar k$ inside the integral, we see that for a free particle the quantity in parentheses in the integrand is $E - p^2/2m = 0$. This implies that $\Lambda = 0$, and hence

$$-\frac{\hbar^2}{2m} \nabla^2 \psi(\mathbf{r}, t) = i\hbar \frac{\partial}{\partial t} \psi(\mathbf{r}, t). \qquad (1.5)$$

This is Schrödinger's equation for a free particle for which we have provided a heuristic derivation here. The reader is referred to the texts by Schiff,[3] Landau and Lifshitz,[4] and Shankar[5] for a complete development of these concepts. A very readable introduction to the early developments in quantum theory is given by Born,[6] and a thorough presentation of the historical developments is given by Jammer.[7]

The above considerations give the recipe for writing Schrödinger's equation for any quantum mechanical system. First, obtain an expression for the total energy of the system. For example, a particle

---

[§]The integral sign in the following equation stands for four integrals as understood from the presence of $d^3k\, d\omega$ in the integrand.

moving in a potential well and having kinetic energy $p^2/2m$ and potential energy $V(\mathbf{r})$ has the total energy

$$\frac{p^2}{2m} + V(\mathbf{r}) = E.$$

The left side of the above equation is the sum of the kinetic energy $T = p^2/2m$ and the potential energy V and is called the Hamiltonian, $\mathcal{H}$. Now substitute

$$E \rightarrow i\hbar \frac{\partial}{\partial t}, \qquad \mathbf{p} \rightarrow -i\hbar \nabla. \qquad (1.6)$$

Then $p^2/2m + V(\mathbf{r}) = E$ is viewed as an *operator* equation acting on $\psi(\mathbf{r}, t)$. Schrödinger's equation for this system is

$$-\frac{\hbar^2}{2m} \nabla^2 \psi(\mathbf{r}, t) + V(\mathbf{r})\psi(\mathbf{r}, t) = i\hbar \frac{\partial}{\partial t}\psi(\mathbf{r}, t)$$

$$\hat{\mathcal{H}}\psi(\mathbf{r}, t) = i\hbar \frac{\partial}{\partial t}\psi(\mathbf{r}, t), \qquad (1.7)$$

with the left hand side being the Hamiltonian operator acting on ψ. We use the method of separation of variables and write

$$\psi(\mathbf{r}, t) = \psi(\mathbf{r})\, e^{-i\omega t}, \qquad (1.8)$$

to obtain the time-independent Schrödinger's equation

$$-\frac{\hbar^2}{2m} \nabla^2 \psi(\mathbf{r}) + V(\mathbf{r})\, \psi(\mathbf{r}) = \hbar\omega\, \psi(\mathbf{r}, t) = E\, \psi(\mathbf{r}). \qquad (1.9)$$

This is an eigenvalue problem with eigenvalues $E = E_\lambda$, and the corresponding eigenfunctions are $\psi_\lambda(\mathbf{r})$. The spectrum of eigenvalues E_λ will be discrete for bound states and continuous for unbound states.

Given a spectrum of eigenvalues and eigenfunctions, a general solution to Schrödinger's equation can be constructed by a linear superposition of the complete set of eigenfunctions. The set of eigenfunctions defines a function space with the "coordinate axes" being the eigenfunctions, and the general solution may be thought of as a vector, the state vector, in this space.

1.1.2 Postulates of quantum mechanics

The general solution of Schrödinger's equation is represented as a linear superposition in the above manner in a complex vector space, called a Hilbert space.¶ This can be stated formally by associating a "normalizable" state vector with each state of the quantum mechanical system. We use Dirac's notation[8] to denote a state vector, called a *ket*, in a complex vector space spanned by $|\alpha_\lambda\rangle$. Its complex conjugate vector is called a *bra* and is denoted by $\langle\alpha_\lambda|$. The α_λ are labels for the observable properties, such as the energy, the position, etc., of the state. These state vectors satisfy the following postulates

Postulate 1: The general state vector is given by the superposition

$$|\alpha\rangle = \sum_\lambda c_\lambda |\alpha_\lambda\rangle.$$

For any two vectors $|\alpha\rangle$ and $|\beta\rangle$ we can define a complex number c called the scalar product of a *bra* with a *ket*, given by

$$\langle\alpha|\beta\rangle = c = (\langle\beta|\alpha\rangle)^*. \tag{1.10}$$

The quantity $\langle\alpha|\alpha\rangle$ is normalized to unity, as in equation (1.2).

Postulate 2: Every observable property A, such as the particle's position, energy, charge, angular momentum, and so on, corresponds to a linear Hermitian operator $\mathbf{A}$ with eigenvalue a

$$\mathbf{A}|\alpha\rangle = a\,|\alpha\rangle. \tag{1.11}$$

A general operator $\mathcal{O}$ is given a matrix representation in the space of $|\alpha\rangle$ through the quantity

$$\mathcal{O}_{\lambda\mu} = \langle\alpha_\lambda|\mathcal{O}|\alpha_\mu\rangle.$$

Every observable is assumed to have a complete set of eigenfunctions which spans a subspace of the Hilbert space of state vectors. We can make use of these eigenfunctions as a basis in that subspace to form a representation or a "coordinate" system in this subspace.

¶Most books on quantum mechanics imply that the Hilbert space is denumerably infinite with a complete basis of square integrable functions. However, the Hilbert space of the spin wavefunctions of a particle with spin j is of dimensionality $(2j + 1)$ and is finite. Also, for a brief discussion on normalizability, see Ref. 5.

For example, the representation in which the position operator **x** is diagonal is called the position representation. The state vector $|\psi\rangle$ is specified in configuration (position) space through its components $\langle x|\psi\rangle$, which is Schrödinger's wavefunction $\psi(x)$ encountered earlier. Again, the physical meaning of the quantity $\langle x|\psi\rangle$ is the following. The probability that a measurement made to determine the position of the particle with state vector $|\psi\rangle$ will be between x and $x + dx$ is given by $|\langle x|\psi\rangle|^2 \, dx$. The eigenfunctions $|x\rangle$ satisfy the relations

$$\mathbf{x}\,|x\rangle = x\,|x\rangle, \tag{1.12}$$

and the spectrum of eigenvalues x corresponds to the points along the x-axis in configuration space. The eigenfunctions further satisfy the normalization and completeness relations

$$\langle x|x'\rangle = \delta(x - x'),$$
$$\int dx'\,|x'\rangle\langle x'| = 1. \tag{1.13}$$

Here $\delta(x)$ is the Dirac δ-function.[‖] In an analogous manner, the momentum representation $|p\rangle$, with an eigenspectrum p, is defined by the relations

$$\mathbf{p}\,|p\rangle = p\,|p\rangle,$$
$$\langle p|p'\rangle = \delta(p - p'),$$
$$\int dp'\,|p'\rangle\langle p'| = 1. \tag{1.14}$$

The state vector $|\psi\rangle$ can be projected into the momentum representation as $\langle p|\psi\rangle \equiv \psi(p)$. As an illustration, the Fourier representation of $\psi(x)$ in momentum space

$$\psi(x) = \frac{1}{2\pi\hbar} \int dp' \, e^{ip'x/\hbar} \, \phi(p')$$

can be written as

$$\psi(x) = \langle x|\psi\rangle$$
$$= \int dp'\,\langle x|p'\rangle\langle p'|\psi\rangle. \tag{1.15}$$

A particle localized in a given region of space has a wavefunction which is a superposition of many plane waves. The physical properties of the particle are represented in the wavefunction. The spread

‖ The Dirac δ-function is treated in detail in Appendix B.

in the position or the uncertainty in x is obtained by evaluating the expectation value of the quantity $\Delta\hat{\mathbf{x}} = \hat{\mathbf{x}} - \langle\hat{\mathbf{x}}\rangle$ defined in terms of the expectation values

$$\langle\psi\,|(\Delta x)^2\,|\psi\rangle = \langle\psi|\,\hat{\mathbf{x}}^2\,|\psi\rangle - (\langle\psi|\,\hat{\mathbf{x}}\,|\psi\rangle)^2. \tag{1.16}$$

We now turn to the commutation properties of operators. Two operators **A** and **B** are said to commute if

$$\mathbf{AB} - \mathbf{BA} \equiv [\mathbf{A},\mathbf{B}] = 0. \tag{1.17}$$

The state vector can be a simultaneous eigenvector of two commuting operators and hence can be labeled by their eigenvalues.

Postulate 3: The position **x** and the canonically conjugate momentum **p** of a particle obey the commutation relation

$$[\mathbf{x},\mathbf{p}] = i\hbar. \tag{1.18}$$

The Heisenberg uncertainty relation: An uncertainty relation of the form

$$\Delta x\,\Delta p \geq \hbar/2 \tag{1.19}$$

holds for all canonically conjugate pairs of physical variables such as x and p. Thus the position and the momentum of a particle cannot be determined exactly in a simultaneous manner. This fundamental aspect of quantum mechanics has profound consequences in all arenas of quantum physics. We note that a similar relation holds for energy and time; however, this is due the time evolution of the state vector which ceases to have its original form after a time interval of the order $\hbar/\Delta E$.[9] A finite lifetime for any quantum mechanical state implies an uncertainty or linewidth for the energy of the state. For further details, the reader is referred to the books on quantum mechanics by Dirac[8] and Sakurai.[9]

In the following, we will be concerned with obtaining solutions for Schrödinger's equation for the coordinate representation of the wavefunction $\psi(x)$.

1.2 Principle of stationary action

Physical quantities, such as the temperature distribution in a thermally conducting system heated at one end, the electrostatic potential in a region occupied by charges, Newton's laws of motion for a particle experiencing forces on it, or the wavefunction of a quantum mechanical particle, *all obey differential equations which can be derived from a principle of stationary action.*[12,10,11] The principle of stationary action represents a truly beautiful synthesis of all the dynamical principles that have been discovered in the physical sciences. Thus, the principle of shortest optical path in the theory of lenses, Hamilton's principle in classical mechanics, the principle of energy minimization, and Schwinger's variational principle[13,14] for quantum mechanics are all manifestations of this single principle, and from this one remarkable principle one can derive the equations of motion governing the physical quantities describing the dynamics of the system. A mathematical framework for the application of this principle was provided in the calculus of variations, developed by Euler and Lagrange.[15,16]

1.2.1   The action integral

Point particles

In classical mechanics,[17] consider the motion of a single particle with kinetic energy $T = m\dot{q}(t)^2/2$ in a region where its potential energy is $V(q)$. Here $q(t)$ is the coordinate position of the particle, and $\dot{q}(t)$ is its time derivative, $dq(t)/dt$. We begin by defining a quantity $L = T - V$, called the Lagrangian. The principle of stationary action states that for a conservative system the integral

$$\mathcal{A} = \int_{t_a}^{t_b} L(q, \dot{q}, t)\, dt \tag{1.20}$$

is stationary for the physically realizable, i.e., the actual, motion of the particle. The quantity $\mathcal{A}$ is called the action integral, or simply the action. Here, the function L should be twice differentiable with respect to $q, \dot{q}$ and t. The second derivative $\ddot{q}$ is also assumed to be continuous.

The variational problem is stated as follows. We wish to determine that function $q_0(t)$, satisfying the boundary conditions (BCs), for which the action $\mathcal{A}$ takes on an extremum value. The extremum of $\mathcal{A}$ has to be determined by varying the function $q(t)$. The quantity $\mathcal{A}$ is called a functional of q and is denoted by $\mathcal{A}[q]$, with q in square brackets rather than in parentheses. The distinction between a function and a functional is that a function $q(t)$ takes on a value (a number) for each value of t, assuming that q is a single-valued

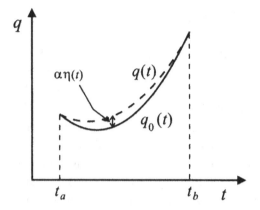

Fig. 1.1 The function $q_0(t)$ (solid curve) is one for which the functional $A[q]$ is an extremum; a neighboring curve $q(t) = q_0(t) + \alpha\eta(t)$ is shown (dashed curve) to illustrate the variation of $q(t)$ over the entire range of t.

function, while a functional A depends on q not at just one value but for *all* values of t.

We can reduce the functional variation to a standard problem in calculus for determining the extremum of a function. We suppose that $q(t) = q_0(t) + \alpha\eta(t)$, where α is a small parameter which can be varied, and $\eta(t)$ is an arbitrary continuous function having a continuous derivative and is such that $\eta(t_a) = 0 = \eta(t_b)$. An example of $q_0(t)$ and $q(t)$ is shown in Fig. 1.1. Writing out the first few terms in the Taylor series for A treated as a function of the parameter α we have

$$A(\alpha) = \int_{t_a}^{t_b} dt\, L(q_0(t) + \alpha\eta(t), \dot{q}_0(t) + \alpha\dot{\eta}(t), t)$$

$$= A(0) + \alpha \int_{t_a}^{t_b} \left(\frac{\partial L}{\partial q}\eta + \frac{\partial L}{\partial \dot{q}}\dot{\eta} \right) dt + \mathcal{O}(\alpha^2).$$

Existence of an extremum at $\alpha = 0$ requires

$$\left. \frac{dA(\alpha)}{d\alpha} \right|_{\alpha=0} = 0,$$

so that

$$\int_{t_a}^{t_b} \left(\frac{\partial L}{\partial q}\eta + \frac{\partial L}{\partial \dot{q}}\dot{\eta} \right) dt = 0, \tag{1.21}$$

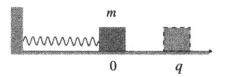

Fig. 1.2 The simple harmonic oscillator in its position of equilibrium at $q = 0$. The displaced mass at q, shown with a dashed outline, is still attached to the stretched spring. The elastic restoring force of the spring on the mass displaced by a distance q is $F = -kq$, where k is the spring constant.

for any $\eta(t)$ we may select. Integrating by parts the second term in equation (1.21), we have

$$\eta \left. \frac{\partial L}{\partial \dot{q}} \right|_{t_a}^{t_b} + \int_{t_a}^{t_b} \left(\frac{\partial L}{\partial q} - \frac{d}{dt} \frac{\partial L}{\partial \dot{q}} \right) \eta(t) \, dt = 0.$$

The first term vanishes since η is zero at the two end points, and for the integral to vanish for arbitrary $\eta(t)$ we must have

$$\frac{\partial L}{\partial q} - \frac{d}{dt} \left(\frac{\partial L}{\partial \dot{q}} \right) = 0. \tag{1.22}$$

Equation (1.22) is called the Euler–Lagrange equation. Together with the BCs $q(t_a) = q_a$ and $q(t_b) = q_b$, it is equivalent to the variational statement of the problem. Performing the derivatives with respect to $q(t)$ and $\dot{q}(t)$ on L, we obtain the equation of motion $-\partial V / \partial q = m\ddot{q}$. This is simply Newton's second law stating that the force $F = -\partial V / \partial q$ acting on the particle equals its mass m times its acceleration $\ddot{q}$.

As an example, consider the simple harmonic oscillator (shown in Fig. 1.2), which is a particle of mass m attached to a spring with a spring constant k, whose opposite end is held fixed. Hooke's law states that the particle experiences a restoring force $F = -kq$ when it is displaced from its equilibrium position. The particle has a potential energy $kq^2/2$ and the corresponding Lagrangian is $L = T - V = m\dot{q}^2/2 - kq^2/2$. The equation of motion for the oscillator obtained using equation (1.22) is then $m\ddot{q} + kq = 0$.

The Lagrangian approach is more general and has several advantages over directly setting up Newton's laws of motion. Newton's method is more natural for point particles and requires the identification of forces and their components in a specially chosen coordinate system. This may not be trivial for a number of extended objects in contact with each other, or for mechanical systems having forces arising through electromagnetic or other interactions. The Lagrangian method begins by defining readily identifiable scalar quantities, the kinetic and potential energies, which can be expressed in generalized coordinates. We can also derive general theorems, such as conservation laws, by using the Lagrangian approach.[10,18] The method is readily extended from particle dynamics to the Lagrangians for continuous fields.[17,19,20]

We can simplify the above variational calculus by employing the rules of functional differentiation so that the Euler–Lagrange equation may be derived without the device of introducing a parameter α as was done above. The functional $\mathcal{A}[q]$, as mentioned earlier, depends not just on one value of $q(t)$ at a point t but on its value over an entire range of the parameter t. On varying q, the variation of $\mathcal{A}[q]$ is defined as (we retain only the first-order term)

$$\delta \mathcal{A}[q] = \mathcal{A}[q + \delta q] - \mathcal{A}[q] \equiv \int dt' \, \frac{\delta \mathcal{A}[q]}{\delta q(t')} \, \delta q(t').$$

Two results useful in functional differentiation are the following:
(i) From the relation

$$q(t) = \int \delta(t - t') \, q(t') \, dt', \tag{1.23}$$

where $\delta(t - t')$ is the Dirac δ-function,** we have

$$\frac{\delta q(t)}{\delta q(t')} = \delta(t - t'). \tag{1.24}$$

(ii) Suppose $\mathcal{A}$ is a functional of a functional $L[q(t)]$. Its functional derivative is

$$\frac{\delta}{\delta q(t)} \mathcal{A}[L[q]] = \int \frac{\delta \mathcal{A}}{\delta L[q(t')]} \frac{\delta L[q(t')]}{\delta q(t)} \, dt'. \tag{1.25}$$

These rules of functional differentiation lead to the lowest order differential

$$\delta \mathcal{A}[q] = \int dt \, \delta L = \int \left[\frac{\partial L}{\partial q(t)} \delta q(t) + \frac{\partial L}{\partial \dot{q}(t)} \delta \dot{q}(t) \right] dt$$

$$= \int \left[\frac{\partial L}{\partial q(t)} - \frac{\partial}{\partial t} \frac{\partial L}{\partial \dot{q}(t)} \right] \delta q(t) \, dt. \tag{1.26}$$

**See Appendix B.

The functional differentiation is recognized as the differentiation of $L[q, \dot{q}, t]$ with respect to the function q. An integration by parts has been carried out, followed by dropping the surface terms since $q(t)$ is fixed and hence $\delta q(t)$ is zero at the end points. Since we seek the extremum of the functional $\mathcal{A}[q]$ we require that $\delta \mathcal{A} = 0$. For arbitrary variations in $q(t)$ we have the desired Euler–Lagrange equation

$$\frac{\partial L}{\partial q(t)} - \frac{d}{dt}\frac{\partial L}{\partial \dot{q}(t)} = 0.$$

Continuous fields

The same considerations carry over to the case of continuous fields. Consider a scalar field $y(x)$, defined at every point in the range $[x_a, x_b]$, representing some property in a physical system with a Lagrangian density $\mathcal{L}$. Let us suppose we can define an integral I given by

$$I = \int_{x_a}^{x_b} dx\, \mathcal{L}\left(y(x), \frac{dy(x)}{dx}, x \right), \qquad (1.27)$$

where $\mathcal{L}$ is a function of $y(x)$ and its first derivative. The variational problem is stated as follows. We wish to determine that function $y_0(x)$ satisfying the boundary values for which the functional $I[y]$ takes on an extremum value. The extremum of I has to be determined by varying the function $y(x)$. Following the rules for functional differentiation we obtain

$$\frac{\partial \mathcal{L}}{\partial y} - \frac{d}{dx}\left(\frac{\partial \mathcal{L}}{\partial y'} \right) = 0. \qquad (1.28)$$

Equation (1.28) is the Euler–Lagrange equation satisfied by the field $y(x)$, and together with the BCs $y(x_a) = y_a$ and $y(x_b) = y_b$, it is equivalent to the variational statement of the problem. More complex BCs and also the variation of end points in the action integral will be treated as the need arises in subsequent chapters.

1.2.2 Examples

Example 1: Poisson's equation

Consider a physical region of volume Ω occupied by a charge distribution $\rho(\mathbf{r})$. The total energy in the volume Ω in terms of the electrostatic potential $\phi(\mathbf{r})$ is given by

$$\mathcal{E}[\phi] = \int_\Omega d^3r\, \mathcal{L}(\phi(\mathbf{r}), \nabla\phi(\mathbf{r}), \mathbf{r})$$
$$= \int_\Omega d^3r \left[\frac{1}{2}\nabla\phi(\mathbf{r}) \cdot \nabla\phi(\mathbf{r}) - 4\pi\phi(\mathbf{r})\rho(\mathbf{r}) \right]. \quad (1.29)$$

The first term corresponds to $\mathbf{E}^2/2$, the energy in the electrostatic field, and the second term represents the potential energy of interaction between the field and the charge distribution. We are assuming that the charges are located in vacuum with dielectric constant $\epsilon = 1$.[††]

The principle of stationary action takes the form of a minimum energy principle. It requires that the actual physical configuration of the potential function be such that the total energy for the system be a minimum, where $\delta\mathcal{E}[\phi] = 0$. Using functional differentiation, we have

$$\delta\mathcal{E}[\phi] = \int_\Omega d^3r \left[\left(\frac{\delta\mathcal{L}}{\delta\nabla\phi(\mathbf{r})} \right) \cdot \delta\nabla\phi(\mathbf{r}) + \left(\frac{\delta\mathcal{L}}{\delta\phi(\mathbf{r})} \right) \delta\phi(\mathbf{r}) \right]$$
$$= \int_\Omega d^3r \left[-\nabla^2\phi(\mathbf{r}) - 4\pi\rho(\mathbf{r}) \right] \delta\phi(\mathbf{r}). \quad (1.30)$$

The surface terms arising from an integration by parts have been dropped under the assumption that the source charges are localized and that the potential vanishes at infinity. The local variation $\delta\phi(\mathbf{r})$ being arbitrary, the minimum energy principle requires that

$$\nabla^2\phi(\mathbf{r}) = -4\pi\rho(\mathbf{r}), \quad (1.31)$$

which is Poisson's equation.

It is worth noting that it is entirely feasible to reverse the variational process and construct the Lagrangian and the action integral from the equation of motion.[21] For example, we multiply equation (1.31) by $\delta\phi(\mathbf{r})$ and integrate over all space to obtain

$$\int_\Omega d^3r\, \delta\phi(\mathbf{r}) \left(-\nabla^2\phi(\mathbf{r}) - 4\pi\rho(\mathbf{r}) \right) = 0. \quad (1.32)$$

Reversing the earlier integration by parts we obtain

$$\delta \int_\Omega d^3r \left(\frac{1}{2}\nabla\phi(\mathbf{r}) \cdot \nabla\phi(\mathbf{r}) - 4\pi\rho(\mathbf{r})\,\phi(\mathbf{r}) \right) = 0.$$

This allows us to identify the Lagrangian density to within a sign, which is then determined by requiring that the energy density be

[††] cgs-units are being used here.

positive. From this differential quantity the functional which is subjected to the variational process is obtained. This exercise has been included here to re-emphasize the importance of the action integral. It will be seen in what follows that the FEM may be thought of as the discretization of the action integral, and this theme will continue to recur throughout this book.

Example 2: Schrödinger's equation

For a nonrelativistic particle moving in a region with potential energy $V(\mathbf{r})$ and having energy E, the time-independent Schrödinger's equation is given by

$$-\frac{\hbar^2}{2m}\nabla^2\psi(\mathbf{r}) + V(\mathbf{r})\,\psi(\mathbf{r}) = E\,\psi(\mathbf{r}). \qquad (1.33)$$

Equation (1.33) can be obtained as the condition under which the action[‡‡]

$$\mathcal{A} = \int d^3r \left[\frac{\hbar^2}{2m}\nabla\psi^*(\mathbf{r}) \cdot \nabla\psi(\mathbf{r}) + \psi^*(\mathbf{r})(V(\mathbf{r}) - E)\psi(\mathbf{r})\right] (1.34)$$

has an extremum. Here ψ^*, the complex conjugate function, and ψ can be considered to be two independent "fields." The second derivative of ψ is assumed to be continuous. On varying $\mathcal{A}$ with respect to $\psi^*(\mathbf{r})$ we obtain Schrödinger's equation, equation (1.33).[§§]

In his initial paper Schrödinger showed that the extremum of the integral

$$I = \int d^3r \left[\frac{\hbar^2}{2m}\nabla\psi^*(\mathbf{r}) \cdot \nabla\psi(\mathbf{r}) + \psi^*(\mathbf{r})V(\mathbf{r})\psi(\mathbf{r})\right], \qquad (1.35)$$

with respect to variations of $\psi^*(\mathbf{r})$ and subject to the subsidiary condition $\int d^3r \, \psi^*(\mathbf{r})\psi(\mathbf{r}) = 1$, leads to equation (1.33). The subsidiary condition ensures that the probability of finding one particle

---

[‡‡]In the case of the time-independent Schrödinger equation I am referring to the Lagrangian as the "action," so that employing the variational method can be designated uniformly throughout the book as invoking the *principle of stationary action.*

[§§]The quantum action principle developed by Schwinger[13] leads to the quantum equations of motion obeyed by fields, as in the example above, and, by varying the end points, also to the canonical equal time commutation rules needed in defining field quantization.

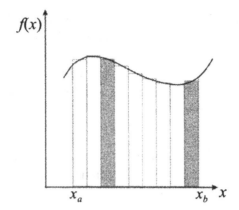

Fig. 1.3 The integral of a function $f(x)$ represented as the sum of rectangular areas under the curve.

in the physical region is unity, and the energy parameter E in equation (1.33) arises from incorporating the subsidiary condition into the integral I by the use of the Lagrange multiplier E.

It should be noted that we have labeled as the action the integral over the Lagrangian density even when it is independent of time. The usual inclusion of an integration over time gives the action the units of energy multiplied by time, which are the same units as for Planck's constant. This notation should cause no confusion since the action is understood as a functional whose functional derivative with respect to the dynamical variable is zero, according to the principle of stationary action.

1.3 Finite elements

Given the differential equation obeyed by a physical observable the natural inclination is to invoke for its solution the mathematical methods developed in classical analysis over the past three centuries. The spatial and temporal evolution of the physical system are then described by that particular solution which satisfies the BCs. The FEM provides an alternative, numerical procedure for the solution of the differential equation.

The concept of the FEM is fairly straightforward. Recall that the integral of a function $f(x)$ over the range $x = a$ to $x = b$, i.e., the area under the curve $f(x)$, can be approximated by a sum of trapezoidal areas, as shown in Fig. 1.3. In each of the trapezoids the integrand can be approximated by a constant, or more generally a polynomial. *The method of finite elements can be characterized as the evaluation*

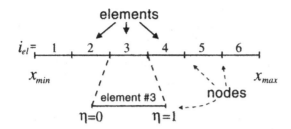

Fig. 1.4 The elements in the range from x_{min}, x_{max}.

of the action integral itself in a similar manner. Consider a 1D physical region $[x_{min}, x_{max}]$, as in Fig. 1.4. The region is divided into smaller regions called *elements*. The action integral is now split up into integrations over each of these elements,

$$\mathcal{A} = \sum_{(i_{el})}^{nelem} \mathcal{A}^{(i_{el})}. \tag{1.36}$$

The unknown solution, say the wavefunction $\psi(x)$, is expressed as a linear combination of interpolation polynomials, $N_i(x)$, multiplied by as-yet-unknown coefficients ψ_i in each of these elements. The spatial integrals in the action integral can now be performed over each element, leaving us with the action given in terms of these unknown coefficients. The basic idea here is to do a variation of the action with respect to these unknown parameters.

Let there be n_{intrp} polynomials defined only over a given element, outside of which they are set to zero. The interpolation polynomials are chosen such that their coefficients ψ_i are the values of the solution at special points within the element called nodes.[¶¶] Thus the wavefunction in the i_{el}th element[‖‖] is given by

$$\psi(x) = \sum_{j}^{n_{intrp}} \psi_j^{i_{el}} N_j^{i_{el}}(x). \tag{1.37}$$

Usually nodes are placed at the beginning and the end points of the element. The first nodal value ψ_1 in a given element will equal

[¶¶]These nodes are not to be confused with zeros of the solution.

[‖‖]Many of the variables will be assigned labels used in typical computer codes written in the Fortran or C programming languages. For example, the running index in a sum over elements will be i_{el} or *iel* and the total number of elements will be labeled by *nelem*.

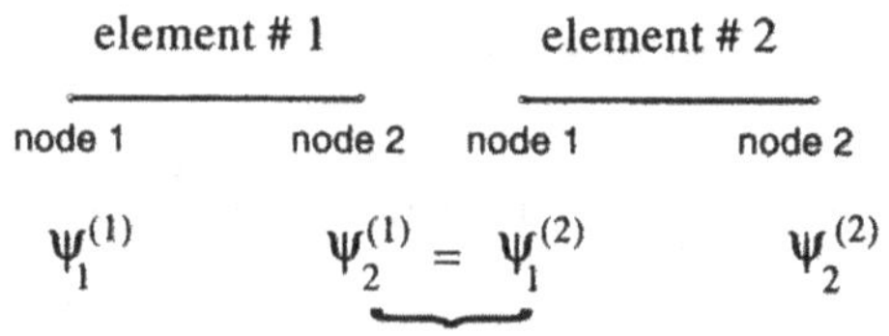

element # 1   element # 2
_____   _____

node 1   node 2   node 1   node 2

$\psi_1^{(1)}$ $\psi_2^{(1)} = \psi_1^{(2)}$ $\psi_2^{(2)}$

Fig. 1.5 The nodal variables in neighboring elements. Here $\psi_2^{(1)}$ and $\psi_1^{(2)}$ are assigned the same value in order to ensure interelement continuity of the solution ψ.

the last nodal value $\psi_{n_{intrp}}$ from the previous element, as shown in Fig. 1.5. This guarantees the interelement continuity of the solution. If the interpolation polynomials of higher degree than the first are desired then additional nodes internal to the element are used.

The interpolation polynomials are limited to within each element and are zero outside it. The spatial dependence in the action functional is now integrated out in each element. The action of equation (1.34) within element i_{el} can be expressed as

$$\mathcal{A}^{(i_{el})} = \sum_{i,j}^{n_{intrp}} \psi_i^* \left[\int dx\, N_i(x)\, \mathcal{O}\, N_j(x) \right] \psi_j$$

$$= \sum_{i,j}^{n_{intrp}} \psi_i^* \mathcal{M}_{ij}^{i_{el}} \psi_j. \tag{1.38}$$

Here $\mathcal{O}$ represents the operators occurring in the Lagrangian density. Clearly, the degree of the polynomial must permit the differentiations present in $\mathcal{O}$. The total action is the sum of the contributions from all the elements. This can be written in a very natural manner in matrix form, obtained on imposing interelement continuity through carefully overlaying the element matrices $\mathcal{M}^{i_{el}}$. The reason is that the continuity of the wavefunction across element boundaries requires the nodal variable $\psi_i^{i_{el}}$ to satisfy the relation

$$\psi_2^{i_{el}} = \psi_1^{i_{el}+1}. \tag{1.39}$$

This is depicted in Fig. 1.6 in the case of linear interpolation.

The above discussion has been focused on an element-by-element evaluation. Very often it is advantageous to switch to a nodal point of view, in that we can consider the interpolation polynomials to be associated with nodes rather than elements. In this case, the

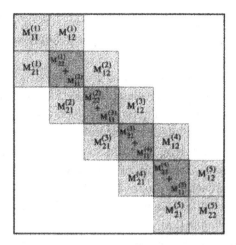

Fig. 1.6 The global matrix constructed from local element matrices using overlap of the matrices is shown for a 1D problem treated with linear interpolation over five elements.

polynomials associated with the element boundary node from two adjacent elements correspond to the interpolation function associated with that node. In the case of linear elements, such an interpolation function will have the shape of a triangular hat and the function is often referred to as a hat-function. This nodal point of view again explains the need for overlap of the element matrices.

With a global matrix $\mathcal{M}$ constructed in this manner, the total action is given by

$$\mathcal{A} = \sum_{\alpha,\beta}^{n_{glob}} \psi_\alpha^* \, \mathcal{M}_{\alpha\beta} \, \psi_\beta. \qquad (1.40)$$

We have replaced the element indices by global indices α and β. The governing dynamical principle, the principle of stationary action, is now invoked on the discretized action integral to derive "equations of motion" which are now algebraic equations. We employ a nodal variational principle and vary the action with respect to the nodal variables ψ_α^* to obtain simultaneous equations for the unknown coefficients ψ_β. We have

$$\frac{\delta\mathcal{A}}{\delta\psi_\alpha^*} = \sum_\beta \mathcal{M}_{\alpha\beta} \, \psi_\beta = 0. \qquad (1.41)$$

This set of simultaneous equations obtained through the breakup of the physical region takes the place of the traditional equation of motion.

Some of the variables ψ_β are known from the BCs. These values are inserted into the simultaneous equations. The rest of the coefficients are solved by standard matrix methods after boundary values are applied to the corresponding coefficients.

In Chapter 2, we apply this method to the problem of the quantum mechanical simple harmonic oscillator and to the solution of the radial differential equation for the radial wavefunctions of the hydrogen atom. These familiar problems allow us to introduce the needed notation for finite element analysis.

The FEM comes into its own when the shape of the physical region is not conducive to the standard methods of attack, such as the method of separation of variables which typically leads to ordinary differential equations in one variable. The entire machinery developed to solve differential equations analytically becomes more of a hindrance when the method of separation of variables fails. Also, when the boundary conditions (BCs) applicable in the problem are complex, as in scattering theory, or when the physical region is inhomogeneous, as in composite materials with interfaces where interface BCs have to be implemented, the problem demands a numerical solution. It is then no longer practical to directly come "down" to the level of differential equations from the "pinnacle" of the action integral.

1.4 Historical comments

The original idea of discretizing the region of physical interest in a variational context is due to Courant[22] who pointed out the advantages of such an approach in a structural engineering problem. Hrenikoff[23] and McHenry[24] used the skeletal and the lattice methods of discretization. Since the early 1940s, the method has evolved hand in hand with each improvement in computational capabilities. The name "finite element method" was first employed by Clough.[25] The pioneering work of Turner *et al.*,[26] Argyris,[27] Zienkiewicz,[28] and others, has led the FEM to be closely associated with structural engineering. An interesting book by Robinson gives further details of the pioneers in the field.[29] However, the general technique of the FEM has since permeated many other branches of engineering and science as a powerful method for the solution of partial and nonlinear differential equations. Perhaps because of its origins in the engineering fields, the typical physicist may at first sight relegate this method as being capable of only very approximate solutions. One of the themes of this book is to emphasize that the FEM can generate results of very high accuracy. The FEM enjoys all the advantages of a variational approach to solving physical problems, so that accuracy of solutions can be systematically improved.

1.5   Problems

It is assumed that the reader has taken a first course on quantum mechanics prior to attempting these problems.

1. Estimate the confinement energy of an electron from Heisenberg's uncertainty principle for an electron in a spherical box of radius $0.528\,\text{Å}$ given that $m_e = 9.109 \times 10^{-28}$ g.

2. In a semiconductor, the many-body interactions between the nearly free electrons in the solid, and also those between the electrons and atomic ions located at lattice sites, alter the electron's effective mass (denoted by m_e^*) from its mass in vacuum (the free-electron mass m_e).

   (a) Estimate the confinement energy of an electron from Heisenberg's uncertainty principle for an electron in a quantum well composed of a GaAs layer sandwiched between barrier layers of AlGaAs. Assume an infinite potential well as a first approximation and suppose that the electron is in the conduction band of the GaAs well layer. Consider quantum wells of widths $1000\,\text{Å}$, $100\,\text{Å}$, and $25\,\text{Å}$ with $m_e^* = 0.0665\,m_e$.

   (b) Estimate the effect of confinement on the electronic energy in GaAs quantum wires of cross-sections $100 \times 100\,\text{Å}$ and $25 \times 25\,\text{Å}$, with $m_e^* = 0.0665\,m_e$. Again assume as a first approximation that the confining potential is infinite.

3. A quantum mechanical particle is limited to move in one dimension from $x = 0$ to $x = L$, with zero potential energy in this region and infinite potential barriers at $x = 0$ and at $x = L$, as in Fig. 1.7.

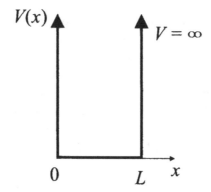

Fig. 1.7 The 1D infinite potential well.

(a) Solve the time-independent Schrödinger equation and obtain the energy spectrum and wavefunctions for this particle confined to the infinite quantum well.

(b) Determine the probability of finding the particle between $L/4$ and $L/2$ in the state $n = 3$.

4. Organic molecules with "conjugated" bonds have essentially free carriers moving along the molecule (as in polyelenes). The conjugated bonds occur in porphyrine rings of molecules such as chlorophyll, hemoglobin, myoglobin, etc. Here we are concerned with the "free" electrons circling in the chlorophyll ring, which has 20 carbon atoms on a ring each of which gives up a π-electron. We are interested in developing an energy-level scheme for the electrons confined to motion on the ring, and to calculate the lowest energy photon which may be absorbed by the system in an electronic transition.***

In order to simplify the problem assume that the 20 electrons are free to propagate on a circular ring of radius $R = 4.21\,\text{Å}$. They move in a potential that is zero along the circumference at $r = R$ and the potential is infinitely high at other radial distances. This is depicted in Fig. 1.8. The classical version of this problem is a bead that is constrained to move on a circular wire held horizontally. These electrons are assumed not to interact with one another.

(a) Solve Schrödinger's equation for the electrons satisfying the

***This problem was suggested to the author by R. Olenick (1977).

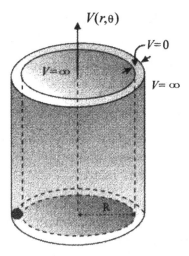

Fig. 1.8 The potential function for a particle on a ring. The particle occupies the region of zero potential between the cylinders. The classical particle is shown in the figure.

periodic BC $\psi(x + L) = \psi(x)$ where $L = 2\pi R$, and x is the distance along the circumference from a specified point on the circumference. Obtain the energy spectrum for the electrons and the normalized wavefunctions.

(b) As mentioned earlier, there are 20 electrons on the ring which are distributed over the lowest few energy levels, consistent with Pauli's exclusion principle (electron spin degeneracy is to be taken into account). Determine the electron distribution in these energy levels. Comment on why the electron distribution is different from the one for a spectrum of a particle in a "box" of length L.

(c) Indicate on an energy-level diagram for the ring the lowest energy transition for the system of 20 electrons. Evaluate the energy difference and determine the wavelength of the photon which would induce this transition.

5. The donor ionization energies in Si and in Ge with dielectric constants $\epsilon_{Si} = 11.7$ and $\epsilon_{Ge} = 15.8$, respectively, vary slightly with the species of the impurity. In Si, the ionization energies are {0.045 eV (P), 0.049 eV (As), 0.039 eV (Sb)}, while in Ge the ionization energies are {0.012 eV (P), 0.0127 eV (As), 0.0096 eV (Sb)}. The symbols in parentheses are the impurity elements

which enter the host lattice substitutionally by replacing a host atom at a lattice site.

(a) Determine the Bohr radius of the ground state of the hydrogenic donor atom in Si and in Ge for each impurity.

(b) Obtain the 2s and 5s energy levels in the two semiconductors.

(c) Determine the temperature at which $k_B T$ equals the ionization energy. At such a temperature the donor has a probability of e^{-1} of being thermally ionized.

(d) Can you suggest reasons for the variation in the ionization energies of the donor electron for the impurities P, As, Sb in the same host semiconductor (Si or Ge)?

6. In the scattering of very slow electrons by noble gases, such as neon and argon, electrons of specific energies are transmitted through the atoms as though the atoms were practically transparent to electrons with these energies. The probability of scattering (rather than direct transmission) in such cases is much less than is obtained with atoms for which this anomalous transmission does not exist. At higher or lower electron energies again the atom is found to be much less transparent. These resonant peaks in the transmission were first observed by Ramsauer and were later explained by quantum theory.[30] A simple model which describes these effects is provided by the 1D square-well potential. Let the attractive potential well be of depth $-V_0$ in the range $-d/2 \le x \le d/2$ and zero outside it. Consider a stream of electrons of velocity v incident from the left on the square well.

(a) Describe what would happen to the incident electrons if the dynamics were governed by classical mechanics.

(b) In quantum theory, Schrödinger's equation is the governing equation satisfied by the electron's wavefunction. State the physical reasons for your choice of BCs, and obtain the solutions for the electron's wavefunction for all values of x in terms of the incident amplitude. Give details of your derivation.

(c) Evaluate the transmission coefficient $T(E)$ and obtain the conditions for perfect transmission. Draw a graph of the behavior of T as a function of the momentum of the electron in the potential well to show the resonant behavior.

(d) Evaluate the "width of the resonance" for the transmission peaks.

(e) Comment on how the analysis of this problem would change if there were a distribution of velocities in the electron beam.

(f) Compare this situation with the Fabry–Pérot interferometer in optics and discuss the parallel features.

(g) Assume a square well of width $d = 2$ Å. How deep would the well have to be to provide a transmission resonance for electrons of 0.1 eV kinetic energy? What is the width of the resonance in this case?

7. The length of an arbitrary curve passing through two points P_1 and P_2 on a plane is given by

$$I = \int_{P_2}^{P_1} ds.$$

Using the relation

$$\frac{ds}{dx} = \left[1 + \left(\frac{dy}{dx} \right)^2 \right]^{1/2},$$

and the variational method obtain an equation for the shortest distance between the two points.

8. Consider a uniform string of length L and linear density ρ held fixed at its two ends. If the displacement of the string is described by $y(x,t)$ then the kinetic energy of a small segment of length δx is given by

$$T = \frac{\rho}{2} \left(\frac{\partial y}{\partial t} \right)^2 \delta x.$$

The string of length δx is stretched due to the displacement of the string by the amount

$$\left[\sqrt{1 + \left(\frac{\partial y}{\partial x} \right)^2} - 1 \right] \delta x \simeq \frac{1}{2} \left(\frac{\partial y}{\partial x} \right)^2 \delta x.$$

If the tension in the string is τ, the potential energy of this segment of string is

$$V = \frac{1}{2}\tau \left(\frac{\partial y}{\partial x}\right)^2 \delta x.$$

(a) Write down the Lagrangian density per unit length for the string, and its action integral.

(b) Derive the equation of motion, the wave equation, for the displacement y of the vibrating string using the variational method.

9. (a) Write a computer program to evaluate the integral

$$\int_0^{\pi/2} d\theta \, \sin\theta \tag{1.42}$$

using 10, 10^2, and 10^3 intervals (or elements) in each of which it is assumed that the integrand is a constant, and compare your numerical result with the analytical value.

(b) Using the same number of elements as in part (a) above, evaluate the integral assuming that the integrand is linearly interpolated between the end points of each element. Represent $\sin\theta$ in the interval $(\theta_{i+1} - \theta_i)$ by

$$\sin\theta = \sin\theta_i \left(\frac{\theta_{i+1} - \theta}{\theta_{i+1} - \theta_i}\right) + \sin\theta_{i+1} \left(\frac{\theta - \theta_i}{\theta_{i+1} - \theta_i}\right)$$

$$= \sin\theta_i \, N_1(\theta) + \sin\theta_{i+1} \, N_2(\theta).$$

Compare your results with the corresponding ones from part (a).

References

[1] D. Bohm, *Quantum Theory* (Prentice Hall, Englewood Cliffs, NJ, 1951).
[2] J. L. Powell and B. Crasemann, *Quantum Mechanics* (Addison-Wesley, New York, 1961).
[3] L. I. Schiff, *Quantum Mechanics* 3rd ed. (McGraw-Hill, New York, 1968).
[4] L. D. Landau and E. M. Lifshitz, *Quantum Mechanics – Nonrelativistic Theory* (Pergamon, London, 1965).

[5] R. Shankar, *Principles of Quantum Mechanics* (Plenum Press, New York, 1994).

[6] M. Born, *Atomic Physics* (Hafner, New York, 1951).

[7] M. Jammer, *The Conceptual Development of Quantum Mechanics* (McGraw-Hill, New York, 1966).

[8] P. A. M. Dirac, *The Principles of Quantum Mechanics*, 3rd ed. (Oxford University Press, Oxford, 1947).

[9] J. J. Sakurai, *Modern Quantum Mechanics* (Benjamin/Cummings, Reading, MA, 1985).

[10] W. Yourgrau and S. Mandelstam, *Variational Principles in Dynamics and Quantum Theory* (Dover, New York, 1968).

[11] R. Weinstock, *Calculus of Variations* (Dover, New York, 1974).

[12] For a very readable introduction to variational principles at the undergraduate level, see: D. S. Lemons, *Perfect Form: Variational Principles, Methods and Applications in Elementary Physics* (Princeton University Press, Princeton, NJ, 1997).

[13] J. Schwinger, *Phys. Rev.* **82**, 914 (1951), *ibid.* **91**, 713 (1953).

[14] B. L. Moiseiwitsch, *Variational Principles* (Interscience, London, 1966).

[15] R. Courant and D. Hilbert, *Methods of Mathematical Physics*, Vol. I (Interscience, New York, 1953).

[16] S. G. Mikhlin, *Variational Methods in Mathematical Physics* (Pergamon, Oxford, 1964).

[17] H. Goldstein, *Classical Mechanics*, 2nd ed. (Addison-Wesley, Reading, MA, 1980).

[18] L. D. Landau and E. M. Lifshitz, *Mechanics* (Addison-Wesley, Reading, MA, 1960).

[19] A. L. Fetter and J. D. Walecka, *Theoretical Mechanics of Particles and Continua* (McGraw-Hill, New York, 1980).

[20] J. J. Sakurai, *Advanced Quantum Mechanics* (Addison-Wesley, Reading, MA, 1967).

[21] J. D. Bjorken and S. D. Drell, *Relativistic Quantum Fields* (McGraw-Hill, New York, 1965).

[22] R. Courant, *Bull. Am. Math. Soc.* **49**, 1 (1943).

[23] A. Hrenikoff, *J. Appl. Mech.* **A8**, 169 (1941).

[24] D. McHenry, *J. Inst. Civ. Eng.* **21**, 59 (1943).

[25] R. W. Clough, *Proceedings of the Second Conference on Electronic Computation*, ASCE (1960).

[26] M. J. Turner, R. W. Clough, H. C. Martin, and L. J. Topp, *J. Aeronaut. Sci.* **23**, 805 (1956).

[27] J. H. Argyris, *Energy Theorems in Structural Analysis* (Butterworths, London, 1960).

[28] O. C. Zienkiewicz, *The Finite Element Method* (McGraw-Hill, New York, 1989).

[29] J. Robinson, *Early FEM pioneers* (Robinson and Associates, Wimborne, England, 1985).

[30] L. I. Schiff, Ref. 3, p123.

2

Simple quantum systems

The aim of this chapter is to introduce the finite element method (FEM) in a familiar setting by treating two problems: the quantum harmonic oscillator and the hydrogen atom. We use linear interpolation functions for the wavefunctions in each finite element, so that all the matrix elements appearing in the discretized action integral are readily obtained analytically. The element matrices are evaluated first and these are inserted into global matrices. The principle of stationary action is used to derive a discretized version of Schrödinger's equation in the form of a generalized eigenvalue problem. The results are compared with the exact analytical answers.

Next, the traditional Rayleigh–Ritz approach using global trial wavefunctions is presented with examples and the method is compared with the FEM.

This chapter concludes with a brief discussion of the essentials of finite element programming, using the radial equation for the hydrogen atom as an example.

2.1 The simple harmonic oscillator

The classical energy, $E = p^2/2m + kx^2/2$, of the simple harmonic oscillator and the prescription $p \to -i\hbar\, \partial/\partial x$ (see equation (1.6)) give for the time-independent Schrödinger equation of a quantum oscillator[1]

$$-\frac{\hbar^2}{2m}\frac{d^2}{dx^2}\,\psi(x) + \frac{1}{2}\,k\,x^2\psi(x) = E\,\psi(x), \qquad (2.1)$$

where k is the spring constant of the oscillator.

It is useful to convert the differential equation to a dimensionless form. The particle undergoing harmonic motion has its energy quantized in units of $\hbar\omega_0$, where $\omega_0 = \sqrt{k/m}$ is the natural frequency of the harmonic oscillator. It is convenient to measure the energy in units of $\lambda = \hbar\omega_0/2$. Writing $x = \alpha\xi$, where $\alpha = \sqrt{\hbar/(m\omega_0)}$, Schrödinger's equation in dimensionless form is given by

$$\frac{d^2}{d\xi^2}\psi(\xi) + (\epsilon - \xi^2)\,\psi(\xi) = 0. \qquad (2.2)$$

The presence of the confining quadratic potential indicates that all the eigenstates will be bound states and that we will have a discrete energy spectrum. Furthermore, the asymptotic form of the differential equation for $|\xi| \to \infty$ suggests that all the solutions will have the factor $\exp(-\xi^2/2)$, and the eigenfunctions will have the form $\psi_n(\xi) = f_n(\xi)\exp(-\xi^2/2)$ where f_n turn out to be polynomials. The wavefunctions ψ_n must converge to zero as $|\xi|$ tends to infinity, as required for bound states.

We wish to obtain an approximate numerical solution to the eigenvalues and eigenfunctions for the oscillator using the FEM. The action integral for the oscillator is

$$\mathcal{A} = \int_{-\infty}^{\infty} d\xi \left[\frac{d}{d\xi}\psi^*(\xi)\,\frac{d}{d\xi}\psi(\xi) + \psi^*(\xi)\left(\xi^2 - \epsilon\right)\psi(\xi)\right]. \qquad (2.3)$$

First, we truncate the range of the integral. With the asymptotic solutions requiring a Gaussian fall-off with increasing $|\xi|$, it should be adequate to limit the range to $[\xi_{min} = -5, \xi_{max} = +5]$. The action integral will have a Gaussian factor $\exp(-x^2)$ which is approximately 1.4×10^{-11} at the two limits of integration. In the following, we will numerically investigate the effect of increasing or decreasing this range on the eigenvalues. The action integral is now estimated by breaking up the region of integration into *nelem* finite elements, as in Fig. 1.4 (Chapter 1). Let the element size be $h = (\xi_{max} - \xi_{min})/nelem$. The beginning and the end points, the *nodal points*, of element i_{el} have the coordinates

$$\xi_1^{(i_{el})} = \xi_{min} + h\,(i_{el} - 1),$$
$$\xi_2^{(i_{el})} = \xi_{min} + h\,i_{el},$$

respectively. The integral over the coordinate in the action integral can now be evaluated by using the discretized form

$$\mathcal{A} = \sum_{i_{el}=1}^{nelem} \mathcal{A}^{(i_{el})}$$

$$= \sum_{i_{el}=1}^{nelem} \int_0^h d\xi \left[\psi^{*\prime}(\xi)\,\psi'(\xi) + \psi^*(\xi)\left(\xi^2 - \epsilon\right)\psi(\xi)\right]. \qquad (2.4)$$

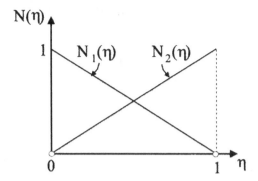

Fig. 2.1 The linear interpolation functions $N_1(\eta)$ and $N_2(\eta)$. The two nodes for the element are at $\eta = 0$ and at $\eta = 1$.

As the first approximation for ψ beyond the element-wise constant form, we attempt a linear interpolation between its as-yet-undetermined values, ψ_1 and ψ_2, at the nodal points of the element i_{el}. We write

$$\psi(\xi) = \psi_1^{(i_{el})} \left(\frac{\xi_2^{(i_{el})} - \xi}{\xi_2^{(i_{el})} - \xi_1^{(i_{el})}} \right) + \psi_2^{(i_{el})} \left(\frac{\xi - \xi_1^{(i_{el})}}{\xi_2^{(i_{el})} - \xi_1^{(i_{el})}} \right)$$

$$= \psi_1^{(i_{el})} (1 - \eta) + \psi_2^{(i_{el})} \eta, \qquad (2.5)$$

where η is a local coordinate in the element having a range $[0, 1]$. It is usual to denote the local interpolation polynomials by $N_1(\eta) = 1 - \eta$, and $N_2(\eta) = \eta$, so that

$$\psi(\xi) = \psi_1^{(i_{el})} N_1(\eta) + \psi_2^{(i_{el})} N_2(\eta). \qquad (2.6)$$

The functions N_i are shown in Fig. 2.1. The derivative of $\psi(\xi)$ is

$$\frac{d\psi(\xi)}{d\xi} = \left(\psi_1^{(i_{el})} \frac{1}{h} \frac{d}{d\eta} N_1(\eta) + \psi_2^{(i_{el})} \frac{1}{h} \frac{d}{d\eta} N_2(\eta) \right)$$

$$= \frac{1}{h} (-\psi_1^{(i_{el})} + \psi_2^{(i_{el})}), \qquad (2.7)$$

the factor h in the denominator arising in the change in coordinates from ξ to η. The range $[0, 1]$ for η is convenient when doing the integrations analytically by hand. It is usual in computer programs to automate the evaluation of element integrals by employing Gauss–Legendre quadrature,[†] in which case the natural range for the local variable η is taken to be $[-1, 1]$.

[†]The theory of Gauss quadrature is discussed in Appendix A.

When written in a 2×2 matrix notation the action integral over one element has the form

$$
\mathcal{A}^{(i_{el})} = \int_0^1 h \, d\eta \, \{\psi_1^{*(i_{el})}, \psi_2^{*(i_{el})}\} \left[\frac{1}{h^2} \begin{pmatrix} 1 & -1 \\ -1 & 1 \end{pmatrix} \right.
$$

$$
+ (\xi_{min} + h\,(i_{el} - 1 + \eta))^2 \begin{pmatrix} (1-\eta)^2 & \eta(1-\eta) \\ \eta(1-\eta) & \eta^2 \end{pmatrix}
$$

$$
\left. - \epsilon \begin{pmatrix} (1-\eta)^2 & \eta(1-\eta) \\ \eta(1-\eta) & \eta^2 \end{pmatrix} \right] \begin{pmatrix} \psi_1^{(i_{el})} \\ \psi_2^{(i_{el})} \end{pmatrix}. \tag{2.8}
$$

Here the transposed row vector is denoted by $\{\Psi\}$. On performing the integrals in the above equation we obtain

$$
\mathcal{A}^{(i_{el})} = \{\psi_1^{*(i_{el})}, \psi_2^{*(i_{el})}\} \left[\frac{1}{h} \begin{pmatrix} 1 & -1 \\ -1 & 1 \end{pmatrix} + h \xi_\ell^2 \begin{pmatrix} \frac{1}{3} & \frac{1}{6} \\ \frac{1}{6} & \frac{1}{3} \end{pmatrix} \right.
$$

$$
+ 2h^2 \xi_\ell \begin{pmatrix} \frac{1}{12} & \frac{1}{12} \\ \frac{1}{12} & \frac{1}{4} \end{pmatrix} + h^3 \begin{pmatrix} \frac{1}{30} & \frac{1}{20} \\ \frac{1}{20} & \frac{1}{5} \end{pmatrix}
$$

$$
\left. - \epsilon h \begin{pmatrix} \frac{1}{3} & \frac{1}{6} \\ \frac{1}{6} & \frac{1}{3} \end{pmatrix} \right] \begin{pmatrix} \psi_1^{(i_{el})} \\ \psi_2^{(i_{el})} \end{pmatrix}. \tag{2.9}
$$

Here $\xi_\ell = \xi_{min} + h(i_{el} - 1)$ is the coordinate of the left side of the element i_{el}. For further discussion, we write this equation in the form

$$
\mathcal{A}^{(i_{el})} = \{\psi_1^{*(i_{el})}, \psi_2^{*(i_{el})}\} \left[\begin{pmatrix} H_{11}^{(i_{el})} & H_{12}^{(i_{el})} \\ H_{21}^{(i_{el})} & H_{22}^{(i_{el})} \end{pmatrix} \right.
$$

$$
\left. - \epsilon \begin{pmatrix} U_{11}^{(i_{el})} & U_{12}^{(i_{el})} \\ U_{21}^{(i_{el})} & U_{22}^{(i_{el})} \end{pmatrix} \right] \begin{pmatrix} \psi_1^{(i_{el})} \\ \psi_2^{(i_{el})} \end{pmatrix}. \tag{2.10}
$$

The basis functions used to interpolate the wavefunctions are nonzero only over a given element, and are zero everywhere else.

Furthermore, the piecewise-polynomial functions are not orthonormal over each element and hence the matrix U is not unit diagonal.

In summing the contributions to the action integral from the elements, we have to recognize that the last nodal variable, $\psi_2^{(i_{el}-1)}$, from the previous element is the same as the first variable, $\psi_1^{(i_{el})}$, of the element i_{el}, as shown in Fig. 1.5 (Chapter 1). In a matrix representation, this interelement continuity is enforced by adding the matrix element $H_{11}^{(i_{el})}$ to $H_{22}^{(i_{el}-1)}$. The same condition is applied to the matrix U. In this fashion we construct the global representation of the action integral, as illustrated in Fig. 1.6 (Chapter 1) where just five elements are shown spanning the physical region.

At this stage, it is useful to convert from element-based indexing into a global index for the nodes and for the nodal values. Nodes 1 and 2 in element i_{el} are renumbered as node numbers $\alpha = i_{el}$ and $\alpha+1$. The corresponding nodal values of the wavefunction are ψ_α and $\psi_{\alpha+1}$. The complete action integral, after the sum over the action in each element is performed, is denoted by

$$\mathcal{A} = \psi_\alpha^* \left[\mathcal{H}_{\alpha\beta} - \epsilon \mathcal{U}_{\alpha\beta} \right] \psi_\beta; \qquad \alpha, \beta = 1, \ldots, nelem + 1, \quad (2.11)$$

where it is understood that repeated indices are summed over. In the context of the discretized action, the principle of stationary action is represented by

$$\delta\mathcal{A} = 0 = \delta\psi_\alpha^* \frac{\delta\mathcal{A}}{\delta\psi_\alpha^*}$$
$$= \delta\psi_\alpha^* \left(\mathcal{H}_{\alpha\beta} - \epsilon \mathcal{U}_{\alpha\beta} \right) \psi_\beta. \quad (2.12)$$

Since $\delta\psi_\alpha^*$ is arbitrary, the discretized action yields Schrödinger's equation in the form of a generalized eigenvalue equation

$$\mathcal{H}_{\alpha\beta} \psi_\beta - \epsilon \mathcal{U}_{\alpha\beta} \psi_\beta = 0. \quad (2.13)$$

We impose the boundary conditions that $\psi(\xi_{min}) = 0 = \psi(\xi_{max})$. Within the framework of our approximation these conditions require the initial and the final nodal values, ψ_1 and $\psi_{nelem+1}$, to be zero. The simultaneous equations defining the eigenvalue problem thus have their first and last terms as zero. This is implemented in the equations by setting the entries in the first and the last columns in the matrices $\mathcal{H}$ and $\mathcal{U}$ to zero. We also remove the first and the last rows, since the values of ψ^* at the two ends are zero. The

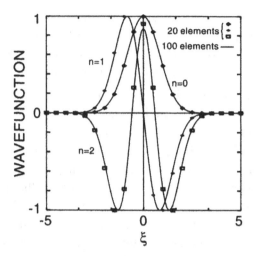

Fig. 2.2 The harmonic oscillator eigenfunctions for the lowest three energies. The data points are those obtained by the FEM using 20 elements with linear interpolation over the range −5 to +5, while the continuous curves are the eigenfunctions obtained using 100 elements with linear interpolation. The wavefunctions have not been normalized to unity.

net effect is to reduce the dimension of the global matrix by two, to $nelem-1$. The generalized matrix eigenvalue equation is solved using diagonalizers from the well-known, and readily available, EISPACK or LAPACK software packages.[2,3] After the nodal eigenvalues are obtained, we reinsert the null nodal values at the first and the last node in the vector arrays representing the complete nodal solutions for the wavefunctions. An alternative procedure that retains the original matrix dimensions is discussed in Section 2.4.

The wavefunctions for the first three eigenstates are shown in Fig. 2.2. The continuous curves in the figure were generated using 100 linear elements over the range −5 to +5. These are visually indistinguishable from the analytically derived wavefunctions. The wavefunctions obtained using 20 elements are shown in the same figure as data points without interconnecting them by straight lines as per linear interpolation. It should be noted that the solutions displayed in Fig. 2.2 have not been normalized to unity. The output from the diagonalizer for wavefunctions is set to unity at the maximum or minimum values.

In Table 2.1, we compare the values for the lowest four eigenvalues obtained by utilizing 20 elements and 100 elements over the

Table 2.1 The energy levels of the 1D simple harmonic oscillator in the FEM with linear interpolation functions. The range used is $-6 \leq x \leq 6$, and the number of elements is 20 (second column) and 100 (third column).

Quantum number n	Energy eigenvalues $(\hbar\omega_0/2)$		Exact eigenvalue
	20 elements	100 elements	
0	1.022 368	1.000 900	1
1	3.109 022	3.004 494	3
2	5.276 233	5.011 675	5
3	7.517 227	7.022 430	7

range $[-6, 6]$. It is clear that as the number of elements is increased we obtain better convergence towards $(2n+1)$, the exact eigenvalues in units of λ. We also obtain better values for the higher eigenvalues on using more elements. This is again seen in Table 2.2, where the dependence on the choice for the range of the physical region is also shown. As the range is increased, we are able to accommodate more eigenfunctions with their oscillations fully taken into account in the physical region, and this can be achieved by employing an adequate number of elements. We are assuming that we have reached the asymptotic region at ξ_{min} or ξ_{max}, where $\psi = 0$. This is clearly not the case for the higher excited states $(n = 6, 7)$ when the range is limited to ± 3. The eigenvalues for $n = 6, 7$ are closer to their natural values when the range is increased to ± 5. Another feature of Table 2.2 is that the ground state energy gets worse as the range is increased while using 20 elements. This is because the elements are distributed uniformly over the given ranges. Consequently less emphasis is placed near the origin and the density of elements is lowered as the range is increased.

In summary, the resulting eigenvalues depend on the number of elements, their placement, and the location of the cutoff which truncates the infinite domain. It is remarkable that linear interpolation gives eigenvalues for the low energy states to an accuracy of 1 part in 10^3 for a modest number of elements. We should keep in mind that we are representing the solution functions in terms of linear interpolation polynomials; on the other hand, the actual wavefunctions also

Table 2.2 The dependence on the range and the number of elements of the lowest few energy levels of the 1D simple harmonic oscillator in the FEM with linear interpolation functions. The exact eigenvalues are $(2n+1)$.

Range $\|\xi_{max}\|$	Quantum number n	Energy eigenvalues $(\hbar\omega_0/2)$	
		20 elements	100 elements
3	0	1.006 322	1.001 004
	1	3.040 006	3.013 286
	2	5.160 238	5.085 410
	3	7.509 500	7.335 732
	4	10.297 768	9.923 888
	5	13.720 976	12.977 212
	6	17.924 670	16.561 161
4	0	1.009 973	1.000 401
	1	3.049 332	3.002 028
	2	5.127 177	5.005 596
	3	7.245 027	7.013 387
	4	9.417 218	9.035 899
	5	11.695 957	11.104 645
	6	14.193 125	13.281 606
	7	17.060 911	15.647 319
5	0	1.015 559	1.000 625
	1	3.076 449	3.003 122
	2	5.195 341	5.008 113
	3	7.369 096	7.015 594
	4	9.594 450	9.025 579
	5	11.868 156	11.038 195
	6	14.188 308	13.054 157
	7	16.560 983	15.076 449

include Gaussian factors shown below,

$$\psi_n(\xi) = H_n(\xi)\,e^{-\xi^2/2}, \tag{2.14}$$

with $H_n(\xi)$ being the Hermite polynomials obtained for each positive integer n by making use of the formula

$$H_n(\xi) = (-1)^n e^{-\xi^2} \frac{d^n}{d\xi^n} e^{-\xi^2}. \qquad (2.15)$$

Further improvement in the accuracy of the eigenvalues requires higher order polynomials together with continuity conditions at the interfaces of adjacent elements, as shown in Chapter 3.

We can provide a mechanical analogy for understanding the above numerical work in physical terms. Consider the bending of a metal bar to fit the shape of one of the wavefunctions of Fig. 2.2. The wavefunction has n zero crossings between $-\infty < \xi < \infty$, where n is the quantum number of the state. The kinetic and potential energy terms can be viewed as the "stiffness matrix" of the bar.[4-6] For a fixed number of elements, a shorter bar is more difficult to bend so as to provide the correct oscillatory shape of a wavefunction for large n. On the other hand, a larger number of elements with adjustable nodal values can have enough flexibility to do so and to minimize the action more effectively.

In the above calculations no attempt was made to optimize the size or the location of the elements. We return to the treatment of adaptive FEM in Chapter 4. A finite element treatment of the quantum oscillator has been given by Burnett[7] using a quadratic interpolation over each element; Searles and von Nagy-Felsobuki[8] employ cubic Hermite interpolation polynomials. We introduce these and other interpolation schemes in Chapter 3.

2.2   The hydrogen atom

The energy levels of the electron in the hydrogen atom are evaluated in the same manner as in the above example. We begin with the time-independent Schrödinger equation[1]

$$-\frac{\hbar^2}{2m} \nabla^2 \Psi_n(\mathbf{r}) - \frac{e^2}{r} \Psi_n(\mathbf{r}) = E_n \Psi_n(\mathbf{r}), \qquad (2.16)$$

where the wavefunction of the electron in a stationary state with energy E_n is denoted by $\Psi_n(\mathbf{r})$. We rescale the radial coordinate r by a_0 (the Bohr radius) and divide equation (2.16) by the energy unit of a Rydberg R_∞

$$R_\infty = \hbar^2/2ma_0^2 = e^2/2a_0 = 13.6 \text{ eV}, \qquad (2.17)$$

to cast equation (2.16) in dimensionless variables. We retain the same symbol r for the radial coordinate after the rescaling, for convenience, but introduce the symbol ϵ for the reduced energy.

We shall specifically concern ourselves here with spherically symmetric bound states with zero angular momentum (the ns-states) with the principal quantum number n. This leads to the equation for the radial wavefunctions $\psi(r)$ given by

$$-\frac{1}{r^2}\frac{d}{dr}\left(r^2\frac{d}{dr}\psi(r)\right) - \frac{2}{r}\psi(r) = \epsilon\,\psi(r). \qquad (2.18)$$

The asymptotic form of the solution at $r \to \infty$ is $\sim e^{-r/n}$, and the physically acceptable solutions to the eigenvalue equation must satisfy the boundary condition $\psi(r) \to 0$ as $r \to \infty$. From the asymptotic form we see that at $r = \infty$ the wavefunction dies exponentially. What are the boundary conditions at $r = 0$? We discuss this below.

To implement the FEM, we first multiply equation (2.18) by the complex conjugate wavefunction $\psi^*(r)$ (together with a Jacobian factor of r^2) and reconstruct the action integral by integrating the differential equation by parts. The "surface" terms generated by this procedure vanish at $r = \infty$ since the wavefunctions and their derivatives vanish there. At $r = 0$, the Jacobian factor of r^2 ensures that the surface term vanishes there assuming that neither ψ nor ψ' behave badly at the origin. Thus the surface terms vanish at both ends. This also implies that we need not assign a value a priori to $\psi(r = 0)$. Recall that the form of the wavefunction is first determined and then it is normalized to unity. This normalization procedure will in any case change any nominal value assigned to it.

We have at hand

$$\mathcal{A} = \int_0^\infty dr\, r^2 \left[\psi^{*\prime}(r)\psi'(r) - \psi^*(r)\left(\frac{2}{r} + \epsilon\right)\psi(r)\right]. \qquad (2.19)$$

To do the above integration numerically, the integral is truncated and the upper limit of the integration is set to a finite value, $r = r_c$. The value of r_c is chosen judiciously so as to reduce the correction term from the integral over the limits r_c to ∞. For example, we could require that the wavefunction be at least six orders of magnitude smaller at r_c than at the origin for the ground state, and use this as the criterion for choosing r_c. We return to the issue of improving accuracy in Chapter 3. The truncated action integral is

$$\mathcal{A} = \int_0^{r_c} dr\, r^2 \left[\psi^{*\prime}(r)\psi'(r) - \psi^*(r)\left(\frac{2}{r} + \epsilon\right)\psi(r)\right]. \qquad (2.20)$$

The integrals in equation (2.20) are discretized into *nelem* small 1D elements. In each element i_{el} we make the simplest assumption and choose a linear interpolation for the wavefunctions, with the nodes located at the two ends of each element. For convenience, we introduce the local coordinate η in each element, and also assume that all the elements are of the same size h. The coordinate r in the element i_{el} is then given by $r = (i_{el} - 1)h + \eta h$. The unknown wavefunction $\psi(r)$ is assumed to be given at the two nodal points of the element i_{el} and linearly interpolated between them in terms of the interpolation polynomials $N_1(\eta) = 1 - \eta$ and $N_2(\eta) = \eta$.

With the assumed linear interpolation for the wavefunction we can directly integrate the action, equation (2.20), in each element; expressing the action in element i_{el} in a 2×2 matrix notation,[9] we have

$$
\begin{aligned}
\mathcal{A}^{(i_{el})} = & \langle \psi_i^{(i_{el})} | H_{ij}^{(i_{el})} | \psi_j^{(i_{el})} \rangle \\
& - \epsilon \langle \psi_i^{(i_{el})} | U_{ij}^{(i_{el})} | \psi_j^{(i_{el})} \rangle,
\end{aligned}
\tag{2.21}
$$

where

$$
\begin{aligned}
H_{11}^{(i_{el})} &= h \left(\frac{1}{3} - n + n^2 \right) + h^2 \left(\frac{1}{2} - \frac{2}{3}n \right), \\
H_{12}^{(i_{el})} &= H_{21}^{(i_{el})} \\
&= h \left(-\frac{1}{3} + n - n^2 \right) + h^2 \left(\frac{1}{6} - \frac{1}{3}n \right), \\
H_{22}^{(i_{el})} &= h \left(\frac{1}{3} - n + n^2 \right) + h^2 \left(\frac{1}{6} - \frac{2}{3}n \right),
\end{aligned}
\tag{2.22}
$$

and

$$
\begin{aligned}
U_{11}^{(i_{el})} &= h^3 \left(\frac{1}{5} - \frac{1}{2}n + \frac{1}{3}n^2 \right), \\
U_{12}^{(i_{el})} &= U_{21}^{(i_{el})} \\
&= h^3 \left(\frac{1}{20} - \frac{1}{6}n + \frac{1}{6}n^2 \right), \\
U_{22}^{(i_{el})} &= h^3 \left(\frac{1}{30} - \frac{1}{6}n + \frac{1}{3}n^2 \right).
\end{aligned}
\tag{2.23}
$$

Usually, the integrations in each element are performed by Gauss quadrature in order to provide the advantages of flexibility and the ability to re-employ subroutines in other computer programs. The expressions given above can be verified with programs implementing symbolic algebra on the computer, such as Mathematica or Maple.

As discussed earlier, the contributions from all the *nelem* elements are assembled to construct the global matrices $\mathcal{H}$ and $\mathcal{U}$ by the overlay of element matrices so as to guarantee interelement continuity. The action integral is now given by

$$\mathcal{A} = \langle \psi_\alpha | \mathcal{H}_{\alpha\beta} | \psi_\beta \rangle - \epsilon \langle \psi_\alpha | \mathcal{U}_{\alpha\beta} | \psi_\beta \rangle;$$
$$(\alpha, \beta = 1, \ldots, nelem + 1), \quad (2.24)$$

where the local index $\{(i_{el}), i\}$ has been replaced by a global index α, and the matrices $\mathcal{H}$ and $\mathcal{U}$ are $(nelem + 1) \times (nelem + 1)$ matrices. The quantity $|\psi_\alpha\rangle$ may be called the "nodal representation" of the wavefunction $\psi(r)$. We now invoke the principle of stationary action and vary the action with respect to the nodal values of the wavefunction ψ_α^*. This leads to the eigenvalue equation

$$\mathcal{H}_{\alpha\beta} | \psi_\beta \rangle = \epsilon \, \mathcal{U}_{\alpha\beta} | \psi_\beta \rangle. \quad (2.25)$$

The vanishing of the wavefunction at infinity is implemented as a boundary condition by requiring that the last nodal value at $r = r_c$ be zero; this is achieved by dropping the row and column with index $nelem + 1$ in both the global matrices. The Jacobian factor of r^2 in the integral together with the fact that both ψ and ψ' are well behaved imply that we can attach any finite value to *psi* at $r = 0$. In other words, we allow the value of *psi* at the first node to "float" to its natural value as obtained from the solution of the eigenvalue problem. The resulting generalized eigenvalue problem is solved using the EISPACK or LAPACK numerical procedures[2,3] for the eigenvalues and eigenfunctions.

The wavefunctions for the lowest three energy levels corresponding to the ground state of the hydrogen atom and the first two excited states are shown in Fig. 2.3. They have been evaluated with 20 finite elements for a range of $20\,a_0$ using linear interpolation functions in each element. The analytical results are represented by the continuous curves and the FEM results are the data points in Fig. 2.3. The eigenfunctions are not normalized; also, their value of unity at $r = 0$ is merely what is provided by the diagonalization subroutine and is of no special significance. The value of the wavefunction at $r = 20\,a_0$ has been set to zero in the calculation. This affects the $n \geq 3$ eigenfunction more than the $n = 1, 2$ states, distorting it to

Table 2.3 The energy levels of the hydrogen atom in the FEM with linear interpolation.

Quantum number n	Energy (Rydberg)	Range (a_0)	Number of elements	Exact eigenvalue
1	−0.941 709 13	20.0	20	−1.000 000
2	−0.238 162 28			−0.250 000
3	−0.092 925 16			−0.111 111
1	−0.997 725 06	20.0	120	−1.000 000
2	−0.249 544 95			−0.250 000
3	−0.099 600 92			−0.111 111
1	−0.941 709 13	100.0	100	−1.000 000
2	−0.238 203 64			−0.250 000
3	−0.106 764 46			−0.111 111
4	−0.060 445 97			−0.062 500
5	−0.038 872 84			−0.040 000
6	−0.027 036 40			−0.027 777
7	−0.018 513 97			−0.024 082
8	−0.008 388 86			−0.015 625
1	−0.997 725 06	100.0	600	−1.000 000
2	−0.249 571 17			−0.250 000
3	−0.110 957 15			−0.111 111
4	−0.062 428 37			−0.062 500
5	−0.039 961 05			−0.040 000
6	−0.027 713 05			−0.027 777
7	−0.019 170 23			−0.024 082
8	−0.009 294 91			−0.015 625

conform to the zero value. The error in the numerically calculated eigenfunction for $n = 3$ is clearly seen in Fig. 2.3. By setting the final nodal value to zero we are, in effect, modifying the problem of the hydrogen atom with its Coulomb potential having an infinite range into the problem of a hydrogen atom enclosed in a spherical box of radius r_c.

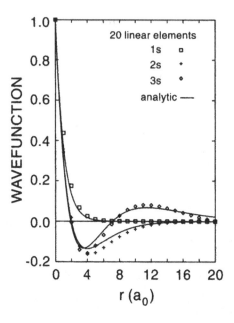

Fig. 2.3 The radial wavefunctions for the lowest three spherically symmetric states of the hydrogen atom. The data points are those obtained by the FEM using linear interpolation in 20 elements, while the continuous curves are the analytical, exact solutions. The wavefunctions have not been normalized to unity.

The lowest three eigenvalues for a range of $20\,a_0$, and the lowest eight calculated eigenvalues for a range of $100\,a_0$, are given in Table 2.3. When the range of the action integral is limited to $20\,a_0$, we note from Table 2.3 that the eigenvalues obtained for 20 elements for the ground and the excited states are higher than the exact values. As the number of elements is increased to 120, these three eigenvalues approach the exact values from above, as is expected in a variational procedure. The $n = 3$ state is more influenced by the short range and by the requirement that $\psi(r_c) = 0$. Once the range is increased to $100\,a_0$ the additional freedom in shaping the eigenfunctions closer to their analytical forms allows eight states to have negative energies. The fact that the ground state has almost the same energy as the ground state calculated with 20 elements over the range of $20\,a_0$ is because the first 20 elements in the longer calculation have the same distribution of elements over the critical range associated with the ground state. When the number of elements is increased to 600, these states have energies closer to their analytical values, again approaching the exact value from above as expected of the variational procedure.

It is clear from Table 2.3 that the linear interpolation functions are providing convergence, with increasing number of elements, to $-1/n^2$, the exact eigenvalues in this problem. For $r_c = 20\,a_0$ the lowest eigenvalue is higher by 6% for 20 elements and off by 0.2% for 120 elements. When r_c is increased to $100\,a_0$ the higher eigenvalues improve considerably. Recall that the exponential factor for the nth eigenfunction is $e^{-r/n}$, and the higher the value of n the larger the value of the cutoff that is needed to properly incorporate the oscillations in the eigenfunctions.

It will be shown in Chapter 3 that in order to improve accuracy it is essential to go to polynomials of higher degree, and, in particular, that imposing derivative continuity at nodal points dramatically improves the accuracy of the calculations. These improvements together with the adaptive placing of more elements where they are essential make the FEM the ideal numerical method for obtaining high accuracy in quantum mechanical calculations.

2.3   The Rayleigh–Ritz variational method

It is important to compare the FEM with the traditional approach to variational calculations for bound states, known as the Rayleigh–Ritz variational method.[10–12] Consider a solution, $\psi(\mathbf{r})$, to Schrödinger's equation defined over the entire physical region (a global wavefunction in the parlance of the FEM). It satisfies the equation

$$\mathcal{H}\,\psi(\mathbf{r}) = E\,\psi(\mathbf{r}).$$

The functional

$$J[\psi] = \frac{\int d^3r\,\psi^*(\mathbf{r})\,\mathcal{H}\,\psi(\mathbf{r})}{\int d^3r\,\psi^*(\mathbf{r})\,\psi(\mathbf{r})} \tag{2.26}$$

has the value $J[\psi] = E$ for the actual solution. A small change $\delta\psi(\mathbf{r})$ in $\psi(\mathbf{r})$ leads to the change $\delta J[\psi]$ in J which is most simply written as

$$\delta J[\psi]\int d^3r\,\psi^*(\mathbf{r})\,\psi(\mathbf{r}) + J[\psi]\int d^3r\,(\delta\psi^*(\mathbf{r})\,\psi(\mathbf{r}) + \psi^*(\mathbf{r})\,\delta\psi(\mathbf{r}))$$

$$= \int d^3r\,(\delta\psi^*(\mathbf{r})\,\mathcal{H}\,\psi(\mathbf{r}) + \psi^*(\mathbf{r})\,\mathcal{H}\,\delta\psi(\mathbf{r})). \tag{2.27}$$

The Hamiltonian operator $\mathcal{H}$ contains derivative operators in the kinetic energy term; it is also Hermitian. Also, the wavefunctions

ψ and $\delta\psi$ for the states are assumed to converge to zero at least as fast as $\sim 1/r$ for increasing values of r. We can then rewrite $\int d^3r\,\psi^*(\mathbf{r})\,\mathcal{H}\,\delta\psi(\mathbf{r})$ as $\int d^3r\,\delta\psi(\mathbf{r})\,\mathcal{H}\,\psi^*(\mathbf{r})$, with the vanishing of the surface terms arising from the integration by parts. We then obtain

$$\delta J[\psi] = \frac{\int d^3r\,\{\delta\psi^*(\mathcal{H}-E)\psi + \delta\psi(\mathcal{H}-E)\psi^*\}}{\int d^3r\,\psi^*\psi}. \qquad (2.28)$$

Since ψ and ψ^* are solutions of Schrödinger's equation, $(\mathcal{H}-E)\psi = 0$, we see that $\delta J[\psi] = 0$. We conclude that $J[\chi]$ with a trial solution χ is a minimum for $\chi = \psi$.

For problems that do not permit an exact solution we can obtain an approximation to the wavefunction by selecting a trial function $\chi(\mathbf{r})$ which attempts to minimize the functional J and thereby obtain an upper bound for the energy of the stationary state. This is seen by the following consideration. Let the complete set of orthonormal eigenfunctions of the Hamiltonian be denoted by $\varphi_i(\mathbf{r})$. Then any function χ can be represented by

$$\chi(\mathbf{r}) = \sum_i a_i\,\varphi_i(\mathbf{r}), \qquad (2.29)$$

where the symbol $\sum$ represents a sum over discrete states and an integral over the continuous spectrum.

If E_0 is the energy of the ground state, we have

$$\int d^3r\,\chi^*(\mathbf{r})\,\mathcal{H}\,\chi(\mathbf{r}) = \sum_i E_i|a_i|^2$$

$$\geq E_0 \sum_i |a_i|^2. \qquad (2.30)$$

Since $\int d^3r\,\chi^*\chi = \sum|a_i|^2$ we find that

$$J[\chi] \geq E_0. \qquad (2.31)$$

The form of the global wavefunction χ may depend on a number of parameters λ_i. In that case, the functional $J[\chi]$ will be a function of these parameters. Then J is minimized with respect to these parameters, and we determine them from

$$\frac{\partial J[\chi;\{\lambda_i\}]}{\partial\lambda_i} = 0. \qquad (2.32)$$

Example 1: The simple harmonic oscillator

Let us calculate the ground state wavefunction for the simple harmonic oscillator. Its Hamiltonian is

$$\mathcal{H} = -\frac{\hbar^2}{2m}\frac{d^2}{dx^2} + \frac{1}{2}m\omega^2 x^2, \tag{2.33}$$

and a trial solution that vanishes as $x \to \pm\infty$ is

$$\chi(x, \alpha) = A\,e^{-\alpha x^2/2}. \tag{2.34}$$

The normalization condition, $\int \chi^* \chi\, dx = 1$, determines the constant $A = (\alpha/\pi)^{1/4}$. The functional $J[\chi, \alpha]$ is

$$\begin{aligned} J[\chi, \alpha] &= \int dx\, \chi^* \mathcal{H} \chi \\ &= \frac{\alpha\hbar^2}{4m} + \frac{m\omega^2}{4\alpha}. \end{aligned} \tag{2.35}$$

The condition $\partial J/\partial\alpha = 0$ leads to $\alpha_0 = m\omega/\hbar$, which yields a ground state energy $E_0 = \hbar\omega/2$. The wavefunction is given by

$$\chi(x) = \sqrt{\frac{m\omega}{\hbar\pi}}e^{-m\omega x^2/\hbar}. \tag{2.36}$$

Example 2: The hydrogen atom

The standard form of the normalized trial wavefunction

$$\chi(\mathbf{r}) = \left(\frac{b^3}{\pi a_0^3}\right)^{1/2} e^{-br/a_0} \tag{2.37}$$

together with the Hamiltonian for the hydrogen atom

$$\mathcal{H} = -\frac{\hbar^2}{2m}\nabla^2 - \frac{e^2}{r} \tag{2.38}$$

gives the functional, equation (2.26),

$$J[\chi, b] = \frac{\hbar^2 b^2}{2ma_0^2} - \frac{e^2 b}{a_0}. \tag{2.39}$$

The minimum of J is attained for $b = 1$, with $E = -e^2/2a_0$. This is the exact solution. In other words, the trial wavefunction had precisely the behavior of the actual solution. A more interesting example[13] is

$$\chi(\mathbf{r}) = A \frac{1}{b^2 + \left(\dfrac{r^2}{a_0^2}\right)}, \tag{2.40}$$

in which the normalization condition determines that

$$A = \frac{1}{\pi} \sqrt{\frac{b}{a_0^3}}. \tag{2.41}$$

The minimization of the functional J for this case gives $b = \pi/4$, and a ground state energy $E_0 = -0.81 e^2 / 2a_0$ which is higher than the actual value by 19%.

For excited states, it is natural to start with a form of the wavefunction that is orthogonal to the ground state wavefunction. For example, let us calculate the energy of the first excited state, the $2s$-state, of the hydrogen atom using a trial wavefunction

$$\chi_{2s}(r) = A \left(1 + c_1 \frac{r}{a_0} \right) e^{-\alpha r/a_0}. \tag{2.42}$$

We require χ_{2s} to be orthogonal to χ_{1s} by imposing the condition

$$\int r^2 dr\, \chi_{2s}^* \chi_{1s} = 0.$$

This determines $c_1 = -\frac{1}{3}(1 + \alpha)$. The normalization condition

$$\int_0^\infty r^2 dr\, d\Omega\, |\chi_{2s}(r)|^2 = 1$$

determines the coefficient A to be

$$A = \left[\frac{3\alpha^5}{\pi a_0^3 (\alpha^2 - \alpha + 1)} \right]^{1/2}. \tag{2.43}$$

The functional $J[\chi_{2s}, \alpha]$ is now given by

$$J[\chi_{2s}, \alpha] = -\frac{e^2}{a_0} \left[\frac{3\alpha + 7\alpha^2}{6} + \frac{\alpha^2/2}{(\alpha^2 - \alpha + 1)} \right]. \tag{2.44}$$

The functional J has a minimum in the variable α for $\alpha = 1/2$, and the eigenvalue at the minimum is $E_{2s} = -e^2/8a_0$.

Thus, the basic rule to get an upper bound on the energy eigenvalue of the nth excited state is to choose a wavefunction

$$\chi(\mathbf{r}) = \varphi(\mathbf{r}) - \sum_{i=1}^{n-1} \int d^3r \, \psi_i(\mathbf{r}) \, \varphi(\mathbf{r}), \qquad (2.45)$$

where $\varphi(\mathbf{r})$ is an arbitrary quadratically integrable function that is orthogonal to all the eigenstates, $\psi_i(r)$, of lower energy. Then $J[\chi]$ has a minimum for a function χ_0 such that $J[\chi_0] \geq E_n$. This procedure presumes that we already have the exact wavefunctions for the states of lower energy.

In this and the above example, fairly simple forms of the trial wavefunction were considered so that the variational procedure could be performed analytically.[‡]

Example 3: Gaussian wavefunctions

As a final example, we consider the consequence of choosing an expansion of the trial wavefunction in terms of global wavefunctions in the evaluation of the action integral, equation (2.19), and the principle of stationary action. In quantum chemistry, it is usual to represent atomic wavefunctions by a sum of Gaussian orbitals over the entire physical range

$$\psi(r) = \sum_{i=1}^{M} c_i e^{-\beta_i r^2}. \qquad (2.46)$$

The coefficients β_i can be chosen as follows.[14] Numerical values are first assigned for the maximum, β_M, and the minimum, β_1. The other $M-1$ values are given by a geometric progression between the extreme values

$$\beta_i = \beta_{i-1} \left(\frac{\beta_M}{\beta_1} \right)^{1/(M-1)} . \qquad (2.47)$$

On integrating over the radial dependence in equation (2.19), we obtain

$$\mathcal{A} = \langle c_i \, | \, \mathcal{H}_{ij} \, | \, c_j \rangle - \epsilon \langle c_i \, | \, \mathcal{U}_{ij} \, | \, c_j \rangle. \qquad (2.48)$$

[‡]Other examples are given in the problems at the end of this chapter.

Here $\mathcal{H}$ and $\mathcal{U}$ are filled matrices. The variational principle is implemented by setting $\delta\mathcal{A}/\delta c_i^* = 0$, leaving us the generalized eigenvalue problem

$$\mathcal{H}_{ij}\,|\,c_j\rangle = \epsilon\,\mathcal{U}_{ij}\,|\,c_j\rangle. \qquad (2.49)$$

The eigenfunctions are the vector array of the coefficients c_i and the full wavefunction can be reconstructed by returning to equation (2.46) and normalizing it to unity. This approach is very similar to the FEM except that at the starting point we assume a global form for the wavefunctions. For about 150 terms in the Gaussian expansion,[14] the lowest few eigenvalues are obtained with an accuracy of 1 part in 10^3–10^4.

In conclusion, we note that the main criteria for selecting the form of a trial global wavefunction χ are that it must be square integrable, it must satisfy the boundary conditions present in the problem, it should be approximately of the same shape as the actual solution in order to provide the least upper bound, and it must give a functional $J[\chi]$ that is amenable to the evaluation of its variation with respect to the parameters in χ. If the Ritz approach is used by employing a large number of orthonormalized global wavefunctions, the global matrix is invariably completely filled.

By contrast, in the FEM we employ interpolation polynomials over each piece (element) of the physical region. While this provides the necessary flexibility in the element-based trial wavefunction, it also gives rise to more parameters in the form of the nodal values for the wavefunctions. Because the interpolation polynomials are nonzero only over specific elements, the global matrices are usually banded, and this banded structure can be exploited during the diagonalization procedure.

2.4 Programming considerations

Consider the problem of determining the energy levels of the electron in the hydrogen atom. The flowchart shown in Fig. 2.4 provides an overview of the programming tasks needed in the finite element analysis. The standard programming language in finite element analysis is still Fortran. Some of the advantages of the C programming language are incorporated into the most recent version, F90, of the language. We shall make use of a more conservative approach, in order to keep the programming transparent, and use the F77 version

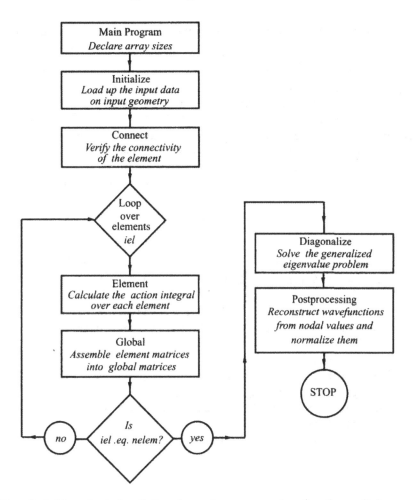

Fig. 2.4 Flowchart for finite element programming for the radial wavefunctions of the hydrogen atom

in the discussion. In the following, we present fragments of Fortran source codes to illustrate key issues in programming.

We begin by deciding on the number of elements, **nelem**, we want to use for the discretization of the action integral, equation (2.20), and the type of interpolation polynomials to be used. We allow ourselves the choice between linear or quadratic interpolation polynomials by assigning the variable **node_elm** the value 2 or 3, respectively. A simple procedure for obtaining interpolation polynomials is given in Chapter 3.

A useful way of allowing changes in key variables which influence the values of a large number of other variables is to assign them values through the parameter assignment at the beginning of the main program. This is illustrated in the following lines of code.

```
c===============================================================
c 1D finite element program to calculate energy
c eigenvalues and wavefunctions for the radial wave
c equation for the h-atom.
c Interpolation  : linear or quadratic interpolation.
c===============================================================
      integer nelem1
      parameter (nelem1  = 10) !The number of elements.

      integer  npts             !(20..100)#Points for graphs.
      parameter (npts = 20)

      real*8 xmin1, xmax1       !Range of the physical region
      parameter (xmin1 = 0.0d0, xmax1 = 20.0d0)

      integer nodelm1          !Number of nodes per element.
      parameter (nodelm1 = 3)!(2/3) for (linear/quadratic)

      integer nelsz1
      parameter (nelsz1 = nodelm1)
      integer ng_nodes1, ngl1
      parameter (ng_nodes1 = nelem1*(nodelm1 - 1) + 1 )
      parameter (ngl1 = ng_nodes1) ! global matrices:
      real*8 agl(ngl1, ngl1), bgl(ngl1, ngl1)
      ...        ...      ...

      nelem = nelem1
      ngl   = ngl1
      ...        ...      ...


c===============================================================
```

The declaration of the dimensions of all the needed arrays, such as the global matrices agl, bgl, can be controlled by just the assigned parameter, nelem1. The disadvantage of this approach to keeping the dimensions of arrays flexible is that Fortran expects the assigned parameters not to change in a program. We can bypass this restriction by assigning the values of the above "constants" to the actual variables we use in the rest of the program, as shown in the last few lines in the above example.

The main program is kept as short as possible and in a separate file, so as to allow faster recompilation, while the subroutines are compiled and converted to object files only once.

The next few steps initialize arrays to zero, include data on Gauss quadrature points and weights, define the beginning and the end coordinates of all the elements, and load this information into common blocks that are invoked in a number of subroutines.

It soon becomes obvious to anyone programming the FEM that keeping track of arrays and inserting element matrices correctly into global matrices are at the core of the programming considerations. Nodes are usually given a local node number in each element. Thus, the two nodes in an element `iel` employing linear interpolation are locally labeled 1 and 2, respectively, whereas their global node numbers are `(iel-1)*(node_elm-1)+1` and `(iel-1)*(node_elm-1)+2`. By including the variable `node_elm` in the expression for the global node number in terms of the local number we can allow for the possibility of either linear or quadratic interpolation in the code. Once the global node number is known the corresponding global coordinate of the node can also be determined. This mapping, called connectivity, from local variables to global variables is critical for precisely fitting element matrices and overlaying them in global matrices. A good discussion of this is available in the book by Hinton and Owen,[15] who define an integer array `lnods(nelem, node_elm)` which returns the global node number when the element number and the local node number are used as input. The first milestone in the programming is reached when we can print out the correct connectivity data for the problem.

The next stage involves the evaluation of the action integral over each element and the calculation of element matrices. The integration is performed using Gauss quadrature. We loop over the Gauss points and evaluate the interpolation (shape) functions and their derivatives at each Gauss point. These are used in the quadrature. We show this calculation in the following subroutine `element`.

```
c===============================================================
      subroutine element(aelm, belm, lnods, nelem,
     $   node_elm, iel)
c===============================================================
c alem, belm    : element matrices -- dimension node_elm1
c lnods         : integer array of dimension
c                     (nelem, node_elm).
c                 Given a node number within an element,
c                 lnods returns the corresponding global
```

```
c                     node number.
c nelem, node_elm : number of elements, size of element
c                     matrices
c iel               : index for the current element being
c                     evaluated.
c This subroutine is where almost all of the physics
c is included.

      integer  iel, nelem, node_elm
      real*8   aelm(node_elm, node_elm),
     $         belm(node_elm, node_elm)
      integer  lnods(nelem, node_elm)

      real*8   xmin, xmax, coord(500)
      common/geometry/ xmin, xmax, coord
      real*8   gauswt(8), gauspt(8)
      integer ngaus
      common/gauss/ gauswt, gauspt, ngaus

c Local variables:
c ----------------
c Here phi and phi_p store the value of shape functions
c and their first derivative at given point xi.

      real*8 potential
      real*8 xbegin, xend, xjacobian, xmid, xglobal,
     $    xi, wt, xintfactor, phi(3), phi_p(3),
     $    akinetic, apotential, afactor, bfactor
      integer ndeg ii, jj, ig, ndim

c Use subroutine ''matzero'' to initialize matrices
c to zero.
      call matzero (aelm, node_elm)
      call matzero (belm, node_elm)

c Calculate the global coordinates of the nodes in
c the element.The local coordinate xi ranges over
c [-1, 1] so that Gauss-Legendre quadrature can be
c performed.

      ii        = lnods(iel, 1) !first node of element iel
      jj        = lnods(iel, node_elm) !last node in iel
      xbegin    = coord (ii) !r-coordinate of first node
      xend      = coord (jj) !r-coordinate  of last node
      xjacobian = (xend - xbegin)/2.0d0
```

```
      xmid        = (xend + xbegin)/2.0d0

c Loop to perform the sum over Gauss points
c The order of the quadrature is ngaus (= 8).

      do ig = 1, ngaus
         xi  = gauspt(ig)        ! Gauss point
         wt  = gauswt(ig)        ! Gauss weight

c The global variable is given by
c       x_global    = xmid  + \xi * xjacobian
c Thus:
c   d x_global = d xi * x_jacobian = d xi * (xbegin-xend)/2

         xglobal = (xmid + xjacobian*xi)
         xintfactor = wt * xjacobian * xglobal**2

c Evaluate the shape functions and their derivatives at
c Gauss quadrature points. phi(i) stores the values of
c N_i(xi), while phi_p(i) stores  values of d N_i(xi)/dr.

         ndeg = node_elm -1 ! degree of polynomial
         ndim = 3 ! The dimension of arrays phi, phi_p
         call shape (phi, ndim, ndeg, xi)
         call deriv (phi_p, ndim, ndeg, xi, xjacobian)

c Calculate all matrix elements of the element matrices
c aelm and belm.

         potential = - 2.0d0/xglobal
         do ii = 1, node_elm    !loop over rows of element
            do jj = 1, node_elm !loop over cols of element

               akinetic    =xintfactor*phi_p(ii)*phi_p(jj)
               apotential =xintfactor*phi(ii)*phi(jj)*
     $                          potential

               afactor     =akinetic+apotential
               bfactor     =xintfactor* phi(ii)*phi(jj)

               aelm(ii,jj) =aelm(ii,jj) + afactor
               belm(ii,jj) =belm(ii,jj) + bfactor
            enddo
         enddo
```

```
     enddo !end loop over Gauss pts in Gauss Quadrature

     return
     end
c=============================================================
```

A subroutine **global** is used to loop over the number of elements **nelem** and call the subroutine **element** each time through the loop, and load the element matrices into global matrices. The starting global matrix indices for the row and column where the first matrix elements of the element matrices **aelm**, **belm** are placed in the global matrices **agl**, **bgl** are given by

```
     igl = (iel -1)*(nodelm -1) + ielm
     jgl = (iel -1)*(nodelm -1) + jelm
```

where **ielm**, **jelm** run over $\{1, \ldots, \text{node\_elm}\}$. This ensures the overlay of element matrices in the global matrices.

We are now ready to impose the boundary conditions on the nodal variables ψ_i. In the present problem, we wish to set to zero the final nodal variable $\psi_{nelem+1}$ located at $r = r_c$. This is done by setting to zero the last row and column of the matrices **agl**, **bgl**.

Fortran 77 is unforgiving in not allowing this reduction of the array size anywhere in the program once it has been declared at the beginning. We can circumvent this limitation by inserting a large number **nlarge** ($\sim 10^5$, say) for the last diagonal entry in **agl**, and unit value for the last diagonal entry of **bgl**. This makes the last entry a block-diagonal 1×1 matrix with an eigenvalue **nlarge**, and the corresponding eigenfunction is orthogonal to all the other nodal wavefunctions. This solution is easily identified and discarded from further consideration after the diagonalization is performed. At this stage, the global matrices are passed into the diagonalizer from the EISPACK mathematical library which performs the diagonalization using the so-called QZ algorithm.[16]

The final phase, *post-processing*, of the programming consists of casting the output of the diagonalization in a useful format. Since typical full-matrix diagonalizers return the eigenvalues in no particular order, we begin by sorting the eigenvalues to select the negative eigenvalues, in the present example, and the corresponding nodal wavefunctions. Now we reconstruct the eigenfunctions at **npts=100** evenly spaced points r_i in the range $[0, r_c]$. The interpolation formula equation (2.6) is used to obtain $\psi(r_i)$ by first determining the element iel in which r_i occurs. For each r_i we loop over all the elements to see if $r_1^{iel} \leq r_i < r_2^{iel}$ in order to identify the correct value of iel. We are comparing real numbers here, as opposed to integers, in this test for the element number. Hence this comparison is best done with r_1^{iel}, r_i, and r_1^{iel+1} being first converted to single precision; we have to account for the round-off errors in dividing the range into elements and comparing "double-precision" numbers with possible round-off will not be effective every time. Using the value of lnods (iel, inode) and the array of nodal coordinates, we can obtain the global coordinates of the element nodes. The wavefunction reconstruction is then straightforward.

The nodal wavefunctions can also be normalized by using Gauss quadrature. The wavefunction is now obtained at Gauss points in each element and the quadrature performed element by element. Once the value $\mathcal{N} = \int |\psi(r)|^2 \, r^2 \, dr$ is calculated, each nodal value of the wavefunction is scaled by $1/\sqrt{\mathcal{N}}$ to obtained the desired normalization. This completes the post-processing, unless computations of matrix elements of specific operators, such as the momentum matrix element evaluated for specific initial and final state wavefunctions, are needed.

We have described the essential points that should be kept in focus during program development for the FEM. As with all computational approaches, such programming should include "hooks" for expanding the program and its applicability to a variety of additional problems. In other words, because sequential programming is so rigid in its flow, it is crucial to think about the next problems one wishes to consider, and build their essential requirements into the program. In this connection, it is important to demand high accuracy in the output at each level of program development. The importance of numerical accuracy at each stage of the calculations is emphasized in the textbook by Acton.[17] We consider ways of improving accuracy in the following chapters.

2.5 Problems

1. Evaluate the energy levels and eigenvalues for bound states in a finite square-well potential using the method of finite elements with linear interpolation in each element.

   (a) Suppose the potential barriers on both sides of a 100 Å well

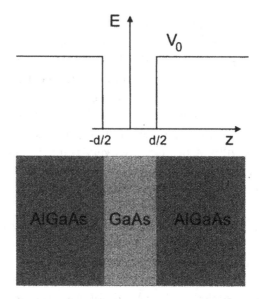

Fig. 2.5 A quantum well with a finite potential barrier V_0. An electron in the conduction band of the GaAs layer in the semiconductor heterostructure shown above experiences a barrier of V_0 at the AlGaAs layers.

are 340 meV, and the effective mass of the particle is $m^* = 0.0667\,m_e$ inside the well as well as in the barrier regions. How many bound states are present within the well? This problem is realized in a semiconductor heterostructure quantum well grown by molecular beam epitaxy, by growing a 100 Å GaAs layer sandwiched between thick layers of Al$_{0.35}$Ga$_{0.65}$As, as shown in Fig. 2.5.

(b) What is the layer thickness of the quantum well so that the highest bound state is just 10 meV below the barrier height?

2. Obtain, using the FEM, the quantum mechanical energy levels and radial wavefunctions for the radially symmetric states of a particle moving in the Morse potential[18]

$$V(r) = V_0 \left(e^{-2\mu x} - 2\,e^{-\mu x} \right).$$

Here $x = (r - r_e)/r_e$ with the potential having a minimum of $-V_0$ at $r = r_e$. For numerical work, let $\mu = 2.4$, with $V_0 = -4$ eV, and $\hbar^2/(2\,m\,r_e^2) = 1.3 \times 10^{-3}$ eV.

3. The radial equation for the hydrogen atom with nonzero orbital angular momentum $\ell\hbar$ is

$$-\frac{\hbar^2}{2m}\left[\left(\frac{d^2}{dr^2}+\frac{2}{r}\frac{d}{dr}\right)-\frac{\ell(\ell+1)}{r^2}\right]\psi(r)-\frac{e^2}{r}\psi(r)=E\,\psi(r).$$

(a) Obtain the corresponding action integral. With convenient reduced variables, express the action in terms of dimensionless variables.

(b) Break up the action integral into $nelem = 100$ elements, and integrate the radial dependence of the action using linear interpolation over each element.

(c) Apply the boundary conditions appropriate to $\ell \neq 0$ states. As usual, at infinity these states vanish exponentially as $e^{-r/n}$, and at $r = 0$ the requirement that the solution be finite leads to the boundary condition that the radial wavefunction vanish there.

(d) Solve the resulting generalized eigenvalue problem using matrix diagonalization subroutines, and obtain the lowest three eigenvalues and eigenfunctions for (i) p-states ($\ell = 1$), (ii) d-states ($\ell = 2$), and (iii) f-states ($\ell = 3$).

4. (a) Evaluate analytically the error in the ground state energy of the simple harmonic oscillator on employing the trial wavefunction $\chi(x) = A\exp(-\alpha|x|)$ in the variational method.

(b) Analytically determine the error in the ground state energy of the hydrogen atom if the variational method employs the trial wavefunction $\chi(r) = A\exp(-\alpha r^2)$.

5. The ground state of the helium atom, with its two electrons bound to the α-particle, is obtained by solving the two-particle Schrödinger equation with the Hamiltonian[§]

$$\mathcal{H} = -\frac{\hbar^2}{2m}(\nabla_1^2 + \nabla_2^2) - 2\,e^2\left(\frac{1}{r_1} + \frac{1}{r_2}\right) + \frac{e^2}{r_{12}}.$$

[§]The books by Schiff,[1] Margenau and Murphy,[19] and Bethe and Salpeter[20] may be consulted.

(a) Use the normalized trial wavefunction

$$\chi(\mathbf{r}_1, \mathbf{r}_2) = \frac{\alpha^3}{\pi a_0^3} e^{-\alpha(r_1 + r_2)/a_0},$$

where α is an arbitrary variational parameter, to show that

$$J[\chi, \alpha] = \frac{e^2}{a_0}(\alpha^2 - 4\,\alpha) + e^2 \left(\frac{\alpha^3}{\pi a_0^3}\right)^2 \left(\frac{5\pi^2 a_0^5}{8\alpha^5}\right).$$

(The term $\langle 1/r_{12}\rangle$ is evaluated by expansion in Legendre polynomials.)

(b) Obtain the minimum of J by varying α and determine an upper bound for the ground state energy of the helium atom. Compare your value with a ground state energy $(-2.903724354\, e^2/a_0)$ obtained with the variational method using a more complex trial function by Freund et al.[21]

References

[1] L. I. Schiff, *Quantum Mechanics*, 3rd ed. (McGraw-Hill, New York, 1968).

[2] B. T. Smith *et al.*, *Matrix Eigensystem Routines - EISPACK Guide*, 2nd ed., 'Lecture Notes in Computer Science', Vol. 6 (Springer-Verlag, New York, 1976).

[3] E. Anderson, Z. Bai, C. Bischof, J. Demmel, J. J. Dongerra, J. Du Croz, A. Greenbaum, S. Hammarling, A. McKenney, S. Ostrouchov, and D. Sorensen, *LAPACK User's Guide* (Society for Industrial and Applied Mathematics, Philadelphia, 1992).

[4] K.-J. Bathe and E. Wilson, *Numerical Methods in Finite Element Analysis* (Prentice Hall, Englewood Cliffs, NJ, 1976); K.-J. Bathe, *Finite Element Procedures in Engineering Analysis* (Prentice Hall, Englewood Cliffs, NJ, 1982); K.-J. Bathe, *Finite Element Procedures* (Prentice Hall, Englewood Cliffs, NJ, 1996).

[5] O. C. Zienkiewicz and Y. K. Cheung, *Finite Element Methods in Structural and Continuum Mechanics* (McGraw-Hill, New York, 1967); O. C. Zienkiewicz, *The Finite Element Method* (McGraw-Hill, New York, 1977).

[6] T. J. R. Hughes, *The Finite Element Method* (Prentice Hall, Englewood Cliffs, NJ, 1987).

[7] D. S. Burnett, *Finite Element Analysis* (Addison-Wesley, Reading, MA, 1987), p429.

[8] D. J. Searles and E. I. von Nagy-Felsobuki, *Am. J. Phys.* **56**, 444 (1988).

[9] A. Askar, *J. Chem. Phys.* **62**, 732 (1975); also see M. Friedman, Y. Rosenfeld, A. Rabinovitch, and R. Thieberger, *J. Comput. Phys.* **26**, 169 (1978).

[10] Lord Rayleigh, *Theory of Sound*, 2nd ed. (Dover, New York, 1945).

[11] W. Ritz, *J. reine angew. Math.* **135**, 1 (1908).

[12] R. Courant and D. Hilbert, *Methods of Mathematical Physics*, Vol. I (Interscience, New York, 1953).

[13] F. Constantinescu and E. Magyari, *Problems in Quantum Mechanics* (Pergamon, New York, 1971), p206.

[14] C. Aldrich and R. L. Greene, *Phys. Status Solidi* (b) **93**, 343 (1979).

[15] E. E. Hinton and D. R. J. Owen, *An Introduction to Finite Element Computations* (Pineridge Press, Swansea, UK, 1979).

[16] G. H. Golub and C. F. van Loan, *Matrix Computations* (Johns Hopkins University Press, Baltimore, MD, 1983), p251.

[17] F. S. Acton, *Numerical Methods That Work* (Harper & Row, New York, 1970), corrected ed. (Mathematical Association of America, Washington, DC, 1990).

[18] P. M. Morse, *Phys. Rev.* **34**, 57 (1928).

[19] H. Margenau and G. M. Murphy, *The Mathematics of Physics and Chemistry* (van Nostrand, Princeton, NJ, 1956), p381.

[20] H. A. Bethe and E. E. Salpeter, *Quantum Mechanics of One- and Two-Electron Atoms* (Plenum Press, New York, 1977).

[21] D. E. Freund, B. D. Huxtable, and J. D. Morgan III, *Phys. Rev.* **29**, 980 (1984).

3

Interpolation polynomials in one dimension

3.1 Introduction

In the finite element method (FEM) in one dimension, we solve for the values of an unknown function $f(x)$ at discrete nodal points in the elements. In this chapter, we consider the representation of $f(x)$ over a given interval using interpolation polynomials, called Lagrange polynomials, assuming we know the values f_i of the function at n discrete nodal points x_i in an interval. With the use of Lagrange interpolation polynomials, we can ensure that $f(x_i) = f_i$ and $f(x)$ is approximated by the polynomials in between. This ensures that the function has continuity at the nodes in the element, and the polynomial representation is said to have C_0 continuity. The end nodes of an element are common to the adjacent elements so that interelement continuity of the function is also readily assured. The Lagrange polynomials are discussed in the next section.

If we desire that the function *and some of its derivatives* be continuous at the nodes, we need a different set of polynomials, the Hermite interpolation polynomials.[†] The continuity of the first derivative at the nodes is referred to as C_1 continuity. Hermite interpolation polynomials are presented in Section 3.3.

Two special-purpose elements are introduced: transition elements and infinite elements. The transition element provides the flexibility to maintain interelement continuity at element boundaries, while the derivative continuity requirements at the two ends of the transition element can be different. The infinite domain "finite" element allows one to include the effects of the infinite domain without introducing cutoffs in the range of integration in the action integral.

[†]To avoid confusion with the solutions of the Hermite equation appearing in the treatment of the quantum harmonic oscillator we will consistently refer to these as "Hermite interpolation polynomials".

In the following, we use a simple numerical method for generating the Lagrange and Hermite interpolation polynomials. We assume without any loss of generality that the range of interpolation is $[-1, +1]$. (This coincides with the range of integration in Gauss–Legendre quadrature.) A simple change of variables can always be used to map the region onto any other region $[a, b]$ for a given element.

Short programs in Mathematica are given in Section 3.6 for generating the Lagrange and Hermite polynomials of any desired degree.

We conclude this chapter by recalculating the energy levels of the quantum oscillator and the radial equation of the hydrogen atom using Hermite interpolation polynomials. The use of these polynomials demonstrate that the FEM can yield the lower eigenvalues in bound state problems to double-precision accuracy with relatively few elements. This is achieved without the need to guess at the form of global wavefunctions.

3.2   Lagrange interpolation polynomials

Any function can be written over a finite interval as

$$f(x) = \sum_i^n f_i \, \phi_i(x) + R_n(x), \tag{3.1}$$

where the polynomials $\phi_i(x)$ are of degree $n - 1$ and the remainder term $R_n(x)$ accounts for the difference between $f(x)$ and its polynomial approximation. The polynomials are chosen so that they have unit value at the specific points x_i. Such polynomials are called Lagrange interpolation polynomials and are given by

$$\phi_i(x) = \prod_{\substack{j=1 \\ j \neq i}}^n \frac{(x - x_j)}{(x_i - x_j)} \tag{3.2}$$

such that

$$\phi_i(x_j) = \delta_{ij}. \tag{3.3}$$

Here δ_{ij} is the Kronecker delta which equals zero when $i \neq j$, and is unity for $i = j$. The above representation, equation (3.1), is exact at these sampling points, with R_n set to zero there. The basic assumption is that for a small enough interval and for large enough

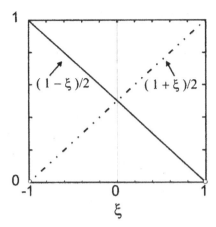

Fig. 3.1 The Lagrange functions for linear interpolation – two nodes per element. The nodes are located at ± 1

n the remainder term can be neglected over the entire interval. This approximation is part of the discretization procedure.

The simplest of the Lagrange interpolation schemes, beyond the piecewise constant interpolation, is linear interpolation within each element. The interpolation functions in the range $[-1, 1]$ are shown in Fig. 3.1. For illustrative purposes, let us evaluate the next set of Lagrange interpolation polynomials using the nodal points $(-1, 0, 1)$. While it is not necessary to have uniformly spaced nodes, it is usual to make this choice and to locate nodes at either end of an element. With three nodal points $(\xi_i = -1, 0, 1)$, we will have three polynomials which will be the quadratic functions

$$\phi_i(\xi) = a_i + b_i\,\xi + c_i\,\xi^2; \qquad (i = 1, 2, 3), \qquad (3.4)$$

where a_i, b_i, c_i are coefficients to be determined from the requirement that

$$\phi_i(\xi_j) = \delta_{ij}. \qquad (3.5)$$

This condition can be written in a matrix form:

$$\begin{pmatrix} 1 & -1 & 1 \\ 1 & 0 & 0 \\ 1 & 1 & 1 \end{pmatrix} \cdot \begin{pmatrix} a_1 & a_2 & a_3 \\ b_1 & b_2 & b_3 \\ c_1 & c_2 & c_3 \end{pmatrix} = \mathbf{I}, \qquad (3.6)$$

or

$$\mathcal{C} \cdot \mathbf{\Phi} = \mathbf{I}. \qquad (3.7)$$

Here the numerical entries of the matrix $\mathcal{C}$ in the ith row refer to the values of $(1, \xi, \xi^2)$ at the nodal points ξ_i, and $\mathbf{I}$ is the unit matrix.

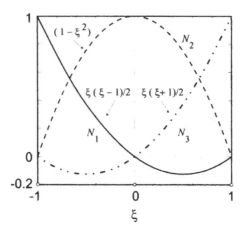

Fig. 3.2 The quadratic Lagrange functions for an element with three nodes. The nodes are located at -1, 0, and 1.

The functions ϕ_i are obtained by solving for the coefficients $(a_i, \ldots)$, and are shown in Fig. 3.2. The coefficients a_i, b_i, c_i are the *column* entries of the ith column in the inverted matrix C^{-1}. The extension to a larger number of nodes in the range $[-1, 1]$ is straightforward.[1,2]

It should be noted that the standard notation for interpolation functions is $N_i(\xi)$. (We reserve x for the global coordinate and ξ for the local coordinate.) We will denote the general polynomials of degree $i - 1$ by ϕ_i, and those particular ϕ_i which satisfy the nodal conditions, equation (3.5), by N_i. Explicit expressions for N_i for the lowest two sets of interpolation polynomials are collected together in Section 3.5.

3.3   Hermite interpolation polynomials

The Hermite interpolation polynomials allow us to impose derivative continuity at the nodes. As an example, let us determine the interpolation functions $N_i(\xi)$, $\overline{N}_i(\xi)$ for an element having two nodes with two degrees of freedom at each node. These functions allow us to represent a function $\Psi(\xi)$, whose values and derivatives at the nodes at ± 1 are known, in the form

$$\Psi(\xi) = \sum_i (N_i(\xi)\, \Psi_i + \overline{N}_i(\xi)\, \Psi_i'); \qquad i = 1, 2, \qquad (3.8)$$

where the Hermite interpolation polynomials $N_i(\xi), \overline{N}_i(\xi)$ must satisfy the conditions

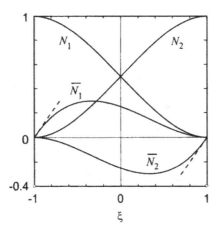

Fig. 3.3 The cubic Hermite interpolation functions for an element with two nodes and two degrees of freedom at each node. The dashed lines indicate the unit slope at the nodes located at ± 1 for the functions $\overline{N}_1$ and $\overline{N}_2$.

$$N_i(\xi_j) = \delta_{ij}; \qquad \overline{N}_i(\xi_j) = 0;$$
$$\frac{d}{d\xi} N_i(\xi_j) = 0; \qquad \frac{d}{d\xi} \overline{N}_i(\xi_j) = \delta_{ij}; \qquad i, j = 1, 2. \qquad (3.9)$$

The degrees of freedom refer to the "freedom" in assigning values for Ψ_i and Ψ_i'. The general polynomial for the case of two nodes with two degrees of freedom is a cubic which can be written as

$$\phi_i(\xi) = a_i + b_i\xi + c_i\xi^2 + d_i\xi^3; \quad (i = 1, \ldots, 4). \qquad (3.10)$$

We let ϕ_1, ϕ_3 equal N_1, N_2, and ϕ_2, ϕ_4 equal $\overline{N}_1, \overline{N}_2$, respectively. Equations (3.9) can be written in a matrix form, as the product of two matrices. The first matrix contains the values of powers of ξ evaluated at ± 1, with the rows arranged alternately for N_i and for $\overline{N}_i'$. The second matrix contains the as-yet-unknown coefficients appearing in equations (3.9). We have

$$\begin{pmatrix} 1 & -1 & 1 & -1 \\ 0 & 1 & -2 & 3 \\ 1 & 1 & 1 & 1 \\ 0 & 1 & 2 & 3 \end{pmatrix} \cdot \begin{pmatrix} a_1 & a_2 & a_3 & a_4 \\ b_1 & b_2 & b_3 & b_4 \\ c_1 & c_2 & c_3 & c_4 \\ d_1 & d_2 & d_3 & d_4 \end{pmatrix} = \mathbf{I}. \qquad (3.11)$$

Again, an inversion of the matrix allows us to obtain the coefficients a_i, etc., as the numerical entries of the inverted matrix along the

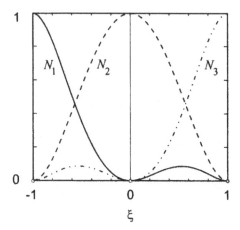

Fig. 3.4 The quintic Hermite interpolation functions, N_1, N_2, and N_3, are shown. They have unit magnitudes at the nodes located at $\xi = -1$, 0, $+1$, respectively, while their derivatives at the three nodes are zero.

ith column. The cubic Hermite interpolation polynomials obtained in this manner are given in Section 3.5. Figures 3.3–3.5 display the resulting cubic and quintic polynomials.

With Hermite interpolation, the representation of a function $\Psi(x)$ over a range $[a, b]$ of a given element, in terms of the interpolation functions which are defined over $[-1, 1]$, requires a further step. We have to map the global coordinate x into the local coordinate ξ using

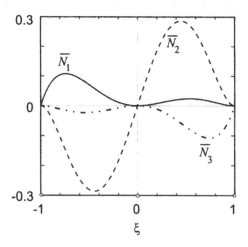

Fig. 3.5 The quintic Hermite interpolation functions, $\overline{N}_1$, $\overline{N}_2$, and $\overline{N}_3$, are shown. They are zero at $\xi = -1$, 0, $+1$. Note that their derivatives have unit magnitude at $\xi = -1$, 0, $+1$, respectively.

$$x = \frac{(a+b)}{2} + \frac{(b-a)}{2}\,\xi,$$

$$\xi(x) = \frac{(2x-a-b)}{b-a}. \tag{3.12}$$

We have

$$\Psi(x) = \sum_i \Psi_i\, N_i(\xi(x)) + \sum_i \Psi_i'\, \overline{N}_i(\xi(x))\,\frac{dx}{d\xi}, \tag{3.13}$$

where Ψ_i are the nodal values of the function and Ψ_i' are the derivatives of $\Psi(x)$, *with respect to x*, at the nodes. The additional factor of $dx/d\xi = (b-a)/2$ does not affect the definition of $\Psi(x)$ at the nodes since the functions $\overline{N}_i(\xi)$ vanish there. However, this factor is essential in reproducing $d\Psi(x)/dx = \Psi'$ at $x = x_i$. Differentiating $\Psi(x)$ given in equation (3.13) with respect to x we obtain

$$\frac{d}{dx}\Psi(x) = \sum_i \Psi_i\left(\frac{dN_i(\xi)}{d\xi}\cdot\frac{d\xi}{dx}\right) + \sum_i \Psi_i'\,\frac{d\overline{N}_i(\xi)}{d\xi}. \tag{3.14}$$

With the functions $N'(\xi)$ vanishing at the nodes, we see that $\Psi'(x)$ is properly reproduced at x_i.

In formulating a physical problem, the notational details, including these scaling factors, are best kept hidden behind the scenes, and it is useful to write the above equations in the compact form

$$\Psi(x) = \sum_\alpha \Psi_\alpha\, \Phi_\alpha(x); \qquad \alpha = 1,\ldots,\{nnode \times ndof\}. \tag{3.15}$$

Here *nnode* is the number of nodes and *ndof* is the number of degrees of freedom per node (in this case, $ndof = 2$), and

$$\Psi_\alpha = (\{\Psi_i\},\ \{\Psi_i'\}),$$

$$\Phi_\alpha(x) = \left(\{N_i(\xi)\},\ \left\{\overline{N}_i(\xi)\frac{dx}{d\xi}\right\}\right). \tag{3.16}$$

The derivative of $\Psi(x)$ is given by

$$\Psi'(x) \equiv \frac{d\Psi(x)}{dx} = \sum_\alpha \Psi_\alpha\,\frac{d\Phi_\alpha(x)}{dx}, \tag{3.17}$$

where

$$\frac{d\Phi_\alpha(x)}{dx} = \left\{\frac{dN_i}{d\xi}\frac{d\xi}{dx},\ \frac{d\overline{N}_i(\xi)}{d\xi}\right\}. \tag{3.18}$$

3.4 Transition elements

On occasion, it is convenient to reduce the number of degrees of freedom across an element which could link, for example, an element
using Hermite interpolation on one side with an element employing Lagrange interpolation on the other side. Another circumstance
could be that we are using higher order Hermite interpolation, with
continuity of $f(x)$, $f'(x)$, $f''(x)$ at the nodes, in homogeneous regions
where the second-order differential equation obeyed by $f(x)$ also
provides an additional continuity condition. At interfaces and at
boundaries the order of derivatives which are specified by physical
considerations could be lower. A "transition element" would be a
natural choice for lowering the degrees of freedom down to the desired number. At interfaces and boundaries we can then implement
the physically appropriate boundary conditions in a natural manner.

Let us consider the specific case of an element with two nodes,
with the left node having two degrees of freedom (f_1, f_1' are specified
there) and the right node having only one (f_2 alone is specified here).
We write

$$f(x) = f_1\, N_1(\xi) + f_1'(x)\, \overline{N}_1(\xi)\, \frac{dx}{d\xi} + f_2\, N_2(\xi), \qquad (3.19)$$

where the nodal subscripts 1,2 refer to the two nodes at $\xi = \mp 1$,
respectively. We can write

$$\begin{aligned}
N_1(\xi) &= a_1 \ + \ b_1\,\xi \ + \ c_1\,\xi^2, \\
\overline{N}_1(\xi) &= a_2 \ + \ b_2\,\xi \ + \ c_2\,\xi^2, \\
N_2(\xi) &= a_3 \ + \ b_3\,\xi \ + \ c_3\,\xi^2.
\end{aligned} \qquad (3.20)$$

Following the same procedure as in earlier examples, we derive the
three quadratic interpolation functions for this case by solving for
the coefficients from the relation

$$\begin{pmatrix} 1 & -1 & 1 \\ 0 & 1 & -2 \\ 1 & 1 & 1 \end{pmatrix} \cdot \begin{pmatrix} a_1 & a_2 & a_3 \\ b_1 & b_2 & b_3 \\ c_1 & c_2 & c_3 \end{pmatrix} = \mathbf{I}. \qquad (3.21)$$

This leads to the interpolation functions

$$N_1'(\xi) = \frac{3}{4} - \frac{1}{2}\xi - \frac{1}{4}\xi^2,$$

$$\overline{N}_1(\xi) = \frac{1}{2} - \frac{1}{2}\xi^2, \tag{3.22}$$

$$N_2(\xi) = \frac{1}{4} + \frac{1}{2}\xi + \frac{1}{4}\xi^2.$$

It can be verified that $N_i(\xi_j) = \delta_{ij}$, $\{i, j = 1, 2\}$, and that N_i have zero slope at $\xi = -1$. On the other hand, $\overline{N}_1$ is zero at the two nodes at $\xi = \pm 1$, while it has unit slope at $\xi = -1$. The derivatives of these functions at $\xi = +1$ are of no concern since no derivative degree of freedom is invoked at $\xi = 1$.

We see that the numerical method of obtaining the Hermite interpolation polynomials presented here is a convenient and flexible procedure. This can be extended in two ways: one is by increasing the number of nodal points in the range $[-1, 1]$, and the other is by increasing the number of degrees of freedom. Hermite interpolation functions of low order are given in the following.

3.5 Low order interpolation polynomials

3.5.1 Low order Lagrange interpolation

1. *Linear interpolation functions for an element with two nodes.* (See Fig. 3.1.)

$$N_1(\xi) = (1 - \xi)/2,$$
$$N_2(\xi) = (1 + \xi)/2; \qquad -1 \le \xi \le 1. \tag{3.23}$$

2. *Quadratic Lagrange interpolation functions for an element with three nodes.* (See Fig. 3.2.)

$$N_1(\xi) = -\xi/2 + \xi^2/2,$$
$$N_2(\xi) = 1 - \xi^2,$$
$$N_3(\xi) = \xi/2 + \xi^2/2; \qquad -1 \le \xi \le 1. \tag{3.24}$$

3.5.2   Low order Hermite interpolation

1. Cubic Hermite interpolation polynomials for elements with two nodes and two degrees of freedom. (See Fig. 3.3.)

$$
\begin{aligned}
N_1(\xi) &= (2 - 3\xi + \xi^3)/4, \\
\overline{N}_1(\xi) &= (1 - \xi - \xi^2 + \xi^3)/4, \\
N_2(\xi) &= (2 + 3\xi - \xi^3)/4, \\
\overline{N}_2(\xi) &= (-1 - \xi + \xi^2 + \xi^3)/4; \qquad -1 \le \xi \le 1.
\end{aligned}
\tag{3.25}
$$

2. Quintic Hermite interpolation polynomials for elements with three nodes and two degrees of freedom. (See Figs. 3.4 and 3.5.)

$$
\begin{aligned}
N_1(\xi) &= (4\xi^2 - 5\xi^3 - 2\xi^4 + 3\xi^5)/4, \\
\overline{N}_1(\xi) &= (\xi^2 - \xi^3 - \xi^4 + \xi^5)/4, \\
N_2(\xi) &= 1 - 2\xi^2 + \xi^4, \\
\overline{N}_2(\xi) &= \xi - 2\xi^3 + \xi^5, \\
N_3(\xi) &= (4\xi^2 + 5\xi^3 - 2\,\xi^4 - 3\xi^5)/4, \\
\overline{N}_3(\xi) &= (-\xi^2 - \xi^3 + \xi^4 + \xi^5)/4; \qquad -1 \le \xi \le 1.
\end{aligned}
\tag{3.26}
$$

3.6   Interpolation polynomials in Mathematica

Symbolic or algebraic manipulation programs such as Mathematica have an important place in developing numerical algorithms. It is useful to employ such symbolic programming methods, particularly for short computations because of the ease in programming. We illustrate this by the following examples in which Mathematica is employed to obtain the Lagrange and Hermite interpolation polynomials of any degree.

3.6.1   Lagrange interpolation

```
(*****************************************)
(* MATHEMATICA  INPUT PROGRAM  lgpoly.m  *)
(*****************************************)


(*****************************************)
(*    MATHEMATICA PROGRAM FOR GENERATING  *)
(*    LAGRANGE INTERPOLATION POLYNOMIALS  *)
```

```
(*    Use the following as the 'lgpoly.m' *)
(*    input file to generate the Lagrange *)
(*    interpolation polynomials.          *)
(*                                        *)
(*    Comment out, using 'parenthesis *', *)
(*    any line not needed for the n-value *)
(*    you need.                           *)
(*****************************************)
(*  The following has been set up for n=3 nodes; *)
(*  it can be extended to any number of nodes in *)
(*  a straightforward manner *)


n = 3


(* DEFINE FUNCTION FOR NODAL COORDINATES   *)
node[i_]:= -1 + (2/(n-1))*(i-1)


(* DEFINE FUNCTION FOR GENERATING X^{I-1}   *)
g[x_,i_]=x^(i-1)


(* MAKE AN ARRAY OF  POWERS  OF X:  X^{I-1} *)
lgpower= Table[g[x,i],{i,n}]

f[1,x_]:=1
f[2,x_]:=x
f[3,x_]:=x^2
f[4,x_]:=x^3
f[5,x_]:=x^4
f[6,x_]:=x^5



(* Enter more equations here for higher polynomials *)
(* as n is increased.                               *)



(* FORM ROWS OF MATRIX USED TO GET COEFFICIENTS IN *)
(* POLYS. AFTER INVERSION OF MATRIX                *)

l1=Table[f[i,node[1]], {i,n}]
l2=Table[f[i,node[2]], {i,n}]
```

```
13=Table[f[i,node[3]], {i,n}]

(* INCLUDE MORE ROWS IF n BECOMES GREATER THAN 3:   *)
(* UNCOMMENT MORE ROWS BELOW AS NEEDED              *)

(* 14=Table[f[i,node[4]], {i,n}]     *)
(* 15=Table[f[i,node[5]], {i,n}]     *)
(* 16=Table[f[i,node[6]], {i,n}]     *)

(* EXTEND ARRAY M IF n IS INCREASED: *)
(* m = {11,12,13,14,15,16}           *)

m={11,12,13}

minv=Inverse[m]
mtr=Transpose[minv]

lgrpoly= mtr.lgpower

(* COMBINE TERMS TO GET COMMON DENOMINATOR   *)
Together[lgrpoly]

(*  Output is saved in file "lagrang.pol"    *)
PutAppend[%y,"lagrang.pol"]
(***********************************************)
(* END OF MATHEMATICA INPUT PROGRAM lgpoly.m *)
(***********************************************)
```

3.6.2   Hermite interpolation

```
(*****************************************)
(*  MATHEMATICA INPUT PROGRAM  hermite.m *)
(*****************************************)
(*****************************************)
(*   MATHEMATICA PROGRAM FOR GENERATING  *)
(*   HERMITE INTERPOLATION POLYNOMIALS   *)
(*   Use the following as the 'hermite.m'*)
(*   input file to generate the Hermite  *)
(*   interpolation polynomials.          *)
(*                                       *)
(*   Comment out, using 'parenthesis *', *)
```

```
(*    any line not needed for the n-value *)
(*    you need.                           *)
(*****************************************)
(* Increase n as needed: *)
n=2

(* Define nodal coordinates *)
node[i_]:= -1 + (2/(n-1))*(i-1)

(* Define a function for x^{i-1} *)
g[x_,i_]=x^(i-1)

(* Make an array of powers of x^{i-1} *)
lgpower= Table[g[x,i],{i,n}]

(* We need n*2 -1 powers; If n .gt. 3 we must    *)
(* define more functions extending the equations *)
(* given below                                   *)
f[1,x_]:=1
fp[1,x_]:=0

f[2,x_]:=x
fp[2,x_]:=1

f[3,x_]:=x^2
fp[3,x_]:=2*x
f[4,x_]:=x^3
fp[4,x_]:=3*x^2
f[5,x_]:=x^4
fp[5,x_]:=4*x^3
f[6,x_]:=x^5
fp[6,x_]:=5*x^4

(* Enter more equations here for higher *)
(* polynomials as n is increased:       *)
h1 = Table[f[i,node[1]],{i,2n}]
h2 = Table[fp[i,node[1]],{i,2n}]
h3 = Table[f[i,node[2]],{i,2n}]
h4 = Table[fp[i,node[2]],{i,2n}]
```

```
(*  h5 = Table[f[i,node[3]],{i,2n}]        *)
(*  h6 = Table[fp[i,node[3]],{i,2n}]       *)
(*  h7 = Table[f[i,node[4]],{i,2n}]        *)
(*  h8 = Table[fp[i,node[4]],{i,2n}]       *)
(*  h9 = Table[f[i,node[5]],{i,2n}]        *)
(* h10 = Table[fp[i,node[5]],{i,2n}]       *)


(* Extend the number of rows in r by putting in   *)
(* more h_i if n.ge.3, as in the following line.  *)
(*  r = {h1,h2,h3,h4,h5,h6}                        *)
r = {h1,h2,h3,h4}
rinv = Inverse[r]
rtr = Transpose[rinv]
hrpow= Table[g[x,i],{i,2n}]
hrpoly = rtr.hrpow


(*  Combine terms to get common denominator   *)
Together[hrpoly]


(* The output is in file "hermite.pol"         *)
PutAppend[%,"hermite.pol"]
(***********************************************)
(* END OF MATHEMATICA INPUT PROGRAM hermite.m *)
(***********************************************)
```

3.7   Infinite elements

In problems with infinite domains it is natural to be concerned about the truncation of the physical region in numerical calculations. It has been shown by Bettess[3] that one can extend the range of the FEM all the way to infinity by attaching an "infinite element" extending from the cutoff value $r = r_c$ to $r = \infty$. The interpolation functions for such an element are taken to be Lagrange interpolation functions multiplied by an appropriate decay factor. For example, we can choose the functions to be

$$N_{\infty,\alpha} = e^{(r_\alpha - r)/L} \prod_{\beta,(\beta \neq \alpha)}^{s} \frac{(r_\beta - r)}{(r_\beta - r_\alpha)}. \tag{3.27}$$

Here the s points r_α are in the range r_c to $r = \infty$. The rate of decay of the exponential factor is governed by the scale parameter

L which is chosen to have some reasonably large value. Other forms are explored in the literature.[4-8]

In some bound state problems, the extension to infinite domains may not be so simple. The eigenfunctions may decay exponentially with the exponential fall-off depending on the principal quantum number. For example, for electronic states in hydrogen, the wavefunctions behave asymptotically as $\exp(-r/na_0)$. This is not conducive to a universal form as in equation (3.27). Nevertheless, in scattering theory we would have a natural candidate for employing such infinite elements. These problems are topics for further exploration.

3.8   Simple quantum systems revisited

In Chapter 2, we noted that the finite element calculations become more accurate with increasing number of elements. To go beyond linear interpolation, we can employ higher order Lagrange polynomials in each element by providing a larger number of nodes in each element. However, one may expect that far fewer elements will be n eeded to obtain double-precision accuracy for the eigenvalues of the lowest few states when Hermite interpolation polynomials are used. This is because the additional condition on the continuity of the derivative of the function provides a very rapid convergence to the minimum of the action integral. For a system with propagating states the continuity of the first derivative of ψ in quantum mechanics means that the "probability current" is conserved across the nodal points and across element interfaces. For bound state problems with states stationary in time, the continuity of the first derivative of the wavefunction across the element boundaries is required by the fact that Schrödinger's equation is of second order. We have shown[9] that cubic Hermite interpolation polynomials (two nodes per element with two degrees of freedom per node) and quintic Hermite interpolation polynomials (three nodes per element with two degrees of freedom per node) provide the necessary accuracy in quantum mechanical calculations.

Typically, for a function obeying a second-order differential equation, there are two boundary conditions on the function or its derivative. These boundary conditions can be implemented very effectively using the C_1-continuous Hermite interpolation polynomials. For homogeneous physical domains, the differential equation itself ensures that the second derivative of the function is also continuous. This suggests that we explore the use of C_2-continuity at nodes, as an experiment to investigate the level of accuracy obtainable in this case. The use of C_2-continuity at nodes corresponds to having three degrees of freedom at each node. For an element with two nodes and three degrees of freedom we work with six quintic Hermite polynomials, while for an element with three nodes and three degrees of freedom we require nine polynomials of degree 8. Such interpolation functions are not usually used in calculations because the element matrices get to be of large dimension. For example, in the three-nodal element with three degrees of freedom at each node, the element matrix will be of dimension 9. However, the flexibility inherent in the FEM is retained while the advantages of derivative continuity and the use of higher degree polynomials are made available. We show below the effect of including three degrees of freedom per node in bound state problems.

Consider the case where continuity of the wavefunction and its first two derivatives across element boundaries is imposed. In the i_{el}th element the wavefunction can be expressed as

$$\psi(r) = N_i(r)\psi_i^{(i_{el})} + \overline{N}_i(r)\psi_i'^{(i_{el})} + \overline{\overline{N}}_i(r)\psi_i''^{(i_{el})} \qquad (3.28)$$

where i is the node index within the element, and $N_i(r)$, $\overline{N}_i(r)$, $\overline{\overline{N}}_i(r)$ are Hermite interpolation functions and $\psi_i'^{(i_{el})}$, $\psi_i''^{(i_{el})}$ are the as-yet-undetermined derivatives of ψ defined at the nodal points i in the element. The contributions of the individual finite elements to the action integral are evaluated as discussed in earlier chapters using Gauss integration. Now each node i in element i_{el} will have three unknown coefficients $\psi_i^{(i_{el})}$, $\psi_i'^{(i_{el})}$, and $\psi_i''^{(i_{el})}$ associated with it, and the element matrices are now $3\,nnode \times 3\,nnode$ in size, where $nnode$ is the number of nodes per element. The continuity of the wavefunction and its derivatives across the element boundary is ensured by overlaying the element matrices while composing the global matrices appearing in the action, such that the global matrix elements corresponding to the common nodes at the boundary obtain contributions from the local element matrices on either side of the boundary nodes.

We have recalculated the eigenvalues of the simple harmonic oscillator using Hermite interpolation polynomials with C_2-continuity. In Table 3.1, we show the results for the eigenvalues obtained using

Table 3.1 The energy levels of the 1D simple harmonic oscillator in the FEM using Hermite interpolation with three degrees of freedom (d.o.f.) at each node. Two and three nodes per element have been used. The range used is $-10 \le \xi \le 10$, and the number of elements is 60. The exact eigenvalues are $(2n+1)$.

Quantum no.: n	Energy 2 nodes & 3 d.o.f.	Energy 3 nodes & 3 d.o.f.
0	1.000 000	1.000 000..
1	3.000 000	3.000 000..
2	5.000 000	5.000 000..
3	7.000 001	7.000 000..
⋮	⋮	⋮
38	77.000 268	76.999 703
39	78.999 934	78.999 173

60 elements with two and three nodes and C_2-continuous Hermite interpolation. With two nodes per element, the lowest four eigenvalues are accurate to better than six significant figures, and the number of eigenvalues obtained to five significant figures is 36. On using Hermite interpolation with three degrees of freedom and three nodes per

Table 3.2 The energy levels of the hydrogen atom using Hermite interpolation using two and three nodes per element with three degrees of freedom (d.o.f.) in the FEM, with a range of $r_c=100a_o$. The number of elements is 50. The exact eigenvalues are $-1/n^2$.

Quantum no.: n	Energy 2 nodes & 3 d.o.f.	Energy 3 nodes with 3 d.o.f
1	−0.999 999 9	−0.999 999 999 999 9
2	−0.249 999 7	−0.250 000 000 000 0
3	−0.111 068 8	−0.111 111 111 111 1
4	−0.062 483 3	−0.062 500 000 000 9
5	−0.039 997 8	−0.040 000 068 132 8

element the first 20 eigenvalues agree with the analytical results to 14 significant figures, and the number of eigenvalues obtained to five significant figures or better increases to 40.[9]

In Table 3.2 we present the results for eigenvalues of the electronic states in the hydrogen atom obtained using Hermite interpolation functions with three degrees of freedom at each node. The results with three nodes per element with three degrees of freedom at each node are clearly in excellent agreement with the analytical values.[9] We note that the improvements in the levels of accuracy we have demonstrated by the use of Hermite interpolation have not been reported in the literature for the energy levels of quantum mechanical systems until recently. Our analysis clearly shows that the FEM is a reliable method for the bound state problems of quantum mechanics.

3.9 Problems

1. Obtain the quantum mechanical energy levels and radial wavefunctions for the radially symmetric states of a particle moving in a modified[‡] Morse potential[10] with a Hamiltonian

$$H = -\frac{1}{2}\frac{d^2}{dx^2} + D\left(e^{-2\alpha x} - 2e^{-\alpha x}\right),$$

using the FEM with quintic Hermite interpolation polynomials. Here $D = 12.00$ and $\alpha = 0.204\,124\,1$, and let $-5 \le x \le 100$. Show that this potential energy function supports 24 bound states and compare your results with those reported by Kimura et al.[11]

2. (a) Examine the energy spectrum and wavefunctions of the Hamiltonian

$$H = -\frac{1}{2}\frac{d^2}{dx^2} + D\left(e^{-\alpha(x-x_1)^2} - e^{-\alpha(x+x_1)^2}\right),$$

with a potential in the form of a double Gaussian well. Here $D = 12.0$, $\alpha = 0.1$, and $x_1 = 5.0$. Use the FEM with quintic Hermite interpolation polynomials to do the numerical analysis.

(b) Show that there are 12 pairs of bound doublet states with even and odd symmetry with energy splitting between pairs that

---

[‡]We modify Problem 2.2 here so that results may be compared directly with numerical values reported in Ref. 11 where quintic Hermite interpolation functions were used and Refs. 12 where Gaussian basis functions were used.

increases as the energy increases. Compare your results with those of Ref. 11.

References

[1] G. Dhatt and G. Touzot, *The Finite Element Method Displayed* (Wiley, New York, 1984).

[2] O. C. Zienkiewicz, *The Finite Element Method* (McGraw-Hill, New York, 1989).

[3] P. Bettess, *Int. J. Numer. Methods Eng.* **11**, 53 (1977); *ibid.*, **15**, 1613 (1980).

[4] S. Pissanetzky, *Int. J. Numer. Methods Eng.* **19**, 913 (1983).

[5] R. W. Thatcher, *SIAM J. Numer. Anal.* **15**, 466 (1978).

[6] G. Beer and J. L. Meek, *Int. J. Numer. Methods Eng.* **17**, 43 (1981).

[7] F. Medina and R. L. Taylor, *Int. J. Numer. Methods Eng.* **19**, 1209 (1983).

[8] O. C. Zienkiewicz, K. Bando, P. Bettess, C. Emson, and T. C. Chiam, *Int. J. Numer. Methods Eng.* **21**, 1229 (1985).

[9] L. R. Ram-Mohan, S. Saigal, D. Dossa, and J. Shertzer, *Comput. Phys.* **4**, 50 (1990).

[10] P. M. Morse, *Phys. Rev.* **34**, 57 (1928).

[11] T. Kimura, N. Sato, and S. Iwata, *J. Comput. Chem.* **9**, 827 (1988).

[12] M. J. Davis and E. J. Heller, *J. Chem. Phys.* **71**, 3383 (1979).

I. P. Hamilton and J. C. Light, *J. Chem. Phys.* **84**, 306 (1986).

4
Adaptive FEM

4.1   Introduction

The discretization of the physical region and the representation of the wavefunction by piecewise polynomials generate errors both of which are intrinsic to the finite element procedure. Since its inception, it has been recognized that a successful application of the finite element method (FEM) requires estimating these errors and developing approaches to minimize them. In Chapters 1–3, it was shown that the FEM is a variational method in which better results can be obtained by the use of more elements and also through the use of higher order interpolation functions. However, no attempt was made to optimize the methods so as to provide a systematic approach for improving the results. We had implicitly assumed that the refinements in the FEM lead to convergence towards the exact results. In this chapter, we provide error estimates for the finite element solutions, derive convergence properties of the FEM, and develop schemes for automating the optimization procedure. There are several approaches to improving results obtained from the FEM, and here we will describe their essentials.

We have seen, in Chapter 3, that a general refinement of the mesh through doubling the number of elements and reducing the element size (length) h improves the accuracy of the eigenvalues in the standard examples. This suggests the first approach to adaptive FEM. The systematic method of *mesh refinement* and a strategic placement of smaller elements is called the h-adaptive approach. In this method the number of elements is increased.

The approach of keeping the same number of elements, but raising the order p of the interpolation polynomials, is said to be p-adaptive. This is also referred to as *enrichment*. A hierarchy of interpolation polynomials of increasing order defined on the same element without increasing the number of nodes can be used to systematically reduce the error in the FEM. Both the h-adaptive and the p-adaptive methods can be combined in an algorithm which can be designated as being *hp*-adaptive. Our use of Hermite interpolation polynomials and the consequent improvement in the solution of Schrödinger's equation is an example of the *p*-adaptive method (see Chapter 3). Only a single step was taken, however, to improve the results obtained by Lagrange interpolation. Here we wish to develop methods that will take a hierarchical approach.

A third approach is to use the *r*-adaptive method, in which the accuracy of the FEM is improved through a *relocation* of the existing nodes; in effect, the size of the elements is changed so as to improve the representation of the solution over each element with the existing interpolation scheme. We do not consider this method any further in this chapter.

We begin with a discussion of the error in the interpolated representation of the solution.

4.2   Error in interpolation

In the FEM, the wavefunction $\psi(x)$ is approximated by an interpolation polynomial in each element. Consider a single element ranging over $[x_0, x_m]$ of length h in which we have $m + 1$ equidistant nodes located at $x_0, x_1, \ldots, x_m$. Let us suppose, for the moment, that the values of $\psi(x)$ at these $m + 1$ nodes are known. A unique polynomial $p_m(x)$ of degree m can be constructed such that $\psi(x_i) = p_m(x_i)$ at the $m + 1$ nodes. It is straightforward to prove the uniqueness of this interpolating polynomial.[1-3] If we use Lagrange interpolation polynomials

$$L_i(x) = \prod_{\substack{j=0 \\ j \neq i}}^{m} \frac{(x - x_j)}{(x_i - x_j)}, \qquad (4.1)$$

satisfying the condition $L_i(x_j) = \delta_{ij}$, we have

$$\psi(x) = \sum_{i=0}^{m} \psi(x_i) \, L_i(x) + E_m(x)$$
$$= \hat{\psi}(x) + E_m(x). \qquad (4.2)$$

The remainder term $E_m(x)$ is the error in approximating the wavefunction with the Lagrange polynomials. Since there is only one unique interpolating polynomial of degree m, we can also represent

$\psi(x)$ by a Taylor expansion with $m + 1$ terms, assuming that $\psi(x)$ can be differentiated $m+1$ times. This allows us to identify the error as the remainder term in the Taylor expansion

$$E_m(x) = \frac{\psi^{(m+1)}(\xi)}{(m+1)!} \prod_{i=0}^{m} (x - x_i), \qquad (4.3)$$

for all x in the interval $[x_0, x_m]$, where ξ is in the open interval. If $|\psi^{(m+1)}(x)|$ is bounded by the finite quantity $C_{(m)}$ in the interval $[x_0, x_m]$, we have

$$|E_m(x)| \leq \frac{C_{(m)}}{(m+1)!} \left| \prod_{i=0}^{m} (x - x_i) \right|. \qquad (4.4)$$

The error vanishes at the nodal points.

Specializing to the case of linear interpolation, we note that $|E_{(lin)}(x)|$ is zero at the two end points x_0 and x_1 of each element, and hence must have a maximum at some point in between. We identify the maximum error by determining the maximum of $(x - x_0)(x - x_1)$ over each element. This occurs at $x = (x_0 + x_1)/2$. The error bound in the linear interpolation is obtained as

$$|E_{(lin)}| \leq \frac{C_{(1)}}{8} \left| (x_1 - x_0)^2 \right| = C_{(1)} \frac{h^2}{8}. \qquad (4.5)$$

For a quadratic element of length $2h$ with equidistant nodes the error function is

$$|E_{(quad)}(x)| \leq \frac{C_{(2)}}{3!} |(x - x_0)(x - x_1)(x - x_2)|. \qquad (4.6)$$

The maximum of the coordinate-dependent factors occurs at $x = x_1 \pm h/\sqrt{3}$, and is given by $2h^2/3\sqrt{3}$. Thus the interpolation error in the quadratic element is bounded by

$$|E_{(quad)}| \leq C_{(2)} \frac{h^3}{9\sqrt{3}}. \qquad (4.7)$$

For Hermite interpolation with the values of $\psi(x)$ and $\psi'(x)$ specified at the nodes, Hildebrand[1] shows that the error in the interpolation polynomial is

$$\overline{E}_{2m}(x) = \frac{\psi^{(2m+2)}(x)}{(2m+1)!} \prod_{i=0}^{2m+1} (x - x_i). \qquad (4.8)$$

The error bounds in this case can be obtained in a similar manner for a given order of the polynomial. For cubic Hermite interpolation it is of $\mathcal{O}(h^5)$, demonstrating the advantages inherent in derivative continuity.

4.3   Error in the discretized action

4.3.1 h-convergence

In the presentation of the FEM in earlier chapters, we have tacitly assumed that the action integral exists and is finite. This implies that the integral over the quadratic form constructed with ψ and its derivative is finite. Consider a quantum mechanical particle over the interval $[a, b]$, moving in a potential $V(x)$ that is finite over the interval. The exact wavefunction is denoted by $\psi(x)$. Let the boundary conditions be $\psi(a) = 0 = \psi(b)$. We make the Schrödinger action dimensionless by replacing $x \to \tilde{x}\ell_0$, $\ell_0 = 1\,\text{Å}$, and dividing the action by $\mathcal{C} = \hbar^2/(2m_0\ell_0^2) = 3.81\,\text{eV}$. Dropping the tilde on x and using reduced units, we see that

$$\mathcal{A} = \int_a^b dx \left[\psi'^*(x)\,\psi'(x) + \psi^*(x)\,(v(x) - \epsilon)\,\psi(x)\right] \qquad (4.9)$$

exists provided the wavefunction and its first derivative are square integrable and the expectation value of the potential is also finite. This requires the first derivative to be piecewise continuous. The existence of the second derivative of $\psi(x)$ is assured in typical physical problems by Schrödinger's equation itself.[†] Functions that satisfy the conditions

$$\langle\psi|\psi\rangle = \int_a^b dx\,\psi^*(x)\,\psi(x) \le \mu,$$

$$\langle\psi'|\psi'\rangle = \int_a^b dx\,\psi'^*(x)\,\psi'(x) \le \nu, \qquad (4.10)$$

and satisfy the boundary conditions at a, b are called admissible functions. Here μ, ν are finite constants.[‡] We usually assume that our wavefunctions belong to the set of admissible functions.

[†]For a related discussion, see Courant and Hilbert.[4]

[‡]The set of admissible functions with norm

$$\|\psi\| = \left[\int_a^b dx\,\psi'^*(x)\psi'(x) + \psi^*(x)\psi(x)\right]^{1/2}$$

are said to belong to a Sobolev space $\mathcal{S}_1$.

We estimate the error in the discretization of the action integral as follows. Assume that we have n elements of equal length h, given by $h = (b-a)/n$. We represent the wavefunction using linear interpolation functions $N_1(x)$ and $N_2(x)$ in each element iel. The interpolated wavefunction is given by

$$\hat{\psi}_{(iel)}(x) = N_1^{(iel)}(x)\,\psi_1(x) + N_2^{(iel)}(x)\,\psi_2(x). \tag{4.11}$$

The global boundary conditions are satisfied if we drop the first interpolation function N_1 in the first element and the second function N_2 in the last element so that the conditions $\psi(a) = 0$ and $\psi(b) = 0$ are enforced. The maximum error $E^{(iel)}$ in each element from the linear interpolation is given by equation (4.5). We have

$$|E^{(iel)}| = |\psi_{(iel)}(x) - \hat{\psi}_{(iel)}(x)| \leq C_{(1)}^{(iel)}\frac{h^2}{8}. \tag{4.12}$$

The square of the error $\mathcal{E}_\mathcal{N}$ in the normalization of the wavefunction is

$$|\mathcal{E}_\mathcal{N}|^2 = \int_a^b dx\,(\psi^*(x) - \hat{\psi}^*(x))\,(\psi(x) - \hat{\psi}(x))$$
$$\leq \{\max C_{(1)}\}^2\frac{nh^5}{64}$$
$$= C_\mathcal{N}\,L\,h^4. \tag{4.13}$$

The error in the normalization is thus proportional to h^2. The square of the error in the action integral $\mathcal{E}_\mathcal{A}$ is defined as

$$|\mathcal{E}_\mathcal{A}|^2 = \sum_i^n \int_i dx(\psi^*(x) - \hat{\psi}^*(x))\,[\overleftarrow{\partial}\overrightarrow{\partial} + (v(x) - \epsilon)]\,(\psi(x) - \hat{\psi}(x))$$
$$\leq \{\max C_{(1)}\}^2\frac{nh^3}{4} + \mathcal{O}(h^4)$$
$$= C_\mathcal{A}\,L\,h^2. \tag{4.14}$$

Here we have employed equation (4.3) in deriving the maximum error. This estimate shows that the rate of convergence of the action integral is $\mathcal{O}(h)$ for linear interpolation. Furthermore, as we add more and more elements, while decreasing the element size h, we are assured of convergence[5,6] in the FEM.

Given the estimated error in the interpolation polynomials we can determine the error in the discretized action for each type of interpolation when the physical region is divided evenly into elements. For example, quadratic interpolation results in $\mathcal{E}_A \sim \mathcal{O}(h^2)$, while cubic Hermite interpolation has $\mathcal{E}_A \sim \mathcal{O}(h^3)$. The quintic Hermite interpolation polynomials employed in the examples in Chapter 3 have $\mathcal{E}_A \sim \mathcal{O}(h^5)$. These so-called a priori error estimates are useful in determining the rate of convergence as the size of the uniform mesh is reduced. The convergence is represented graphically by replacing h by the number of degrees of freedom N in the FEM. In linear interpolation, the total number of degrees of freedom is essentially the total number of elements, $N \simeq n$. For quintic Hermite interpolation the number of degrees of freedom is $N \simeq 4n$. The logarithmic error can be expressed as

$$\log \mathcal{E}_A = \Lambda - \beta \log N, \qquad (4.15)$$

where β is proportional to the degree p of the interpolating polynomial and Λ is a constant. This relation describes the rate of convergence with the number of degrees of freedom.[5,7,8]

In practice, the actual finite element solution is used to determine an a posteriori error, which is found to be a good measure of the accuracy of the method. This error can then be employed in refining the breakup of the physical region. The reader is referred to Refs. 9–12 for further elaboration. This is taken up again in Section 4.4

4.3.2 p-convergence

The p-adaptive method retains the initial grid while additional higher order interpolation polynomials are introduced in each element. We wish to set up a hierarchical scheme for a systematic improvement of the results. The following features show the advantages of this method:

(i) In this scheme, we include an additional interpolation polynomial of degree one higher than in the previous level of refinement and orthogonal to all the previous polynomials, through each level of iteration. This ensures a faster rate of convergence than in the h-refinement method.

(ii) We wish to avoid the introduction of additional nodes so that the previous solutions and computations can be reused in the next level of refinement. The new shape functions should not be associated with nodal values; instead, we require them to be zero at the nodes in the element. The initial representation of the solution in a given element is

$$\psi(x) = \sum_{i=1}^{n} \psi_i N_i(x), \qquad (4.16)$$

and the additional shape functions are used to write

$$\psi(x) = \sum_{i=1}^{n} \psi_i N_i(x) + \sum_{j=n+1}^{n+m} a_j P_j(\xi). \qquad (4.17)$$

The new polynomials are chosen to be *integrals* of the Legendre polynomials, so that at least the derivative (kinetic energy) terms in the action integral accrue only diagonal elements in the global matrix from them. (In the usage of computational mechanics this term leads to the stiffness matrix.[7,12]) As we have seen, the derivative terms in the action actually control the level of convergence. Thus the polynomial P_{j+1}, of degree j, is defined as

$$
\begin{aligned}
P_{j+1}(\xi) &= \sqrt{\frac{2j-1}{2}}\, \frac{1}{(j-1)!2^{j-1}} \frac{d^{j-2}}{d\xi^{j-2}} (\xi^2 - 1)^{j-1} \\
&= \frac{1}{\sqrt{2(2j-1)}} (P_j(\xi) - P_{j-2}(\xi)); \qquad j \geq 2, \quad (4.18)
\end{aligned}
$$

where P_j are the usual Legendre polynomials. Using the recursion relation obeyed by the Legendre polynomials:

$$(2n+1)P_n(\xi) = P'_{n+1}(\xi) - P'_{n-1}(\xi); \qquad n = 1, 2, \ldots, \quad (4.19)$$

we can show that the polynomials P_j satisfy the relation

$$\int_{-1}^{1} d\xi \frac{dP_j}{d\xi} \frac{dP_\ell}{d\xi} = \delta_{j\ell}. \qquad (4.20)$$

From equation (4.18) we see that the P_j are zero at the two end points ± 1, and their derivatives are orthogonal to each other and to all the lower order nodal shape functions as well. In the absence of the potential energy terms, the additional interpolation functions lead to a diagonal enlargement of the element matrices. All such diagonal terms can be attached to the global matrix of the previous iteration fairly easily.

(iii) While the potential energy terms in the action will contribute to off-diagonal terms also, the diagonal dominance of the kinetic terms leads to a significantly better condition number for the global matrices. This is because the entries in the derivative terms (the stiffness matrix) increase with decreasing element size h, while the potential energy terms are proportional to h.

(iv) At each stage in the iteration process the solution vector of nodal values and the coefficients of the $\mathcal{P}_j$ from the previous iteration, extended suitably to the dimension of the new global matrices, provides an excellent seed vector for an iterative solution of the global set of the finite element equations.

(v) Finally, at each stage we can use local error criteria to decide the level of p-refinement desired in each element. This issue is considered in the context of h-refinement in the following section; the same considerations apply to p-refinement.

The additional interpolation terms introduced in the p-refinement can be considered as part of a series of terms in a truncated Fourier-like expansion in polynomials with corresponding Fourier-like coefficients which are determined by solving the matrices generated in the FEM.

The first nine p-hierarchical functions starting with linear interpolation are given below:

$$N_1(\xi) = \frac{1 - \xi}{2}, \qquad -1 \leq \xi \leq 1;$$

$$N_2(\xi) = \frac{1 + \xi}{2},$$

$$\mathcal{P}_3(\xi) = \sqrt{\frac{3}{2}} \frac{1}{2} (\xi^2 - 1),$$

$$\mathcal{P}_4(\xi) = \sqrt{\frac{5}{2}} \frac{1}{2} (\xi^3 - \xi),$$

$$\mathcal{P}_5(\xi) = \sqrt{\frac{7}{2}} \frac{1}{8} (5\xi^4 - 6\xi^2 + 1), \qquad (4.21)$$

$$\mathcal{P}_6(\xi) = \sqrt{\frac{9}{2}} \frac{1}{8} (7\xi^5 - 10\xi^3 + 3\xi),$$

$$\mathcal{P}_7(\xi) = \sqrt{\frac{11}{2}} \frac{1}{16} (21\xi^6 - 35\xi^4 + 15\xi^2 - 1),$$

$$\mathcal{P}_8(\xi) = \sqrt{\frac{13}{2}} \frac{1}{16} (33\xi^7 - 63\xi^5 + 35\xi^3 - 5\xi),$$

$$\mathcal{P}_9(\xi) = \sqrt{\frac{15}{2}} \frac{1}{128} (429\xi^8 - 924\xi^6 + 630\xi^4 - 140\xi^2 + 5).$$

The first five $\mathcal{P}_j$ polynomials are shown in Fig. 4.1.

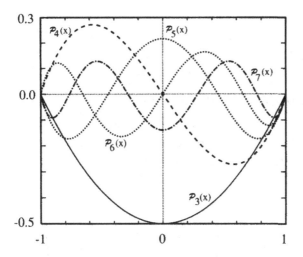

Fig. 4.1 The first five hierarchical p-adaptive interpolation polynomials are displayed. The overall sign of the functions is not relevant.

The p-refinement approach was initiated in the pioneering work of Peano.[13] The reader is directed to a very clear presentation by Robinson.[14] The mathematical aspects of p-refinement and error analysis are discussed in Refs. 7, 15, and 16.

The combination of both h- and p-adaptive methods into a single finite element procedure provides a very powerful algorithm for very accurate solutions of differential equations. It has been shown by Babuska and colleagues[7,17] that in the hp-adaptive FEM one has exponential convergence to the exact solution. The relative error when a hierarchy of hp operations is performed,

$$\mathcal{E} = \frac{|\psi - \hat{\psi}_{FEM}|}{|\psi|},$$

is given by

$$\mathcal{E} \leq k \cdot \exp(-\beta N^{\gamma}),$$

where k, β, and γ are positive constants, and N is the number of degrees of freedom.

4.4   The action in adaptive calculations

The FEM, based on the variational principle, may be considered to be the discretization of the action integral itself. Here we show how the action integral can again provide criteria for the adaptive application of the FEM to improve accuracy. We provide two one-dimensional calculations of adaptive h-refinement as examples illustrating the simplicity and the universality of the method.[18] We discuss briefly the adaptive p-refinement approach for eigenvalue problems.

4.4.1 An ordinary differential equation

As a first example, consider the Helmholtz differential equation, obtained from Schrödinger's equation for scattering after appropriate scaling of length and energy variables,

$$\frac{d^2}{dx^2} u(x) + 100\, u(x) = 0; \quad 0 \le x \le 1. \tag{4.22}$$

For illustrative purposes, let us consider the simple Dirichlet boundary conditions $u(0) = 1$, and $u(1) = -0.839\,071$. The analytic solution is obtained as $u(x) = \cos(10x)$. The corresponding action integral is

$$\mathcal{A} = \int_0^1 dx \left[-\frac{1}{2} u'(x) u'(x) + 100\, \frac{1}{2} u^2(x) \right]. \tag{4.23}$$

Given the form of the differential equation, we anticipate a trigonometric solution which will vary fairly rapidly. We begin by dividing the interval [0,1] into a modest number of intervals, say 10.

Linear interpolation is used to obtain a first approximation to the actual solution. As usual, we replace $u(x)$ by its interpolated form in terms of unknown nodal values u_i. The integration over the elements is performed, and the resulting action is bilinear in the nodal values. On minimizing the action with respect to the nodal values we obtain the approximate solution.

At this stage, the action can be considered to be the sum of the action integrals over the individual elements. The approximate solution provides us with a rough location of the extremum of the action in the parameter space defined by the nodal values. We now view each integral in this sum as an action integral with boundary values set by the approximate solution. Our strategy will be to subdivide further the individual elements into two elements each, and subject the corresponding action integrals to the variational principle. While the exact nodal values are not known, the approximate solution at the nodes will be used as the boundary conditions for these "elemental" actions. This may prove to be adequate for the next step. Naturally, such a fortuitous condition requires that an adequate number of elements are included in the first attempt at the solution so as to approximately reproduce the general shape of the solution.

Now consider any one of the elements, iel. The action integral over this element is split into two integrals by inserting a mid-node. We evaluate the action integral over each of the sub-elements, and overlay the element matrices to obtain the action over the parent element. The variational problem is solved once again for this element in order to determine the solution at the mid-node. Once this is obtained we explicitly re-evaluate the locally minimized action integral in each sub-element and compare the sum $\mathcal{A}_1^{iel} + \mathcal{A}_2^{iel}$ with the original value of the action integral over the element, $\mathcal{A}^{iel}$. If the relative error defined by

$$\delta = |\mathcal{A}^{iel} - (\mathcal{A}_1^{iel} + \mathcal{A}_2^{iel})|/|\mathcal{A}^{iel}|$$

is less than a set tolerance we do not subdivide further, and go on to the next element. An additional refinement would be to scale the local tolerance δ proportionately by the ratio of the length of the parent element to the entire physical region. In this manner, we can distribute the final error uniformly over the entire physical region. We retain the location of the mid-node for a final global re-evaluation of the action integral, just to improve matters. This ability to redistribute and refine the error over the physical region is one of the remarkable strengths of the FEM.

If the difference is larger than the tolerance we take up each of the sub-elements in turn, and subdivide it into the next "generation" of two sub-elements. Again, the solutions for the nodal values from the previous subdivision are used as known, if approximate, boundary values. This process is continued till the local change in the action due to bisection of each sub-element is below the tolerance level. The locations of the new mid-nodes are stored in an array. We have now determined the optimal placement of nodes over the entire region for an optimal solution. The new nodes are arranged according to their coordinate values and a new connectivity relation is established between the nodes and the corresponding elements.

The next step is to repeat the global calculation with the new nodal structure. At this stage, we can decide whether a linear interpolation in each new element is adequate, or whether we wish to further improve the solution by employing higher order (Lagrange or

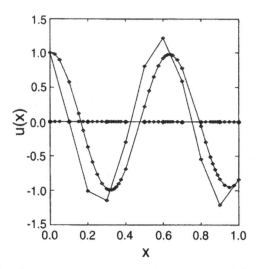

Fig. 4.2 The first approximation to the solution $u(x)$ is shown with 10 linear segments. The result of adaptive bisection of the intervals for minimizing the action is the smooth curve with 39 linear elements. The nodes are marked on the x-axis.

Hermite) interpolation. Note that this is precisely the adaptive algorithm suggested in Appendix A for evaluating any integral numerically using the procedure of adaptive Gauss–Legendre quadrature.

In Fig. 4.2, the approximate initial solution with just 10 elements is compared with the result obtained after adaptive calculations. The global tolerance was set at 10^{-2} for $\delta\mathcal{A}$, and proportionately scaled downward for the daughter elements. The new nodal locations are marked along the x-axis in the figure. It should be emphasized that the global problem has been solved only twice, the second time with the new elements determined by the adaptive calculations. In some problems this may not be adequate, especially when the initial approximate solution varies substantially from the final one. The entire adaptive procedure may have to be repeated more than once in order to overcome any deficiency in the boundary conditions used in the first attempt. For example, if the number of elements used in the first calculation is too small, the approximate solution will be too far off the mark to provide any guidance for the adaptive procedure.

An alternative criterion for bisection is to compare the average of the two nodes with the variational solution at the new node located at the midpoint. The variational principle is applied on the local action that is evaluated with the two sub-elements for this purpose.

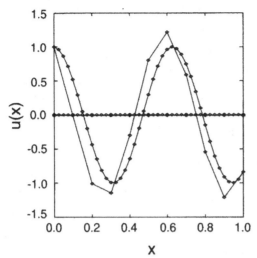

Fig. 4.3 The first approximation to the solution $u(x)$ is shown with 10 linear segments. The result of adaptive bisection of the intervals for minimizing the difference in the wavefunction at the midpoint of each interval is the smooth curve with 40 linear elements. The nodal locations are indicated on the x-axis.

Again, if the difference between the linearly interpolated value at the mid-node and the one from the solution with two elements is smaller than a tolerance value we stop the bisection for that segment of the physical interval. The results are shown in Fig. 4.3. The local calculations, for deciding whether to bisect a given element or not, proceed very quickly since the local element matrices are very small.

Results from the adaptive calculation with both the criteria are shown in Table 4.1. The starting number of linear elements is 10 in both sets of calculations. The computations were performed on a Digital Equipment Corporation Alpha machine running at 500 MHz. The times for the calculations are shown in order to compare the actual CPU time required for different levels of tolerances, and the numbers correspond to the time taken for a full recalculation of the final global problem. The matrices were stored in a sparse matrix format. The variation of the global action gives rise to a set of simultaneous equations for the unknown nodal values u_i, which can be written in the form

$$\mathbf{M} \cdot \mathbf{u} = \mathbf{r}. \tag{4.24}$$

An LU decomposition of the matrix $\mathbf{M}$ was performed to obtain the solution. With larger matrices it would be advantageous to use

Table 4.1 Number of elements generated by the adaptive improvement of the extremal value of the action integral and approximate run times on a 500 MHz DEC Alpha computer are given. The starting number of linear elements is 10.

Tolerance	Minimizing local $\delta\mathcal{A}$		Minimizing local $\delta\psi$	
	Final elements	Time (s)	Final elements	Time (s)
1.0×10^{-2}	39	0.08	40	0.08
1.0×10^{-3}	123	0.12	160	0.19
1.0×10^{-4}	361	0.33	640	0.74
1.0×10^{-5}	1198	1.57	1 280	1.71
1.0×10^{-6}	3875	9.09	5 121	13.93
2.0×10^{-7}	5282	31.50	20 480	155.00

conjugate gradient methods to solve iteratively the matrix equation.

The treatment given above differs in detail from the widely used Zienkiewicz–Zhu[11] criterion for adaptive calculations. The focus here has been on employing the change in the local action integral as an effective estimator for h-adaptive calculations.

4.4.2 The H atom again

The h-adaptive procedure can be applied to eigenvalue problems in a manner similar to the solution of regular differential equations. Let us treat the problem of determining the lowest eigenvalues of the radial equation for the spherically symmetric states of the hydrogen atom. We limit the range of the physical region to $[0, r_c]$. A cut-off range of $r_c = 20a_0$, where a_0 is the Bohr radius, is used in the calculations reported here. The action integral in reduced variables is

$$\mathcal{A} = \int_0^{r_c} r^2 dr \left[\psi'^* \psi' - \frac{2}{r} \psi^* \psi - \epsilon \psi^* \psi \right]. \qquad (4.25)$$

We start the calculations with 10 equal elements over the physical region. The spatial integral is performed using linear interpolating functions; the action is then reduced to a bilinear form in the nodal variables ψ_i. We have

$$\mathcal{A} = \psi_i^* H_{ij} \psi_j - \epsilon \psi_i^* B_{ij} \psi_j. \qquad (4.26)$$

The variation of the action with respect to the nodal values ψ_i^* leads to the generalized eigenvalue problem $H\psi = \epsilon B\psi$. The final value of the wavefunction at the cutoff radius is set to zero, and correspondingly the last row and column of H and B are deleted. With 10 linear elements in use, the matrix equation has 10×10 matrices. The eigenvalue problem is solved using a diagonalizer to obtain a first approximation to the lowest eigenvalues and the corresponding eigenfunctions. With just 10 elements we obtain very approximate values for the lowest three eigenvalues: $\epsilon_1 = -0.853\,36$, $\epsilon_2 = -0.219\,05$, and $\epsilon_3 = -0.079\,59$, instead of -1, $-1/2^2$ and $-1/3^2$, respectively. In the following, we consider the ground state only, though the calculations could proceed for any other eigenstate in a parallel manner. The cutoff range we have used accommodates only three eigenstates with negative eigenvalues, all other eigenvalues having been pushed above zero.

The approximate ground state wavefunction is used to evaluate the actual value of the action in each of the 10 elements. In the following we use these nodal values to provide us with approximate boundary conditions for each element.

Consider any one element iel and divide it into two sub-elements iel_1 and iel_2. The action integral is evaluated over the two sub-elements. The element matrices are determined using linear interpolation again, and are assembled to form 3×3 local matrices, $h_{ij}^{(iel)}$ and $b_{ij}^{(iel)}$. The action is now a bilinear function of the nodal values ψ_1, ψ_2, and ψ_3 and their complex conjugates. Here ψ_1 and ψ_3 are fixed since they are the local boundary values for the element as determined from the first iteration. We are then left with a variation over ψ_2^* giving us a single equation of the form

$$h_{21}\psi_1 + h_{22}\psi_2 + h_{23}\psi_3 = \epsilon\,(b_{21}\psi_1 + b_{22}\psi_2 + b_{23}\psi_3). \qquad (4.27)$$

Using the global value of ϵ and the known values ψ_1 and ψ_3, we determine the solution at the midpoint in the element.

We consider three alternate criteria for improving the solution over each element. The first uses the local action integral itself as a measure for the h-adaptive procedure. Given the local solutions $\psi_{1,2,3}$, we evaluate the local action in each of the sub-elements and sum the two contributions. We compare this sum with the value obtained initially with linear interpolation over the element iel. Let δ be the error tolerance for the global action integral. This is typically chosen to be 10^{-2}–10^{-4}. In order to define a local tolerance level we let the local value of the tolerance be proportional to the actual range of the element. With 10 starting elements of equal size, we define the local error as $\tilde{\delta} = (\delta/r_c) \times (r_c/10)$. If

$$\frac{|\mathcal{A}^{(iel)} - (\mathcal{A}^{(iel_1)} + \mathcal{A}^{(iel_2)})|}{|\mathcal{A}^{(iel)}|} < \tilde{\delta},$$

we stop subdividing the daughter elements any further, and proceed to the next element for adaptive improvement. Otherwise, we take the sub-elements iel_1 and iel_2 in turn and subdivide these into two elements each. We carry out the same procedure in order to determine whether these sub-elements require further bisection. The corresponding local error is rescaled appropriately, as mentioned above.

In each new level of iteration we use the eigenvalue from the previous iteration in equation (4.27). This requires some explanation. The eigenvalue is defined as the Rayleigh quotient

$$\rho(\epsilon) = \epsilon = \frac{\langle \psi | K + V | \psi \rangle}{\langle \psi | \psi \rangle}$$
$$= \frac{\langle \psi_i | H | \psi_i \rangle - S}{\langle \psi_i | B | \psi \rangle},$$

where K and V are the kinetic and potential energy terms in the Schrödinger Hamiltonian (as opposed to the Schrödinger action) and H is the matrix obtained for the symmetrized derivative terms together with the potential energy term in the action integral. The difference between the expectation of the Hamiltonian and the extremum value of the action integral is represented by the surface terms $S = \psi^* \psi'|_{x_{(iel)}}^{x_{(iel+1)}}$. In typical bound state problems, the surface terms vanish at the global end points since the confined wavefunctions vanish at these points. In the present case, the wavefunction is approximated to zero at $r = r_c$, with no constraint imposed on the wavefunction at $r = 0$. We are considering here only a small portion of the physical region with the wavefunction taking on specified boundary values. Once the wavefunction at the midpoint is solved using equation (4.27), we use this solution to obtain the next approximation for the local eigenvalue by constructing the Rayleigh quotient. This is used in the next bisection as the input eigenvalue while solving for the wavefunction at the midpoint.

Once all the elements have been passed through this adaptive procedure, we determine the new set of nodal coordinates for a final global evaluation of the problem. The eigenfunctions for the lowest

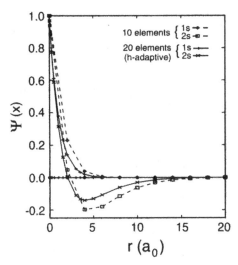

Fig. 4.4 The first approximation, with 10 linear segments, to the ground and first excited state wavefunctions of the hydrogen atom are shown. The results from the adaptive bisection of the intervals for minimizing the action are the smooth curves with 37 linear elements. The tolerance was set at 10^{-3} for $\delta\mathcal{A}$ in the starting segment and proportionately scaled downward for the daughter elements. The marks on the x-axis show the location of nodes.

two eigenvalues obtained using this method are shown in Fig. 4.4. As is evident, the additional nodes are placed closer to the origin where the wavefunctions contribute the most to the action integral. A substantial improvement is achieved if we employ higher order Lagrange polynomials, or better yet, Hermite interpolation functions for a final calculation in the newly defined elements.

If the final number of elements is on the order of 20–50, it is not advantageous to use the method of subspace iteration to determine the eigenvalues and eigenfunctions. This is shown by the run-times listed in Table 4.2. We then use the full-matrix diagonalizer based on the QZ algorithm.[19] It is seen from the first half of the data presented in this table that we steadily improve the eigenvalues by lowering the global error tolerance on the action integral. As we have noted earlier, with linear elements the advance towards double-precision accuracy is slow; the small value of the cutoff for the range is also a factor in limiting the accuracy of the final result.

We used the method of subspace iteration for diagonalizing the

Table 4.2 Adaptive improvement of the discretized action integral. This table compares the accuracy obtained on using the local change in action and the result of requiring the local change in the wavefunction for tolerance criteria.

Tolerance	Final elements	Eigen-values	Time (s)	Diagonalizer
Minimizing the local $\delta\mathcal{A}$				
1.0×10^{-2}	20	−0.970 913	0.09	Full-matrix: QZ
		−0.242 521		
		−0.093 779		
1.0×10^{-4}	67	−0.992 241	1.39	Full-matrix: QZ
		−0.248 655		
		−0.099 070		
1.0×10^{-6}	229	−0.999 427	78.34	Full-matrix: QZ
		−0.249 872	26.84	Subspace
		−0.099 771		
1.0×10^{-8}	735	−0.999 927	187.79	Subspace
		−0.249 961		
		−0.099 828		
Minimizing the local $\delta\psi$				
1.0×10^{-2}	40	−0.985 201	0.37	Full-matrix: QZ
		−0.246 787	2.00	Subspace
		−0.098 013		
1.0×10^{-4}	156	−0.998 742	11.80	Subspace
		−0.249 715		
		−0.099 682		
1.0×10^{-6}	518	−0.999 856	174.33	Subspace
		−0.249 942		
		−0.099 817		
1.0×10^{-8}	1652	−0.999 981	1342.41	Subspace
		−0.249 971		
		−0.099 834		

larger matrices generated by this method.[§] An error tolerance of 10^{-7} was used in this procedure, and the subspace was restricted to 14 so as to fully accommodate the lowest (negative) three eigenfunctions of interest to us. The method required a total of 85–90 iterations for

[§]The method of subspace iteration is discussed in Part IV of this book.

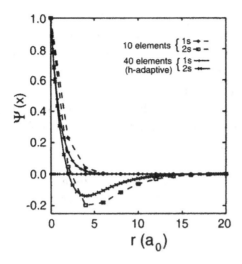

Fig. 4.5 The first approximation, with 10 linear elements, to the ground state and first excited state wavefunctions of the hydrogen atom are shown (dashed lines). The result of adaptive bisection of the intervals for minimizing the difference in the wavefunction at the midpoint of each interval are the smooth curves with 40 linear elements. The nodes are shown on the x-axis.

determining the 14 eigenvalues and eigenfunctions in all cases. Of these, only the lowest three were negative eigenvalues.

In the second approach, we compare the value for the mid-node ψ_{mid} with the average ψ_{avg} of the values for ψ_1 and ψ_2. If

$$\frac{|\psi_{avg} - \psi_{mid}|}{|\psi_{avg}|} \leq \tilde{\delta}, \tag{4.28}$$

we stop the h-adaptive division of the element. Otherwise each sub-element is once again subjected to bisection, and the wavefunction is determined at the midpoint of each portion for comparison with the average of the two end values. In Fig. 4.5, the results based on reducing the tolerance on the error in the wavefunction are shown. The tolerance for that calculation was set at 10^{-2} for $\delta\psi$ over the global domain and it was proportionately scaled downward for the bisected elements as the element size was diminished.

The results from the use of the wavefunction criterion are shown in the second half of Table 4.2. It is seen that the number of final elements in this case is double those obtained when the local action is used as a criterion for improving the solution. The improvements in the eigenvalues and in the eigenfunctions suggest that an adaptive approach to breaking up the physical region does indeed yield superior results; we can see this quantitatively by comparing the numbers in Table 2.3 of Chapter 2 with those in Table 4.2.

A third alternative for the adaptive improvement of solutions is to use the convergence to a given energy eigenvalue as the criterion. In this case, we compare the eigenvalue obtained at each iteration with the Rayleigh quotient from the previous iterate. The details of implementing this method are very similar to the two criteria considered above. We return to this issue in the context of adaptive p-refinement below.

4.4.3   Adaptive p-refinement

Let us consider the previous eigenvalue calculation in the context of adaptive p-refinement. We start again with a small number of elements which provide a rough breakup of the physical region. We make use of linear interpolation functions $N_1(\xi)$ and $N_2(\xi)$ of equations (4.21) in the elements and solve the global eigenvalue problem as usual. This gives us approximate eigenvalues and the corresponding nodal values for the eigenfunctions. At this point, we treat each element as a physical region in which the physical considerations of the global problem hold, and consider the same eigenvalue problem with the boundary conditions given by the nodal values at the two ends of the element. The action is given by

$$\mathcal{A} = \int_a^b r^2 dr \left[\psi^{*\prime}(r)\psi'(r) - \psi^*(r)\frac{2}{r}\psi(r) - \epsilon\psi^*(r)\psi(r) \right]. \qquad (4.29)$$

We improve the solution in each element by employing the additional hierarchical shape functions of equations (4.21). We add the polynomials $\mathcal{P}_i$, one by one, to the interpolation at each iteration. Consider a particular element ranging over the radial interval $[a, b]$. At the ith iteration we have

$$\psi(x) = \sum_{j=1,2} \psi_j^{(i)} N_j(\xi) + \sum_{j=3,i-1} c_j \mathcal{P}_j(\xi). \qquad (4.30)$$

At each stage of the iteration, the addition of a new polynomial can be thought of as a perturbation on the previous wavefunction. We insert the new wavefunction into the action integral and perform the spatial integration. The resulting action is varied with respect to the parameters c_j in equation (4.30). In this instance, $\psi^{(i)}$ are not varied since they are fixed by the given Dirichlet boundary conditions.

The variation of $\mathcal{A}$ with respect to c_j^* leads to the set of simultaneous equations

$$H_{j,1}\psi_1 + H_{j,2}\psi_2 + H_{jk}c_k = \epsilon^{(i-1)}(B_{j1}\psi_1 + B_{j2}\psi_2 + B_{jk}c_k),$$
$$j = 1, i-1; \quad k = 3, i-1. \qquad (4.31)$$

The eigenvalue from the $(i-1)$th iterate is used to solve for the coefficients c_k.

We use the wavefunction determined at this level to obtain the new approximation to the eigenvalue $\epsilon^{(i)}$. As mentioned above, the relation between the action integral and the expectation value of the Hamiltonian for an arbitrary wavefunction is given by

$$\mathcal{A} = \int_a^b r^2 dr \left[-\psi^*\psi'' - \psi^*\frac{2}{r}\psi - \epsilon\psi^*\psi \right] + r^2\psi^*\psi'|_a^b$$
$$= \langle \psi\, |H|\, \psi \rangle - \epsilon\langle \psi\, |\psi \rangle + S$$
$$= \mathcal{J} + S, \qquad (4.32)$$

where S is the surface term generated by an integration by parts. The usual calculation for the local element matrices is modified slightly to include the surface terms so that the functional $\mathcal{J} = (\mathcal{A} - S)$ is determined. The variational extremum of $\mathcal{J}$ gives the Rayleigh quotient

$$\rho(\epsilon^{(i)}) = \frac{\langle \psi^{(i)}\, |\mathcal{H}|\, \psi^{(i)} \rangle}{\langle \psi^{(i)}\, |\psi^{(i)} \rangle}$$
$$= \frac{\mathcal{A}^{(i)} - S}{\langle \psi(i)\, |\psi^{(i)} \rangle}$$
$$= \epsilon^{(i)}, \qquad (4.33)$$

which gives the ith approximation to the energy eigenvalue. The procedure, called inverse vector iteration,[¶] for determining the improved eigenpair can be repeated once again to obtain local convergence before going to the next level of iteration.

Here again we have three options for deciding when the iteration has arrived at a satisfactory conclusion. We can use (i) the change in the action integral itself obtained by the inclusion of yet

---

[¶] The method of inverse vector iteration is presented in Part IV of this book.

another hierarchical interpolation polynomial at every iteration, or (ii) the change in the wavefunction over each iteration, as defined by $\langle(\delta\psi)^2\rangle = \int r^2 dr |\psi^{(i)}(r) - \psi^{(i-1)}(r)|^2$, or (iii) the change in the eigenvalue $|\epsilon^{(i)} - \epsilon^{(i-1)}|$, as the criterion for stopping the iteration.

If the p-refinement through, say, seven to eight higher order polynomials does not lead to convergence with respect to a specified tolerance, the element is bisected and the p-refinement over each sub-element is pursued to a proper convergence. In Ref. 20, this approach is used in conjunction with triangular elements for 2D problems. In Ref. 21, hierarchical interpolation functions are given for derivative continuity.

4.5 Concluding remarks

We note that when adaptive and iterative techniques actually achieve convergence, we invariably perceive such happy endings as a reiteration of our faith in the "physics" of the problem. In the examples presented, we have employed the action integral itself to provide criteria for convergence. While the adaptive finite element approach has been explored extensively in the computational engineering literature the criteria have been presented there in a somewhat less transparent manner.

In one dimension we are able to keep track of the breakup of the physical region without excessive book-keeping. The desired adaptive changes in 2D meshes are straightforward for triangular elements so that no new interpolation polynomials are needed. On the other hand, in the case of rectangular elements, the breaking up of an element leads to sub-elements with nodes at the midpoints of some of the sides, and consequently a need arises for interpolation polynomials over five-noded, six-noded, and seven-noded rectangles.[12]

In problems with a negative potential energy, the optimization based on the energy minimization principle naturally exploits the partial cancellation between the kinetic and potential energy terms at a global level. A given choice of a global wavefunction could give the energy to a high degree of accuracy without correspondingly providing accurate values for the expectation values for other quantum mechanical operators. For example, a global wavefunction for the hydrogen atom could give a highly accurate energy eigenvalue while yielding poor values for $\langle(1/r)\rangle$, $\langle r \rangle$, or $\langle r^2 \rangle$. The cancellation in the energy minimization may occur globally. This situation can be altered by demanding a *local* minimization of the energy norm in the adaptively selected finite elements. Now the cancellation cannot include contributions from elements other than the one under consideration. This is one of the attractive features of the FEM and its use of local wavefunctions extending only over each element. Another source of error is the use of Gauss quadrature for functions that are varying rapidly in a given element. This error can be minimized by going to higher order rules for Gauss quadrature, or by employing adaptive integration schemes for such elements. The FEM also has the advantage of being able to easily accommodate local additions or deletions of degrees of freedom.

Allowing the computer to make the final determination of the interpolated form of the solution through adaptive methods has its own appeal. It is clearly important for nonlinear problems in which our guesses at breaking up the physical region into finite elements may have no intuitive guidelines.

References

[1] F. B. Hildebrand, *Introduction to Numerical Analysis*, 2nd ed. (Dover, New York, 1987).

[2] S. D. Conte and C. de Boor, *Elementary Numerical Analysis, an algorithmic approach* (McGraw-Hill, New York, 1980).

[3] P. M. Prenter, *Splines and Variational Methods* (Wiley, New York, 1975).

[4] R. Courant and D. Hilbert, *Methods of Mathematical Physics*, Vol. I (Interscience, New York, 1953), p199.

[5] G. Strang and G. J. Fix, *An Analysis of the Finite Element Method* (Prentice Hall, Englewood Cliffs, NJ, 1973).

[6] P. G. Ciarlet, *The Finite Element Method for Elliptic Problems* (Elsevier North-Holland, New York, 1978).

[7] B. Szabo and I. Babuska, *Finite Element Analysis* (Wiley, New York, 1991).

[8] M. Mori, *The Finite Element Method and its Applications* (Macmillan, New York, 1986).

[9] E. Rank and O. C. Zienkiewicz, *Commun. Appl. Numer. Methods* **3**, 1986 (1987).

[10] O. C. Zienkiewicz and A. Craig, in *Accuracy Estimates and Adaptive Refinements in Finite Element Computations*, ed. I. Babuska, O. C. Zienkiewicz, J. Gago, and E. R. de Oliveira (Wiley, New York, 1986), p25.

[11] O. C. Zienkiewicz and J. Z. Zhu, *Int. J. Numer. Methods Eng.* **24**, 337 (1987).

[12] O. C. Zienkiewicz and R. L. Taylor, *The Finite Element Method*, 4th ed. (McGraw-Hill, New York, 1994).

[13] A. G. Peano, *Comput. Math. Appl.* **2**, 3 (1976); A. G. Peano, A. Pasini, R. Riccioni, and L. Sardella, *Comput. Struct.* **10**, 333 (1979).

[14] J. Robinson, *Finite Elements Anal. Des.* **2**, 377 (1986).

[15] I. Babuska, in *Accuracy Estimates and Adaptive Refinements in Finite Element Computations*, ed. I. Babuska, O. C. Zienkiewicz, J. Gago, and E. R. de Oliveira (Wiley, New York, 1986), p3.

[16] B. A. Szabo, in *Accuracy Estimates and Adaptive Refinements in Finite Element Computations*, ed. I. Babuska, O. C. Zienkiewicz, J. Gago, and E. R. de Oliveira (Wiley, New York, 1986) p61.

[17] I. Babuska and M. R. Dorr, *Numer. Math.* **37**, 257 (1981); W. Gui and I Babuska, *Numer. Math.* **48**, 557, 613, 658 (1986); B. Guo and I. Babuska, *Comput. Mech.* **1**, 21, 203 (1986).

[18] L. R. Ram-Mohan, J. E. Moussa, J. A. Gagnon, J. M. Sullivan, and D. Dossa, "The action integral and adaptivity in finite element applications to quantum mechanics" (unpublished, 2002).

[19] G. H. Golub and C. F. van Loan, *Matrix Computations* (Johns Hopkins University Press, Baltimore, MD, 1983), p251.

[20] J. Ackermann and R. Roitzsch, *Chem. Phys. Lett.* **214**, 109 (1993); J. Ackermann, B. Erdmann, and R. Roitzsch, *J. Chem. Phys.* **101**, 7643 (1994); J. Ackermann, *Phys. Rev.* A **52**, 1968 (1995).

[21] S. Franchiotti and M. A. Rubin, *Phys. Rev.* D **43**, 3524 (1991).

Part II

Applications in 1D

5
Quantum mechanical tunneling

5.1 Introduction

A distinguishing feature of quantum mechanics, as compared with classical mechanics, is the phenomenon of matter exhibiting wave-like properties. A direct consequence of this wave property of particles is the tunneling of particles through potential energy barriers.[1-3] The classical particle is forbidden from penetrating the barrier region. In this chapter, the problem of one-dimensional (1D) quantum mechanical tunneling of particles through potential barriers of arbitrary shape is treated. We show that tunneling is a mixed boundary value problem, and obtain the action integral that leads to Schrödinger's equation and at the same time gives the equations defining the boundary conditions (BCs).

We use the tunneling problem to introduce the Galerkin method of weighted residuals for solving differential equations and to show its relation to the variational method. The use of the Galerkin method to obtain the transmission coefficient for the usual rectangular barrier is displayed in some detail as an example. Results are shown as well for tunneling through other standard potential barriers. These investigations demonstrate that very high levels of accuracy can be achieved with the finite element method (FEM) using quintic Hermite interpolation polynomials. The examples presented here also highlight the fact that this numerical approach provides results where the semiclassical or the Wentzel–Kramers–Brillouin approximation fails.

We further discuss the occurrence of quasibound states in asymmetric quantum wells (AQWs) and the conditions under which *no* bound states need occur in an *asymmetric* quantum well, and review the experimental evidence for such quantum states in semiconductor heterostructures. The formation of barrier-localized states above the energy barrier and surface quantum-well states are treated in the problems at the end of this chapter.

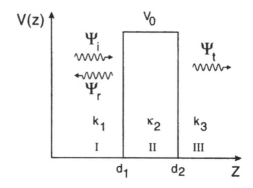

Fig. 5.1 A 1D rectangular potential barrier of height V_0. The particle incident with energy $E < V_0$ from the left has wavevectors k_1, κ_2, and k_3 in the three regions labeled I, II, and III. The incident, reflected and transmitted wavefunctions are denoted by ψ_i, ψ_r, and ψ_t, respectively.

5.2 Mixed BCs: redefining the action

The 1D Schrödinger equation governing the wavefunction $\psi(x)$ of a particle incident on a potential barrier $V(x)$ is

$$-\frac{\hbar^2}{2m}\frac{d^2}{dx^2}\psi(x) + V(x)\psi(x) = E\psi(x), \tag{5.1}$$

where the potential $V(x)$ is positive in the barrier regions.[†] The simplest potential barrier, shown in Fig. 5.1, is given by

$$V_I(x) = \begin{cases} V_0, & \text{for } d_1 \le x \le d_2, \\ 0, & \text{otherwise.} \end{cases} \tag{5.2}$$

We wish to solve equation (5.1) for the wavefunction and the transmission and reflection coefficients. This simple example is used to illustrate the issues in the application of the FEM to tunneling. The particle in region I is incident from the left with energy E, and it is reflected with an amplitude r and transmitted through the barrier with amplitude t into region III. By solving equation (5.1) with zero potential in the two exterior regions, the wavefunctions can be shown to be of the form

$$\begin{aligned} \psi_{(I)}(x) &= A\,e^{ik_1x} + r\,e^{-ik_1x}, & x < d_1, \\ \psi_{(III)}(x) &= t\,e^{ik_3x}, & x > d_2, \end{aligned} \tag{5.3}$$

[†]A potential well of finite depth also will scatter quantum mechanical particles of energy greater than the barrier height incident on it. See Problem 1.6 in Chapter 1, and also Problem 5.2 at the end of this chapter.

where $k_1 = k_3 = \sqrt{2mE/\hbar^2}$, and the coefficients A, t and r are complex numbers in general. It is assumed that the incident amplitude A is known. For the region within the barrier, the wavefunction is given by

$$\psi(x) = C\,e^{\kappa_2 x} + D\,e^{-\kappa_2 x}, \qquad d_1 < x < d_2, \qquad (5.4)$$

with $\kappa_2 = \sqrt{2m(V_0 - E)/\hbar^2}$ for $E < V_0$. In the above equation, κ_2 is replaced by ik_2 for $E > V_0$.

The transmission coefficient is defined as the ratio of the transmitted probability current to the incident current. With the transmitted probability current given by

$$j(x) = \frac{\hbar}{2\,m\,i}\left[\psi_t^*(x)\left(\frac{d}{dx}\psi_t(x)\right) - \left(\frac{d}{dx}\psi_t^*(x)\right)\psi_t(x)\right], \quad (5.5)$$

we have

$$T = (k_3\,|t|^2)/(k_1\,|A|^2). \qquad (5.6)$$

The usual expressions[1] for the transmission coefficient with $k_3 = k_1$ for the two cases $E < V_0$ and $E > V_0$ are given by

$$T = \cfrac{1}{1 + \cfrac{V_0^2}{4\,E(|V_0 - E|)}\begin{cases}\sinh^2(\kappa_2 d)\\ \sin^2(k_2 d)\end{cases}}, \qquad (5.7)$$

where $d = d_2 - d_1$. It can be shown in a similar manner that the reflection coefficient defined as the ratio of the reflected current to the incident current is

$$R = |r|^2/|A|^2. \qquad (5.8)$$

As a consequence of the conservation of the probability current, which requires that the incident probability current equal the sum of the reflected and transmitted currents, we must have $R + T = 1$.

The BCs on the wavefunction at d_1 and at d_2, equations (5.3), can be expressed as

$$\begin{aligned}\psi'(d_1) + ik_1\,\psi(d_1) &= 2ik_1\,A\,e^{ik_1 d_1},\\ \psi'(d_2) - ik_3\,\psi(d_2) &= 0.\end{aligned} \qquad (5.9)$$

The first relation identifies the incoming current, while the second defines the condition that we have an outgoing current on the right hand side. These are said to represent a mixed boundary value problem since the BCs are linear relations between the solution and its derivative. The first of the equations is an inhomogeneous BC while the second equation is homogeneous.

We wish to construct the action integral for this problem. We begin by converting equation (5.1) to a dimensionless form through the substitutions

$$x = \ell_0 \tilde{x}; \qquad \mathcal{C} = \frac{\hbar^2 c^2}{2 m \ell_0^2}. \qquad (5.10)$$

If the incident particle is an electron with mass $m = m_e$ and the length scale is $\ell_0 = 10^{-8}$ cm ($= 1$ Å), the scale of energy is given as $\mathcal{C} \simeq 3.81$ eV. In the following, for convenience, we shall continue to indicate the scaled coordinate by x and the reduced wavevectors by k_i. The potential energy is written as $v(x) = V(x)/\mathcal{C}$ and the energy of the particle as $\epsilon = E/\mathcal{C}$. With these substitutions we have

$$-\psi''(x) + (v(x) - \epsilon)\,\psi(x) = 0. \qquad (5.11)$$

The mixed boundary values require a modification of the form of the action from that encountered in earlier chapters. We begin with the functional

$$\mathcal{A}[\psi^*, \psi] = \int_{d_1}^{d_2} dx \, \left[\psi'^*(x)\psi'(x) \; + \; \psi^*(x)\,(v(x) - \epsilon)\,\psi(x) \right]$$
$$+ \psi^*(x)[\,-ik_3\,]\,\psi(x)|_{x=d_2}$$
$$- \psi^*(x)[+ik_1\,\psi(x) - 2ik_1\,A\,e^{ik_1 x}]\Big|_{x=d_1}. \qquad (5.12)$$

As shown below, this modified action will lead to the usual differential equation as well as the BCs appropriate for the tunneling problem. The variational principle requires that the variation of $\mathcal{A}$ with respect to $\psi^*(x)$ be zero. When applied to equation (5.12), and followed by an integration by parts, we have

$$\delta\mathcal{A}[\psi^*, \psi] = 0 = \int_{d_1}^{d_2} dx \, \delta\psi^*(x) \left[-\psi''(x) \; + \; (v(x) - \epsilon)\,\psi(x) \right]$$
$$+ \delta\psi^*(x)[\,\psi'(x) - ik_3\,\psi(x)]|_{x=d_2}$$
$$- \delta\psi^*(x)[\,\psi'(x) + ik_1\,\psi(x) - 2ik_1\,A\,e^{ik_1 x}]\Big|_{x=d_1}. \qquad (5.13)$$

The variational principle requires that for all possible choices for the function $\delta\psi^*(x)$ we must have $\delta\mathcal{A} = 0$. When we select $\delta\psi^*$ arbitrarily, except that these functions vanish at the two end points, we again obtain Schrödinger's equation (5.11), over the region $d_1 \leq x \leq d_2$. So the term containing the integral can be dropped. We now enlarge the set of $\delta\psi^*$ by removing the restriction that $\delta\psi^*$ is zero at the end points. We do this in two different ways. First, we choose $\delta\psi^*(x)$ with $\delta\psi^*(d_2) = 0$ and $\delta\psi^*(d_1) = 1$. This yields the first BC in equations (5.9). Next, we let $\delta\psi^*(d_2) = 1$ while $\delta\psi^*(d_1) = 0$, to obtain the second of the BCs in equations (5.9). In appreciating this argument it must be remembered that the sets of choices for the function $\delta\psi^*(x)$ made in the above are otherwise arbitrary.[4] Thus the action integral, equation (5.12), yields Schrödinger's equation and the two BCs equations (5.9).

In a variational formulation within the framework of the FEM, we would proceed as in Chapters 2 and 3, by discretizing the action integral, equation (5.12), using suitably placed finite elements. The wavefunctions are expressed in terms of interpolation polynomials multiplied by coefficients representing the nodal values of the wavefunctions, and the spatial dependence of the action is integrated. The resulting action is subjected to a nodal variational procedure to obtain a set of simultaneous equations for the unknown coefficients of the interpolation polynomials which are the nodal values of the wavefunction. This, however, is not the technique that we use in this chapter. In the following, we alter this procedure and introduce the Galerkin approach to solving differential equations.

5.3   The Galerkin method

Consider the differential equation (5.11) together with the BCs given in equations (5.9). The Galerkin method[5] assumes that the solution $\psi(x)$ can be approximated by

$$\psi(x) = \phi_0(x) + \sum_{i=1}^{n} a_i\,\phi_i(x). \tag{5.14}$$

Here $\phi_0(x)$ is assumed to satisfy the BCs (5.9), while the functions $\phi_i(x)$ satisfy the homogeneous BC at $x = d_2$, and the *homogeneous* part of the BC at $x = d_1$ obtained by setting $A = 0$ in equation (5.9). The functions are assumed to be orthogonal over the physical region. The approximation (5.14) satisfies the BCs at the two ends exactly for any choice of a_i. However, when substituted into equation (5.11), this approximate form for $\psi(x)$ leads to a *residual* $R(x)$ given by

$$R(x) = -\phi_0''(x) + (v(x) - \epsilon)\,\phi_0(x)$$

$$+ \sum_{i=1}^{n} a_i \left[-\phi_i''(x) + (v(x) - \epsilon)\,\phi_i(x) \right], \quad (5.15)$$

which in general will not be zero. We approach the actual solution by selecting a_i such that the residual $R(x)$ is as small as possible over each point in the physical region. The best choice would be a zero residual throughout the physical region. This will rarely be possible except through a happy choice of just the right set of approximation functions; we therefore insist on the weaker condition that the *integral* of the residual multiplied by some known and suitably chosen weighting functions $w_i(x)$ be zero, i.e., that the residual vanish in the sense that

$$\int_{d_1}^{d_2} w_j^*(x)\,R(x)\,dx = 0, \quad j = 1, 2, \ldots, n. \quad (5.16)$$

By selecting n linearly independent weighting functions we obtain n conditions for determining the unknown coefficients a_i. This is known as the method of weighted residuals.[6] In the Galerkin method, the weight functions are chosen to be the trial functions $\phi_j(x)$ themselves, and we have the n simultaneous equations

$$\int_{d_1}^{d_2} \phi_j^*(x)\,R(x)\,dx = 0, \quad j = 1, 2, \ldots, n. \quad (5.17)$$

An integration by parts of the second-order derivative terms in $R(x)$ gives surface terms. Retaining terms with the unknown coefficients a_i on the left side we obtain

$$\sum_{i=1}^{n} a_i \left[\int_{d_1}^{d_2} dx \left\{ \phi_j'^*(x)\,\phi_i'(x) + \phi_j^*(x)\,(v(x) - \epsilon)\,\phi_i(x) \right\} \right.$$

$$\left. - [\phi_j^*(x)\,\phi_i'(x)] \Big|_{x=d_1}^{x=d_2} \right]$$

$$= -\int_{d_1}^{d_2} dx \left[\phi_j'^*(x)\,\phi_0'(x) + \phi_j^*(x)\,(v(x) - \epsilon)\,\phi_0(x) \right]$$

$$+ [\phi_j^*(x)\,\phi_0'(x)] \Big|_{x=d_1}^{x=d_2}. \quad (5.18)$$

The integration over x can be performed since the trial functions are known functions. The above relation may be cast in the form of a matrix equation, and in an obvious notation it is expressible as

$$G_{ji}\,a_i = b_j, \quad (5.19)$$

with the column vector b_j containing all the terms with the known coefficients from the BCs. This set of n simultaneous equations is solved for the unknown coefficients a_i.

We note that the substitution of equation (5.14) into the action integral, equation (5.12), leads to the same set of simultaneous equations. The details have been relegated to Problem 5.1. The above description of the Galerkin method is thus *equivalent* to the Rayleigh–Ritz variational method, discussed in Chapter 2, which employs global wavefunctions. The Galerkin method has one advantage over the variational approach in that it can also be implemented *for differential equations which are not derivable from a variational principle.*[7]

The Galerkin method can also make use of test functions defined over finite elements which discretize the physical domain. In the next section we use the barrier tunneling problem to demonstrate in detail how the Galerkin method is used in practice. In order to illustrate the implementation of the mixed BCs in the most natural manner we shall employ Hermite interpolation polynomials defined over each finite element. This procedure also enjoys the advantage of giving results of high accuracy.

5.4   Tunneling calculations in the FEM

5.4.1   Evaluation of the residual

As a test of applying the Galerkin FEM for solving Schrödinger's equation, let us consider the constant potential V_I given by equation (5.2). We divide the range $[d_1, d_2]$ into *nelem* elements, each with three nodes: a node at the two ends and one in the middle of the element. For ensuring C_1 derivative continuity within the element we use two degrees of freedom at each node corresponding to the function and its derivative, i.e., we use *nnode* $= 3$ and *ndof* $= 2$. The desired polynomials are the quintic Hermite interpolation polynomials given in Section 3.5. These interpolation functions, defined over each element, are connected element by element to approximate the solution over the entire range of interest. If the potential has discontinuities, these can be accommodated by placing the element boundaries (nodes) at the discontinuities in the potential.

The three nodes in a given element i_{el} have the global coordinates

$$x_1 = d_1 + (i_{el} - 1)\, h,$$

$$x_2 = d_1 + (i_{el} - 1/2)\,h, \qquad (5.20)$$
$$x_3 = d_1 + i_{el}\,h,$$

with $h = (d_2 - d_1)/nelem$. The global coordinate x and the local coordinate ξ which ranges over $[-1, 1]$ in the element i_{el} are related by

$$x = x_2 + \frac{h}{2}\,\xi,$$
$$\xi(x) = \frac{2\,(x - x_2)}{h}. \qquad (5.21)$$

The wavefunction in the element i_{el} is written as

$$\Psi(x) = \sum_{i=1}^{3} \Psi_i\, N_i(\xi(x)) + \sum_{i=1}^{3} \Psi_i'\, \overline{N}_i(\xi(x))\, \frac{dx}{d\xi}, \qquad (5.22)$$

where the interpolation functions satisfy the conditions

$$N_i(\xi_j) = \delta_{ij}; \qquad \overline{N}_i(\xi_j) = 0;$$
$$\frac{d}{d\xi} N_i(\xi_j) = 0; \qquad \frac{d}{d\xi}\overline{N}_i(\xi_j) = \delta_{ij}; \qquad i, j = 1, 2, 3, \quad (5.23)$$

at the nodal points $\xi_1 = -1$, $\xi_2 = 0$, and $\xi_3 = +1$. We use the compact notation for equation (5.22)

$$\Psi(x) = \sum_{\alpha} \Psi_\alpha\, \Phi_\alpha(x); \qquad \alpha = 1, \ldots, 6, \qquad (5.24)$$

with the odd values of α corresponding to N_i and the even values α indexing $\overline{N}_i$ multiplied by the derivative $dx/d\xi$. This interpolated form is now substituted into Schrödinger's equation, equation (5.11).

In the Galerkin FEM the weight functions $w_i(x)$ are the same as the interpolation polynomials $\Phi_\alpha(x)$, and these local polynomials are truncated to zero outside the given element. Thus the weighted residual is obtained by multiplying equation (5.11) by $\Phi_\alpha(x)$, which are real polynomials, and integrating over the range of an element. We recognize that the nodal variables $\Psi_{5,6}$ of element i_{el} are the same as the variables $\Psi_{1,2}$ of element $i_{el} + 1$ as required to guarantee interelement continuity of the wavefunction and also of the probability current. In view of this equivalence, the interpolation functions from adjacent elements are added together in order to generate weight functions associated with the nodal variables at the boundaries. To illustrate this, consider the range $[d_1, d_2]$ split up into just two elements. The total number of unknown coefficients, before any BCs are imposed, is $4\,nelem + 2 = 10$. The first four weight functions are $\Phi_\alpha^{(1)}$, $\alpha = 1, \ldots, 4$, which are not zero only in the first element. The next two weight functions are $w_5 = \Phi_5^{(1)}(x) + \Phi_1^{(2)}(x)$, and $w_6 = \Phi_6^{(1)}(x) + \Phi_2^{(2)}(x)$, respectively. The last four weight functions are $\Phi_\alpha^{(2)}$ with $\alpha = 3, \ldots, 6$. The range of integration is $[-1, 1]$ in the variable ξ, and Gauss–Legendre quadrature is employed. This is also true for each of the two polynomials constituting the weight functions for the nodal variables at the boundary since each is nonvanishing only over one element. We integrate the coordinate dependence using a six-point Gauss–Legendre quadrature; for our quintic polynomials and for the potential V_I this is more than adequate to obtain exact (double-precision) results for the numerical integrations.[‡] Higher order Gauss quadrature is used for more complex potentials.

The second derivative term appearing in Schrödinger's equation is eliminated by an integration by parts, which leads to the usual surface terms corresponding to the functional values of Ψ at only the beginning and the end of the entire physical region. (If there are physical boundaries such as interfaces between materials, we will have to account for wavefunction and probability current continuity across these interfaces. It is best to begin with the integration by parts done analytically, rather than after the FEM implementation, to avoid any problems associated with discontinuous or ill-defined derivatives.) We are thus led to the $4\,nelem + 2$ equations for the unknown coefficients ψ_i and ψ_i' at each of $2\,nelem + 1$ nodes in the physical region.

When expressed as a matrix equation, the procedure corresponds to an "overlay" of the element matrices, exactly in the same manner as that discussed in Chapters 2 and 3. We write

$$
M_{\mu\nu} = \sum_{i_{el}}^{nelem} \left[\sum_{\alpha=1}^{6} \int_{-1}^{1} d\xi \left(\frac{dx}{d\xi} \right) \left\{ \Phi_\beta'^{(i_{el})}(x)\, \Phi_\alpha'^{(i_{el})}(x)\, \Psi_\alpha^{(i_{el})} \right.\right.
$$
$$
\left.\left. + \Phi_\beta^{(i_{el})}(x)\, (v(x) - \epsilon)\, \Phi_\alpha^{(i_{el})}(x)\, \Psi_\alpha^{(i_{el})} \right\} \right]
$$

[‡]We still have to deal with errors generated by the discretization used in the calculation.

$$- \left[\Phi_5^{(nelem)}(x) \, \Phi'^{(nelem)}_6(x) \, \Psi_6^{(nelem)} \right] \Big|_{d_2}$$
$$+ \left[\Phi_1^{(1)}(x) \, \Phi'^{(1)}_2(d_1) \, \Psi_2^{(1)} \right] \Big|_{d_1}. \qquad (5.25)$$

Here μ ranges over 1 to $4\,nelem+2$, while $\nu = 4\,i_{el}-4+\alpha$. The surface terms have only $\Phi_1(x)$ and $d\Phi_2(x)/dx$ not being zero at $x = d_1$ in the first element, and only $\Phi_5(x)$ and $d\Phi_6(x)/dx$ are not zero at $x = d_2$ in the last element. Only these terms have been retained in the above.

While we have described the Galerkin FEM using a *node-oriented* indexing of the variables and weight functions, it is also useful to keep in mind the usual *element-based* indexing for uniformity of the book-keeping in the computer codes; both points of view have advantages which can be exploited for realizing error-free programming.

5.4.2 Applying mixed BCs

We have two mixed BCs in equations (5.9) which have to be imposed on the system of $4\,nelem + 2$ equations (5.25). An example will best serve to demonstrate the method used to insert the BCs. For this purpose, let us suppose that we employ just one element over the entire region $[d_1, d_2]$, in which we make use of quintic Hermite interpolation. On multiplying equation (5.11) by the six polynomials in turn and integrating, we obtain the weighted residuals giving six equations for the nodal values $\psi_{1,2,3}$ and $\psi'_{1,2,3}$. Equation (5.25) takes the form

$$M_{i1}\,\psi_1 + (M_{i2} + \delta_{i1})\,\psi'_1 + M_{i3}\,\psi_2 + M_{i4}\,\psi'_2$$
$$+ M_{i5}\,\psi_3 + (M_{i6} - \delta_{i5})\,\psi'_3 = 0. \quad (5.26)$$

Here M_{ij} is the matrix element obtained for the weight function Φ_j with terms proportional to ψ_i and ψ'_i in the differential equation having been grouped together. The surface terms generated on inte-gration by parts are included through the δ_{ij} terms.

We now use equations (5.3) to substitute for the wavefunctions and their derivatives at d_1 and at d_2 in equation (5.26). Collecting like terms we obtain

$$[M_{i1} + ik_1(M_{i2} + \delta_{i1})]\,e^{ik_1 d_1}\,A + [M_{i1} - ik_1(M_{i2} + \delta_{i1})]\,e^{-ik_1 d_1}\,r$$
$$+ M_{i3}\,\psi_2 + M_{i4}\,\psi'_2 + [M_{i5} + ik_3(M_{i6} - \delta_{i5})]\,e^{ik_3 d_2}\,t = 0. \quad (5.27)$$

We have used the mixed BC $\psi'_t - ik_3\psi_t = 0$ to eliminate ψ'_3. The amplitude A of the incoming wave is assumed to be given in the ini-tial conditions for the problem. The terms in the above equations which have the known coefficient A are moved to the right side of the equations with a change of sign. This gives a set of four simultaneous equations for the remaining unknown quantities r, ψ_2, ψ_2', and t. Without loss of generality, we can set the amplitude A of the incoming wave to unity. This system of simultaneous equations is solved using standard numerical routines, and the transmission coefficient is calculated using equation (5.6). The actual number of elements required will be somewhat larger than one and will depend on the energy of the incident particle, as discussed below.

In anticipation of similar applications in the following chapters, we wish to introduce notation here which will provide cleaner programming for the above implementation of BCs. Let us suppose that we have *nelem* elements, having a total of *ng_nodes* nodal points, and that the global matrix, $\tilde{\mathcal{M}}$, has been formed by overlaying element matrices; the surface terms are also included in $\tilde{\mathcal{M}}$. The matrix is of dimension $4\,nelem + 2$ before the BCs are imposed. This global matrix acting on the array of nodal variables is equated to a null vector, as required for a zero-weighted residual. It is simplest to allocate a single matrix dimension in a computer program and retain it throughout the given calculation. The following procedure will allow us to keep the matrix size as $4\,nelem + 2$. We transform the first two nodal variables ψ_1, ψ_1' into the variables A and r using a matrix notation

$$\begin{pmatrix} \psi_1 \\ \psi_1' \end{pmatrix} = \begin{pmatrix} e^{ik_1 d_1} & e^{-ik_1 d_1} \\ ik_1 e^{ik_1 d_1} & -ik_1 e^{-ik_1 d_1} \end{pmatrix} \begin{pmatrix} A \\ r \end{pmatrix}. \qquad (5.28)$$

Let $\nu_0 = 2\,nelem + 1$, so as to allow for a compact subscript. We have

$$\begin{pmatrix} \psi_{\nu_0} \\ \psi_{\nu_0}' \end{pmatrix} = \begin{pmatrix} e^{ik_3 d_2} & 0 \\ ik_3 e^{ik_3 d_2} & 0 \end{pmatrix} \begin{pmatrix} t \\ 0 \end{pmatrix}. \qquad (5.29)$$

In effect, we have expressed the finite element variables in terms of the propagating modes in regions I and III. The transformation matrices converting ψ_1, ψ_1', and ψ_{ν_0}, ψ_{ν_0}', into amplitudes of the propagating modes can be embedded in a matrix, $\mathcal{P}$, of the same dimension as the global matrix to form a block-diagonal matrix, in which the 2×2 matrices given above appear at the beginning and the end of the diagonal, with all other diagonal elements in between being unity.

We now have

$$\tilde{\mathcal{M}} \cdot \mathcal{P} \,\tilde{\Psi} \equiv \mathcal{Q} \,\tilde{\Psi} = 0, \qquad (5.30)$$

in which the first two elements of the vector $\tilde{\Psi}$ are A and r, while the last two elements are t and some dummy variable (eventually to be assigned the value $ik_3\,t$). The product $\mathcal{Q}$ of the two matrices $\tilde{\mathcal{M}} \cdot \mathcal{P}$ performs precisely the column recombinations done in equation (5.27). Since A is known (and has been set to unity), we multiply the first column of the matrix $\mathcal{Q}$ by A and move the terms $\mathcal{Q}_{i1} \times A$ to form a vector column on the right side of the equation. Once this is done, we set to zero the elements of the first column *and also the first row* of $\mathcal{Q}$, and replace the first diagonal element by unity. Correspondingly, we insert the value of A in the first element of the vector on the right hand side. Setting the row elements to zero is equivalent to our defining the shape function to be zero at that node. Mathematically, this requirement corresponds to admitting only those test functions (shape functions here) that satisfy the BCs. The result of this procedure of inserting zeros along the row and column with a given index and assigning a unit value for the diagonal allows us to retain the original matrix size throughout the numerical calculations. Also, the "variable" A solves to its assigned value during the numerical solution of the simultaneous equations. This procedure may be called giving *benediction* to the matrix for the particular row/column index.[§] In the case of tunneling, we apply benediction to the last row and column of $\mathcal{Q}$ and set to zero the last element of the right hand side. Since t is unknown as yet, it is premature to assign any value to the right side for t'. The result of these changes in the matrix elements of $\mathcal{Q}$ is shown in Fig. 5.2 for a single element.

Once the simultaneous equations are solved using standard linear equation solvers,[8,9] the second element of the solution vector gives the value of the reflection amplitude r and the penultimate element gives the value of the amplitude t. The nodal wavefunction is reconstructed simply by multiplying the solution vector by the matrix $\mathcal{P}$ and the full wavefunction is obtained by interpolation.

[§]This rather suggestive nomenclature can be appreciated by the reader if he or she were to physically move his or her arm down the column of the matrix and across the row to be set to zero, and insert unity at the diagonal element.

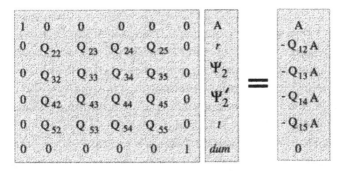

Fig. 5.2 The matrix equation obtained after applying BCs on using one element with quintic Hermite interpolation.

The additional matrix $\mathcal{P}$, of the same dimension as the global matrix $\mathcal{M}$, is very sparsely occupied. Thus this method is particularly applicable to problems which require small global matrices. For finite element applications generating large matrices the usefulness of this procedure would require implementing sparse matrix methods to reduce the memory requirements on a computer. The transmission coefficient obtained by this procedure for the rectangular potential barrier of Fig. 5.1 for just two elements is shown in Fig. 5.3a; the result with 30 elements is shown in Fig. 5.3b. The comparison between the exact and the calculated transmission using the FEM with

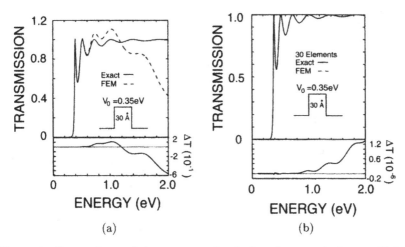

Fig. 5.3 Comparison of the exact and calculated transmission coefficient (a) using just two elements, and (b) with 30 elements in the region of the potential barrier V_I.

30 elements is excellent, and the results cannot be distinguished visually. The bottom curve in Figs. 5.3a and 5.3b are plots of the difference between the exact and the FEM-calculated transmission. With 30 elements, the resulting accuracy is of the order of 10^{-6}, which for many applications is more than adequate. Further improvement can be generated by the use of additional elements. A perusal of Figs. 5.3a and 5.3b indicates that as the energy of the incident particle is increased, it becomes desirable to include more elements. This is understandable, since we are representing more rapidly oscillating wave functions with increasing E (for $E > V_0$) using quintic interpolation polynomials.

We draw attention to earlier work on tunneling calculations using finite elements by Laloyaux et $al.$,[10] and by Nordholm and Bacskay.[11] Tunneling in quantum semiconductor structures has been investigated independently by Nakamura et $al.$[12]

The transmission coefficient for the potential barrier

$$V_{II}(x) = V_0/\cosh^2 x, \qquad (5.31)$$

where V_0 is again the maximum height of the barrier, is also analytically solvable,[2] and we have

$$T = \frac{\sinh^2(\pi k)}{\sinh^2(\pi k) + \cosh^2(\frac{1}{2}\pi\sqrt{8MV_0/\hbar^2 - 1})} \qquad (5.32)$$

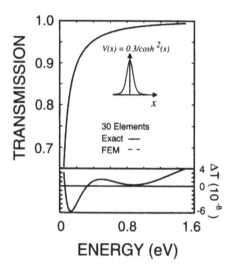

Fig. 5.4 Comparison of the exact and calculated transmission using 30 elements for the potential barrier V_{II}.

when $8MV_0/\hbar^2 > 1$, or

$$T = \frac{\sinh^2(\pi k)}{\sinh^2(\pi k) + \cos^2(\frac{1}{2}\pi\sqrt{1 - 8mV_0/\hbar^2})} \qquad (5.33)$$

when $8MV_0/\hbar^2 < 1$. We apply the Galerkin FEM to the potential V_{II} by a simple replacement of V_I by V_{II} in equation (5.11). The results are similar to those mentioned for the previous example. Figure 5.4 shows a comparison between the exact and the numerically evaluated transmission for 30 elements. The error is of the order of 10^{-8} up to an energy E which is five times the barrier height.

5.5   Comparing Galerkin FEM with WKB

When the transmission coefficient cannot be determined analytically, the phase-integral approach of Wentzel, Kramers, and Brillouin (WKB)[13] is commonly used for calculating the tunneling coefficient for a particle incident on the potential barrier with an energy below the barrier height. However, this approximation is not valid for energies above the barrier because the potential has no classical turning points. A path-integral approach is advocated by Holstein for this case.[14] The WKB method requires that the potential vary slowly in comparison with the wavelength of the particle. The method exploits the fact that the wavefunction does not change drastically from that for a constant potential for any small region of the potential. A clear treatment of the method and its classical connection is presented in Refs. 1 and 3, and the reader's attention is directed to these textbooks. Here we restrict ourselves to comparing the WKB method with the Galerkin FEM.

As an example, we take the potential function[15]

$$V_{III}(x) = V_0\left(1 - x^2/x_0^2\right). \qquad (5.34)$$

The lowest approximation within the WKB method gives the transmission coefficient

$$T = \exp\left[-\pi x_0\sqrt{\frac{2mV_0}{\hbar^2}}\left(1 - \frac{E}{V_0}\right)\right]. \qquad (5.35)$$

Equation (5.35) is valid when

$$\pi x_0\sqrt{\frac{2mV_0}{\hbar^2}}\left(1 - \frac{E}{V_0}\right) \gg 1. \qquad (5.36)$$

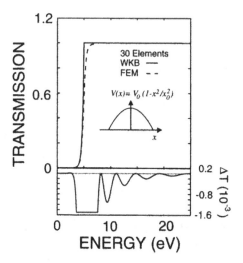

Fig. 5.5 The calculated transmission (dashed line) using 30 elements for the potential barrier V_{III} is compared with the WKB result. Note that the lower plot has ΔT which is the difference between the calculation using the WKB approximation and the finite element result.

Note, however, that the WKB method results in $T = 1$ for $E > V_0$, where no classical turning points exist in this case.[15] In Fig. 5.5 we compare the WKB and the calculated transmission for 30 elements. For energies above the barrier height V_0 the FEM calculation continues to give oscillations in the transmission, as is seen in the lower graph in Fig. 5.5, which plots the difference between the WKB and the FEM results. Note that the FEM continues to provide accurate results where the WKB method fails, and the accuracy is comparable to the earlier examples since numerical considerations here are not different in any way from those in earlier examples.

In summary, the FEM provides two main improvements over the WKB approximation. The first is the ability of the FEM to accommodate any arbitrary potential with a finite quantum mechanical expectation value, whereas the WKB approximation requires that the potential change slowly as a function of position, as compared to the wavelength of the particle. The second improvement resides in the inability of the WKB approximation to address cases where the energy of the particle is greater than the maximum potential of the barrier. Both these issues are accounted for in a rather straightforward manner in the FEM implementation. The need for a given

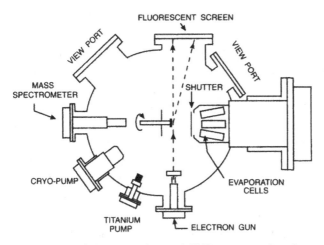

Fig. 5.6 Schematic for an MBE vacuum chamber.

minimum number of elements in the FEM is governed by the spatial frequency of oscillation of the wavefunction, which in turn depends on the energy (or wavevector) of the particle.[16]

5.6 Quantum states in asymmetric wells

Layered semiconductor heterostructures provide a natural testing ground in one dimension for quantum mechanical concepts.[17] In this section, we consider the physics of asymmetric quantum wells and the variety of states within them, and discuss how the FEM can be used to numerically model the properties of such quantum wells. The practical motivation for such investigations in semiconductors is driven by the fact that the new compound semiconductor heterostructures hold promise for a number of new devices which will provide the next generation of fast switching ($\leq 1\,\mathrm{ps}$) and high frequency devices in electronics and optoelectronics.[18] Recent advances in semiconductor growth technology, as represented by the methods of molecular beam epitaxy (MBE) and metallo-organic chemical vapor deposition (MOCVD), have led to the growth of layered semiconductors with near-atomic perfection in the crystalline growth of layers and of interfaces between layers. A schematic of an MBE vacuum chamber is shown in Fig. 5.6. The layered structures are grown atomic layer by atomic layer on a substrate with an unprecedented control over the composition of the atomic layers.

Consider a planar heterostructure consisting of a layer of Al-

GaAs grown on top of a layer of GaAs. The energy of the conduction bandedge in the AlGaAs layer is higher than in GaAs. Hence, an electron in GaAs incident on the GaAs/AlGaAs interface experiences a potential energy barrier in the AlGaAs layer. The height of the barrier depends on the concentration of Al in AlGaAs. By placing a thin layer of GaAs between two AlGaAs layers it is possible to construct a quantum well structure, in which an electron in the GaAs layer is confined to the quantum well layer in its motion along the crystal's "growth direction." The electron is free to propagate in the direction parallel to the layer. A quantum well structure is simultaneously formed for holes at the top of the valence band, since the energy bandgap in AlGaAs is larger than that in GaAs. As the thickness of the well layer is lowered below about 1000 Å the effect of carrier confinement along the growth direction begins to influence the dynamics of the carriers in the conduction band. The Heisenberg uncertainty principle leads to an increase in the energy of the carriers, and the effective conduction bandedge now corresponds to the position of the lowest energy level in the quantum well. A similar shift in energy takes place with the holes in the valence band of the quantum well material. The effective energy bandgap is thus raised depending on the confinement effect which in turn depends on the layer thickness of the quantum well. The optical properties of the heterostructure are altered in the process, and by adjusting the geometry of the layered structure we can tailor these properties, within limits imposed by the material properties of the constituent semiconductors. This is the basic idea of *bandgap engineering*.

Let us consider quantum states in layered semiconductor structures localized within the well region. Problem 2.1 at the end of Chapter 2 addresses the question of determining energy levels and wavefunctions of bound states below a finite barrier height in symmetric quantum wells. As is well known, single symmetric quantum wells must have at least one bound state, regardless of the barrier height.[2,3] However, potential wells that are asymmetric due to the barrier heights on either side of the well region being unequal do not have this requirement that a bound state must exist.[2] Furthermore these structures in general contain not only bound but also quasibound states as will be explained presently. AQWs can be constructed with layered GaAs/AlGaAs structures simply by altering the concentration of Al in the barrier regions. We have mentioned that quantum wells can be asymmetric because of unequal barrier

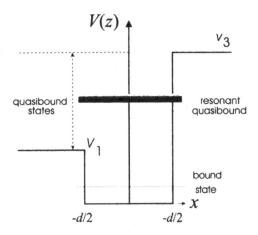

Fig. 5.7 A plot of the potential profile of an asymmetric single quantum well indicating a bound state with energy below V_1 and quasibound states for $V_1 < E < V_3$. Some quasibound states can resonantly occupy the well region.

heights on either side of the well. This may be called geometrically (compositionally) induced asymmetry. We can also induce an asymmetry in a symmetric quantum well by applying an electric field across it. In this section, we consider the application of BCs for the wavefunctions in a compositionally asymmetric well in a finite element calculation.

The potential energy function for the AQW is given by

$$V_a(x) = \begin{cases} V_1, & \text{for } x < 0, \\ 0, & \text{for } 0 < x < d, \\ V_3, & \text{for } x > d, \end{cases} \qquad (5.37)$$

where $0 < V_1 < V_3$, as in Fig. 5.7. We solve equation (5.11) with scaled variables and a length scale of $\ell_0 = 1$ Å. The well thickness d is usually specified in angstrom units in quantum semiconductor structures. For energies $E < V_1$ the wavefunctions fall off exponentially in the barrier regions for bound states

$$\psi(x) = \begin{cases} c_1\, e^{+\kappa_1(x-d/2)} & \text{for } x \le -d/2, \\ c_3\, e^{-\kappa_3(x-d/2)} & \text{for } x \ge -d/2. \end{cases} \qquad (5.38)$$

Here the wavenumbers are

$$\kappa_{1,3} = \sqrt{\frac{2\,m\,\ell_0^2\,(V_{1,3} - E)}{\hbar^2}}. \qquad (5.39)$$

The BC $\psi(|x|) \to 0$ has to be satisfied in the asymptotic region $|x| \to \infty$ in the two barriers. Within the framework of the FEM we set the wavefunction and its derivative to zero at some suitably large cutoff value $|x_c|$, where the wavefunction is 10^{-6} times smaller, say, than its value at the origin. We have at hand an eigenvalue problem similar to the examples treated in Chapters 2 and 3. This finite element implementation is now a matrix generalized eigenvalue problem of the form

$$\mathcal{H}\cdot\Psi = \epsilon\,\mathcal{U}\cdot\Psi, \tag{5.40}$$

with global matrices of dimension $4\,nelem - 2$ when quintic Hermite interpolation polynomials are used. The acceptable solutions corresponding to the bound state eigenvalues ϵ_i are selected in the energy region $\epsilon_i < V_1$, where the above BCs are operative. Bound states in arbitrary potential profiles in quantum semiconductor structures have been investigated earlier using the FEM.[19]

States with energy $V_3 > E > V_1$ are unbound on the left side and fall off exponentially into the barrier on the right side. These states are referred to as quasibound states. Electrons incident from the left on the quantum well are completely reflected, and the reflection coefficient continues to be $R = 1$ for all energies in this range. At certain energies such states satisfy the condition $k_2 d = n\pi$ in the quantum well region, where $k_2 = \sqrt{2\,m\,\ell_0^2\,E/\hbar^2}$. These states may be thought of as remnants of the bound states of a symmetric quantum well with both barrier heights being V_3. As the barrier height on the left side is lowered these states continue to be confined on the right side while becoming unconfined in the left region. A useful way of identifying such "resonant states" is by considering the probability of finding the particle in the quantum well region, which we call the occupancy $\Omega(E)$. Following Messiah's treatment[20] one can obtain an analytic expression for the occupancy

$$\Omega(E) = \int_{-d/2}^{d/2} |\psi(x)|^2 dx. \tag{5.41}$$

We take the energies at which $\Omega(E)$ has local maxima to be those corresponding to quasibound states.

To apply the FEM for the evaluation of the wavefunctions of quasibound resonant states we use the mixed BC

$$\psi'(-d/2) + ik_1\,\psi(-d/2) = 2ik_1\,A, \tag{5.42}$$

at $x = -d/2$ and require the vanishing of the wavefunction and its derivative at some large value of the coordinate $x_b \gg d/2$. Here the energy of the incident wave is assumed to be given, so that the bound state BC can be made more precise. At each energy E we can determine the imaginary wavevector κ_3 in the right barrier region, and hence determine the explicit exponential fall-off of the wavefunction

$$\psi(x) = c_1 \, e^{-\kappa_3(x-d/2)}. \qquad (5.43)$$

Now the coefficient c_1 plays a role analogous to the transmission amplitude t for tunneling wavefunctions considered earlier, and using quintic Hermite interpolation we can apply the BCs as in the tunneling problem. The BC on the right side can then be written as

$$\psi'(d/2) + \kappa_3 \, \psi(d/2) = 0. \qquad (5.44)$$

Note that we no longer have an eigenvalue problem, and the set of equations for the nodal values of the wavefunctions is solved with standard linear equation solvers. Once the wavefunction is reconstructed using the nodal wavefunction values and the interpolation polynomials, we can directly calculate the occupancy at the given energy E. This calculation is repeated to obtain $\Omega(E)$ over the range $V_1 < E < V_3$.

In the above discussion, we considered only the continuity of the wavefunction and its derivative at quantum well interfaces. In the actual calculations, the BCs are more complex for two reasons. First, the single Schrödinger equation is inadequate for the description of the carrier dynamics in semiconductors. The equations for the valence and the conduction bands are coupled in the so-called k·P model for envelope functions. Second, the current continuity condition requires a more elaborate treatment which will not be considered here. A first approximation is achieved by accounting for the fact that Schrödinger's equation requires the use of the carrier's effective mass. The many-body effects arising from the energy band formation in semiconductors alter the free carrier mass. The electron's effective mass in GaAs is $m^* = 0.0665 \, m_0$, while the effective mass in the Al-GaAs barriers is somewhat higher, depending on the concentration of Al.[21,22] The resulting current continuity condition, in a one-band calculation,[17] at the interface located at $x = -d/2$ is given by

$$\left. \frac{1}{m_1^*} \, \psi'(x) \right|_{x=-d/2-\delta} = \left. \frac{1}{m_2^*} \, \psi'(x) \right|_{x=-d/2+\delta}, \qquad (5.45)$$

with δ being a small positive quantity. This BC was first proposed by Ben Daniel and Duke.[23]

We now turn to the experimental verification of the existence of resonant quasibound states. The layered crystals of semiconductor quantum wells were investigated optically while subjected to alternate compression and expansion by attaching a lead zirconate–titanate transducer to the GaAs substrate in the heterostructure. A modulated reflectivity spectrum was observed which is the derivative of the usual reflectivity features. This derivative spectroscopy technique provides spectra of exceptional quality,[24,25] as seen in Figure 5.8. The details of the experimental procedure have been reported in the literature.[26,27] Samples of symmetric and compositionally asymmetric quantum wells of the same thickness with $d = 20\,\text{Å}$, $33\,\text{Å}$, $50\,\text{Å}$, and $100\,\text{Å}$ were grown to investigate the role of the quasibound states in interband optical transitions in the energy range $E > 1.52\,\text{eV}$, i.e., above the bandgap of GaAs.

If the barrier height V_3 is very large and at the same time the well width d is small enough, it is possible that no bound states exist with energies $\epsilon_i < V_1$. The condition for the expulsion of the bound state from the quantum well is[2,20]

$$k_2\, d > \arccos\left(\sqrt{V_1/V_3}\,\right). \tag{5.46}$$

The $20\,\text{Å}$ AQW was specifically grown with the purpose of verifying that at this well thickness and with Al concentrations of $x_1 = 0.1$ and $x_3 = 0.3$ in the left and the right barriers of $Al_x Ga_{1-x} As$ we do not have any light-hole bound states in the AQW.

We now review the experimental observations for asymmetric wells.

Samples of 100 Å wide quantum wells

(i) We begin by first considering the well-known case of a symmetric single quantum well. Figure 5.8a shows the piezomodulated reflectivity of the $100\,\text{Å}$ symmetric quantum well. Note that the insets shown in this figure, and in all figures to follow, are not a magnification of the main plot but instead are separate scans where the spectrometer slits and the amplifier have been optimized for weak signals.

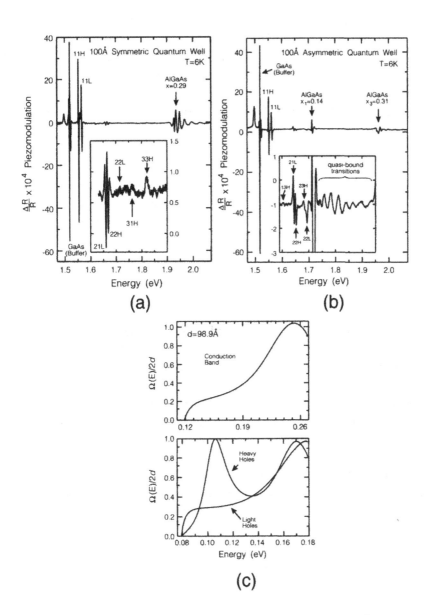

Fig. 5.8 The piezomodulated reflectivity spectra of a $100\,\text{Å}$ wide (a) $Al_{0.29}Ga_{0.71}As/GaAs/Al_{0.29}Ga_{0.71}As$ symmetric quantum well, and (b) of a $100\,\text{Å}$ $Al_{0.14}Ga_{0.86}As/GaAs/Al_{0.31}Ga_{0.69}As$ AQW at $T=6\,\text{K}$ are shown. In (c), a plot of the occupancy $\Omega(E)/2d$ as a function of energy for the conduction and valence bands of the $100\,\text{Å}$ wide AQW is shown.

Table 5.1 Transition energies (eV) for the 100 Å symmetric well. The theoretical values are uncorrected for excitonic binding energies. A well width of 101.76 Å was used for the calculations.

Experiment	Theory	Identification
1.551	1.558	11H
1.563	1.573	11L
1.656	1.658	21L
1.663	1.665	22H
1.712	1.716	22L
1.761	1.757	31H
1.817	1.814	33H

The strongest feature at 1.5156 eV ± 0.0004 eV is the free exciton from the 0.5 μm thick GaAs buffer layer on which the quantum well was grown. (The exciton is the bound state of the electron (optically) excited to the conduction band from the valence band interacting with the hole left in the valence band.) The AlGaAs barrier signature at 1.931 eV allowed calibration of the Al content in the barriers. Both the experimentally determined and the theoretically predicted transition energies according to a three-band bound state model are presented in Table 5.1. The calculations were done using three coupled Schrödinger equations which were solved by the FEM. Three quintic Hermite elements in each layer were adequate to obtain an accuracy for the eigenvalues of the order of 10^{-5} eV. The results were compared with highly accurate calculations based on the transfer-matrix method developed by the author.[28] For labeling the transitions we use a notation in which the first number indicates the conduction band state and the second the valence band state, with "H" and "L" identifying the heavy-hole (hh) and light-hole (lh) states. Any transitions which involve quasibound states are denoted by a superscript "q" after the number. All the theoretical values are uncorrected for excitonic binding energies. The calculations were carried out with the assumption that the well width is an integral multiple of half the lattice constant of GaAs (1/2 of $a = 5.65$ Å) and with the conduction band offset taking up 60% of the total band offset. The theory predicted five hh states of which three were ob-

Table 5.2 Transition energies (eV) for the $100\,\text{Å}$ asymmetric well. Note that the theoretical values are uncorrected for excitonic binding energies. A well width of $98.9\,\text{Å}$ was used for the calculations.

Experiment	Theory	Identification
1.548	1.554	$11h$
1.559	1.568	$11l$
1.601	1.608	$13h$
1.641	1.645	$21l$
1.648	1.652	$22h$
1.678	1.685	$23h$
1.691	1.695	$22l$
1.732	1.725	$13^q l$
	1.730	$24^q h$
1.742	—	—
1.739	1.779	$3^q 1h$
1.782	1.793	$3^q 1l$
	1.794	$25^q h$
	1.800	$3^q 2h$
	1.802	$23^q l$
1.809	1.833	$3^q 3h$
1.835	1.843	$3^q 2l$
1.859	1.878	$3^q 4^q h$
1.894	—	—

served and three lh states of which we observed two. Transitions involving all three predicted conduction band states were observed. In Fig. 5.8a, the AlGaAs barrier signature above $1.9\,\text{eV}$ contains many features; this complex structure may arise from a number of possibilities, and further details are given elsewhere.[27]

(ii) The $100\,\text{Å}$ asymmetric single quantum well was grown with unequal barrier heights corresponding to Al concentrations of $x_1 = 0.14$ and $x_3 = 0.31$ as determined from the barrier signatures shown in Fig. 5.8b. The lowering of one barrier causes a few of the bound states in the symmetric case to become quasibound. The number of bound states is reduced to two conduction band states, three *hh* states, and two *lh* states. The plot of the occupancy $\Omega(E)$ in Fig. 5.8c, in which the energy is measured with respect to the GaAs bandedge and ranges from the lower barrier energy to the upper barrier energy, shows two quasibound *hh* states and one quasibound *lh* state, thus accounting for the missing bound states from the symmetric quantum well due to the lowered barrier height. Each of the quasibound states appears at a slightly lower energy than its bound counterpart in the symmetric quantum well. The experimental and theoretical transition energies are shown in Table 5.2.

The transitions involving quasibound states are easily seen between the two barrier signatures. For this particular sample, bound to quasibound state transition energies are expected below the lower barrier signature. However, they are expected to appear too close to the more intense bound-to-bound transitions to allow observation.

Samples of 20 Å wide quantum wells

(i) The smallest quantum well examined was 20 Å wide. The symmetric quantum well has just one bound state each for the conduction band and the *hh* and *lh* states as shown in Fig. 5.9a. The 11H and 11L transition energies are at 1.754 eV and 1.786 eV respectively. The transition from the GaAs buffer layer spin–orbit split-off band to the conduction band is visible at 1.859 eV.

(ii) A symmetric well must have at least two transitions corresponding to 11H and 11L. However, an AQW with unequal barrier heights is not required to possess a bound state. The piezomodulated reflectivity spectrum shown in Fig. 5.9b shows just one transition from a bound *hh* state to the bound conduction band state. There is no bound *lh* state in this sample. This is demonstrated by the experimental results as well as by the theoretical predictions. This is verified by the occupancy plot shown in Fig. 5.9c. The 11H transition is just 14 meV below the lower AlGaAs valence band barrier. The missing *lh* state has become a quasibound state 17 meV above the lower valence band barrier as shown by the plot of $\Omega(E)$ in Fig. 5.9c. Note that in this sample the *hh* lacks a true maximum in $\Omega(E)$. In order to adequately explain all the features in Table 5.3, the second *hh* state (2^qH) was assigned to the valence band edge where $\Omega(E)$ reaches its highest value in Fig. 5.9c.

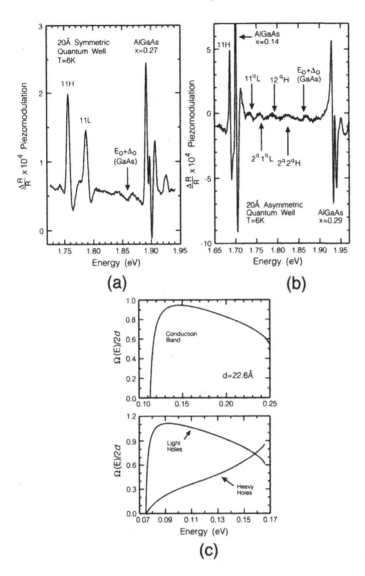

Fig. 5.9 The piezomodulated reflectivity spectra of a 20 Å wide (a)
$Al_{0.27}Ga_{0.73}As/GaAs/Al_{0.27}Ga_{0.73}As$ symmetric quantum well, and (b)
$Al_{0.14}Ga_{0.86}As/GaAs/Al_{0.29}Ga_{0.71}As$ AQW at $T=6$ K are shown. Note the
absence of the 11L transition. The GaAs buffer layer signature has been
omitted for clarity. In (c), a plot of the occupancy $\Omega(E)/2d$ as a function
of energy for the conduction and valence bands of the 20 Å wide AQW is
shown.

Table 5.3 Transition energies (eV) for the 20 Å
asymmetric well. Note that the theoretical values
are uncorrected for excitonic binding energies. A
well width of 19.8 Å was used for the calculations.

Experiment	Theory	Identification
1.685	1.688	11H
1.737	1.722	11^qL
1.763	1.759	2^q1^qL
1.789	1.798	12^qH
1.829	1.833	2^q2^qH

For completeness, we note that interest in AQWs has also been based on the fact that they have second-order optical nonlinear susceptibilities, $\chi^{(2)}$, which can be three orders of magnitude larger than those of bulk GaAs ($\chi^{(2)}_{GaAs} \sim 2 \times 10^{-10}$ m/V, at 10.6 μm). The enhancement is easily appreciated when we realize that the transition dipole matrix elements are of the order of the unit cell ($\sim$3–5 Å) in a bulk III–V semiconducting material with inversion asymmetry, whereas the dipole matrix elements in AQWs are of the order of 30–50 Å, the size of typical quantum wells being $\sim$100 Å. Recent work has focused on (i) symmetric single quantum wells with inversion asymmetry induced by an electric field applied across the quantum well,[31] (ii) single quantum wells with a step potential at the bottom of the well from a raised bandedge in the well arising from the introduction of Al in the GaAs layer,[31,32] and (iii) asymmetric double quantum wells.[33,34] These investigations have focused on inter-subband as well as interband transitions which contribute to the nonlinear susceptibility. The compositionally asymmetric single quantum well is the simplest of structures with a "geometrically induced" inversion asymmetry. Our estimate for the interband second-order nonlinear susceptibility for a 100 Å asymmetric well with Al concentrations of 0.4 and 0.1 on the two sides is: $\chi^{(2)}_{asym}(SHG) \sim 6 \times 10^{-7}$ m/V for an AQW.[35]

We have shown here that finite element calculations can be used with advantage to *design* heterostructures for investigating experimentally the quantum mechanical features exhibited by quantum semiconductor structures. The finite element calculations provide the means for *interpreting* the experimental results; the presence of heterostructure interfaces necessitates numerical computations before the experimental features can be adequately understood. Novel features arise in the electronic energy level scheme of a quantum well structure as a result of the nature of the quantum well profile. The AQW structures which exhibit quasibound states are worth exploring further in the context of photodetectors or optical modulators.[29,30] The differences in the photo-ionization cross-section between the symmetric and the AQWs may be significant in such phenomena. The Stark effect of the bound versus the quasibound states is another aspect worthy of consideration.

5.7 Problems

1. Substitute the approximate form of the solution to Schrödinger's equation, equation (5.14), into equation (5.12) for the action integral and implement the variational principle by determining the condition for the extremum of the action by varying the parameters a_i^*. Show that the Galerkin equations, equation (5.18), are obtained on substituting the BCs equation (5.9).

2. A particle with positive energy E is incident on a potential well

$$V(x) = \begin{cases} -V_0, & \text{for } -d/2 \leq x \leq d/2, \\ 0, & \text{for } x > |d/2|. \end{cases}$$

(a) Show analytically that the transmission coefficient is

$$T = \cfrac{1}{1 + \cfrac{V_0^2}{4E(V_0 + E)} \sin^2(k_2 d)}.$$

(b) Evaluate the transmission coefficient by the Galerkin FEM. Plot the calculated transmission obtained by employing quintic Hermite interpolation functions in (i) *nelem* $= 2$, and (ii) *nelem* $= 30$ elements. Compare the computed transmission coefficients with the above exact expression.

3. One area of current interest is the localization of states with energies above the barrier height. There exist energies at which the probabilities of finding the particle in the *barrier region* are local

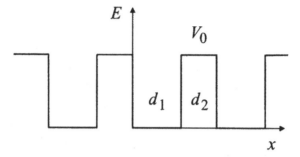

Fig. 5.10 The periodic square-well potential for the conduction band in a semiconductor superlattice.

maxima. Such localized states have been found to exist experimentally, and to participate in optical transitions in quantum semiconductor heterostructures.[36-38] These occupancy maxima arise through Fabry–Pérot interference effects between the forward and backward propagating waves between the boundaries of the barrier layer. To model this effect, it is necessary to calculate the wavefunction as described in the text.

Use the nodal variables and the interpolation functions and integrate the probability density over the width of the barrier and

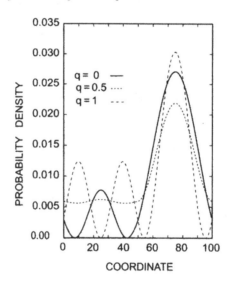

Fig. 5.11 Localized states in a superlattice structure for energy above the barrier height $V_0 = 300\,\text{eV}$ in a $50\,\text{Å}/50\,\text{Å}$ GaAs/AlGaAs superlattice . Here q is in units of $\pi/(d_1 + d_2)$.

plot this integral versus the incident energy. Determine the energies at which these occupancy plots have maxima. Consider potential barriers of the type V_I, V_{II}, and V_{III} discussed in the text.

4. A semiconductor superlattice structure[18] is formed by growing a periodic arrangement of alternating layers of GaAs and AlGaAs. Let the two layers be of thicknesses d_1 and d_2. The periodic square-well potential, shown in Fig. 5.10, is referred to as the Kronig–Penney potential.[39] Use a barrier height $V_0 = 300$ meV and an effective mass $m^* = 0.0665\,m_0$ for the electron in all the layers. The periodicity of the potential gives rise to the Bloch periodicity condition on the wavefunction[40]

$$\psi(d_1 + d_2) = \psi(0)\,e^{iq(d_1+d_2)}.$$

Here q is the wavevector of the superlattice. The energy of the electron displays a dispersion $E(q)$ with energy bandgaps induced by the periodic potential.

(a) Within the FEM, impose the periodic BC for the superlattice by requiring that the variables corresponding to the last node be equivalent to those of the first node multiplied by a phase factor $e^{iq(d_1+d_2)}$. Describe the rearrangement of rows and columns required to ensure this periodicity.

(b) Determine the localization in the barrier region of a conduction electron with energy above the barrier height. Show that carrier localization occurs for all values of q, as shown in Fig. 5.11.

5. A surface quantum well[41] consists of a layer of AlGaAs with thickness $d_b \simeq 800$ Å, grown on a GaAs substrate, with the growth terminated by a cap layer of GaAs of thickness $d_w \simeq 100$ Å. The resulting potential profile is shown in Fig. 5.12. An electron in the conduction band of such a structure experiences a very large confining potential barrier at the vacuum–GaAs interface, and a barrier height of $V_b = 300$ meV in the AlGaAs barrier region.

By using the FEM, show that the single barrier and the quantum well at the surface give rise to states localized above the barrier energy. Show that the localized states above the barrier can be

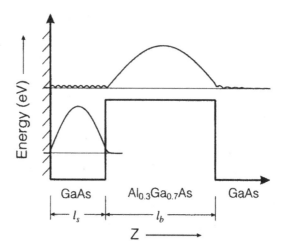

GaAs | Al$_{0.3}$Ga$_{0.7}$As | GaAs

l_s | l_b

Z ⟶

Fig. 5.12 Localized states in a surface quantum well are formed in the surface layer as well as in the barrier region. See Ref. 41.

assigned quantum numbers $\nu = 1, 2, \ldots$, in a manner analogous to the usual quantum number assignment for states localized within quantum wells.

6. The potential energy of an electron in symmetric quantum wells in the presence of an external electric field has an additional contribution $V_E = +|e|Ex$. The potential well gets tilted and the original bound states in the quantum well become resonant quasibound states. Consider a quantum well of GaAs/AlGaAs with a well thickness $d_w = 100$ Å and very wide barriers on either side of thickness $d_b = 500$ Å. Assume a barrier height $V_0 = 300$ meV in the absence of any external field and a common effective mass of $m^* = 0.0665\, m_0$ for the electron throughout the three layers in the quantum well structure.

(a) For an external electric field of $1.0\,\mathrm{V\,cm^{-1}}$ across the three layers, determine the number of quasibound states which resonantly occupy the quantum well layer. Employ the method of finite elements and implement the BCs on the wavefunctions in a manner similar to that for the AQW treated in the text.

(b) Define the energy width of the resonances as the full width at half-maximum, and estimate the lifetime of such a resonant state.

References

[1] D. Bohm, *Quantum Theory* (Prentice Hall, New York, 1951), p233.

[2] L. D. Landau and E. M. Lifshitz, *Quantum Mechanics – Nonrelativistic Theory* (Pergamon, New York, 1958), pp78–80.

[3] L. I. Schiff, *Quantum Mechanics*, 3rd ed. (McGraw-Hill, New York, 1968).

[4] R. Courant and D. Hilbert, *Methods of Mathematical Physics*, Vol. I (Interscience, New York, 1953), p208.

[5] B. G. Galerkin, *Vestn. Inz. Tech.* **19**, 897 (1915). A very readable introduction to Galerkin methods together with a brief history is given in C. A. J. Fletcher, *Computational Galerkin Methods* (Springer-Verlag, New York, 1984). Also see, for example, T. J. R. Hughes, *The Finite Element Method* (Prentice Hall, Englewood Cliffs, NJ, 1987).

[6] B. A. Finlayson, *The Method of Weighted Residuals and Variational Principles, with Application in Fluid Mechanics, Heat and Mass Transfer* (Academic Press, New York, 1972).

[7] G. Strang and G. J. Fix, *An Analysis of the Finite Element Method* (Prentice Hall, Englewood Cliffs, NJ, 1973).

[8] W. H. Press, S. A. Teukolsky, W. T. Vetterling, and B. P. Flannery, *Numerical Recipes* (Cambridge University Press, Cambridge, UK, 1992).

[9] E. Anderson, Z. Bai, C. Bischof, J. Demmel, J. J. Dongerra, J. Du Croz, A. Greenbaum, S. Hammarling, A. McKenney, S. Ostrouchov, and D. Sorensen, *LAPACK User's Guide* (Society for Industrial and Applied Mathematics, Philadelphia, 1992).

[10] T. Laloyaux, P. Lambin, J. P. Vigneron, and A. A. Lucas, *J. Comput. Phys.* **83**, 398 (1989).

[11] S. Nordholm and G. Bacskay, *Chem. Phys. Lett.* **42**, 259 (1976).

[12] K. Nakamura, A. Shimizu, M. Koshiba, and K. Hayata, *IEEE J. Quantum Electron.* **27**, 1189 (1991).

[13] Sometimes referred to as the JWKB method, developed by Jefferys, Wentzel, Kramers, and Brillouin. See Refs. 1 and 3, above. H. Jeffreys, *Proc. London Math. Soc.* **23**, 428 (1923); G. Wentzel, *Z. Phys.* **38**, 518 (1926); H. A. Kramers, *Z. Phys.* **39**, 828 (1926); L. Brillouin, *CR Acad. Sci. Paris* **183**, 24 (1926).

[14] B. R. Holstein, *Am. J. Phys.* **52**, 321 (1984); *Topics in Advanced Quantum Mechanics* (Addison-Wesley, Redwood City, CA, 1992).

[15] F. Constantinescu and E. Magyari, *Problems in Quantum Mechan-*

ics, trans. V. V. Grecu, ed. J. A. Spiers (Pergamon, New York, 1971), p97.

[16] R. Goloskie, J. W. Kramer, and L. R. Ram-Mohan, *Comput. Phys.* **8**, 679–686 (1994). Portions of this article were part of the undergraduate thesis of JWK at Worcester Polytechnic Institute (1992).

[17] G. Bastard, *Wave Mechanics Applied to Semiconductor Heterostructures* (Les Editions de Physique, Les Ulis, France, 1988).

[18] R. Tsu and L. Esaki, *Appl. Phys. Lett.* **22**, 562 (1973).

[19] K. Nakamura, A. Shimizu, M. Koshiba, and K. Hayata, *IEEE J. Quantum Electron.* **25**, 889 (1989).

[20] A. Messiah, *Quantum Mechanics*, Vol. I (North-Holland, New York, 1964), p88.

[21] *Landolt-Börnstein Numerical Data and Functional Relationships in Science and Technology*, ed. O. Madelung Group III (Springer-Verlag, Berlin, 1982), Vol. 17.

[22] I. Vurgaftman, J. R. Meyer, and L. R. Ram-Mohan, "Band Parameters for III-V Compound Semiconductors and Their Alloys," Applied Physics Reports, Review Section of *J. Appl. Phys.* **89**, 5815–5875 (2001).

[23] D. J. Ben Daniel and C. B. Duke, *Phys. Rev.* **152**, 683 (1966).

[24] Y. R. Lee, A. K. Ramdas, F. A. Chambers, J. M. Meese, and L. R. Ram-Mohan, *Appl. Phys. Lett.* **50**, 600 (1987); Y. R. Lee, A. K. Ramdas, F. A. Chambers, J. M. Meese, and L. R. Ram-Mohan, *SPIE Proc.* **794**, 105 (1987).

[25] Y. R. Lee, A. K. Ramdas, A. L. Moretti, F. A. Chambers, G. P. Devane, and L. R. Ram-Mohan, *Phys. Rev. B* **41**, 8380 (1990).

[26] D. Dossa, L. C. Lew Yan Voon, L. R. Ram-Mohan, C. Parks, R. G. Alonso, A. K. Ramdas, and M. R. Melloch, *Appl. Phys. Lett.* **59**, 2706 (1991).

[27] C. Parks, R. G. Alonso, A. K. Ramdas, L. R. Ram-Mohan, D. Dossa, and M. R. Melloch, *Phys. Rev. B* **45**, 14215 (1992); D. Dossa, Ph.D. thesis, WPI (1996).

[28] L. R. Ram-Mohan, K. H. Yoo, and R. L. Aggarwal, *Phys. Rev. B* **38**, 6151 (1988).

[29] B. F. Levine, C. G. Bethea, K. K. Choi, J. Walker, and R. J. Malik, *J. Appl. Phys.* **64**, 1591 (1988).

[30] G. Hasnain, B. F. Levine, D. L. Sivco, and A. Y. Cho, *Appl. Phys. Lett.* **56**, 770 (1990).

[31] M. M. Fejer, S. J. B. Yoo, R. L. Byer, Alex Harwit, and J. S. Harris Jr., *Phys. Rev. Lett.* **62**, 1041 (1989); S. J. B. Yoo, M. M.

Fejer, R. L. Byer, and J. S. Harris Jr., *Appl. Phys. Lett.* **58**, 1724 (1991).

[32] P. Bois, E. Rosencher, J. Nagle, E. Martinet, P. Boucaud, F. H. Julien, D. D. Yang, and J-M. Lourtioz, *Superlattices Microstruct.* **8**, 369 (1990); E. Rosencher and Ph. Bois, *Phys. Rev.* B **44**, 11315 (1991).

[33] J. Khurgin, *Appl. Phys. Lett.* **51**, 2100 (1987); *Phys. Rev.* B **38**, 4056 (1988).

[34] Y. L. Xie, Z. H. Chen, D. F. Cui, S. H. Pan, D. Q. Deng, Y. L. Zhou, H. B. Lu, Y. Huang, S. M. Feng, and G. Z. Yang, *Phys. Rev.* B **43**, 12477 (1991).

[35] L. C. Lew Yan Voon, Ph.D. thesis, WPI (1993); L. C. Lew Yan Voon and L. R. Ram-Mohan, *Phys. Rev.* B **50**, 14421 (1994).

[36] F. C. Zhang, N. Dai, H. Luo, N. Samarth, M. Dobrowolska, J. K. Furdyna, and L. R. Ram-Mohan, *Phys. Rev. Lett.* **68**, 3220 (1992).

[37] H. Luo, L. R. Ram-Mohan, G. L. Yang, and J. K. Furdyna, *J. Electron. Mater.* **22**, 1103 (1993).

[38] Y. Xuan, H. Luo, F. C. Zhang, M. Dobrowolska, J. K. Furdyna, and L. R. Ram-Mohan (unpublished work).

[39] R. de L. Kronig and W. G. Penney, *Proc. R. Soc. (London)* **A130**, 499 (1931).

[40] C. Kittel, *Introduction to Solid State Physics*, 6th ed. (Wiley, New York, 1986).

[41] C. Parks, A. K. Ramdas, M. R. Melloch, G. Steblovsky, L. R. Ram-Mohan, and H. Luo, *Solid State Commun.* **92**, 563 (1994); *J. Vac. Sci. Technol.* **B 13**, 657 (1995).

6

Schrödinger–Poisson self-consistency

6.1 Introduction

In Group IV bulk semiconductors such as Si, it is possible to replace
some of the atoms in the lattice by impurity atoms belonging to
group V of the periodic table, such as P. The group V impurity atom
has an additional electron which is not used in the covalent bonding
in the crystal. Due to the effective dielectric constant of the host
and its reduced effective mass the electron is weakly bound to the
impurity atom forming a hydrogenic structure. In III–V compound
semiconductors such as GaAs an analogous impurity doping can be
accomplished by using Si which replaces a Ga atom in the GaAs lattice. These impurity atoms, called donors, ionize easily, giving off the
bound electron to enable electrical conduction; the binding energy of
an Si donor in GaAs is about $5 \, \text{meV}$. The conduction is limited by
scattering from the ionized impurity atoms, besides other scattering
mechanisms. With the advent of molecular beam epitaxy (MBE) it
was recognized that the conduction channel and the location of impurity doping can be separated to improve the mobility of carriers.
With the remarkable control over the composition and almost atomic
perfection in the heterostructure crystals grown by MBE, a quantum
well structure can be grown with selective doping only in specified
sections of the barrier regions. If the barrier regions alone are doped
with impurity atoms, the electrons initially bound to the impurity
atoms seek lower energy states and fall into the quantum well. This
reduces their probability of being in the barrier region with the ionized impurities, resulting in a reduction in ionized impurity scattering
in transport in the in-plane direction. This selective doping of the
barrier, or any particular layer for that matter, is called modulation
doping. The mobility defined as the drift velocity per unit electric
field is a measure of the collision time for the scattering, and mobili-
ties of $4 \times 10^6 \, \text{cm}^2/\text{V s}$ have been exceeded in transport of electrons
in quantum wells in the in-plane direction. This may be compared

with typical mobilities of $10^5 \, \text{cm}^2/\text{V s}$ in bulk semiconductors.

In this chapter, we wish to calculate the energy levels in a carrier-confining heterostructure by accounting for the additional potential ϕ_{ch} experienced by a given electron due to the presence of other electrons and due to the ionized impurities. We limit ourselves to the specific case of a single quantum well, though the theory we develop here will apply to any heterostructure having a confining potential. Similar strategies are given in Refs. 1 and 2. We employ the finite element method (FEM) to solve Schrödinger's equation in the self-consistent potential defined by the Poisson equation. We shall work in the Hartree approximation so that the electrostatic potential function ϕ_{ch} is defined only by the carrier density and the ionized donor density.[†]

Consider a quantum well of width d and barrier height V_0, with the barriers on both sides having n_D impurities per cm^3. Figure 6.1a shows the un-ionized impurities with the Fermi level, E_F, located at the ground state energy of the donors. The donor electrons, being in bound states, will have a ground state energy E_F that is less than the bandedge energy V_0 by the donor binding energy E_D, so that $E_F = V_0 - E_D$. This is true for $T = 0 \, \text{K}$. At $T \neq 0 \, \text{K}$, the Fermi level will have to be determined by accounting for thermal ionization of the donors. We shall assume that E_F is essentially at its zero temperature value.

This "instantaneous" picture of the initial arrangement of carriers at time $t = 0$ changes into that shown by Fig. 6.1b, as the carriers start occupying the lowest energy levels in the quantum well. This cascade of carriers into the well cannot continue indefinitely since the number of available states in the quantum well is finite, and because the presence of electrons already in the well provides a repulsive Coulomb potential limiting further intrusion by more electrons. The former can be estimated by summing over the states given by the in-plane energy dispersion of the quantum well. While doing so we can fairly readily account for any nonparabolic behavior which usually occurs in the energy dispersion; however, we will ignore this non-parabolicity in the following. A calculation of the influence of other electrons requires the simultaneous solution of (i) Schrödinger's equation for the electron's energy levels and the wavefunctions that are employed in the calculation of the spatial distribution of carriers, and

---

[†]The inclusion of electronic exchange and correlation effects is mentioned later.

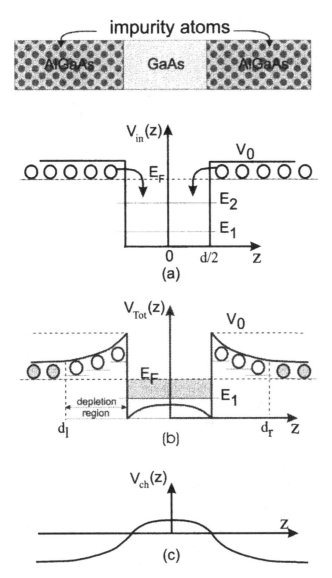

Fig. 6.1 (a) A modulation doped quantum well, of width d, with impurity doping in the barrier regions is shown. At its initial configuration, all the carriers are bound to donors and are at energy $E_F = V_0 - E_D$. (b) The quantum well bandedge after a rearrangement of the charges. The Fermi level E_F can be considered to be at the same position as in (a), while the bandedges move relative to it. The depletion regions correspond to $-d/2 - d_\ell$ and $d_r - d/2$. (c) The net change in the potential energy function V_{ch} due to charge redistribution.

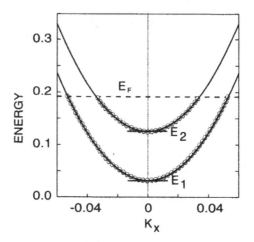

Fig. 6.2 The in-plane parabolic dispersion in the quantum well. The carriers occupy the subband energy levels up to the Fermi level E_F. The k_y-dependence is not shown here.

(ii) the Poisson equation for the electrostatic potential, which determines the potential energy of the electrons. The Poisson equation involves the charge density, which is given by the electronic charge density and by the distribution of ionized impurities in the barrier region. This coupling of Schrödinger and Poisson equations results in a nonlinear problem and requires an iterative approach for its solution.

We shall assume that the system is electrically neutral at all times, with the total electronic charge equaling the total positive charge from the ionized impurities in the system.

It is useful to discuss Fig. 6.1b further. For the moment, let us adopt the point of view that the Fermi level E_F is always held at its initial value shown in Fig. 6.1a while the bandedges change in energy. The un-ionized donors located beyond the "depletion region" on either side of the well in the final configuration again specify the location of E_F at its former value. The depletion region contains the positively charged, i.e., ionized, donors which have provided the carriers entering the well region. The electrons in the quantum well occupy energy levels from the sub-bandedges, defined by the quantum well levels E_ν, up to the common Fermi level, shown in Fig. 6.2. This rearrangement of charges leads to the band bending shown in Fig. 6.1b. The initially flat bandedge at the bottom of the well is now raised up, because the electrons already occupying the well increase the potential energy of the well region for the next electron entering the well region. The band offset V_0 continues to be present at the edges of the quantum well as it is intrinsic to the materials on either side of the interfaces. Finally, the downward curve of the bandedges in the barrier regions is due to the attractive potential associated with the positive charges on the ionized impurities in the barriers. In Fig. 6.1c the curved portions of the bandedges are pieced together to show the net potential energy change V_{ch} as a function of the coordinate. Beyond the depletion region the potential energy again has zero slope, representing the fact that there is no electric field there. Note that all the electric field lines will emanate from the positive charges in the barriers and terminate at the electrons in the well region.

We solve for V_T, the total potential, given by the sum of the initial bandedge profile $V_{in}(z)$ of Fig. 6.1a and the electrostatic potential energy contribution V_{ch} from the rearrangement of the charges.

6.2   Schrödinger and Poisson equations

Schrödinger's equation

We begin by solving Schrödinger's equation

$$-\frac{\hbar^2}{2m_i^*}\nabla^2\Psi(\mathbf{r}) + V_T(z)\Psi(\mathbf{r}) = E\Psi(\mathbf{r}). \qquad (6.1)$$

Each layer i is considered to be of uniform composition with a constant effective mass m_i^*. We make use of the reduced variables $\tilde{z}$ and ϵ, introduced in Chapter 5, defined by the relations $z = \ell_0\tilde{z}$ and $E = \epsilon\mathcal{C}$, where $\mathcal{C} = \hbar^2/(2m\,\ell_0^2)$. We assume that the in-plane motion can be factored out, and that the energy associated with it is given by $E_\perp = \hbar^2k_\perp^2/(2m_i^*)$. The three-dimensional wavefunction in the effective-mass approximation, called the envelope function, is given by

$$\Psi(\mathbf{r}) = \frac{1}{(\sqrt{L})^2}\,e^{ik_x x}\,e^{ik_y y}\,f(\tilde{z}). \qquad (6.2)$$

The envelope function represents the slowly varying part of the total wavefunction in the presence of the periodic arrangement of atoms.[3] The normalization of the wavefunction is defined by an in-plane quantization area L^2. Here, the function $f(\tilde{z})$ in the wavefunction has to be determined by solving the equation

$$-\frac{1}{m_i^*/m_0} f''(\tilde{z}) + (V_T(\tilde{z})/\mathcal{C} - \epsilon) f(\tilde{z}) = 0. \tag{6.3}$$

The natural scale of length for heterostructure geometry is $\ell_0 = 1\,\text{Å},^\ddagger$ for which $\mathcal{C} \simeq 3.81\,\text{eV}$.

We determine the quantum well energy levels E_ν and the corresponding envelope wavefunctions $f_\nu(\tilde{z})$ by the FEM.

The Poisson equation

The Poisson equation for the electrostatic potential $\phi_{ch}(z)$ is given by

$$\nabla\left[\varepsilon_i(z)\nabla\phi_{ch}(z)\right] = -4\pi\,\rho_{ch}(z). \tag{6.4}$$

Here ε_i is the static dielectric constant, assumed to be a constant in each layer i.§ We employ reduced variables and express energies in units of eV in the following. The electrostatic potential $\phi(z)$ is converted to the electron's potential energy by multiplying it by the charge of the electron, $-|e|$, to write $V_{ch}(z) \equiv -|e|\,\phi(z)$. The potential energy is not a function of the in-plane coordinates if the barrier layers have uniform doping. Setting $z = \ell_0\tilde{z}$, equation (6.4) rephrased in terms of the electron's potential energy V_{ch} is given by

$$\frac{d}{d\tilde{z}}\left(\varepsilon_i\frac{d}{d\tilde{z}}V_{ch}(\tilde{z})\right) = +4\pi\,\ell_0^2\,|e|(\,\rho_e(\tilde{z}) + \rho_D(\tilde{z})\,)$$
$$\equiv \mathcal{S}_D + \mathcal{S}_e. \tag{6.5}$$

It is useful to express the charge densities in terms of the number densities

$$\rho_e(\tilde{z}) = -|e|\,n_e(\tilde{z}),$$
$$\rho_D(\tilde{z}) = |e|\,n_D(\tilde{z})\,\theta(V_T(\tilde{z}) - E_D - E_F). \tag{6.6}$$

The Heaviside step-function¶ $\theta(V_T(\tilde{z}) - E_D - E_F)$ determines the regions where donor ionization can take place: the donor electrons require energies larger than the Fermi energy for ionization to occur. We consider below the individual terms in equation (6.5) separately.

‡In the following we use this value: $\ell_0{=}1\,\text{Å}= 10^{-8}\,\text{cm}$.
§The factor of 4π is present since we are using cgs units.
¶See Appendix B for properties of the step-function.

6.3 Source terms

The source term arising from the presence of ionized donors takes
the form

$$S_D = +4\pi\, \ell_0^2\, |e|^2\, n_D(\tilde{z})\, \theta(V_T(\tilde{z}) - E_D - E_F). \qquad (6.7)$$

We scale the number density n_D by a typical carrier density of
$n_{D,16} = 10^{16}\,\mathrm{cm}^{-3}$. Using the dimensionless Sommerfeld fine-structure constant (in cgs units) $\alpha = e^2/\hbar c \simeq 1/137$ and the quantity
$\hbar c \simeq 1.97327 \times 10^{-5}\,\mathrm{eV\,cm}$, we express the source term in units of
eV, to obtain

$$S_D = [\,1.809\,51 \times 10^{-6}\,\mathrm{eV}\,]\left(\frac{n_D(\tilde{z})}{n_{D,16}}\right)\theta(V_T(\tilde{z}) - E_D - E_F). \qquad (6.8)$$

We return to the issue of determining the Fermi level and the depletion region in the following.

Consider the contribution of the electron density to the source
term in equation (6.5). From

$$S_e = -\,4\pi\,|e|^2\,\ell_0^2\,n_e(\tilde{z}), \qquad (6.9)$$

we obtain

$$S_e = -\,[\,1.809\,51 \times 10^{-22}\,\mathrm{eV\,cm}^3\,]\,n_e(\tilde{z}). \qquad (6.10)$$

We evaluate the number density of electrons n_e at any temperature
T by counting the total number of occupied states. Assuming a
uniform in-plane carrier density we write

$$n_e(z) = 2\sum_{n_x, n_y, n_z} |\Psi(x, y, z)|^2\, f(E, E_F, T), \qquad (6.11)$$

where

$$f(E, E_F, T) = \frac{1}{e^{(E-E_F)/k_B T} + 1}, \qquad \text{for } T \neq 0,$$

$$= \theta(E_F - E), \qquad\qquad \text{for } T = 0\,\mathrm{K}, \qquad (6.12)$$

is the Fermi distribution function. The factor of 2 on the right of
equation (6.11) takes account of the spin degeneracy of each energy
level, and the sum extends over all the allowed quantum numbers.

The carriers have a continuous energy spectrum for motion in the plane of the quantum well, and the sums over n_x, n_y can be converted to continuous integrals over k_x, k_y using the relation

$$k_{x,y} = \frac{2\pi\, n_{x,y}}{L},$$

arising from the periodic boundary conditions (BCs) of Born and von Karman for traveling waves in the solid.[3] The sum over n_z is replaced by the sum over the energy levels in the quantum well and is labeled by the index ν.

Substituting equation (6.2) into equation (6.11) we have

$$n_e(\tilde{z}) = 2\sum_\nu \frac{L^2}{4\pi^2} \int dk_x\, dk_y \left\{ \frac{1}{L^2} |f_\nu(z)|^2 \right\} f(E_\perp + E_\nu, E_F, T).$$

With $dk_x\, dk_y = \pi\, dk_\perp^2$ and $dE_\perp = \hbar^2 dk_\perp^2/2m^*$ we obtain

$$n_e(\tilde{z}) = \left[\frac{m^* k_B T}{\pi \hbar^2} \right] \sum_\nu |f_\nu(z)|^2 \ln(1 + e^{(E_F - E_\nu)/k_B T}). \quad (6.13)$$

In our numerical computations we determine the eigenfunctions $f_\nu(\tilde{z})$ and normalize them over the heterostructure in the reduced coordinate. Since we require $\int dz\,|f(z)|^2 = 1$, and we have employed $\int d\tilde{z}\,|f(\tilde{z})|^2 = 1$, we have the definition

$$|f(z)|^2 \equiv |f(\tilde{z})|^2/\ell_0. \quad (6.14)$$

The factor in the square brackets in equation (6.13) has units of inverse area and our normalization adds an extra factor of a length scale so that the units of n_e are correctly specified to be cm^{-3}. For computational purposes, we scale m^* by m_0, the rest mass of the electron, and use the fact that $k_B T = (T/11\,604.5)\,\mathrm{eV}$, with T on the Kelvin scale. The final expression for S_e is

$$S_e = -[6.513\,80 \times 10^{-4}\,\mathrm{eV}] \left(\frac{m_i^*}{m_0} \right) \frac{T}{1\,\mathrm{K}} \sum_\nu |f_\nu(\tilde{z})|^2$$

$$\times \ln(1 + e^{(E_F - E_\nu)/k_B T}). \quad (6.15)$$

The expression $1\,K$ is present in the denominator in order to make the temperature factor dimensionless. At $T = 0K$, the Fermi distribution function is the step-function, $\theta(E_F - E_\perp - E_\nu)$, and the phase space integrals reduce to simple expressions. We have

$$
\begin{aligned}
n_e(\tilde{z}) &= 2\frac{L^2}{(2\pi)^2}\sum_\nu \int dk_x\, dk_y \left\{\frac{1}{L^2}|f_\nu(z)|^2\right\}\ \theta(E_F - E_\perp - E_\nu)\\
&= \frac{m_i^*}{\pi\hbar^2\ell_0}\sum_\nu |f_\nu(\tilde{z})|^2 \int dE_\perp\, \theta(E_F - E_\perp - E_\nu)\\
&= \frac{m_i^*}{\pi\hbar^2\ell_0}\sum_\nu |f_\nu(\tilde{z})|^2\ (E_F - E_\nu).
\end{aligned}
\tag{6.16}
$$

Thus, the source term for electrons is

$$
\begin{aligned}
\mathcal{S}_e &= -[1.809\,51 \times 10^{-22}\,\mathrm{eV\,cm^3}]\left(\frac{m_i^*}{m_0}\right)\frac{m_0 c^2}{\pi\hbar^2 c^2\ell_0}\\
&\qquad\times \sum_\nu |f_\nu(\tilde{z})|^2\ (E_F - E_\nu)\\
&= -7.558\,90\left(\frac{m_i^*}{m_0}\right)\sum_\nu |f_\nu(\tilde{z})|^2\ (E_F - E_\nu).
\end{aligned}
\tag{6.17}
$$

Here $|f_\nu(\tilde{z})|^2$ is dimensionless and *its integral over $\tilde{z}$ is normalized to unity*. The energies E_F and E_ν are in units of eV.

6.4   The Fermi energy and charge neutrality

As yet we have not specified the Fermi level, E_F, or the thickness of the depletion region. While we may consider E_F to be always fixed at its initial value, it is practical to use charge neutrality to determine its actual value while allowing the bandedge to move in energy through each iteration.

We begin by choosing an initial seed value for E_F. With this value we require the total number of positive and negative charges over the entire heterostructure to be equal:

$$
\int d^3r\, n_D(z) = \int d^3r\, n_e(z).
\tag{6.18}
$$

Let us assume a uniform doping density of n_D in the barrier regions. The depletion region is actually determined[4,8] by finding the zeros of $g_{depl}(\tilde{z}) = (V_T(\tilde{z}) - E_D - E_F)$ since the impurity atoms in the region with $V_T(\tilde{z}) - E_D \geq E_F$ get ionized. This determination of the zeros of $g_{depl}(\tilde{z})$ is a necessary step in the overall iteration scheme since the function V_T changes in each iteration. Within the FEM, the location of the zero of $g_{depl}(\tilde{z})$ could occur anywhere within any element in the barrier regions. In each element, the function $V_T(\tilde{z})$ is determined as follows. The function V_{ch} is obtained by interpolation from its nodal values obtained in the previous iteration. Then $V_T(\tilde{z})$ is obtained by adding V_{in} to V_{ch}. Next, Brent's algorithm[9] is used to test for a zero crossing of $g_{depl}(\tilde{z})$ within the given element. Once such zeros are determined and the edges of the depletion regions are identified, we can curtail the integration of $n_D(\tilde{z})$ over the regions. Let the depletion regions add up to a thickness $\mathcal{D}$ in angstrom units. With lateral dimensions being $L \times L$ we have

$$
\begin{aligned}
\mathcal{N}_D &\equiv \int d^3r \, n_D(z) \\
&= L^2 \, \ell_0 \int_{depl.reg.} d\tilde{z} \, n_D(\tilde{z}) \\
&= \mathcal{D} \, L^2 \, \ell_0 \, n_D \, \frac{10^{16} \, \mathrm{cm}^{-3}}{n_{D16}}.
\end{aligned}
\tag{6.19}
$$

We have scaled n_D by a standard value $n_{D16} = 10^{16} \, \mathrm{cm}^{-3}$.

The total number of negative charges is given by

$$
\mathcal{N}_e = 2 \int d^3r \, \frac{L^2}{(2\pi)^2} \int dk_\perp^2 \, \pi \sum_\nu |\Psi_\nu(x,y,z)|^2 \, f(E_F, E_\perp, E_\nu).
$$

Here the spatial integration over the normalized wavefunctions no longer gives unity because of the variation of the m^* factor with layers. We obtain

$$
\mathcal{N}_e = \frac{L^2}{\ell_0} \int \ell_0 \, d\tilde{z} \, \frac{m_i^*}{m_0} |f(\tilde{z})|^2 \left(\frac{m_0 c^2}{\pi \hbar^2 c^2} \right) \sum_\nu (E_F - E_\nu),
$$

$$
\text{for } T = 0\,\mathrm{K}; \tag{6.20}
$$

$$
= \frac{L^2}{\ell_0} \left(\frac{k_B T}{\pi \hbar^2} \right) \sum_\nu \int d\tilde{z} \left(\frac{m_i^*}{m_0} \right) |f_\nu(\tilde{z})|^2 \ln(1 + e^{(E_F - E_\nu)/k_B T}),
$$

$$
\text{for } T \neq 0\,\mathrm{K}.
$$

The integrations are performed using Gauss quadrature over each element, and the full range for the integration is the entire physical

Table 6.1 Approximate depletion region for a
100 Å GaAs well with a 320 meV barrier. The
depletion region extends on either side of well.

Donor density (cm^{-3})	Depletion region (Å)
10^{16}	2000
10^{17}	600
10^{18}	200
10^{19}	80

region. We equate the right sides of equations (6.19) and (6.20) to
verify charge neutrality. If the two quantities differ significantly then
the value of E_F is adjusted towards attaining charge neutrality. We
iterate through the charge neutrality calculation once again in order
to verify that the new value of E_F has converged to its correct value
to within a tolerance of say $\sim 10^{-5}$ eV. If, for the moment, we ignore
the variation with layer composition for the effective masses, we can
integrate out the wavefunction dependence. For the calculation at
$T = 0\,\text{K}$ we obtain

$$\{\mathcal{D}\ 10^{-8}\}\,\frac{n_D}{n_{D16}} = \frac{m^*}{m_0}\sum_\nu \frac{(E_F - E_\nu)}{2\pi\,C}. \qquad (6.21)$$

This approximate charge neutrality condition can be expressed in
the form

$$\sum_\nu (E_F - E_\nu) = \mathcal{D}\left(\frac{n_D}{n_{D16}}\right)\left(\frac{m_0}{m^*}\right)[2.3939 \times 10^{-7}\,\text{eV}]. \qquad (6.22)$$

This approximate formula can be used to check the numerical solutions for the size of the depletion region. In Table 6.1, the numerical
estimations are given for the depletion regions for impurity concentrations from 10^{16} to $10^{19}\,\text{cm}^{-3}$ for a 100 Å GaAs quantum well.

6.5 The Galerkin finite element approach

6.5.1 Boundary conditions

The BCs for Schrödinger's equation at the interfaces of the layers
present in the planar heterostructure are the continuity of the function $f(\tilde{z})$ and the continuity of the probability current across interfaces. Current continuity now corresponds to continuity of the "mass

derivative" $\{\frac{1}{(m_i^*/m_0)} f'(\tilde{z})\}$ across interfaces, since the effective mass of the electron depends on the layer. We also require the vanishing of the wavefunctions in the barriers as $\tilde{z} \to \pm\infty$ for states confined by the quantum well. In the FEM, we use quintic Hermite interpolation to implement these BCs in a straightforward manner, and use a cutoff value of $|\tilde{z}|$ at which we set the wavefunction to zero in the barrier region. The elements are located so that the layer interfaces are congruous with element boundaries. Element matrix insertions into the global matrix are performed in the usual manner except at interfaces. Here the derivative nodal variable f_r' from the element on the right of an interface is replaced by the derivative nodal variable f_l' from the element on the left after an appropriate scaling by the ratio of the masses

$$f_r' = \frac{m_r^*}{m_l^*} f_l'. \tag{6.23}$$

In the Schrödinger action, this rescaling is included as follows. Consider a quantum well structure modeled with three elements, one for each layer. The Hermite interpolation polynomials are used in each element with the nodal variables f_i and f_i' arranged alternately. With this choice the element matrix for the second element representing the well layer has its second row *and* column multiplied by the mass ratio m_w^*/m_b^* for the well and barrier masses, before the element matrix is overlayed into the global matrix. Similarly, the third element representing the right barrier layer of the quantum well has its second row and column multiplied by the mass ratio m_b^*/m_w^* before the element matrix is placed in the global matrix. The BCs requiring the vanishing of the wavefunction $f(\tilde{z})$ at $\pm\infty$ are imposed by eliminating the first and the last rows and columns which correspond to the nodal variables at the first and last nodes. The eigenvalue problem resulting from the nodal variational principle is solved along the lines discussed in Chapter 3.

We use the Galerkin approach to solve the Poisson equation

$$\frac{d}{d\tilde{z}} \left(\varepsilon(\tilde{z}) \frac{dV_{ch}(\tilde{z})}{d\tilde{z}} \right) = \mathcal{S}_D + \mathcal{S}_e, \tag{6.24}$$

in reduced variables. Let the coordinate to the left of the depletion region, on the left side of the well, be d_ℓ and the corresponding coordinate on the right side be d_r. The physical region is broken up into *nelem* finite elements in each of which the function $V_{ch}(\tilde{z})$

is expressed again in terms of quintic Hermite interpolation.[4-6] The potential energy function in element iel is

$$V_{ch}^{(iel)}(\tilde{z}) = \sum_{\alpha=1}^{6} \Phi_\alpha(\tilde{z}) (V_{ch}^{(iel)})_\alpha. \tag{6.25}$$

We arrange the nodal variables pairwise so that at the first node the first variable is $V_{ch,1}$ and the second is $V'_{ch,1}$, etc. It is simplest to use the same element distribution for the Poisson calculation as in solving Schrödinger's equation.

In the Galerkin method, we multiply equation (6.24) by weighting functions which are the Hermite interpolation polynomials and are local to each element, and integrate the equation over the physical domain. An integration by parts of equation (6.24) gives

$$-\int_{d_\ell}^{d_r} d\tilde{z}\, \Phi'_\alpha(\tilde{z})\, \varepsilon(\tilde{z})\, \Phi'_\beta(\tilde{z})\, V_{ch,\beta}$$

$$+ \left[\left(\Phi_\alpha(\tilde{z})\, \varepsilon(\tilde{z})\, \Phi'_\beta(\tilde{z})\, V_{ch,\beta} \right) \Big|_{d_\ell}^{-d/2} + \Big|_{-d/2}^{+d/2} + \Big|_{+d/2}^{d_r} \right]$$

$$= \int_{d_\ell}^{d_r} d\tilde{z}\, \Phi_\alpha(\tilde{z})\, (\mathcal{S}_D + \mathcal{S}_e). \tag{6.26}$$

For the Poisson equation, the first BC is the continuity of the electrostatic potential ϕ, or equivalently the potential energy $V_{ch}(\tilde{z})$. The second is the continuity across interfaces for the normal component of the displacement field $\mathbf{D} = \varepsilon \mathbf{E}$. The latter condition is expressible as the continuity of $\varepsilon_i V'_{ch}(z)$ across the material interfaces. This eliminates the interface surface terms at $\tilde{z} = \pm d/2$ in equation (6.26) leaving us with the surface terms at the physical boundary. At d_ℓ only the term with $\alpha = 1$, $\beta = 2$ survives, while at d_r the term with $\alpha = 5$, $\beta = 6$ is nonzero. The integral on the right hand side of equation (6.26) is readily evaluated since the source terms are known. The weighting functions associated with each element give rise to six simultaneous equations for the unknown pairs of parameters, V_{ch} and V'_{ch}, at each of the three nodes. We can represent these simultaneous equations in a matrix form

$$M_{\alpha,\beta}^{(iel)} (V_{ch}^{(iel)})_\beta = R_\alpha^{(iel)}. \tag{6.27}$$

Here the matrix M^{iel} is thought of as an element matrix. Again, suppose that just three elements are used to model the three layers.

We insert the first element matrix into a global matrix. In the second element matrix, representing the well region, we replace $V'_{ch,1}$ at the first node by $V'_{ch,3}$ from the last node of the first element, and rescale the element matrix by multiplying its second row *and* column by $\varepsilon_1/\varepsilon_2$. This rescaled element matrix is now inserted into the global matrix partially overlaying the first element. This procedure accounts for the interlayer boundary condition for the electric displacement vector. The third element matrix representing the right barrier region is treated in a similar fashion. With more than one element per layer, the element-overlay procedure guarantees the continuity of V_{ch} and V'_{ch} across *element* boundaries in the same layer. Once the global matrix is built up the surface terms are inserted at the matrix elements with row and column indices (1,2) and the $(4nelem+1, 4nelem+2)$ in the global matrix. We also require BCs at $\tilde{z} \to \pm\infty$ for V_{ch} and/or V'_{ch}.

It is clear from Fig. 6.1 that the change in the bandedge beyond the depletion region is zero. We either set (a) $V_{ch}(d_\ell) = 0$ and (b) $V_{ch}(d_r) = 0$, or alternatively, employ (a') $V_{ch}(d_\ell) = 0$ and (b') $V'_{ch}(d_r) = 0$.[2,7] We are concerned here only with the change in the potential energy and not with the total potential energy. Clearly, either of the two sets of BCs are acceptable when there are no additional external electric fields applied across the quantum well. In the presence of an external field, we can still use these BCs by invoking the superposition principle for the potentials and again focusing on the change in the potential energy. Now the effect of the external field can be incorporated into V_{in}.

The final form of the weighted residuals leads to the equation

$$-\mathcal{M}_{\mu,\nu} (V_{ch})_\nu = R_\mu, \qquad (6.28)$$

with $\mu, \nu = 1, \ldots, (4nelem + 2)$. The BCs at the left and the right of the physical region are imposed by substituting the nodal values of the potential energy in the above equations, and these known terms are transferred to the right side with a change of sign. The nodal values of the potential energy are obtained by solving the set of equations, as discussed in Chapter 5. We use 25–40 elements depending on the expected size of the depletion region (see Table 6.1). More elements are needed for lower donor densities.

6.5.2 The iteration procedure

We solve Schrödinger's equation and the Poisson equation in dimensionless variables, with the source terms S_D and S_e given by equations (6.8) and (6.15) or (6.17), using the following steps in each iteration:

Step 1: Schrödinger's equation is solved for the energy levels and wavefunctions for the quantum well. We need a seed potential to initiate the iteration. Just for the first iteration, we assume that we can represent V_{ch} by three sections in which the function can be approximated by quadratic functions. (If the source terms were constants we would have a quadratic function for V_{ch}.) For example, in the well region we assume a starting potential energy V_{ch} given by

$$V_{ch}(\tilde{z}) = c_1 \left[(d/2)^2 - \tilde{z}^2 \right], \quad \text{for} \ -d/2 \le \tilde{z} \le d/2. \qquad (6.29)$$

Here $c_1(d/2)^2$, the value of the potential at the center of the well,

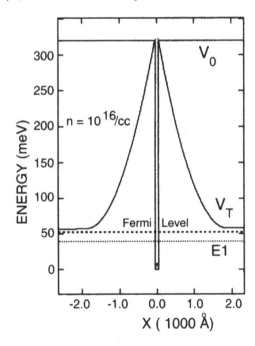

Fig. 6.3 The self-consistent conduction bandedge in a GaAs/AlGaAs quantum well, of width $100\,\text{Å}$, with $10^{16}\,\text{cm}^{-3}$ doping in the barrier regions. Note that the depletion regions extend to $\sim 2000\,\text{Å}$ on either side of the well. E_F represents the Fermi level in the system. Here V_T represents the final bandedge configuration.

Table 6.2 Approximate potential energy at the center of a 100 Å GaAs well relative to the edge of the well.

Donor density (cm^{-3})	Energy offset (meV)
10^{16}	8
10^{17}	23
10^{18}	55
10^{19}	105

can be chosen to be about 25–50 meV, depending on the impurity concentration. The increase in the potential energy at the center of a 100 Å quantum well for a range of impurity doping densities is shown in Table 6.2, showing that the barrier height is effectively reduced to one-third its initial value when the donor density is 10^{19}cm^{-3}. Similarly, the seed function for potential energy V_{ch} in the barrier on the right side of the well can be chosen as

$$V_{ch} = V_{ch}(d_r) \left(1 - \frac{(d_r - \tilde{z})^2}{(d_r - d/2)^2} \right), \quad \text{for } d/2 \leq \tilde{z} \leq d_r. \quad (6.30)$$

The value of $V_{ch}(d_r)$ can be about $-(25$–$50)$ meV as an initial guess. This potential is used in the calculation of the energy levels and wavefunctions only for the first iteration. In each subsequent cycle of iteration Schrödinger's equation will be solved with the potential obtained from the Poisson equation in the previous iteration.

Step 2: The band bending assumed above allows us to define a reasonable depletion region. In this step, we solve for $\mathcal{D}$. A Fermi energy E_F is chosen and we determine the number of ionized donors and the number of electrons in the quantum well occupying states below the Fermi level. If the two are not equal, we alter E_F by a small amount towards achieving charge neutrality. This self-consistent determination of E_F also determines the appropriate depletion region for this iteration.

Step 3: The Poisson equation is solved with the electron density in the well region given in terms of the wavefunctions and energy levels obtained from *Step 1*, and in terms of the number of ionized donors given by the thickness $\mathcal{D}$ of the depletion region and the impurity concentration, as identified in *Step 2*. This leads to a new solution of the total potential energy function, and the result is reinserted back

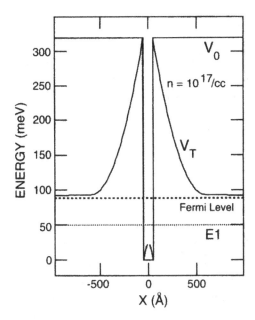

Fig. 6.4 The self-consistent conduction bandedge in a GaAs/AlGaAs quantum well, of width $100\,\text{Å}$, with $10^{17}\,\text{cm}^{-3}$ doping in the barrier regions. The depletion regions extend nearly $600\,\text{Å}$ on either side of the well. Here V_T represents the final bandedge configuration.

into *Step 1*. The procedure is repeated till solutions that are stable under iteration are obtained.

It is useful to emphasize here that one can achieve a monotonic evolution to the final result if one uses a *very conservative* approach to this feedback. The iterative process can be made adaptive by using only a fraction (e.g., 0.1 or less) of the change between the old potential and the new solution in the feedback, with the percentage of the change allowed being governed by the rate of change of the potential in some fashion. (See Refs. 2, 4, 5 for further elaboration.) Results from a calculation based on the FEM for a $100\,\text{Å}$ GaAs quantum well are shown in Figs. 6.3–6.6. The depletion region is large ($\sim2000\,\text{Å}$) when the donor concentration is low ($10^{16}\,\text{cm}^{-3}$), and vice versa.

6.5.3 Numerical issues

In order to attain convergence for the numerical solution we have to take account of the following issues; these have not been considered at length in the literature.

At the edges of the structure, the self-consistent potential is flat and a miniband of quasibound states forms immediately above the self-consistent potential energy edge. The upward-sloping potential energy curve provides an effective potential barrier; this miniband formation has been noted by others.[10] These unbound states present difficulties in automating the calculations since their wavefunctions are not localized within the well, and hence care must be taken to see that they are excluded from the sum over true bound states. For the 100 Å GaAs quantum well, we find that there are two bound states in the well contributing to the sum over states. It is interesting to note that the energy difference $E_2 - E_1$ decreases with increasing donor number densities, as shown in Table 6.3.

The self-consistent potential energy emerging from the Poisson equation is almost completely flat deep within the barrier regions, as required by the BCs. However, since the FEM only solves for the nodal points and approximates the result in between them with a polynomial, a closer inspection shows that the interpolated solution has small oscillations between the nodal points. These small oscillations introduce noise and instability into the iterative process, especially in the determination of the depletion regions. We could

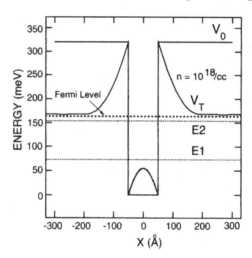

Fig. 6.5 The self-consistent conduction bandedge in a GaAs/AlGaAs quantum well, of width 100 Å, with 10^{18} cm^{-3} doping in the barrier regions. The depletion regions extend to $\sim$200 Å on either side of the well. V_T is the final configuration of the bandedge, obtained after an iterative calculation.

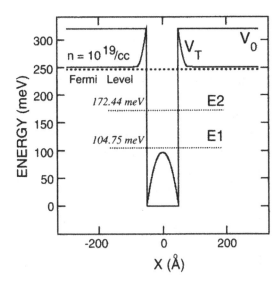

Fig. 6.6 The self-consistent conduction bandedge in a GaAs/AlGaAs quantum well, of width 100 Å, with 10^{19} cm^{-3} doping in the barrier regions. The depletion regions extend to less than 100 Å on either side of the well.

control them by using extra elements, but it is computationally more efficient to filter out these oscillations from the solution, particularly because it does not require bigger matrices. We filter our solution by insisting that all points below the bandedge potential energy be set to the value of the energy at the bandedge, and that their derivatives are set to zero. This filtering is not strictly necessary; the solution will converge without it, but it converges more slowly and less smoothly.

The convergence criterion must be carefully managed. Even small changes of the potential profile significantly alter the number of charges in the quantum well. We therefore apply a fraction of

Table 6.3 Difference between first and second energy levels in a 100 Å GaAs quantum well for high donor densities.

Donor density (cm^{-3})	$E_2 - E_1$ (meV)
No impurities	97
10^{18}	80
10^{19}	68

the change in the solution at each iteration to allow the system to damp down during the initial iterations.[5] However, we can afford to be less cautious and increase the convergence factor when the solution is close to convergence in order to speed up the algorithm. We thus employ adaptive, cautious convergence to promote stability in a deliberate manner. Even with these conservative measures our algorithm is found to be quite efficient. With 25 elements and a donor density of 10^{19} cm^{-3} the algorithm converges in 36 steps. As the donor density decreases the solution takes longer, until at a concentration of 10^{16} cm^{-3} 150 iterations are required for convergence. Computer run-times are less than a few minutes on a workstation. Earlier use of this approach to the problem of calculating the band bending in modulation doping was made by Laux *et al.*,[5] and by Inoue *et al.*[6] The FEM has proved to be a stable, rapidly converging algorithm for this problem.

6.5.4 Essential and natural boundary conditions

In Section 5.2, it was shown that the Galerkin method and the variational method are equivalent for the 1D Schrödinger equation with the mixed BCs for tunneling. This equivalence in the case of the Poisson equation, equation (6.26), can be demonstrated in a similar manner. Consider the Neumann BCs

$$
\begin{aligned}
V'_{ch}(\tilde{d}_\ell) &= 0, \\
V'_{ch}(\tilde{d}_r) &= 0,
\end{aligned}
\tag{6.31}
$$

at the two ends d_ℓ and d_r. Setting the variation of the functional

$$
\mathcal{A}[V_{ch}] = -\int_{d_\ell}^{d_r} d\tilde{z} \left(\frac{1}{2} \varepsilon(\tilde{z}) \frac{dV_{ch}(\tilde{z})}{d\tilde{z}} \frac{dV_{ch}(\tilde{z})}{d\tilde{z}} \right.
$$
$$
\left. + V_{ch}(\tilde{z}) \left(\mathcal{S}_D + \mathcal{S}_e\right) \right)
\tag{6.32}
$$

to zero, after performing an integration by parts, leads to

$$
\delta\mathcal{A}[V] = 0 = \int_{d_\ell}^{d_r} d\tilde{z} \frac{d}{d\tilde{z}} \left(\varepsilon(\tilde{z}) \frac{dV_{ch}(\tilde{z})}{d\tilde{z}} - \left(\mathcal{S}_D + \mathcal{S}_e\right) \right) \delta V_{ch}(\tilde{z})
$$
$$
- \left[\delta V_{ch}(\tilde{z}) \varepsilon(\tilde{z}) \frac{dV_{ch}(\tilde{z})}{d\tilde{z}} \right]_{\tilde{z}=d_r} + \left[\delta V_{ch}(\tilde{z}) \varepsilon(\tilde{z}) \frac{dV_{ch}(\tilde{z})}{d\tilde{z}} \right]_{\tilde{z}=d_\ell}.
\tag{6.33}
$$

With the derivatives at the boundary being zero we directly obtain the Poisson equation, equation (6.24). In this instance, replacing

$\delta V_{ch}(\tilde{r})$ by weighting functions (interpolation polynomials) in equation (6.33) gives the Galerkin weighted residual that is set to zero.

The BCs, equation (6.31), are called *natural boundary conditions*, reflecting the fact that the action integral is not modified in any way with additional surface terms, the values of V_{ch} are not specified at the two extremities, and the BC is automatically satisfied if we evaluate $\mathcal{A}$ and find its stationary value. The final FEM global matrices are also not altered to impose any of these BCs, since they are naturally (automatically or implicitly) satisfied.

On the other hand, the Dirichlet BCs

$$V_{ch}(\tilde{d}_\ell) = 0,$$
$$V_{ch}(\tilde{d}_r) = 0, \qquad (6.34)$$

require that the vanishing of the trial functions V_{ch} be imposed in the variational procedure used on the action integral. These are called *essential boundary conditions*. In this case the BCs are explicitly satisfied.

6.6 Further developments

Exchange and correlation effects

The effect of exchange and correlation[11] due to electron–electron interactions can be taken into account through numerical calculations. The full electrostatic potential energy

$$V_{ch} = -|e|\,\phi_{ch}(z) + V_{ex}(z)$$

has the electron–electron exchange and correlation effects in the local density approximation[12,13] given by

$$V_{ex}(z) = -\left[1 + 0.0545\, r_s \ln\left(1 + \frac{11.4}{r_s}\right)\right] (18/\pi^2)^{1/3} R_y^*, \qquad (6.35)$$

where

$$r_s = \left[\frac{4}{3}\pi a^{*3} n_e(z)\right]^{-1/3},$$
$$a^* = \varepsilon\hbar^2/m^* e^2,$$
$$R_y^* = m^* e^4/(2\varepsilon^2\hbar^2),$$

and $n_e(z)$ is the electron charge density. Numerical calculations which have included the effects of V_{ex} show[13,10,14,6] that these effects can be neglected at least in GaAs heterostructures with doping levels of about $10^{18}\,\mathrm{cm}^{-3}$.

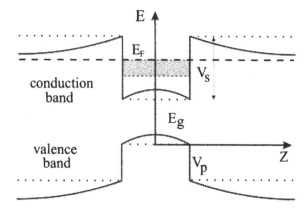

Fig. 6.7 The bandedges for both the conduction and the valence bands shown together.

Thermal excitation across the bandgap

Since we had been concerned with n-doped barriers in the quantum well heterostructure, our treatment has neglected the effects of thermal excitation across the bandgap. The bandedge functions for both the conduction and the valence band are shown together in Fig. 6.7. At finite temperature, there will be carriers in the conduction and the valence bands due to thermal excitation across the bandgap in intrinsic heterostructures. This would lead to an accumulation of electrons with carrier density $n_e(z)$, as well as holes with density $n_p(z)$, in the quantum well region. If we assume that the carriers are all limited to the well region with uniform spatial distribution, the problem of calculating the new bandedge profile becomes straightforward. These concepts are developed in Problem 6.2.

Electrons in a parabolic well

There has been recent interest in studying electrons in parabolic quantum wells. The motivation for this is that, with set-back doping, the carriers that fall into the well raise the potential energy near the center of the parabolic well, with the effect that the final bandedge can be made nearly flat within the quantum well. This leads to a nearly uniform, and dense, electron gas.[15,16] (The valence bandedge becomes even more parabolic when the conduction bandedge in the well becomes horizontal.) Theoretical calculations suggest that, in the presence of an in-plane magnetic field, such an electron gas should display a spin-density-wave state.[17]

Other calculations

A number of modulation-doping geometries of interest can be modeled using the iterative method described in this chapter. For example, the modeling of selectively doped II–VI superlattices has been studied fairly recently.[18] The effect of high magnetic fields on the energy levels in modulation doped quantum wells has been investigated.[19] Optical absorption and inter-subband transitions in the presence of carriers accounting for band bending in superlattices has been studied in Ref. 20. These are representative of the calculations appearing in the literature.

In summary, modulation doping provides a simple, and at the same time remarkably effective, means of controlling impurity scattering in heterostructures. The final form of the potential profile, after charge rearrangement, influences the energy levels of the system. It is thus important to account for it, and we have shown that this can be done in a systematic manner by numerical calculations using the FEM.

6.7   Problems

1. Consider a bulk semiconductor with a bandgap $E_g \gg k_B T$ at temperature T. Given an energy spectrum $E_e = E_g + \hbar^2 k^2 / 2m_c^*$, and $E_h = -\hbar^2 k^2 / 2m_h^*$, determine n_e, the equilibrium concentration of electrons in the conduction band, and n_p, the concentration of holes in the valence band. Denote the Fermi level by E_F.

   (a) Show that the product $n_e n_p$ is independent of E_F, and that

   $$n_e n_p = 4 \left(\frac{k_B T}{2\pi \hbar^2} \right)^3 (m_e^* m_h^*)^{3/2} \, e^{-E_g / k_B T}.$$

   (b) For intrinsic semiconductors we have $n_{e0} = n_{p0}$; using this condition derive the result

   $$E_F = E_g/2 + \frac{3}{4} k_b T \, \ln(m_h^*/m_e^*).$$

   Obtain an expression for n_{e0} or n_{p0} with E_F given by the above expression.

(c) Show that the number densities in the presence of an additional *electron potential energy* $V(z)$ in an intrinsic semiconductor are given by

$$n_e = n_{e0} \times e^{-V(z)/k_B T}; \qquad n_p = n_{p0} \times e^{+V(z)/k_B T}.$$

Note: When $V(z)$ is positive it repels electrons, while attracting holes.

(d) Solve the Poisson equation for the potential energy in a semiconductor

$$\frac{d^2}{dz^2} V(z) = \frac{4\pi |e|}{\varepsilon_\infty} (n_p - n_e),$$

given n_e and n_p from part (c).

Hint: Multiply by V' to write the left side as a perfect differential of V'^2 and obtain the first integral of the equation. Use the BCs $V = 0$ and $V' = 0$ at $z = \pm\infty$, to obtain expressions for V' valid for $z > 0$ and for $z < 0$. Next, integrate V' again to obtain an expression for $V(z)$.

Hint: Once again the trick of expressing the left side as a total differential works here – this time multiply through by $V(z)$.

2. We wish to study the electric fields generated by ionized donors and electron distributions in a quantum well heterostructure in which the well width is $d_w = 100\,\text{Å}$. Suppose the two barriers on either side of the well have constant but different donor impurity concentrations, with the left barrier having $n_{D\ell} = 10^{17}\,\text{cm}^{-3}$, and $n_{Dr} = 10^{18}\,\text{cm}^{-3}$. The depletion regions on the two sides will then differ in thickness. Let the regions have thicknesses $\mathcal{D}_\ell = 600\,\text{Å}$ and $\mathcal{D}_r = 180\,\text{Å}$. The following parts are designed to obtain analytic solutions for the electric field and the potential energy using approximations.

(a) Let us first consider only the ionized impurities in the depletion regions. From the Poisson equation evaluate the electric field everywhere and the potential energy profile generated by the ionized impurities as a function of the coordinate, assuming that the electrons are absent. Ignore variations in the dielectric constant in the different layers.

(b) Solve for the electric field and the potential energy profile as a function of the coordinate for the case when the electrons are present in the quantum well region, and are *uniformly distributed and confined to the quantum well layer only*. The system as a whole is electrically neutral.

(c) Suppose the effective masses in the three regions of the quantum well are identical. Determine the number of carriers per cm^2 present in the well using charge neutrality.

(d) Obtain a rough estimate for the Fermi level assuming that the electrons are distributed in the energy levels of the quantum well. Suppose, for the sake of this calculation, that the energy levels are at $E_1 = 32\,meV$, $E_2 = 135\,meV$, $E_3 = 250\,meV$, and that the effective mass of the electron is $m^* = 0.067\,m_0$. Let the barrier height V_0 be $\sim 320\,meV$.

References

[1] G. Bastard, *Wave Mechanics Applied to Semiconductor Heterostructures* (Les Editions de Physique, Les Ulis, France, 1988).

[2] F. Stern, *J. Comput. Phys.* **6**, 56 (1970).

[3] C. Kittel, *Introduction to Solid State Physics*, 6th ed. (Wiley, New York, 1986).

[4] D. J. Rostcheck, "Selfconsistent calculations of electronic energy levels in quantum heterostructures," Senior Thesis, Worcester Polytechnic Institute, 1992. (The FEM was employed for both the Schrödinger and the Poisson equations, with quintic Hermite interpolation polynomials.)

[5] S. E. Laux, D. J. Frank, and Frank Stern, *Surf. Sci.* **196**, 101 (1988).

[6] K. Inoue, H. Sakaki, J. Yoshino, and T. Hotta, *J. Appl. Phys.* **58**, 4277 (1985).

[7] F. Stern and W. E. Howard, *Phys. Rev.* **163**, 816 (1967).

[8] B. C. Duncan, "Selfconsistent analysis of energy levels in modulation doped heterostructures," Senior Thesis, Worcester Polytechnic Institute, 1990. (A transfer matrix method was used for Schrödinger's equation, and discretization with the "shooting method" for the Poisson problem – this procedure was observed to be unstable for high impurity concentrations $\geq 10^{18}\,cm^{-3}$.)

[9] R. P. Brent, *Algorithms for Minimization Without Derivatives*, Chapters 3, 4 (Prentice Hall, Englewood Cliffs, NJ, 1973).

[10] T. Ando, *J. Phys. Soc. Jpn* **51**, 3893 (1982); *ibid.*, **51**, 3900 (1982).

[11] For the many-body theory of exchange and correlation effects see, for example, S. Raimes, *The Wave Mechanics of Electrons in Metals* (North-Holland, Amsterdam, 1961); A. L. Fetter and J. D. Walecka, *Quantum Statistical Mechanics of Many Particle Systems* (McGraw-Hill, New York, 1971); G. D. Mahan, *Many Particle Physics*, 2nd ed. (Plenum Press, New York, 1990); G. D. Mahan and K. R. Subbaswamy, *Local Density Theory of Polarizability* (Plenum Press, New York, 1990).

[12] O. Gunnarson and B. I. Lundquist, *Phys. Rev.* B **13**, 4274 (1976).

[13] T. Ando and S. Mori, *J. Phys. Soc. Jpn* **47**, 1518 (1979).

[14] F. Stern and S. Das Sarma, *Phys. Rev.* B **30**, 840 (1984).

[15] A. J. Rimberg and R. M. Westervelt, *Phys. Rev.* B **40**, 3970 (1989).

[16] J. H. Burnett, H. M. Cheong, W. Paul, P. F. Hopkins, E. G. Gwinn, A. J. Rimberg, R. M. Westervelt, M. Sundaram, and A. C. Gossard, *Phys. Rev.* B **43**, 12033 (1991).

[17] L. Brey and B. I. Halperin, *Phys. Rev.* B **40**, 11634 (1989).

[18] I. Suemune, *J. Appl. Phys.* **67**, 2364 (1990).

[19] J. Sanchez-Dehesa, F. Meseguer, F. Borondo, and J. C. Mann, *Phys. Rev.* B **36**, 5070 (1987).

[20] K. T. Kim, S. S. Lee, and S. L. Chuang, *J. Appl. Phys.* **69**, 6617 (1991).

7

Landau states in a magnetic field

7.1 Introduction

A uniform magnetic field changes the energy levels of charged carriers in an intrinsic, fundamental way. For this reason, magnetic fields have been used as an externally controlled perturbation in the investigation of the electronic energy levels in quantum systems. We solve Schrödinger's equation by including the interaction energy between the external field and the dipole moment of the electron due to orbital motion in the field, and also the interaction energy for the spin magnetic moment of the electron interacting with the external magnetic field. We investigate the effect of a magnetic field on carriers in a layered quantum semiconductor heterostructure. The finite element method (FEM) is shown to be a natural computational approach for layered structures that easily accounts for the interface boundary conditions (BCs).

7.1.1 Landau levels

First consider bulk semiconductors with electrons in the conduction band obeying an energy dispersion relation $E = \hbar^2(k_x^2 + k_y^2 + k_z^2)/2m^*$. Let $\mathbf{B} = B_0 \hat{z}$ be the applied magnetic field directed along the $\hat{z}$-direction. Then, as in the classical case, the motion parallel to the magnetic field lines remains unaltered. In the plane perpendicular to the magnetic field, the classical picture corresponds to electrons undergoing circular motion due to the Lorentz force $\mathbf{F} = (q/c)\mathbf{v} \times \mathbf{B}$. In the quantum mechanical description, the energy spectrum changes into that of a simple harmonic oscillator. We work in the Coulomb gauge[†] $\nabla \cdot \mathbf{A} = 0$, with $\mathbf{A} = (0, B_0 x, 0)$. The magnetic field is obtained as $\mathbf{B} = \nabla \times \mathbf{A} = B_0 \hat{z}$. The magnetic field gives rise to

---

[†]Appendix C includes a discussion of the gauge invariance of Maxwell's equations.

a velocity-dependent potential energy for a charged particle. This leads to the replacement

$$\mathbf{p} \to (\mathbf{p} - (q/c)\mathbf{A}),$$

in the free-particle Hamiltonian, referred to as the minimal gauge substitution. The electron's intrinsic magnetic dipole moment also contributes a term $-\boldsymbol{\mu}_e \cdot \mathbf{B}$ arising from the interaction energy of the dipole with the external magnetic field. Schrödinger's equation takes the form

$$\frac{p_x^2}{2m^*} \Psi(\mathbf{r}) + \frac{1}{2m^*} \left(p_y + \frac{|e| B_0 x}{c} \right)^2 \Psi(\mathbf{r}) + \frac{p_z^2}{2m^*} \Psi(\mathbf{r})$$
$$+ g^* \mu_B \frac{\sigma_z}{2} B_0 \, \Psi(\mathbf{r}) = E \, \Psi(\mathbf{r}). \qquad (7.1)$$

Here $q = -|e|$ is the charge on the electron. Also, the Bohr magneton is given by $\mu_B = |e|\hbar/2m_0 c$, the electron's effective mass in the conduction band is m^*, and g^* is the electron's effective g-factor. The 2×2 Pauli spin matrix σ_z appearing in equation (7.1) is present due to the spin magnetic moment of the conduction electron with the dipole moment expressed in terms of the electron spin $\boldsymbol{\mu}_e = -g^* \mu_B \boldsymbol{\sigma}/2$.

Following Landau, we assume a solution of the form

$$\Psi(\mathbf{r}) = \frac{1}{\sqrt{(L_y L_z)}} e^{ik_z z} e^{ik_y y} \, \Xi(x) \, \mathcal{X}_s. \qquad (7.2)$$

The spin part of the wavefunction is represented by $\mathcal{X}_s$ The form of the wavefunction in equation (7.2) ensures that the particle displays the "free-particle" behavior in its motion along the $\hat{z}$-direction. Substituting equation (7.2) into equation (7.1) we obtain

$$-\frac{\hbar^2}{2m^*} \Xi''(x) + \frac{1}{2m^*} \left(\hbar k_y + \frac{e B_0 x}{c} \right)^2 \Xi(x)$$
$$+ \left(\frac{\hbar^2 k_z^2}{2m^*} \pm \frac{g^* \mu_B B_0}{2} \right) \Xi(x) = E \, \Xi(x). \qquad (7.3)$$

The $\pm$ sign in the term arising from the magnetic dipole moment of the electron corresponds to the electron spin polarization being either parallel to the magnetic field or antiparallel to it. We define

$$\omega_B = \frac{eB}{m_0 c}, \qquad \omega_B^* = \frac{eB}{m^* c}, \qquad g^* \mu_B B_0/2 = g^* \hbar \omega_B/4,$$

where ω_B^* is the cyclotron frequency of the orbits in the magnetic field. Let us further define

$$x_0 = \frac{\hbar k_y c}{eB_0}$$
$$= \left(\frac{\hbar}{m_0 \omega_B}\right) k_y \equiv \mathcal{R}_0^2 \, k_y. \tag{7.4}$$

Here $\mathcal{R}_0$ is called the cyclotron radius or Landau radius for the ground state of the electron in the magnetic field. With these definitions we have

$$-\frac{\hbar^2}{2m^*} \Xi''(x) + \left(\frac{\hbar^2 k_z^2}{2m^*} + \frac{e^2 B_0^2}{2 \, m^* c^2}(x + x_0)^2\right) \Xi(x)$$
$$\pm (g^* \hbar \omega_B / 4) \, \Xi(x) = E \, \Xi(x). \tag{7.5}$$

We scale the coordinate x by the Landau radius and define

$$(x + x_0) = \mathcal{R}_0 \xi.$$

Here the energy scale is

$$\hbar^2 / (2m^* \mathcal{R}_0^2) = \hbar \omega_B^* / 2.$$

Using these two definitions equation (7.5) becomes

$$\left(\Xi''(\xi) - \xi^2 \Xi(\xi)\right) + \frac{2}{\hbar \omega_B^*} \left(E - \frac{\hbar^2 k_z^2}{2m^*}\right) \Xi(\xi)$$
$$\mp \frac{2}{\hbar \omega_B^*} \left(\frac{g^* \hbar \omega_B^*}{4} \frac{m^*}{m_0}\right) \Xi(\xi) = 0. \tag{7.6}$$

The solution to equation (7.6) is given by the Hermite functions[1]

$$\Xi(\xi) \sim e^{-\xi^2/2} H_n(\xi) \equiv h_n(\xi), \quad n = 0, 1, 2, \ldots, \tag{7.7}$$

with $\xi = (x + x_0)/\mathcal{R}_0$. We normalize $h_n(\xi)$ to unity. The corresponding eigenvalues are

$$E = \frac{\hbar^2 k_z^2}{2m^*} + \frac{\hbar \omega_B^*}{2} \left[(2n + 1) \pm \frac{g^*}{2} \frac{m^*}{m_0}\right]$$
$$= \frac{\hbar^2 k_z^2}{2m^*} + E_{n,\pm}. \tag{7.8}$$

The three-dimensional wavefunction becomes

$$\Psi(x, y, z) = \frac{1}{\sqrt{(L_y L_z)}} e^{ik_z z} e^{ik_y y} h_n(\xi). \tag{7.9}$$

7.1.2 Density of states

It is instructive to consider the problem of distributing N electrons at $T = 0\,\mathrm{K}$ over the available energy levels in the absence, and also in the presence, of the magnetic field in order to clarify the roles played by the various quantum numbers appearing in equation (7.9). The electrons will occupy the energy levels from the lowest value up to a Fermi energy E_F, with each level having a spin degeneracy of 2.

In the absence of a magnetic field we have the spectrum $E = (\hbar^2/2m^*)(k_x^2 + k_y^2 + k_z^2)$, and the total number of electrons, N, occupying energy levels up to the Fermi energy is given by

$$N = 2\,[L^3/(2\pi)^3]\int d^3k\,\theta(E_F - E), \qquad (7.10)$$

and the number density is

$$N/L^3 \equiv n = \int_0^{E_F} dE\,\left[\frac{1}{2\pi^2}\left(\frac{2m^*}{\hbar^2}\right)^{3/2} E^{1/2}\right]. \qquad (7.11)$$

The quantity dn/dE, given in the square brackets, is called the density of states, $\rho(E)$. In the absence of the magnetic field we have the density of states behaving as $E^{1/2}$.

In the presence of the magnetic field, we determine the density of states again by counting the total number of states

$$N = 2\sum_{n_x,n_y,n_z}\int d^3r\,|\Psi(x,y,z)|^2\,\theta(E_F - E). \qquad (7.12)$$

The quantum number n_z can be replaced by the wavevector $k_z = (2\pi n_z)/L_z$ using the usual periodic BC in the z-direction. The sum over n_y can again be replaced by the sum (integral) over the wavevector $k_y = (2\pi n_y)/L_y$. Note that the quantity k_y does not appear in the energy spectrum, equation (7.8), at all! It does, however, appear in the wavefunction, since $\mathcal{R}_0\xi = (x + x_0)$ and $x_0 = \mathcal{R}_0^2 k_y$. The coordinate $-x_0$ is the location of the orbit center for the Landau orbits, and k_y defines for us the orbit center through this relation. The sum over n_x survives as the sum over the Landau quantum numbers n. These substitutions and the integrations over y, z lead to

$$N = 2\frac{L_z L_y}{(2\pi)^2}\sum_n\int dk_z\int dk_y\int_{-L/2}^{L/2} dx$$
$$\times\,[h_n(x/\mathcal{R}_0 + k_y\mathcal{R}_0)]^2\,\theta(E_F - E). \qquad (7.13)$$

Integration over k_y can be performed first, using the orthonormality of the Hermite functions, to obtain

$$N = L_x L_y L_z \frac{2}{(2\pi \mathcal{R}_0)^2} \sum_n \int dk_z\, \theta(E_F - E). \qquad (7.14)$$

We can replace the integral over the wavevector k_z by an integral over the energy since $\hbar^2 k_z^2 / 2m^* = E - E_{n,\pm}$, where $E_{n,\pm}$ is the harmonic oscillator energy together with the energy of the dipole interaction with the external magnetic field. Thus the number density of electrons $\bar{n}$ is given by

$$\bar{n} = \int dE \sum_n \left[\frac{1}{(2\pi \mathcal{R}_0)^2} \left(\frac{2m^*}{\hbar^2} \right)^{1/2} \frac{1}{\sqrt{(E - E_n)}} \right]. \qquad (7.15)$$

We have used the index n for the Landau quantum number, so the carrier density is labeled here by $\bar{n}$. The quantity inside the square brackets is the density of states $\rho_n(E)$, the number of states for each Landau quantum number n per unit energy per unit volume, in a magnetic field. Note that the step-function $\theta(E - E_n)$ is understood to be present as a multiplicative factor within the summation. This ensures that the density of states obtains contributions from the nth term only if the energy is larger than E_n.

It is also possible to evaluate the density of states by using the following method. Let the total number of electrons be N. Then

$$N = 2 \sum_{\nu_x, \nu_y, \nu_z} \{\text{occupied states}\}.$$

Now insert a factor of unity: $1 = \int dE\, \delta(E - E_{\nu_i})$ in the sum on the right side. Then the density of states per unit volume is

$$\frac{1}{V} \frac{dN}{dE} = \rho(E)$$
$$= \frac{2}{V} \sum_{\nu_x, \nu_y, \nu_z} \delta(E - E_{\nu_i}).$$

The density of states from equation (7.11) and equation (7.15) are compared in Fig. 7.1. In the figure we have ignored the magnetic dipole energy. Notice the singularities in the density of states at $E = E_n$; these singularities are generated in the density of states in the presence of a magnetic field however small its magnitude. In

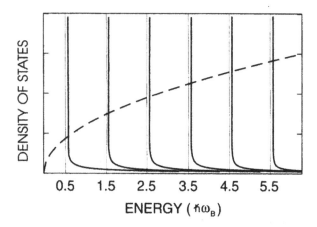

Fig. 7.1 The Landau density of states for an electron in an external magnetic field. Note the $E^{-1/2}$ singularity at the Landau energy levels. This is compared with the bulk density of states $E^{1/2}$ (dashed curve) in the absence of the field.

this sense the spectrum has undergone a profound change from the three-dimensional density of states behaving as $\sim E^{1/2}$ into one which behaves as $E^{-1/2}$. This fall-off with increasing energy is typical of quasi-one-dimensional electron distributions. The electron is bound by the quadratic potential in the plane perpendicular to the direction of the field, whereas it moves as a free particle parallel to the direction of the field. The energy dispersion is $E \sim \hbar^2 k_z^2/(2m^*) + E_n$ and is thus one dimensional in wavevector space. The energy $E_{n=0}$ is $\hbar\omega_B^*/2$ and can be thought of as the "confinement energy" of the ground state in the magnetic field.

For numerical work it is useful to obtain estimates for various physical quantities. We express magnetic fields in units of tesla (T). The cyclotron frequency per unit magnetic field is given by $\omega_B/B_0 = |e|/m_0 c$ with a numerical value $1.75882 \times 10^{11}\,\mathrm{rad\,s^{-1}\,T^{-1}}$. The Landau radius is given by $\mathcal{R}_0^2 = \hbar/(m_0\omega_B)$, with a numerical value given by $6.582 \times 10^{-12}\,\mathrm{cm^2}/[B_0]$ where B_0 is given in tesla. This leads to $\mathcal{R}_0 = 256.56\,\text{Å}\,[B_0]^{1/2}$ with B_0 in tesla. This shows that the cyclotron (or Landau) radius becomes comparable to typical layer thicknesses in quantum heterostructures for magnetic fields of a few Tesla. At $10\,\mathrm{T}$, for example, the Landau radius is approximately $81\,\text{Å}$. Finally, the cyclotron energy $\hbar\omega_B^*$ is given by $\hbar\omega_B^* = 1.15768 \times 10^{-4}(m_0/m^*)\,[B_0]\,\mathrm{eV}$, again with B_0 in tesla.

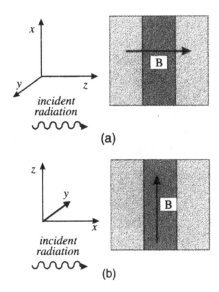

Fig. 7.2 The coordinate systems used for (a) the Faraday configuration and (b) the Voigt configuration.

7.2 Heterostructures in a B-field

7.2.1 Faraday configuration

In a layered heterostructure we can consider two special situations. First, we have $\mathbf{B} \parallel \hat{z}$ where $\hat{z}$ is along the growth direction. This will be called the Faraday geometry, since the usual optical investigations have light being incident in the z-direction. We then have $\mathbf{B}$ parallel to the direction of propagation of light.[‡] Secondly, we have the Voigt configuration, in which $\mathbf{B}$ is perpendicular to the direction of propagation of radiation in the material. Here, we consider $\mathbf{B}$ in the plane of the layers in this situation. The coordinate systems for these two configurations are shown in Fig. 7.2. The more general case with $\mathbf{B}$ along an arbitrary direction is treated in the references given at the end of this chapter.

In this section, we consider the case of Faraday geometry. Schrödinger's equation, equation (7.1), now has an added potential energy term $V(z)$ due to the band offsets in the heterostructure. In the Faraday geometry, the magnetic field being parallel to the growth

---

[‡]The Faraday effect concerns the rotation of the plane of polarization induced by the external field with the light propagating parallel to the magnetic field.

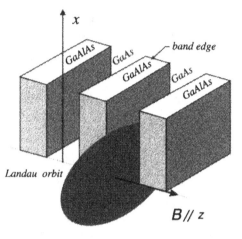

Faraday geometry

Fig. 7.3 The orientation of the Landau orbit in the Faraday geometry.

direction, the Landau orbits are in the planes of the layers, as shown in Fig. 7.3. We now assume a wavefunction of the form

$$\Psi(x, y, z) = f(z) \, e^{ik_y y} h_n((x + x_0)/\mathcal{R}_0). \tag{7.16}$$

Substituting this form of the wavefunction in Schrödinger's equation with the band offset potential energy we have

$$\left(-\frac{\hbar^2}{2m^*} f''(z) + (V(z) - E) \, f(z) \right) h_n(\xi)$$

$$+ f(z) \frac{\hbar\omega_B^*}{2} \left(-h_n''(\xi) + \xi^2 h_n(\xi) \pm \frac{g^* m^*}{2m_0} h_n(\xi) \right) = 0. \tag{7.17}$$

We use the property of the Hermite functions

$$h_n''(\xi) - \xi^2 h_n(\xi) = -(2n + 1) \, h_n(\xi) \tag{7.18}$$

to eliminate $h_n''(\xi)$; next we cancel a common factor of $h_n(\xi)$ in equation (7.17). Furthermore, we rescale the z-coordinate, setting $z = \tilde{z}\ell_0$, and let the energy scale be $\mathcal{C} = \hbar^2/2m_0\ell_0^2$:[§]

$$-\frac{1}{m^*/m_0} f''(\tilde{z}) + \frac{(V(\tilde{z}) - E)}{\mathcal{C}} f(\tilde{z})$$

$$+ \frac{\hbar\omega_B^*}{\mathcal{C}} \left(n + \frac{1}{2} \pm \frac{g^* m^*}{4m_0} \right) f(\tilde{z}) = 0. \tag{7.19}$$

This equation is solved using the FEM.

[§]For $\ell_0 = 1$ Å, a natural length scale for heterostructures, we have $\mathcal{C} = 3.81$ eV.

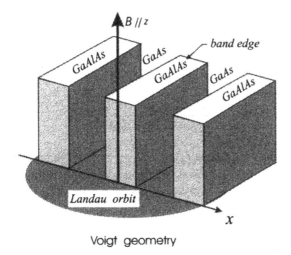

Voigt geometry

Fig. 7.4 The orientation of the Landau orbit in the Voigt geometry.
conditions

7.2.2   Voigt configuration

In the Voigt geometry, the magnetic field is parallel to the layers of
the heterostructure. It is then convenient to take the quantization
axis, $\hat{z}$, to lie in the plane of the layers, and we define the $\hat{x}$-axis to
be along the growth direction for this geometry. This is shown in
Fig. 7.4. The Landau orbits in this configuration cross the interfaces
of the layers, and the interface BCs play a significant role in the
determination of the solution. With $\mathbf{A} = (0, B_0 x, 0)$, and $\mathbf{B} \parallel \hat{z}$, the
Schrödinger Hamiltonian has the form

$$H = \frac{\hbar^2 k_x^2}{2m^*} + \frac{1}{2m^*}(\hbar k_y + |e|B_0 x/c)^2 + \frac{\hbar^2 k_z^2}{2m^*} + g^* \mu_B \mathbf{B} \cdot \mathbf{s} + V(x). \quad (7.20)$$

The form of the wavefunction is chosen as

$$\Psi(x, y, z) = e^{ik_y y} e^{ik_z z} \mathcal{F}(x)\, \mathcal{X}_s, \quad (7.21)$$

where $\mathcal{X}_s$ is the spinor wavefunction for the electron. Substituting
equation (7.21) into equation (7.20) we obtain an equation for $\mathcal{F}$:

$$-\frac{\hbar^2}{2m^*}\frac{d^2}{dx^2}\mathcal{F}(x) + (V(x) - E)\mathcal{F}(x) + \frac{\hbar \omega_B^*}{2}(x/\mathcal{R}_0 + k_y \mathcal{R}_0)^2 \mathcal{F}(x)$$

$$+ \left(\frac{\hbar^2 k_z^2}{2m^*} \pm \frac{\hbar \omega_B^* g^*}{4}\frac{m^*}{m_0}\right)\mathcal{F}(x) = 0. \quad (7.22)$$

Here again setting $x = \tilde{x}\ell_0$ we have

$$-\frac{1}{m^*/m_0}\mathcal{F}''(\tilde{x}) + \frac{(\ell_0/\mathcal{R}_0)^4}{m^*/m_0}\left(\tilde{x} + \frac{\mathcal{R}_0^2}{\ell_0}k_y\right)^2\mathcal{F}(\tilde{x})$$

$$+\left[\frac{V(\tilde{x}) - E}{\mathcal{C}} + \frac{(\ell_0 k_z)^2}{m^*/m_0} \pm \frac{g^*\hbar\omega_B^* m^*}{4m_0\mathcal{C}}\right]\mathcal{F}(\tilde{x}) = 0. \qquad (7.23)$$

All terms here are dimensionless, and terms with k_z and the spin magnetic energy are constants for given values of the quantum number k_z and B_0. Now the effective potential energy term is given by

$$\mathcal{V}_{\text{eff}}(\tilde{x}) = \frac{V(\tilde{x})}{\mathcal{C}} + \frac{(\ell_0/\mathcal{R}_0)^4}{(m^*/m_0)}\left(\tilde{x} + \frac{\mathcal{R}_0^2}{\ell_0}k_y\right)^2, \qquad (7.24)$$

with a quadratic term added to the potential energy arising from the band offsets. In principle, if $V(\tilde{x})$ is a constant in each layer in the heterostructure we could solve the above Hermite equation in each layer and match solutions at each boundary. The Hermite equation, being a second-order differential equation, has two types of solutions only one of which is well behaved at infinity, namely, $h_n(\xi)$. In general, when the range of the coordinate is finite, both types of solutions are needed to represent a general function over the given range. The solutions are given by the confluent hypergeometric functions, and this approach has been used by Altarelli and colleagues[2,3] in studying the energy levels of carriers. The solutions are expressed with an analytic continuation in the Landau quantum number. While this is a mathematically compact description of the solutions to the differential equation, in practice it does not provide any additional physical understanding of the problem beyond what is obtained from a straight numerical attack on the problem. More recently, the use of cylinder functions has been espoused[4] in a transfer-matrix approach to the problem with one energy band. There are other flexible, but computationally intensive, variational procedures for the solution of this problem. One can assume the global wavefunctions to be of some given form with unknown coefficients or parameters, and determine energy eigenvalues and the wavefunction parameters by minimizing the energy. Such a Rayleigh–Ritz technique was used fairly recently with an expansion of the wavefunction in terms of trigonometric functions (truncated Fourier series).[5] This has provided a good physical picture of the behavior of the wavefunctions.

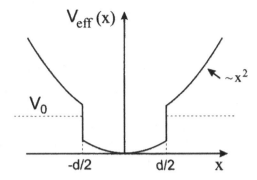

Fig. 7.5 The effective potential experienced by an electron in a quantum well in the presence of an external magnetic field in the Voigt geometry. The orbit center is assumed to be at the center of the well.

This problem has been solved using the FEM,[6] in which the region of interest is divided into a number of elements and the wavefunctions are represented by linear combinations of Hermite interpolation polynomials defined in each of the elements. The quadratic potential ensures that all states are confined in the direction perpendicular to the magnetic field. The interface boundary conditions require the continuity of the wavefunction and the continuity of the "mass derivative" $\frac{1}{m^*/m_0}\mathcal{F}'(\tilde{x})$ across material interfaces. Using the principle of least action, we define a generalized eigenvalue problem and solve it numerically for the energy eigenvalues and eigenvectors which are the coefficients of the Hermite interpolation polynomials in each element. The full wavefunction is reconstructed from the nodal coefficients by interpolation. The FEM is best suited for implementing BCs at interfaces and is a natural choice for the quantum mechanical calculations in semiconductor heterostructures. We shall assume that the eigenvalues and wavefunctions have been obtained by one of the above-mentioned methods and discuss the results in the following section.

7.3 Comparison with experiments

7.3.1 Interband transitions

Magneto-optical experiments on a $45\,\text{Å}/11\,\text{Å}$ GaAs/AlGaAs superlattice with an Al concentration of 40% were performed by Maan.[7] Both Faraday and Voigt configurations were investigated. Theoretical predictions, in a one-band calculation, for the energy levels for the conduction and the heavy-hole valence bands allow one to determine the magnetic field dependence of the $e1-hh1$ transitions between Landau levels with the same quantum number ($\Delta n = 0$ selection rule). The calculated transition energies conform to the experimental data. The calculations of Maan were repeated in the one-band model with the FEM for both the Faraday as well as the Voigt geometries. The effective potential experienced by the electron in the Voigt configuration is shown in Fig. 7.5 and the wavefunctions of the ground state and the first excited states are displayed in Fig. 7.6. The Landau wavefunctions, which correspond to harmonic oscillator wavefunctions, are clearly modified in the presence of the periodic potential of the superlattice. The additional features due to the localizations in the superlattice well regions are clearly visible. Higher excited states can be expected to show occupancy of wells farther away from the orbit center. The excited conduction and valence band states need not have their localization in the same wells because of the differing masses for the conduction electrons and the heavy holes. Hence, with increasing magnetic field, the interband transitions between excited states will become spatially indirect.

The transition energies as a function of magnetic field for the $e1-hh1$ transitions with $\Delta n = 0$, for the lowest few Landau quantum numbers, are shown in Fig. 7.7a for the Faraday geometry, and in Fig. 7.7b for the Voigt geometry, for the above-mentioned superlat-

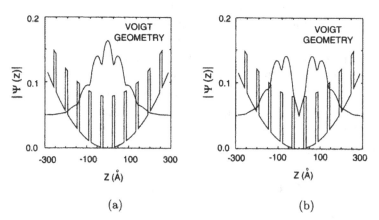

(a)   (b)

Fig. 7.6 The effective potential experienced by an electron in a 45 Å/11 Å GaAs/AlGaAs superlattice in the Voigt geometry for $B_0 = 10$ T. The wavefunction for the ground state at 57 meV is shown in (a), and the modulus of the wavefunction for the first excited state at 73 meV is shown in (b). The orbit center is assumed to be at the middle of one of the wells.

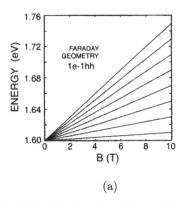
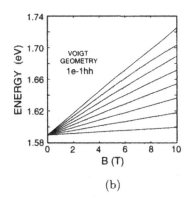

(a)   (b)

Fig. 7.7 The interband transitions between the heavy-hole valence states and conduction states with Landau quantum numbers $n = 0, 1, \ldots$, calculated[6] using a one-band model, for (a) the Faraday geometry, and (b) the Voigt geometry. The calculations are in reasonable agreement with the results quoted in Ref. 7, for a 45 Å/11 Å GaAs/AlGaAs superlattice.

tice. Experimentally,[7] it was found that the transitions are seen up to a transition energy of about 1.8 eV in the Voigt geometry. Above this energy, transitions become unobservable in the Voigt geometry, whereas they continue to be observed for higher energies in the Faraday geometry. This is explained in the following.

7.3.2   Energy dependence on the orbit center

In the Faraday geometry the orbits are in the plane of the layers, and the orbit center can be any point in the plane perpendicular to the magnetic field. We therefore have translational symmetry with respect to the location of the orbit center. With the magnetic field parallel to the layers, the orbits are perpendicular to the layers, and the presence of the interfaces makes a choice of the orbit center uniquely different from any other location for it. This then removes the periodicity of the superlattice potential. The superposed quadratic potential leads to what may be called Voigt localization.

The periodicity of the superlattice potential now manifests itself in the periodicity of the energy as a function of the orbit center. This is demonstrated in Fig. 7.8, where the dependence of the energy on the location of the orbit center is shown.[6] The energies of the various levels vary as the orbit center moves from the center of one well, progresses through the barrier region, and enters the next well. The calculation for the energy levels in the Voigt configuration proceeds

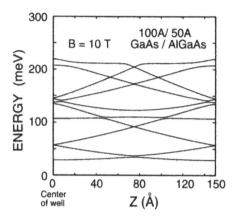

Fig. 7.8 The dependence of the energy on the location of the orbit center in a 100 Å/50 Å GaAs/AlGaAs superlattice in the Voigt geometry. The orbit-center dependence is weaker when the levels have energies corresponding to the superlattice miniband energies in the absence of a magnetic field. After Ref. 5; recalculated by Staveley using FEM in Ref. 6.

by picking a location for the orbit center. Next, the superlattice is treated as a multiple quantum well, with the range of the physical region for the calculation being governed by the Landau radius of the state under consideration. For example, for the lowest Landau level, at $B_0 = 10\,\mathrm{T}$, we have $\mathcal{R}_0 \simeq 81$ Å. So the multiple quantum well system is assumed to occupy the region from about $-2\mathcal{R}_0$ to $+2\mathcal{R}_0$. For higher energy levels this range has to be increased and it extends over $\pm 2\sqrt{2n+1}\mathcal{R}_0$ where n is the Landau quantum number of the excited state. It is clear that at low magnetic fields, the range over which the superlattice system has to be modeled as a multiple quantum well increases as $\mathcal{R}_n$ extends further and further.

A surprising result of the calculations for the energy dependence on the orbit center in the Voigt configuration is that the "Landau" levels vary very little over the energy ranges which correspond to the superlattice miniband width. The separation between the flat Landau levels is somewhat less than the cyclotron energy provided many Landau levels fall within the subband width.[8] This behavior can be understood by noting that at zero field the subband width corresponds to the energy range for effective carrier tunneling through the barriers to occur. For these energies the carriers in a magnetic field can complete their orbits through several barriers, and therefore these states do not have energies strongly dependent on where the center of the orbit is located. In the minigaps, the carriers are reflected by the barriers, and hence these states have position-dependent Landau levels.

This picture is further confirmed by the experimental data. We expect sharp peaks in the transitions between flat Landau levels. With increasing field the highest Landau level moves out of the allowed energy range, develops a dispersion (in terms of the orbit center), and becomes unobservable.[8]

7.3.3   Level mixing in superlattices with small band offsets

The cyclotron energy for an electron in the conduction band in GaAs is 1.73 meV/T. This suggests that if we have a superlattice with wide wells and low barrier heights we should see the effects of level mixing with increasing applied field since the cyclotron energy would become comparable to the minigaps. This effect was proposed by Xia and Fan.[5] A recalculation of their result using finite element analysis is shown in Fig. 7.9. It is clear from the above examples that the analysis of magneto-optical data in the Voigt configuration presents

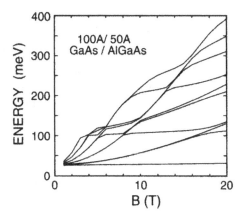

Fig. 7.9 Magnetic field dependence of energy levels in the Voigt geometry for the same superlattice as in Fig. 7.8. Note the remarkable level crossings appearing in this example with the superlattice having small barriers. Calculations performed by Staveley[6] using the FEM to reproduce results in Ref. 5.

very interesting challenges, as well as new effects not seen in the bulk material.

7.3.4   Density of states in the Voigt geometry

The energy spectrum in the Voigt configuration is dependent on the location of the orbit center, with energy levels periodic in the coordinate variable of the orbit center. This spectrum can be approximated, for a given band with index n, by

$$E = E_n - E_{n0} \cos\left(\frac{2\pi x_0}{D}\right) + \frac{\hbar^2 k_z^2}{2m^*}, \qquad (7.25)$$

where $2E_{n0}$ represents the energy "bandwidth" or the variation of energy as the orbit center moves over a superlattice period given by $D = d_1 + d_2$; here E_n corresponds to the cyclotron energy in the middle of the nth mini-bandwidth. The magnitude and sign of E_{n0} can vary depending on the energy level (see Fig. 7.9). This approximation is used only for illustrative purposes, and a parallel development can be made for the numerically determined energy dispersion, which may be obtained in a functional form by fitting the energy band to, say, a polynomial.

We can derive the density of states for the above spectrum following Ref. 5. We begin by distributing N electrons into the energy spectrum given above. It is understood that the energy spectrum begins at $E_n - E_{n0}$ and extends beyond $E_n + E_{n0}$ because of the kinetic energy arising from motion in the $\hat{z}$-direction. In the following we insert a step-function to account for it. Replacing the k_z integration by the energy variable we have

$$N = 2 \frac{L_z L_y}{4\pi^2} \int dk_y \sqrt{\frac{2m^*}{\hbar^2}} \int \frac{dE}{2} \frac{\theta(E - E_n + E_{n0})\,\theta(E_F - E)}{\sqrt{E - E_n + E_{n0}\cos(2\pi x_0/D)}}.$$

Here k_y ranges over positive and negative values. Using $x_0 = k_y \mathcal{R}_0^2$ and the fact that the k_y integration is over a periodic function, we can replace the k_y integral by $\mathcal{N}$ times the integral over one period, with $\mathcal{N}$ being the number of periods of the superlattice present in the quantization volume. Defining $\xi = 2\pi k_y \mathcal{R}_0^2/D$, we have the carrier density $\bar{n}$ given by

$$
\bar{n} = \frac{N}{L_z L_y \mathcal{N} D}
$$

$$
= \left[\frac{1}{(2\pi)^3} \frac{eB}{\hbar c} \sqrt{\frac{2m^*}{\hbar^2}} \right] \int dE \int_0^{2\pi} d\xi \frac{\theta(E - E_n + E_{n0})}{\sqrt{E - E_n + E_{n0} \cos \xi}}.
$$

Let us denote the quantity in square brackets in the above equation by Λ. The density of states is identified as

$$
\rho(E) = \Lambda \int_0^{2\pi} d\xi \frac{1}{\sqrt{E - E_n + E_{n0}(1 - 2\sin^2(\xi/2))}}
$$

$$
= \Lambda \frac{1}{\sqrt{E - E_n + E_{n0}}} \int_0^{2\pi} d\xi \frac{1}{\sqrt{1 - Q^2 \sin^2(\xi/2)}},
$$

where $Q^2 = 2E_{n0}/(E - E_n + E_{n0})$. The integral over ξ is replaced by one over $\xi' = \xi/2$ so that

$$
\rho(E) = \Lambda \frac{1}{\sqrt{E - E_n + E_{n0}}} \int_0^{\pi/2} d\xi' \frac{1}{\sqrt{1 - Q^2 \sin^2 \xi'}}. \qquad (7.26)
$$

The integral in equation (7.26) is a complete elliptic integral of the first kind, $K(Q)$, with $Q^2 < 1$, i.e., with $E > E_n + E_{n0}$.

When the energy E lies in the "miniband", $E_n - E_{n0} \le E \le E_n + E_{n0}$, we have $Q^2 > 1$, and the integral has its limits of integration given by $E_n - E_{n0}$ and α where

$$
\alpha = \arcsin \left\{ \sqrt{(E - E_n + E_{n0})/2E_{n0}} \right\}.
$$

We obtain[¶]

$$
\rho(E) = \Lambda \frac{1}{\sqrt{2E_{n0}}} \int_0^{\alpha} d\xi' \frac{1}{\sqrt{1/Q^2 - \sin^2 \xi'}}
$$

$$
= \left[\frac{1}{2\pi^3} \frac{eB}{\hbar c} \sqrt{\frac{2m^*}{\hbar^2}} \right] \frac{1}{\sqrt{2E_{n0}}} F(\alpha, 1/Q). \qquad (7.27)
$$

Below the energy $E_n - E_{n0}$, there is no contribution. We thus see that we have step-like behavior at the beginning of the "mini-bandedge,"

[¶]The elliptic integral and other related integrals are given in I. S. Gradshteyn and I. M. Ryzik, *Table of Integrals, Series, Products*, trans. and ed. A. Jeffrey (Academic Press, New York, 1980), pp160, 172, and 904.

with a monotonic increase till the singularity in the density at the location of the upper bandedge. This is analogous to the quasi-two-dimensional (quantum well) density of states. Above the upper bandedge we again have a fall-off in the density of states which is similar to the Landau density of states with the characteristics of a 1D density of states. Note that in the limit E_{n0} tends to zero, and the density of states equation (7.26) reverts to the 1D density of states, equation (7.15).

We thus see a complex picture emerging for the Voigt geometry in terms of new features in the density of states, the energy spectrum, and in the transition spectra.

7.4   Voigt geometry and a semiclassical model

7.4.1   Landau orbit theory

The discussion presented so far is an exact one-particle quantum mechanical description of the electronic states. For the Voigt configuration with a low value of the external magnetic field, the Landau radius is large, and a range corresponding to a large number of superlattice periods is needed to model the Landau states. This requirement can make the numerical methods computationally intensive.

It is precisely at such low fields that it is advantageous to consider a simplified model of electronic behavior in the presence of a static magnetic field. The so-called semiclassical model provides a physical picture of the electron orbits in an external magnetic field, and uses the semiclassical Bohr–Sommerfeld quantization rule to obtain energy levels. It is a fairly accurate description as long as the Landau radius is much larger than the confinement length (the superlattice period) appropriate to the heterostructure.

In the following, we describe the semiclassical method for a bulk system,[9] and then use it for layered heterostructures in the Voigt configuration at low fields. We follow the development given by Zilberman,[10] Altarelli,[11] and by Xia and Huang.[12]

The Bohr–Sommerfeld quantization condition for an oscillator is given by

$$\oint \mathbf{p} \cdot d\mathbf{r} = \left(n + \frac{1}{2} \right) 2\pi \hbar. \tag{7.28}$$

In the presence of a magnetic field, the momentum conjugate to $\mathbf{r}$ is $\mathbf{p}$; however, the mass times velocity for an electron is

$$m\dot{\mathbf{r}} = \hbar\mathbf{k} = \mathbf{p} + \frac{|e|}{c}\mathbf{A}.$$

The equation of motion

$$\hbar\frac{d\mathbf{k}}{dt} = -\frac{|e|}{c}\mathbf{v} \times \mathbf{B} \tag{7.29}$$

can be integrated over time, and for periodic orbits in the magnetic field we have

$$\hbar\mathbf{k}(t) = -\frac{|e|}{c}\mathbf{r}(t) \times \mathbf{B}. \tag{7.30}$$

Substituting these relations into the Bohr–Sommerfeld condition we obtain

$$\oint \mathbf{p}\cdot d\mathbf{r} = +\frac{|e|}{c}\mathbf{B}\cdot\oint \mathbf{r}\times d\mathbf{r} - \frac{|e|}{c}\int \mathbf{B}\cdot d\mathbf{S}$$

$$= \frac{|e|}{c}\mathbf{B}\cdot\mathbf{S}$$

$$= \left(n + \frac{1}{2}\right)2\pi\hbar. \tag{7.31}$$

Here, Stokes' theorem and the relation $\nabla \times \mathbf{A} = \mathbf{B}$ has been used, and $\mathbf{S}$ is the surface area vector of the closed orbit in configuration space. According to equation (7.30), the area S in configuration space is related to the area A_k in wavevector space by the relation

$$S = \left(\frac{|e|B_0}{\hbar c}\right)^2 A_k. \tag{7.32}$$

Hence, closed orbits in momentum space must satisfy the quantization condition[9]

$$A_k \equiv \oint k_y\, dk_x = 2\pi\frac{|e|B}{\hbar c}\left(n + \frac{1}{2}\right). \tag{7.33}$$

We now apply this relation to orbits in heterostructures to determine the energy spectrum at very low fields. Consider the case of superlattices with a magnetic field directed parallel to the plane of the layers as in the Voigt configuration. We suppose that we are given the in-plane dispersion and the superlattice dispersion for the

nth miniband in the growth direction (parallel to $\hat{x}$ here), in the *absence* of a magnetic field:

$$E = \frac{\hbar^2 k_y^2}{2m^*} + E_\nu + \frac{\Delta_\nu}{2}\, \varepsilon(q_x), \qquad -\pi/D \le q_x \le \pi/D. \qquad (7.34)$$

We let $k_z = 0$ since we will be concerned with electron orbits in the plane perpendicular to the applied magnetic field. It is usual to label the superlattice wavevector by q_x, rather than by k_x. In equation (7.34), the dispersion $\varepsilon(q_x)$ in the growth direction has been expressed in dimensionless form by extracting out the factor $\Delta/2$. The quantity E_ν defines the midpoint of the mini-bandwidth. We now construct constant energy surfaces in the q_x-k_y plane. The Brillouin zone in the q_x-direction is narrow (of width $2\pi/D$) due to zone folding arising from the periodicity of the superlattice in that direction, while it extends to π/a_0 in the in-plane direction. Note that the superlattice period D is much greater than the lattice constant a_0. In the energy range corresponding to a superlattice miniband, the constant energy surfaces form closed contours. For energies above this allowed energy subband the contour lines are open "orbits" (see Fig. 2 of Ref. 11, or Fig. 1 of Ref. 12). When a magnetic field is switched on in the $\hat{z}$-direction, the closed electron orbits are subject to the Bohr–Sommerfeld quantization conditions discussed above. These quantized orbits correspond to Landau levels which are independent of x_0. The open and closed contours correspond to the possible trajectories that an electron can follow in k-space. We replace k_y by energy variables using equation (7.34):

$$\oint dq_x \sqrt{E - E_\nu + \frac{\Delta_\nu}{2}\, \varepsilon(q_x)} = 2\pi \frac{|e|B}{\hbar c}\left(n + \frac{1}{2}\right). \qquad (7.35)$$

The integral can be performed by parameterizing the dispersion function $\varepsilon(q_x)$ by a polynomial or by a truncated cosine series. The quantity on the left side of equation (7.35) is the area enclosed by the orbit of energy E. This equation is solved for various integer values of the quantum number n to determine the quantized energy eigenvalues. This semiclassical approach is valid only for the Landau radius $\mathcal{R}_0$ being much larger than superlattice period D. In this method, once the superlattice energy dispersion is determined, in the absence of a magnetic field, no further reference is made to the periodic band offset potential of the superlattice in configuration space. In the presence of the applied field, the continuous energy spectrum in q_x and k_y variables is broken up into the discrete Landau states which satisfy the above quantization condition.

7.4.2   Envelope functions and the FEM in k-space

In the semiclassical formulation, we can consider the electron in k-space to move in a periodic effective potential defined by $\varepsilon(q_x)$. This provides an alternative to the approach of implementing the Bohr–Sommerfeld quantization for obtaining energy levels and wavefunctions. We follow the presentation of Xia and Huang,[12] with slight modifications in notation. The 2D Hamiltonian

$$H = \frac{\hbar^2 k_y^2}{2m^*} + E_\nu + \frac{\Delta_\nu}{2}\,\varepsilon(q_x) \qquad (7.36)$$

is modified in the presence of the field B_0 parallel to the $\hat{z}$-direction. In the Landau gauge, we have

$$H = \frac{1}{2m^*}\left(\frac{|e|B_0}{c}\right)^2 (x + \mathcal{R}_0^2 k_y)^2 + E_\nu + \frac{\Delta_\nu}{2}\,\varepsilon(q_x). \qquad (7.37)$$

Within the spirit of the semiclassical approximation, the effect of the periodic superlattice potential is included in the superlattice dispersion in q_x. To obtain the wavefunctions and the corresponding energies, we Fourier transform Schrödinger's equation to momentum space using

$$\phi(q_x) = \int dx\,\phi(x)\,e^{-iq_x x} \qquad (7.38)$$

and the relation $x = +i\,\hbar\,\partial/\partial q_x$ to obtain

$$\frac{m^*\omega_B^{*2}}{2}\left(i\hbar\frac{d}{dq_x} + \mathcal{R}_0^2 k_y\right)^2 \phi(q_x) + \left(E_\nu + \frac{\Delta_\nu}{2}\,\varepsilon(q_x)\right)\phi(q_x)$$
$$= E\,\phi(q_x). \qquad (7.39)$$

Substituting $\phi(q_x) = \eta(q_x)\,e^{i\mathcal{R}_0^2 k_y q_x/\hbar}$ in the above equation and letting $q_x \to \tilde{q}_x/D$ we obtain

$$-\left(\frac{m^*\omega_B^{*2}D^2}{2}\right)\frac{d^2}{d\tilde{q}_x^2}\,\eta(\tilde{q}_x) + \frac{\Delta_\nu}{2}\,\varepsilon(\tilde{q}_x)\,\eta(\tilde{q}_x)$$
$$= (E - E_\nu)\,\eta(\tilde{q}_x). \qquad (7.40)$$

Multiplying and dividing the first term by $\mathcal{R}_0^2$ and dividing through-out by $\hbar\omega_B^*/2$, we obtain

$$-\left(\frac{D^2}{\mathcal{R}_0^2}\right)\frac{d^2}{d\tilde{q}_x^2}\eta(\tilde{q}_x) + \frac{\Delta_\nu}{\hbar\omega_B^*}\varepsilon(\tilde{q}_x) = \frac{2\,(E-E_\nu)}{\hbar\omega_B^*}\eta(\tilde{q}_x). \qquad (7.41)$$

Here $\varepsilon(\tilde{q}_x)$ is a periodic function of period 2π and we can expand it in a cosine series, and use a plane-wave expansion for $\eta(\tilde{q}_x)$. It is sufficient to retain a few terms in the series expansions. In this basis, the matrix elements of $\varepsilon(\tilde{q}_x)$ are readily obtained in terms of Kronecker-delta functions. The secular determinant is solved to obtain the eigenvalues. The wavefunctions $\phi(\tilde{q}_x)$ can then be reconstructed and Fourier transformed back to configuration space. The wavefunctions obtained by this method are similar in form to the ones obtained using the full quantum mechanical calculations, except that they do not show the "lumpiness" seen in Figs. 7.6a and 7.6b. In this sense, the envelope functions obtained by this semiclassical method may be considered as the envelopes of the envelope functions obtained by the quantum mechanical calculation.

   This is a very versatile approach for obtaining eigenvalues and eigenfunctions in the Voigt geometry for the low field region where $\mathcal{R}_0 \gg D$. As emphasized earlier, it comes into its own precisely where the full quantum mechanical calculation is not computationally cost effective, while providing fairly accurate eigenvalues and reasonably good eigenfunctions. For further details and a numerical comparison with the quantum mechanical calculations, see Ref. 12. An obvious extension is to use the FEM in k-space to solve for wavefunctions and eigenvalues for complex functional forms for $\varepsilon(\tilde{q}_x)$. It is then evident that using the FEM for problems better defined in k-space is feasible and is a very promising approach to such problems.

7.5 Problems

1. A sample of n-InSb, with $n_e = 1.2 \times 10^{17}\mathrm{cm}^{-3}$, is placed in a uniform magnetic field $\mathbf{B} = B_0\hat{z}$. The carriers have an effective mass $m^* = 0.015\,m_0$, and an effective g-factor $g^* = -50$.

   (a) Suppose the lowest spin state (1/2, or spin-up, in this case) and the spin-down state of the Landau level with quantum number $n = 0$ are occupied. Determine the energy levels of these states for $B_0 = 6\,\mathrm{T}$.

(b) Show that the number density n_e at $T = 0\,\mathrm{K}$ is given by

$$n_e = \frac{1}{(2\pi R_0)^2}[2\,p_{F,0+} + 2\,p_{F,0-}]$$

where

$$p_{F,0\pm} = \sqrt{2m^*/\hbar^2} \times \sqrt{E_F - E_{0,\pm}}.$$

(c) Evaluate the Fermi velocities of the electrons given by

$$v_{F,0\pm} = \hbar k_{F,0\pm}/m^*.$$

(d) Evaluate E_F by solving for it from part (b). Is E_F greater than $E_{1,\pm}$? What does your answer signify in terms of occupied states?

2. Calculate the Landau orbit radii R_0, for the ground state, and R_9 for the 10th Landau level, for a range of magnetic fields from $B_0 = 0$ to $B_0 = 10\,\mathrm{T}$, and tabulate your results.

3. A quantum well heterostructure has three confined states with energy levels below its barrier.

(a) Obtain an expressions for the density of states, at $T = 0\,\mathrm{K}$ for the quantum well, including the phase space available due to in-plane motion.

(b) Consider the case of a $100\,\text{Å}$ GaAs quantum well with barriers of $Al_xGa_{1-x}As$ having $x = 0.37$. Plot the density of states at $T = 0\,\mathrm{K}$ as a function of energy.

4. A superlattice has minibands whose energy dispersion can be approximated by

$$E(q) = E_\nu + (-1)^\nu \frac{\Delta_\nu}{2} \cos qD, \quad -\pi/D \le q \le \pi/D,$$

where E_ν defines the center of the νth mini-bandwidth, Δ_ν is the bandwidth of the νth miniband, and D is the superlattice period.

(a) Taking account of the in-plane kinetic energy, determine the density of states for a superlattice at zero temperature.

(b) Represent graphically the density of states for the lowest two minibands, taking typical numbers for $E_{1,2}$ and $\Delta_{1,2}$ for a superlattice of GaAs/Al$_{0.37}$Ga$_{0.63}$As with a well width of 75 Å and a barrier width of 25 Å. How well does the above cosine formula represent the actual dispersion obtained numerically by the one-band FEM?

5. A superlattice, of bilayer periodicity D, is placed in an external magnetic field with the field parallel to the layers (Voigt geometry). It has its energy spectrum for the lowest energy "band" described in an approximate form by $E = E_1 - E_{1,0} \cos(2\pi x_0/D) + \hbar^2 k_z^2/2m^*$. Here $x_0 = k_y \mathcal{R}_0^2$ is the position of the orbit center. Let $E_1 = 50$ meV and $E_{10} = 5$ meV at $B_0 = 10$ T. Plot the density of states for the range in energy $0 \leq E \leq 100$ meV using the expression given in equations (7.26) and (7.27). Determine the behavior of $\rho(E)$ as a function of E near the singularity in the density of states. Assume that there are no other energy bands in the vicinity of this range.

References

[1] L. D. Landau and E. M. Lifshitz, *Quantum Mechanics – Non-relativistic Theory* (Pergamon, New York, 1965), pp65–66.
[2] M. Altarelli and G. Platero, *Surf. Sci.* **196**, 540 (1988).
[3] G. Platero and M. Altarelli, *Phys. Rev. B* **39**, 3758 (1989).
[4] K.-M. Hung and G. Y. Wu, *Phys. Rev. B* **45**, 3461 (1992).
[5] J.-B. Xia and W.-J. Fan, *Phys. Rev. B* **40**, 8508 (1989).
[6] B. S. Staveley, "Energy eigenvalues of charged carriers in quantum heterostructures in an externally applied magnetic field," Senior Thesis, Worcester Polytechnic Institute, 1991.
[7] J. C. Maan, *Superlattices Microstruct.* **2**, 557 (1986).
[8] G. Belle, J. C. Maan, and G. Weimann, *Surf. Sci.* **170**, 611 (1986).
[9] C. Kittel, *Quantum Theory of Solids* (Wiley, New York, 1963), p225.
[10] G. E. Zilberman, *Sov. Phys. – JETP* **5**, 208 (1957); *ibid.*, **6**, 299 (1958).
[11] M. Altarelli, in *Interfaces, Quantum Wells, and Superlattices*, ed. C. R. Leavens and R. Taylor, NATO ASI Series B, Vol. 179 (Plenum Press, New York, 1988), p43.
[12] J.-B. Xia and Kun Huang, *Phys. Rev. B* **42**, 11884 (1990).

8

Wavefunction engineering

8.1 Introduction

Over three decades ago, Esaki and Tsu[1] proposed that one can impose a super-periodicity in the crystalline growth of semiconductor heterostructures, and that this would lead to very fundamental changes in the electronic and optical properties of the composite structure as compared with those of the constituent layers. This proposal was an important milestone in the evolution of the field of semiconductor heterostructures. Since then, advances in the technology of MBE and metallo-organic chemical vapor deposition (MOCVD) have enabled the growth of layered semiconductor structures of any binary, ternary, or quaternary combination of semiconducting materials with control at essentially the atomic level. Essentially any of the compound semiconductors from the group III–V or II–VI elements can be arranged, layer by layer, to form a heterostructure. When ~100 Å layers of different materials, such as GaAs and Al-GaAs, are laid one on top of another, the natural or bulk energy bandgaps of the individual layers will generally not align with the adjacent ones. An electron in the conduction band of the GaAs layer will then confront a barrier in the adjacent AlGaAs layer. We have mentioned in earlier chapters[†] how such confinement, through the Heisenberg uncertainty principle, leads to a raising of the conduction bandedge of the heterostructure. A similar effect occurs for the carriers in the valence bands, leading to an overall change in the bandgap of the heterostructure.

Since the degrees of freedom in forming heterostructures include material composition, layer thicknesses, strain, growth direction, and so on, it is clear that these properties can be manipulated within a range of possibilities. This led Capasso[2,3] to propose the concept of "bandgap engineering" which has provided the *modus operandi*

[†]See Chapter 5.

in designing and growing multiple quantum wells (QWs), superlattices, graded energy gap materials, and the like, and in envisioning novel applications for such structures. The manipulation of the sub-bandedges through the appropriate selection of layer thickness and material parameters, and the control this provides in altering the optical properties of the heterostructure, have been a central focus of research in semiconductor heterostructures over the past two decades.

Over the past decade, ongoing improvements in growth that allow control down to the atomic scale, coupled with rapid advances in device fabrication technologies, have encouraged a trend towards increasingly complex structural configurations. Fortunately, recently developed modeling techniques provide the flexibility to explore such complexities through numerical simulations. This has led to the crystallization of the concepts that may be referred to as "wavefunction engineering."[4,5] This third milestone, in terms of fundamental shifts in paradigms, is providing a lodestone for directing efforts towards formulating new electronic mechanisms and concepts, exploring basic physics, and designing new optoelectronic devices. We describe here the computational aspects and present applications of this paradigm. We should note that for over a decade there have been sporadic references to the idea and potential of wavefunction engineering.[6-11] However, a full realization of the concept has required an implementation of advanced software tools that would allow one to explore issues associated with optimizing heterostructure design for specific applications.[12-22] This advance has not been merely in modeling; in fact, heterostructures grown to specifications for individual applications have performed as promised. We provide examples of this remarkable synergism between modeling, growth, and experimental characterization, which are the three facets of the efforts towards bringing new devices from conceptual development to the feasibility stage, and towards commercialization.

The interpretation of the experimentally observed optical properties of heterostructures requires that we understand the effects of layer thicknesses and material properties of the layers. The analysis of the data is usually done by modeling the structure using computer calculations for the actual energy levels and energy bands in the system. This intertwining of modeling and simulation with experimental characterization is a feature that has become vital as the structures being studied have increased in complexity. Having adequate computational resources has thus become imperative for progressing on both the experimental and the theoretical fronts. Accurate theoretical modeling of the electronic dispersion relations and optical properties of the new multi-layered quantum heterostructures entails a serious computational investment. Here we discuss a multiband finite element implementation of the k·P model, which is particularly well suited for taking into account the details of the geometry and the complex BCs imposed on the wavefunctions of the carriers at material interfaces. The energy levels and wavefunctions of the carriers can then be obtained in a routine manner.

This has profound implications for both the design of optoelectronic devices and the exploration of fundamental physical issues. By tailoring the wavefunctions we can control optical selection rules, optical matrix elements, carrier lifetimes, overlap integrals, tunneling currents, electro-optical and nonlinear optical coefficients, and so on. We have already presented some examples which may be restated in the light of wavefunction engineering. The discovery that carriers with energies above the barrier height in superlattices are localized in the barrier layers and have quantum-well-like spectra is leading to a new spectroscopy of above-barrier states. In compositionally asymmetric wells, the confined ground states, when they exist, in the conduction and the valence bands, are very close in energy to the corresponding ground state energies in compositionally symmetric quantum wells. Thus, there is not much difference between them in so far as bandgap engineering is concerned. However, the wavefunctions of the higher bound/quasi-bound states are substantially different in the two cases. This is an example of wavefunction engineering in which the carrier localizations are altered to induce asymmetry. These examples from Chapter 5 clearly illustrate the basic shift in the way we envisage quantum heterostructures, in having progressed from bandgap engineering to wavefunction engineering.

8.2   k·P theory of band structure

We consider[‡] planar layered semiconductor heterostructures with the planes perpendicular to the growth direction z. The layers are taken to be composed of compound III–V or II–VI semiconductors with their conduction and valence bandedges located at the Γ-point in

---

[‡]This section requires some knowledge of semiconductor bandstructure at the center of the Brillouin zone, at the level presented in Kittel (Ref. 23).

the Brillouin zone. The periodic components of the Bloch functions, $u_{j,\mathbf{k}=0}(\mathbf{r})$, with j being the band index, at the bandedges are assumed not to differ much as we traverse the layer interfaces.[24-26] We assume that the original bulk crystal translational symmetry is maintained in the transverse direction.

We consider the zone-center bulk bandstructure of the constituent semiconductors, within the spirit of the $\mathbf{k} \cdot \mathbf{P}$ model. While additional bands (such as a 14-band model) may be treated analogously within our formalism, we find the eight-band $\mathbf{k} \cdot \mathbf{P}$ model to have sufficient sophistication for most applications. The usual eight-band model consists of the Γ_6 conduction band (c), the Γ_8 heavy-hole (hh) and light-hole (lh) bands, and the Γ_7 spin–orbit split-off band (s.o.), with their spin degeneracies. The complete 8×8 Hamiltonian is displayed in Ref. 27. However, for the sake of clarity in the presentation, we will consider here the simplified example of a layered system with no external electric or magnetic fields or built-in strain, and the limit of vanishing in-plane wavevector ($k_\parallel \equiv (k_x^2 + k_y^2)^{1/2} \to 0$). We note that the following considerations also hold for the more general case, except that the dimensions of the matrices correspond to the full eight-band model, and with external perturbations or finite $k_\parallel$ the Kramers degeneracy of the bands is lifted.

With these simplifications, the problem reduces to a three-band model, with the hh band factoring out. Within the envelope function approximation, the problem then reduces to the solution of a set of three simultaneous second-order differential equations for the envelope functions $f_i(z)$ of the constituent layers. We have

$$H_{ij}(k_\parallel = 0, k_z)\, f_j(z) \;=\; E\, f_i(z), \qquad (8.1)$$

where k_z has to be replaced by the differential operator $-i\partial/\partial z$. The eigenvalues of the 3×3 matrix are given by the secular equation (in atomic units)

$$\begin{vmatrix} E_c + (F + \tfrac{1}{2})k_z^2 - E & -\sqrt{E_P/3}\,k_z & -\sqrt{E_P/6}\,k_z \\[2ex] -\sqrt{E_P/3}\,k_z & E_v - \tfrac{1}{2}(\gamma_1 + 2\gamma_2)k_z^2 - E & -\sqrt{2}\gamma_2 k_z^2 \\[2ex] -\sqrt{E_P/6}\,k_z & -\sqrt{2}\gamma_2 k_z^2 & E_s - \tfrac{1}{2}\gamma_1 k_z^2 - E \end{vmatrix} = 0.$$

$$(8.2)$$

Here E_c, E_v, and E_s are the bandedge energies of the conduction, lh, and s.o. bands. The three coupled second-order differential equations represented by equation (8.2) can be written as

$$\left[-\mathcal{A}_{ab}\frac{\partial^2}{\partial z^2} - i\mathcal{B}_{ab}\frac{\partial}{\partial z} + \mathcal{C}_{ab} \right] f_b(z) \;=\; E\, f_a(z). \qquad (8.3)$$

While the matrix coefficients $\mathcal{A}, \mathcal{B}, \mathcal{C}$ in equation (8.3) are usually assumed to be constant in each layer, there are cases for which each material parameter should more generally be considered a function of coordinate, e.g., in a graded-bandgap system or a structure for which the strain is a function of position within a given layer. In a heterostructure, the differences in the bandedge energies give rise to the confining potentials experienced by the carriers. In the full eight-band formalism, the above equation consists of eight coupled Schrödinger equations, as discussed in Ref. 27.

The finite element method (FEM) may be used to solve the above coupled Schrödinger equations for symmetric as well as asymmetric QWs, for superlattices with two or more than two constituents in each period, for resonant tunneling structures, etc. In fact, this method can accommodate the possibility of every material parameter being a function of coordinate. We have already shown in Chapters 3 and 5 that the FEM can be adapted to yield very accurate eigenvalues for bound state problems, and for obtaining solutions in quantum semiconductor heterostructures with complex geometries. In the multi-band scheme, one begins by writing out the appropriate symmetrized (Hermitian) Lagrangian that would generate the above coupled equations through a variational procedure. The integral over the physical region of the Lagrangian density, the action integral, is then split up into a number of "cells" or elements, in each of which the physical considerations of the problem hold. The wavefunctions are assumed to be given locally in each element by fifth-order Hermite interpolation polynomials. The global wavefunctions $f_i(z)$ are constructed by joining the locally defined interpolation functions and matching the function and its derivative across the element boundary for each of the bands included in the analysis. The heterointerface BCs consisting of continuity of the envelope functions and of the probability current, and the BCs for the bound states as well as for scattering states[28] at $z = \pm\infty$, are readily incorporated into the FEM. It is useful to derive the probability current density that is conserved across the interfaces by employing a gauge variational approach for the multi-component wavefunctions using an extension of the gauge variational method of Gell-Mann and Levy.[29] The spatial dependence of the wavefunctions, manifested through the interpolation polynomials, is next integrated out, leaving the action integral dependent on the unknown nodal values of the wavefunction. As described in earlier chapters, the usual variational principle is then implemented as a variation of the nodal values of $f^*(z)$ under which the action integral is a minimum. This nodal variational principle leads to "Schrödinger's equation" for the nodal values. The integration of the action integral is performed element by element, giving rise to element matrices which are then overlaid into a global matrix in a manner consistent with the interelement BCs. This results finally in a generalized eigenvalue problem, which may be solved for the eigenenergies and wavefunctions with a standard diagonalizer on a computer workstation.

In the eight-band model, the element matrices are of dimension 48 for quintic Hermite interpolation. The element matrix overlays used to construct the global matrix require considerable care in the book-keeping since we are now dealing with eight simultaneous equations. Use of the FEM with three elements per layer leads to very accurate quantum well and superlattice energies and eigenfunctions. In the limit of simple geometries the eigenvalues agree with those obtained from the eight-band transfer-matrix method[27,30–32] to within 10^{-6} eV, and double-precision accuracy can be obtained by employing more elements in the computation. We should mention that a two-band finite element calculation of superlattice bandstructures was developed independently by Nakamura *et al.*[33]

8.3   Designing mid-infrared lasers

In this section, we illustrate the application of the FEM to the design of optoelectronic devices. It will be used to determine energy levels, dispersion relations, wavefunctions, and optical properties for specific wavefunction-engineered materials that, following fabrication and testing, have shown promising performance in the laboratory.

8.3.1   The type-II W-laser

Type-I heterostructures, such as GaAs/AlGaAs systems, have interfaces at which both the conduction and valence bandedges of one of the adjacent layer materials lie within the energy bandgap of the other. In type-II interfaces, only one of the bandedges of an adjacent layer is outside the bandgap of the other layer. Until 1994, all interband QW lasers employed well and barrier constituents having a type-I band alignment, since the achievement of gain requires strong optical coupling between the conduction and valence band states. However, recent work has demonstrated that type-II InAs–Ga$_{1-x}$In$_x$Sb structures produce substantial gain as long as the layers are thin enough to allow significant' interpenetration of the electron and hole wavefunctions.[34,35] In fact, mid-wave infrared (mid-IR) lasers based on the InAs–GaSb–AlSb family of type-II heterostructures not only are feasible, but exhibit some significant advantages[36,37] over the type-I systems that have been investigated for this application.[38]

Fundamental limitations of many of the type-I structures include inadequate electrical confinement due to the small conduction and/or valence band offsets (which leads to escape from the active region at higher temperatures) and the increasing predominance of Auger recombination when the energy gap is lowered and the temperature raised. The nonradiative decay in narrow-gap III–V systems currently under investigation is often dominated by the so-called CHHS Auger process, in which the conduction-to-heavy-hole (CH) recombination is accompanied by a heavy-to-split-off-hole (HS) transition. In InAs-rich alloys such as InAsSb, InAsSbP, and InGaAsSb, this process easily conserves both momentum and energy because the energy gap is nearly equal to the split-off gap Δ_0.[39,40] Grein et al. theoretically discussed the minimization of Auger rates in type-II InAs–Ga$_{1-x}$In$_x$Sb superlattices for mid-IR laser applications.[36]

The simplest type-II structure is a two-constituent type-II superlattice. In Fig. 8.1, for illustrative purposes, we show the conduction, valence, and split-off band profiles of a multiple QW studied experimentally by Hasenberg et al.[41] Also shown are the corresponding energy levels and wavefunctions calculated using the eight-band FEM algorithm. Note first that even though the electron wavefunctions (solid curves) have their maxima in the InAs layers and the hole wavefunctions (dashed curves) are centered on the Ga$_{1-x}$In$_x$Sb layers, there is significant overlap because each penetrates into the adjacent layers. Hence the optical matrix element is nearly as large as values typically obtained for type-I QWs. We also find that the resonance between E_g (the separation of E1 and H1) and Δ_0 (the difference between H1 and S1) is completely removed by the type-II

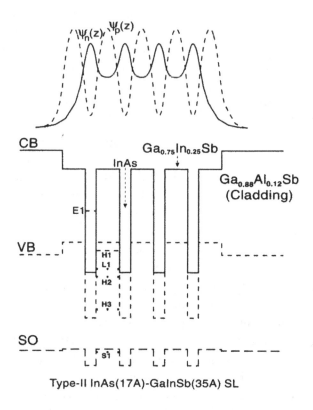

Fig. 8.1 Conduction, valence (heavy-hole), and split-off band profiles for the type-II InAs–$Ga_{0.75}In_{0.25}Sb$ multiple QW.

band alignment, even though it is present in bulk InAs and GaSb, and is potentially an issue in $Ga_{1-x}In_xSb$. Furthermore, Grein *et al.* pointed out that for these particular layer thicknesses, the energy gap does not resonate with any intervalence transitions involving H1 near its maximum (it falls approximately halfway between H1–H2 and H1–H3).[36] Hence all multi-hole Auger processes are energetically unfavorable. Moreover, the rate for CCCH events (in which the CH recombination is accompanied by an electron transition to a higher energy conduction band state (CC)) is suppressed by the small in-plane effective mass for holes near the band extremum ($\approx 0.047m_0$). It has been demonstrated experimentally that type-II quantum heterostructures can have significantly smaller Auger coefficients than analogous type-I structures with the same energy gap.[42,43]

However, the structure shown in Fig. 8.1 is nonoptimal in that the electron dispersion is effectively three dimensional. It is well known that QW lasers (2D) usually significantly outperform[44,45] double-heterostructure lasers (3D) once a given fabrication technology has matured, primarily because the more concentrated 2D density of states yields much higher gain per injected carrier at threshold. While the holes in the type-II superlattice shown in Fig. 8.1 have minimal dispersion along the growth axis (i.e., they are quasi-2D), the strong penetration of the electron wavefunctions into the thin $Ga_{1-x}In_xSb$ barriers leads to a nearly isotropic electron mass $(m_{nz}/m_{n\|} \approx 1.2)$.

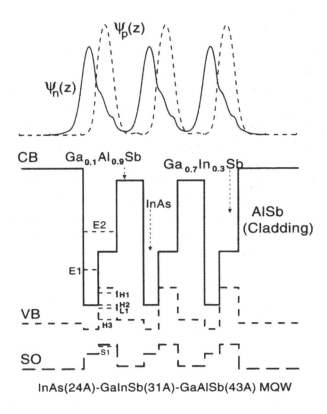

Fig. 8.2 Band profiles and FEM wavefunctions and energy levels for a type-II InAs–$Ga_{0.70}In_{0.30}Sb$–$Ga_{0.1}Al_{0.9}Sb$ three-constituent multiple QW which is almost exactly lattice matched to the AlSb cladding layers. Although the electrons now have 2D dispersion due to the $Ga_{0.1}Al_{0.9}Sb$ barrier layers, the wavefunction overlap is reduced from that in Fig. 8.1.

The most straightforward approach to reducing the electron dimensionality is to convert the superlattice into a multiple QW. Figure 8.2 illustrates the incorporation of additional AlSb layers, which
serve as barriers for both electrons and holes, into each period of the
structure. The FEM calculation for the structure in Fig. 8.2 yields 2D
electron and hole dispersion relations (the electron mini-bandwidth
is < 0.3 meV), since both wavefunctions fall nearly to zero in the
barrier layers and the coupling between successive periods is quite
weak. We emphasize here that while most multiband heterostruc-
ture $\mathbf{k} \cdot \mathbf{P}$ calculations can treat simple two-constituent superlattices
such as that shown in Fig. 8.2, they are not easily adapted to more
complicated geometries such as the three-constituent multiple QW
shown in Fig. 8.2. By contrast, the multiband FEM algorithm readily yields bandstructures, wavefunctions, and optical properties for
layered structures of arbitrary complexity.

Unfortunately, a by-product of the three-constituent asymmetric
geometry in Fig. 8.2 is that the electron–hole wavefunction overlap is
somewhat smaller than in the superlattice of Fig. 8.1. Electrons from
a given InAs QW penetrate only about halfway into the adjacent
$Ga_{1-x}In_xSb$ layer, and now there are no InAs electrons from the
other side to overlap the other half of the hole wavefunction. As
a consequence, the interband optical matrix element is somewhat
smaller and the hole effective mass is more than twice as heavy as
it was in the superlattice ($0.11m_0$). The repulsion of the E1 and
H1 dispersion relations is proportional to the overlap integral; hence
both masses in this narrow-gap system are quite sensitive to the
spatial distribution of the wavefunctions.[46]

Fortunately, we can eliminate these drawbacks by performing
some further wavefunction engineering. Figure 8.3 illustrates the
consequences of incorporating an additional InAs layer into each period, on the other side of the $Ga_{1-x}In_xSb$, to form a "W" laser,[37]
which is so named because of the shape of the conduction band profile. The entire hole wavefunction for this four-constituent quantum
structure is now seen to lie in a region of strong overlap with the electron wavefunction. Hence the in-plane hole mass is virtually equal
to that in the superlattice ($0.048m_0$) and the interband optical matrix element is nearly as large, yet the electrons and holes both have
2D dispersion due to the AlSb barriers. Because of the coupled-well
nature of the conduction band profile, the electron states split into
symmetric (E1S) and antisymmetric (E1A) levels. However, only

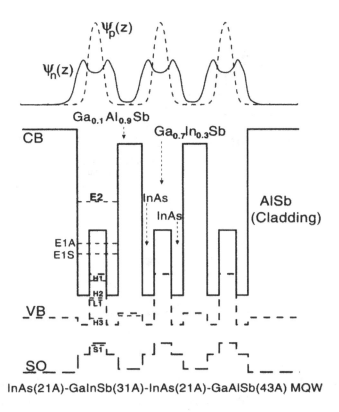

Fig. 8.3 Band profiles and FEM wavefunctions and energy levels for a type-II InAs–$Ga_{0.70}In_{0.30}Sb$–InAs–$Ga_{0.1}Al_{0.9}Sb$ four-constituent multiple QW which is almost exactly lattice matched to the AlSb cladding layers. This structure combines both the 2D electron dispersion of Fig. 8.2 and the large wavefunction overlap of Fig. 8.1.

E1S will be populated at the laser operating temperatures of interest, since the energy separation between the two is 125 meV. The structure in Fig. 8.3 is illustrated with AlSb cladding layers, which provide excellent optical confinement ($\Delta n > 0.2$) and provide a nearly exact lattice match to the active QW region. Note also that we still expect intervalence Auger processes to be weak because the energy gap between E1S and H1 again falls halfway between the gaps for H1–L1 and H1–H3 (and is of course far out of resonance with the split-off gap). We finally point out that electrical confinement ceases to be an issue in type-II W structures because the AlSb cladding layers provide large offsets for both the conduction and valence bands.

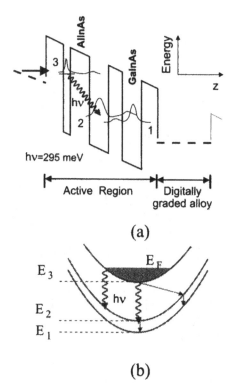

Fig. 8.4 A schematic of the inter-subband QCL showing (a) the conduction bandedge profile, and (b) the relaxation mechanisms that an electron can undergo.

The quantum structural design considered here has quite literally taken the form of wavefunction engineering. Optically pumped type-II W structures were the first interband lasers emitting beyond $3\,\mu$m to operate at room temperature[47] (up to 360 K), and they also hold the record for maximum continuous wave operating temperature (220 K)[48] for all III–V semiconductor lasers in that wavelength range.

8.3.2   The interband cascade laser

In the quantum cascade laser (QCL),[49] unipolar electron injection into a series of coupled QWs produces a subband population inversion and lasing owing to stimulated inter-subband transitions. A schematic of this heterostructure is shown in Fig. 8.4. A key feature is that in contrast to conventional diode lasers, in which a maximum of one photon is emitted for every injected electron and hole,

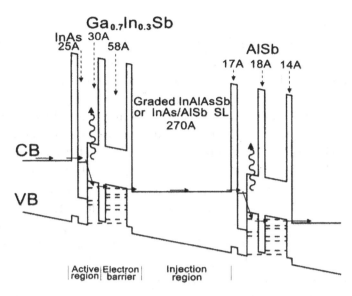

Fig. 8.5 Conduction and valence band profiles and calculated energy levels for the type-II ICL structure under bias $(100\,\mathrm{kV\,cm^{-1}})$.

in the QCL each injected electron can in principle produce an additional photon for each period of the structure. High threshold current densities are required to obtain population inversion, however, since the nonradiative lifetime of electrons in the upper lasing subband is only $\approx 1\,\mathrm{ps}$ due to inter-subband relaxation via optical phonon emission. This shortcoming can be overcome if type-II interband rather than inter-subband transitions are used in conjunction with a cascade geometry.[50] In the interband cascade laser (ICL), the phonon relaxation path indicated in Fig. 8.4 is eliminated while the advantages of electron recycling are retained. Detailed simulations of optimized designs generated using wavefunction engineering are predicted to yield high continuous wave output powers for high temperature operation.[14,51,52]

Conduction and valence band profiles along with quantized energy levels for one period of a type-II ICL designed to emit at $3.15\,\mu\mathrm{m}$ are shown in Fig. 8.5. Electrons are injected into the InAs electron QW from an adjacent period on the left side and emit mid-IR photons by making type-II radiative transitions to the valence band of the GaInSb hole QW. This hole well is optimized so as to avoid possible resonances between the energy gap and any of the valence inter-subband splittings, which is again necessary to minimize losses due to Auger recombination and free carrier absorption. Holes tunnel into the GaInSb well from the adjacent well. The second hole well provides a substantial barrier to electron leakage from the active InAs well (i.e., negligible wavefunction overlap between the active region and the electrons in the first QW of the injection region) while assuring a large indirect optical matrix element (i.e., large overlap of the active electron and hole wavefunctions). This is followed by an n-doped InAs/Al(In)Sb superlattice, whose well thicknesses are graded such that under the appropriate bias the bottom subbands merge into a miniband of width ≈ 65 meV to insure a rapid transfer of electrons across the injection region and into the active InAs QW of the next period. Resonances between the lasing photon energy and the inter-subband transition energies have been avoided in all electron wells to prevent absorption, although even near resonance the inter-subband absorption processes are strongly suppressed for the TE polarization of the lasing mode. The lattice constant for the entire period can be matched to that of the cladding layers by using $Al_{0.82}In_{0.18}Sb$ rather than AlSb for some of the barriers. All layers are lattice matched to the GaSb substrate.

The internal electric fields generated by the hole populations in the undoped GaInSb and GaSb wells are gradually compensated by the excess electron concentrations near the left end of the n-doped injection region. Under the biasing conditions required for near-threshold operation at room temperature, the steady-state field is ≈ 70 kV cm^{-1} towards the middle of the injection region, while that value is substantially reduced in most of the active region. In fact, the internal field across the GaSb well is so strong that the net field is in the opposite direction from the applied bias.

A simulation of the operation of a 15-period ICL structure surrounded by optical cladding layers 1.5 μm thick consisting of n-doped InAs/AlSb superlattices has been performed using the FEM. The optical confinement factor for the active region is $\approx 78\%$, and we estimate a net loss of ≈ 52 cm^{-1}. Energy dispersion relations and optical matrix elements for this complex structure were derived from the eight-band FEM algorithm. Fermi levels and electronic heat capacities were then calculated as a function of the carrier concentration for each well of a given period, and the corresponding optical gain determined as a function of electron and hole densities in the active wells. These dependencies were then input to the time-dependent equations for interwell carrier transfer, which are coupled to the photon propagation equation and the heat balance equation for the carrier temperature in the active wells. As the calculation progresses, the internal electric fields were self-consistently adjusted according to the spatial build-up of excess charge in the various QWs.

Electron–hole recombination arises from spontaneous and stimulated radiative emission, as well as Shockley–Read and Auger nonradiative processes. The energy relaxation time is taken to be 1 ps and it is assumed that frequent interwell carrier–carrier scattering events lead to a common carrier temperature throughout the structure. For operation at room temperature, the calculations yield that the carrier heating does not exceed $\approx 4\,K$ near threshold and increases to $\approx 65\,K$ at a current density of $4\,kA\,cm^{-2}$. The simulation yields that at 300 K, a threshold current of $1.1\,kA\,cm^{-2}$ (due almost entirely to Auger recombination) induces lasing when a net bias of 6.6 V is applied across the 15-period structure.

The ICL threshold currents are found to be much lower, since for a cascade geometry the contacts need only to supply enough carriers to establish a population inversion in the first period, from which the carriers are then injected into the next period, etc. Especially encouraging is the prediction that output powers exceeding 1 W may be feasible, although the maximum is limited by the carrier and lattice heating at high injection currents. The theoretical predictions for the threshold current density of the quantum cascade laser are $\approx 3\,kA\,cm^{-2}$ and $1\,kA\,cm^{-2}$ at 300 and 100 K, respectively.[53] Further wavefunction engineering of the structure of Fig. 8.5 can be introduced to improve the laser performance by putting a W structure at the beginning of each period.[52]

ICLs with designs similar to those illustrated in Fig. 8.5 have now been demonstrated experimentally.[43,54–56] Even at this preliminary stage of development, external quantum efficiencies greater than one photon per injected electron have been confirmed.[43,54,55] This implies that true cascading has been successfully realized. A W ICL has generated up to 170 mW peak power at 180 K,[56] which is the highest for any electrically pumped interband mid-IR semiconductor laser at that temperature. The same device operated up to 225 K, and at that temperature had a slightly lower threshold than any previous mid-IR result.

Wavefunction engineering has also been used recently to design active regions[52] for the first III-V vertical-cavity surface-emitting lasers to emit beyond $2.2\,\mu m$. The optically pumped devices emitting at 2.9–$3.0\,\mu$m have been operated continuous wave up to $280\,$K pulsed[57] and $160\,$K.[58] Inter-subband QCLs based on antimonide QWs have been designed as well and are predicted to have lower thresholds than InGaAs–InAlAs QCLs.[53]

8.4   Concluding comments

Recent developments in what we call wavefunction engineering of the electronic and optical properties of quantum heterostructures have been reviewed in this chapter. We have emphasized in particular the evolution of the field beyond the more restrictive concept of bandgap engineering, in which the confinement of carriers over the physical dimensions of the layers controls the energy gap as a manifestation of Heisenberg's uncertainty relation. Technological advances in the crystal growth of semiconductor heterostructures of III–V and II–VI materials, as well as masking and pattern etching, now allow us to alter much more directly the shape of the carrier wavefunctions to suit specific applications. Thus, to a remarkable degree we can control the probability density, or occupancy, of the carriers in any region of the heterostructure. Through the appropriate insertion of thin (even atomic) layers of a different material, by employing asymmetric wells, through the use of type-II interfaces, through localization in the barrier layers for carriers with above-barrier energy, through control over the dynamics of intervalley transfer via externally applied fields and optical phonon emission, through the use of strain to tailor the band mixing and valence band properties, through the use of diluted magnetic semiconductor layers, and so on, we can significantly alter the carrier wavefunctions. Wavefunction engineering provides a greater appreciation of the mechanisms governing carrier dynamics, since overlap integrals, optical matrix elements, density of states, tunneling times, lifetimes for carrier recombination, optical detection efficiencies, nonlinear optical properties, etc., can now all be altered through a judicious use of "heterostructure architecture."

In creating the new generation of advanced devices based on detailed control over the electronic and optical properties of quantum heterostructures with novel geometries, the designer must necessarily rely to a considerable extent on computer modeling of the energy bands and wavefunctions. We have demonstrated that the FEM is well suited to the task of efficiently calculating the required properties for structures of arbitrary complexity. Although our emphasis has been on integration of the FEM into the $\mathbf{k} \cdot \mathbf{P}$ framework, it should be mentioned that tight-binding models may also be viewed in the same context, with individual atoms functioning as the "finite elements."

FEM computations have allowed us to explore some fundamental aspects of the band structure of quantum semiconductor systems. The investigations of quasibound states and above-barrier states in III–V heterostructures were driven by modeling and feasibility studies performed with the FEM. Additional examples of wavefunction engineering and applications to quantum wires are presented in Chapter 12, where the level degeneracy issues in square and rectangular quantum wires with finite barrier height are revealed through finite element calculations. That we can compute in detail the optical properties of the checkerboard superlattices, as in Chapter 12, is based on the ability of the FEM to model complex geometries. By extrapolation, exciting results revealing new physics can clearly be anticipated with explorations into the properties of systems with 3D confinement. Wavefunction engineering, made feasible by finite element modeling, will play a vital role in the exploration of new physical concepts and in the development of new devices based on quantum heterostructures.

References

[1] L. Esaki and R. Tsu, *IBM J. Res. Dev.* **14**, 61 (1970).

[2] F. Capasso, *J. Vac. Sci. Technol.* B **1**, 457 (1983).

[3] F. Capasso, *Surf. Sci.* **142**, 513 (1984).

[4] L. R. Ram-Mohan and J. R. Meyer, *J. Nonlinear Opt. Phys. Mater.* **4**, 191 (1995).

[5] L. R. Ram-Mohan, I. Vurgaftman, J. R. Meyer, and D. Dossa, "Wavefunction engineering: a new paradigm in quantum nanostructure modeling", in *Handbook of Nanostructured Materials and Nanotechnology*, Vol. 2, Ed. H. S. Nalwa, (Academic Press, New York, 1999), Chapter 15.

[6] H. Sakaki, *IEEE J. Quantum Electron.* **22**, No. 9, 1845 (1986).

[7] K. Ploog, *J. Cryst. Growth* **79**, 887 (1986); *Proc. Workshop on Semiconductor Interfaces: Formation and Properties*, ed. G. Le Lay, J. Derrien, and N. Boccara, p10 (1987); *Surf. Interface Anal.* **12**, 279 (1988); in *Physics, Fabrication and Applications of Multilayered Structures*, Proceedings of NATO Advanced Study Institute, ed. P. Dhez and C. Weisbuch, p33 (1988).

[8] W. Frensley, *TI Tech. J.* **6**, 4 (1989).

[9] S. Bhobe, W. Porod, and S. Bandyopadhyay, *Solid-State Electron.* **32**, 1651 (1989); S. Bhobe, W. Porod, S. Bandyopadhyay, and D. J. Kirkner, in *Nanostructure Physics and Fabrication*, Proceedings of the International Symposium, ed. M. A. Reed and W. P. Kirk, p201 (1989); S. Bhobe, W. Porod, and S. Bandyopadhyay, *Phys. Status Solidi* A, **125**, 375 (1991).

[10] P. Helgesen, R. Sizmann, S. Lovold, and A. Paulsen, *SPIE Proc.* **1675**, 271 (1992).

[11] R. Q. Yang, *Phys. Rev.* B **52**, 11958 (1995); *Appl. Phys. Lett.* **66**, 959 (1995).

[12] L. R. Ram-Mohan, *Proceedings of the 8th Annual IEEE-LEOS Meeting, San Francisco, IEEE*, Vol. **2**, p17 (1995).

[13] I. Vurgaftman, J. R. Meyer, and L. R. Ram-Mohan, *IEEE J. Quantum Electron.* **32**, 1334 (1996).

[14] J. Meyer, I. Vurgaftman, R. Q. Yang, and L. R. Ram-Mohan, *Electron. Lett.* **32**, 45 (1996).

[15] L. R. Ram-Mohan and J. R. Meyer, *Proceedings of the NASA Semiconductor Device Modeling Workshop*, NASA Ames Research Center, ed. S. Saini, p147 (1996).

[16] I. Vurgaftman, J. R. Meyer, and L. R. Ram-Mohan, *IEEE J. Quantum Electron.* **32**, 1334 (1996).

[17] I. Vurgaftman, J. R. Meyer, and L. R. Ram-Mohan, *Photonics Technol. Lett.* **9**, 170 (1997).

[18] L. R. Ram-Mohan and J. R. Meyer, Conference on Experimental and Simulation Challenges in Nanostructured Materials, Baton Rouge, LA, February (1996).

[19] I. Vurgaftman, J. R. Meyer, and L. R. Ram-Mohan, *J. Quantum Electron.* **34**, 147 (1998).

[20] J. R. Meyer, J. I. Malin, I. Vurgaftman, C. A. Hoffman, and L. R. Ram-Mohan, in *Strained-Layer Quantum Wells and their Applications*, ed. M. O. Manasareh (Gordon and Breach, Newark, NJ, 1997), Chapter 6, pp235–272.

[21] J. R. Meyer, C. A. Hoffman, F. J. Bartoli, and L. R. Ram-Mohan, in *Novel Optical Materials and Applications*, ed. I. C. Khoo, F. Simone, and C. Umeton (Wiley, New York, 1996), pp205–237.

[22] I. Vurgaftman, J. Meyer, and L. R. Ram-Mohan, *Solid State Commun.* **100**, 663 (1996).

[23] C. Kittel, *Quantum Theory of Solids* (Wiley, New York, 1963).

[24] G. Bastard, *Phys. Rev.* B **24**, 5693 (1981).

[25] G. Bastard, *Phys. Rev.* B **25**, 7584 (1982).

[26] G. Bastard, *Wave Mechanics Applied to Semiconductor Heterostructures* (Les Editions de Physique, Les Ulis, 1988).

[27] L. R. Ram-Mohan, K. H. Yoo, and R. L. Aggarwal, *Phys. Rev.* B **38**, 6151 (1988).

[28] R. Goloskie, J. W. Kramer, and L. R. Ram-Mohan, *Comput. Phys.* **8**, 679 (1994).

[29] M. Gell-Mann and M. Levy, *Il Nuovo Cim.* **16**, 53 (1960).

[30] K. H. Yoo, R. L. Aggarwal, and L. R. Ram-Mohan, *J. Vac. Sci. Technol.* A **7**, 415 (1989).

[31] K. H. Yoo, L. R. Ram-Mohan, and D. F. Nelson, *Phys. Rev.* B **39**, 12808 (1989).

[32] B. Chen, M. Lazzouni, and L. R. Ram-Mohan, *Phys. Rev.* B **45**, 1204 (1992).

[33] K. Nakamura, A. Shimizu, M. Koshiba, and K. Hayata, *IEEE J. Quantum Electron.* **27**, 2035 (1991).

[34] R. H. Miles, D. H. Chow, J. N. Schulman, and T. C. McGill, *Appl. Phys. Lett.* **57**, 801 (1990).

[35] J. I. Malin, J. R. Meyer, C. L. Felix, C. A. Hoffman, L. Goldberg, F. J. Bartoli, C.-H. Lin, P. C. Chang, S. J. Murry, R. Q. Yang, and S. S. Pei, *Appl. Phys. Lett.* **68**, 2976 (1996).

[36] C. H. Grein, P. M. Young, and H. Ehrenreich, *J. Appl. Phys.* **76**, 1940 (1994).

[37] J. R. Meyer, C. A. Hoffman, F. J. Bartoli, and L. R. Ram-Mohan, *Appl. Phys. Lett.* **67**, 757 (1995).

[38] H. K. Choi, G. W. Turner, and M. J. Manfra, *Electron. Lett.* **32**, 1296 (1996).

[39] M. Aidaraliev, N. V. Zotova, S. A. Karandashev, B. A. Matveev, N. M. Stus, and G. N. Talalakin, *Fiz. Tekh. Poluprov.* **27**, 21 (1993) (*Sov. Phys. Semicond.* **27**, 10 (1993)).

[40] J. R. Lindle, J. R. Meyer, C. A. Hoffman, F. J. Bartoli, G. W. Turner, and H. K. Choi, *Appl. Phys. Lett.* **67**, 3153 (1995).

[41] T. C. Hasenberg, D. H. Chow, A. R. Kost, R. H. Miles, and L. West, *Electron. Lett.* **31**, 275 (1995).

[42] E. R. Youngdale, J. R. Meyer, C. A. Hoffman, F. J. Bartoli, C. H. Grein, P. M. Young, H. Ehrenreich, R. H. Miles, and D. H. Chow, *Appl. Phys. Lett.* **64**, 3160 (1994).

[43] C. L. Felix, W. W. Bewley, I. Vurgaftman, J. R. Meyer, D. Zhang, C.-H. Lin, R. Q. Yang, and S.-S. Pei, *IEEE Photonics Technol. Lett.* **9**, 1433 (1997).

[44] N. Holonyak Jr., R. M. Kolbas, R. D. Dupuis, and P. D. Dapkus, *IEEE J. Quant. Electron.* **16**, 170 (1980).

[45] W. T. Tsang, *Appl. Phys. Lett.* **39**, 786 (1981).

[46] J. R. Meyer, C. A. Hoffman, F. J. Bartoli, and L. R. Ram-Mohan, *Phys. Rev.* B **49**, 2197 (1994).

[47] J. I. Malin, C. L. Felix, J. R. Meyer, C. A. Hoffman, J. F. Pinto, C.-H. Lin, P. C. Chang, S. J. Murry, and S.-S. Pei, *Electron. Lett.* **32**, 1593 (1996).

[48] E. H. Aifer, W. W. Bewley, C. L. Felix, I. Vurgaftman, L. J. Olafsen, J. R. Meyer, H. Lee, R. U. Martinelli, J. C. Connolly, and A. R. Sugg, *Electron. Lett.* **34**, 1587 (1998).

[49] J. Faist, F. Capasso, D. L. Sivco, C. Sirtori, A. L. Hutchinson, and A. Y. Cho, *Science* **264**, 553 (1994).

[50] R. Q. Yang, *Superlattices Microstruct.* **17**, 77 (1995).

[51] I. Vurgaftman, J. R. Meyer, and L. R. Ram-Mohan, *IEEE Photonics Technol. Lett.* **9**, 170 (1997).

[52] I. Vurgaftman, J. R. Meyer, and L. R. Ram-Mohan, *IEEE J. Quantum Electron.* **34**, 147 (1998).

[53] J. Faist, F. Capasso, C. Sirtori, D. L. Sivco, J. N. Baillargeon, A. L. Hutchinson, S.-N. G. Chu, and A. Y. Cho, *Appl. Phys. Lett.* **68**, 3680 (1996).

[54] R. Q. Yang, B. H. Yang, D. Zhang, C.-H. Lin, S. J. Murry, H. Wu, and S. S. Pei, *Appl. Phys. Lett.* **71**, 2409 (1997).

[55] B. H. Yang, D. Zhang, R. Q. Yang, C.-H. Lin, S. J. Murry, and S. S. Pei, *Appl. Phys. Lett.* **72**, 2220 (1998).

[56] L. J. Olafsen, E. H. Aifer, I. Vurgaftman, W. W. Bewley, C. L. Felix, J. R. Meyer, D. Zhang, C.-H. Lin, and S. S. Pei, *Appl. Phys. Lett.* **72**, 2370 (1998).

[57] C. L. Felix, W. W. Bewley, I. Vurgaftman, J. R. Meyer, L. Goldberg, D. H. Chow, and E. Selvig, *Appl. Phys. Lett.* **71**, 3483 (1997).

[58] W. W. Bewley, C. L. Felix, I. Vurgaftman, E. H. Aifer, J. R. Meyer, L. Goldberg, D. H. Chow, and E. Selvig, *IEEE Photonics Technol. Lett.* **10**, 660 (1998).

Part III

2D Applications of the FEM

9

2D elements and shape functions

9.1 Introduction

In progressing from one-dimensional to two-dimensional problems we encounter a number of new issues in the formulation of the finite element method (FEM). The physical region has to be discretized into finite elements, which are now 2D elements, such that these elements must tessellate the physical region in its entirety. For example, if straight-edged elements are used along a curved boundary, the elements would not faithfully represent the physical regions along the boundary, and this could lead to incorrect boundary conditions (BCs) being implemented for the problem at hand. A large number of elements with straight edges would have to be invoked on the boundary to overcome this. A similar problem could arise when elements with curved sides are used in the interior since all adjacent elements must then have compatible sides with no gaps or overlapping coverages in the tessellation. We assume that such geometrical issues are taken into account adequately and focus on other aspects of the 2D FEM.

In this chapter, we derive the shape functions for a few typical 2D elements that are in general use. The basic idea is to obtain a polynomial representation for a function $f(x,y)$ over the area of any element using interpolation polynomials, given the values of the function at nodes in the element. A very large variety of shape functions for 2D elements have been considered in the application of the FEM to structural mechanics over the past several decades.

We consider rectangular and triangular elements and treat the case of elements with curved sides. We then describe the essentials of the evaluation of the discretized action and the application of BCs. The last section is devoted to Gauss quadrature over the 2D elements.

9.2 Rectangular elements

9.2.1 Lagrange elements

Any rectangular area on the x, y plane can be mapped, using linear mapping, into a standard square element with local variables (ξ, η), each ranging over the interval $[-1, 1]$. If the element ranges over (x_{min}, y_{min}) to (x_{max}, y_{max}), the mapping is given by the relations

$$x = \frac{x_{max} + x_{min}}{2} + \frac{x_{max} - x_{min}}{2}\xi,$$

$$y = \frac{y_{max} + y_{min}}{2} + \frac{y_{max} - y_{min}}{2}\eta. \qquad (9.1)$$

The element nodes are usually located at the corners, along the sides, and in the interior of the element. Let there be n nodes. If the coordinates of the nodes are labeled by (ξ_j, η_j), $j = 1, 2, \ldots, n$, the n shape functions must satisfy the conditions

$$N_i(\xi_j, \eta_j) = \delta_{ij},$$

$$\sum_i^n N_i(\xi, \eta) = 1, \qquad (9.2)$$

and the interpolated function is given by

$$f(\xi, \eta) = \sum_{i=1}^n f_i N_i(\xi, \eta). \qquad (9.3)$$

In Chapter 3, we had obtained the Lagrange interpolation polynomials for one variable. We can simply use products of these polynomials in ξ and η to construct the Lagrange interpolation polynomials for the rectangle. Linear elements are obtained as the product of linear Lagrange polynomials leading to

$$N_i(\xi, \eta) = a_i + b_i\xi + c_i\eta + d_i\xi\eta. \qquad (9.4)$$

The product of two linear Lagrange polynomials allows us to identify the coefficients. We have

$$N_1(\xi, \eta) = \frac{1}{4}(1 - \xi)(1 - \eta),$$

$$N_2(\xi, \eta) = \frac{1}{4}(1 + \xi)(1 - \eta), \qquad (9.5)$$

$$N_3(\xi, \eta) = \frac{1}{4}(1 - \xi)(1 + \eta),$$

$$N_4(\xi, \eta) = \frac{1}{4}(1 + \xi)(1 + \eta).$$

The linear element is shown in Fig. 9.1a. This polynomial represen-

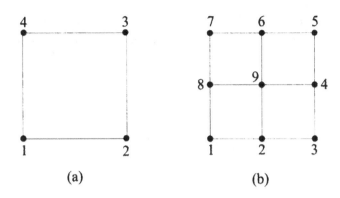

Fig. 9.1 (a) A four-noded rectangular linear element and (b) a nine-noded quadratic rectangular element.

tation does not include terms with ξ^2 and η^2. If all powers of ξ and η appearing above a horizontal line drawn across the Pascal triangle, shown in Table 9.1, are included in the polynomial representation, the polynomial is said to be complete.

A function represented over adjacent elements by such polynomials will have interelement continuity, but with possible discontinuous derivatives, the polynomials being C_0-continuous. Each node has one degree of freedom associated with it.

Quadratic elements correspond to the use of a product of quadratic Lagrange functions dependent on ξ and on η. As shown in Fig. 9.1b, The quadratic element has the four corner nodes, four mid-side nodes, and the node at the center. With higher order poly-

Table 9.1 The Pascal triangle for two variables is shown here through quintic variables.

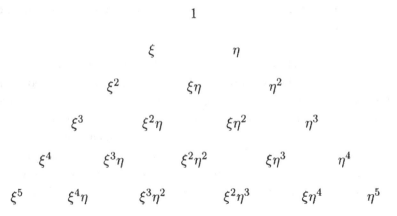

Table 9.2 The terms appearing in shape functions for serendipity elements with up to 16 nodes only on the periphery of the element.

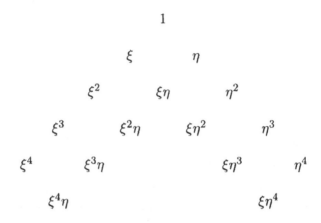

nomials the number of internal nodes increases rapidly, as can be expected with product functions. The Lagrange shape functions have more terms than are needed for completeness for a given degree.

Minimizing the number of internal nodes could have some advantages in finite element programming, and for this reason interpolation polynomials appropriate to such elements are tabulated in the literature. The elements with interior nodes require additional computation to determine their location during mesh generation. Also, the Lagrange shape functions are "overcomplete." These additional terms do not contribute towards faster convergence. For these reasons elements with nodes only on their periphery were invoked. The shape functions for such elements were discovered by chance and this class of rectangular elements are called serendipity elements. It is possible to eliminate the central node in a nine-noded quadratic element and convert it into an eight-noded element. This can be accomplished by suitably modifying the polynomials associated with the boundary nodes and increasing the order of the polynomials while dropping other terms in the expansion. The terms in the interpolation polynomials appearing in these shape functions are shown in Table 9.2. (The notion of exploiting the completeness of the polynomials is given up when we go beyond cubic shape functions.) The polynomial shape functions for the serendipity elements can be derived in

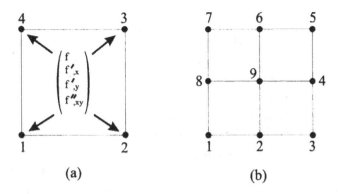

Fig. 9.2 Rectangular four-noded (a) and nine-noded (b) Hermite elements. At each node we have four degrees of freedom corresponding to the four coefficients for the function, its first derivatives along the two rectangular axes, and the cross-derivative at the node.

a systematic manner.[3,4]. The reader is directed to Zienkiewicz and Taylor,[1] Schwarz,[5] and Mitchell and Wait[6] for further details.

9.2.2 Hermite elements

The rectangular Hermite elements are constructed once again as the product of the 1D Hermite interpolation polynomials derived in Chapter 3. Consider a four-noded square element with ξ and η ranging over $[-1, 1]$. We construct products of the cubic Hermite interpolation polynomials in each of these variables defined in the 1D element, to obtain bicubic polynomials in ξ, η for the shape functions. We have 16 shape functions corresponding to the degrees of freedom associated with

$$f(\xi, \eta), \quad \frac{\partial}{\partial \xi} f(\xi, \eta), \quad \frac{\partial}{\partial \eta} f(\xi, \eta), \quad \frac{\partial^2}{\partial \xi \partial \eta} f(\xi, \eta),$$

at each node. The new feature here is the appearance of the cross-derivative degrees of freedom at each node. This is indicated in Fig. 9.2. The element has derivative, C_1-continuity across the edges.

The occurrence of the cross-derivatives requires some care in applying BCs at interfaces between two materials and also at the physical boundary in the global problem. Since Schrödinger's differential equation is of second order in the derivatives, physical considerations directly define the continuity of the wavefunction and the probability current. The cross-derivative terms are either allowed to "float," and be determined from the FEM analysis, or assigned continuity conditions consistent with the first-derivative continuity in the x, y-directions. Examples of the applications of such BCs are given in Chapters 11 and 12.

In one dimension, we have seen that the use of Hermite interpolation polynomials in the FEM leads to convergence to the solutions with fewer elements than with equivalent Lagrange functions. This is also true in two dimensions, and the use of bi-quintic Hermite interpolation polynomials leads to very accurate determination of eigenvalues for bound systems in two dimensions. The physical advantage of being able to directly specify either the function or its derivative as BCs in computations is also a consideration in favor of Hermite interpolation.

It is also possible to define rectangular elements with lower derivative interelement continuity. We can define an element with nodal degrees of freedom corresponding to just the function, $f(\xi, \eta)$, and its first derivatives, $\partial_\xi f$ and $\partial_\eta f$, being specified at the nodes.[7] The terms in the polynomials for a four-noded square element with just 12 degrees of freedom are complete through the cubic terms and include in addition two terms dependent on $\xi^3\eta$ and $\xi\eta^3$. The shape functions are thus incomplete fourth-order polynomials. The matrix inversion method described in Chapter 3 can be used to obtain the coefficients in the polynomials representing the shape functions associated with each of the four nodes.

With rectangular elements the Jacobian of transformation from (x, y) to (ξ, η) is easily obtained as the product of the 1D Jacobian factors. Also, the 2D action integral is readily computed using nested do-loops for the individual ξ and η Gauss quadratures. We can extend the 1D formulation to calculations in two dimensions within the FEM in a straightforward manner.

9.3 Triangular elements

Linear triangular elements were first considered by Courant[8] in the context of vibration analysis and computational mechanics.

Let us suppose that the triangular element has vertices located at coordinates (x_i, y_i), $i = 1, 2, 3$, in the global coordinate system. This triangle is linearly mapped into a standard right-angled triangle on the ξ, η plane. The usual labeling of the nodes is shown in Fig. 9.3. In linear triangular elements the nodal shape functions are given by

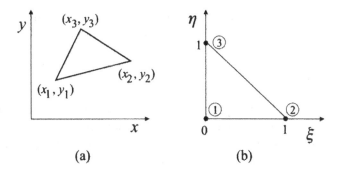

(a)   (b)

Fig. 9.3 The global triangle and its mapping into the standard triangle.
The circled numbers are the node numbers.

$$N_1(\xi, \eta) = 1 - \xi - \eta,$$
$$N_2(\xi, \eta) = \xi,$$
$$N_3(\xi, \eta) = \eta.$$

(9.6)

These shape functions are sometimes expressed in area coordinates
for a more symmetric representation.[1]

A general function $f(x, y)$, with given values f_i at the three vertices of the triangle, can be linearly interpolated inside the triangle
using

$$f(x, y) \simeq \sum_i^3 f_i N_i(\xi, \eta).$$

(9.7)

We can also represent the global coordinates x and y themselves in
terms of shape functions:

$$x = x_1 N_1(\xi, \eta) + x_2 N_2(\xi, \eta) + x_3 N_3(\xi, \eta),$$
$$y = y_1 N_1(\xi, \eta) + y_2 N_2(\xi, \eta) + y_3 N_3(\xi, \eta).$$

(9.8)

This is an example of isoparametric mapping, in which the coordinate transformation to the standard triangle is the same for the
coordinates as for the function being interpolated. Equation (9.8)
can be written in a 2×2 matrix form

$$\begin{pmatrix} x - x_1 \\ y - y_1 \end{pmatrix} = \begin{pmatrix} x_2 - x_1 & x_3 - x_1 \\ y_2 - y_1 & y_3 - y_1 \end{pmatrix} \begin{pmatrix} \xi \\ \eta \end{pmatrix},$$

$$(\mathbf{r} - \mathbf{r}_1) \equiv \mathbf{J} \cdot \boldsymbol{\rho}.$$

(9.9)

This is a linear mapping between the variables (ξ, η) and (x, y). The
determinant of the matrix $\mathbf{J}$ defined above in equation (9.9) is the

Jacobian of the transformation of the area element from $dx\,dy$ to $d\xi\,d\eta$. The Jacobian is twice the area of the global triangle

$$\mathbf{S} = \frac{1}{2}(\mathbf{r}_2 - \mathbf{r}_1) \times (\mathbf{r}_3 - \mathbf{r}_1).$$

The integrand in the quantum mechanical action will in general contain kinetic energy terms. We therefore determine the gradient of the wavefunction expressed in terms of the shape functions. Consider the partial derivative $\partial_x \psi(x, y)$. We have

$$
\begin{aligned}
\frac{\partial}{\partial x}\psi(x,y) &= \frac{\partial}{\partial x}\left(\sum_i \psi_i N_i(\xi, \eta)\right) \\
&= \sum_i \psi_i \left(\frac{\partial N_i}{\partial \xi}\frac{\partial \xi}{\partial x} + \frac{\partial N_i}{\partial \eta}\frac{\partial \eta}{\partial x}\right).
\end{aligned} \tag{9.10}
$$

The inverse of the relation

$$
\begin{pmatrix} dx \\ dy \end{pmatrix} = \begin{pmatrix} \dfrac{\partial x}{\partial \xi} & \dfrac{\partial x}{\partial \eta} \\ \dfrac{\partial y}{\partial \xi} & \dfrac{\partial y}{\partial \eta} \end{pmatrix} \begin{pmatrix} d\xi \\ d\eta \end{pmatrix}
$$

$$
= \mathbf{J} \begin{pmatrix} d\xi \\ d\eta \end{pmatrix} \tag{9.11}
$$

is given by

$$
\begin{pmatrix} d\xi \\ d\eta \end{pmatrix} = \frac{1}{2S} \begin{pmatrix} \dfrac{\partial y}{\partial \eta} & -\dfrac{\partial x}{\partial \eta} \\ -\dfrac{\partial y}{\partial \xi} & \dfrac{\partial x}{\partial \xi} \end{pmatrix} \begin{pmatrix} dx \\ dy \end{pmatrix}. \tag{9.12}
$$

We can therefore identify, for example, $\partial \xi/\partial x = (1/2S)\partial y/\partial \eta$. Proceeding in this manner we find the derivatives of the wavefunction:

$$
\frac{\partial}{\partial x}\psi(x,y) = \frac{1}{2S}\sum_i \psi_i \left(\frac{\partial N_i}{\partial \xi} \cdot \frac{\partial y}{\partial \eta} - \frac{\partial N_i}{\partial \eta} \cdot \frac{\partial y}{\partial \xi}\right)
$$

$$
\frac{\partial}{\partial y}\psi(x,y) = \frac{1}{2S}\sum_i \psi_i \left(-\frac{\partial N_i}{\partial \xi} \cdot \frac{\partial x}{\partial \eta} + \frac{\partial N_i}{\partial \eta} \cdot \frac{\partial x}{\partial \xi}\right). \tag{9.13}
$$

Since the coordinates (x, y) themselves are also expressed in terms of shape functions, we have a systematic way of obtaining the gradient

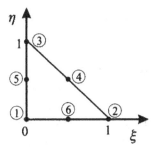

Fig. 9.4 The quadratic triangular element. The node numbers (shown in circles) conform to the usual usage.

of the wavefunction in terms of products of the derivatives of the shape functions,

$$\frac{\partial}{\partial x}\psi(x,y) = \frac{1}{2S}\sum_{ij}\psi_i y_j \left(\frac{\partial N_i}{\partial \xi}\cdot\frac{\partial N_j}{\partial \eta} - \frac{\partial N_i}{\partial \eta}\cdot\frac{\partial N_j}{\partial \xi} \right)$$

$$\frac{\partial}{\partial y}\psi(x,y) = \frac{1}{2S}\sum_{ij}\psi_i x_j \left(-\frac{\partial N_i}{\partial \xi}\cdot\frac{\partial N_j}{\partial \eta} + \frac{\partial N_i}{\partial \eta}\cdot\frac{\partial N_j}{\partial \xi} \right). \quad (9.14)$$

Quadratic triangular elements contain three additional nodes located at the midpoints of the three sides. The standard element is shown in Fig. 9.4, and the corresponding shape functions are

$$N_1(\xi,\eta) = (1 - \xi - \eta)(1 - 2\xi - 2\eta),$$
$$N_2(\xi,\eta) = \xi(2\xi - 1),$$
$$N_3(\xi,\eta) = \eta(2\eta - 1),$$
$$N_4(\xi,\eta) = 4\xi(1 - \xi - \eta), \quad\quad\quad (9.15)$$
$$N_5(\xi,\eta) = 4\xi\eta,$$
$$N_6(\xi,\eta) = 4\eta(1 - \xi - \eta).$$

These shape functions can be used to define triangles with curved sides through an isoparametric mapping. In this case the triangular edges correspond to quadratic functions. In general, the elements must be such that the transformation from the standard triangle to the actual element should not distort the edges, generating multiple crossings of curves representing the sides of the elements.

Triangular Hermite elements have been defined in the literature. These elements have additional derivative degrees of freedom defined at the nodes. Since the mapping from the global triangle with vertices at (x_i, y_i) to the standard right triangle will entail the transformation of the derivatives as well, we have to provide for all cross-derivatives to a given order. For example, if we allow the degree of freedom $\partial^2_{xy} f(x, y)$ at a node, we must also provide for the degrees of freedom corresponding to the values of $\partial^2_{xx} f(x, y)$ and $\partial^2_{yy} f(x, y)$ at the nodes. It is possible to define 21 complete quintic polynomials for shape functions associated with the six degrees of freedom at each vertex for

$$f(x, y), \quad \partial_x f, \quad \partial_y f, \quad \partial^2_{xx} f, \quad \partial^2_{xy} f, \quad \partial^2_{yy} f,$$

together with the normal derivatives $\partial_n f$ at nodes located at the midpoints of the sides. Such an interpolation scheme is C_1-continuous across the triangular network. Now the shape functions $N_i(\xi, \eta)$ are given by

$$N_i(\xi, \eta) = \sum_{p+q \leq 5} a_{i,pq} \, \xi^p \, \eta^q, \quad i = 1, \ldots, 21. \tag{9.16}$$

The degrees of freedom at the mid-side nodes, the normal derivatives at those points, can be eliminated without losing the derivative C_1-continuity by imposing constraints on the shape functions. These additional conditions correspond to equating the coefficients of $\xi^4 \eta$ and $\xi \eta^4$ and requiring that the coefficients of $\xi^5, \xi^3 \eta^2, \xi^2, \eta^3, \eta^5$ add up to zero:

$$a_{i,41} = a_{i,14},$$
$$5a_{i,50} + a_{i,32} + a_{i,23} + 5a_{i,05} = 0. \tag{9.17}$$

The normal derivative now has a cubic variation along the sides. We thus have six degrees of freedom located only at the three vertices. The C_1-continuous polynomials[6,2] are given in Table 9.3.

A final note about triangular elements concerns the use of hierarchical shape functions for triangular elements. This has been investigated by Webb and Aboucharcra,[9] who employ Jacobi polynomials for the development of the hierarchical shape functions. Besides the usual lower order shape functions associated with nodes, we have additional shape functions which vanish at the nodes, but provide additional degrees of freedom for the element in order to represent a function over the triangle. In two dimensions, we have (i) edge

Table 9.3 The shape functions for the Hermite interpolation polynomials for a triangle with 18 degrees of freedom. At the three nodal vertices the ordering of the shape functions corresponds to the degrees of freedom corresponding to $f(\xi,\eta)$, $\partial_\xi f$, $\partial_\eta f$, $\partial^2_{\xi\xi} f$, $\partial^2_{\xi\eta} f$, $\partial^2_{\eta\eta} f$. Here $\sigma = 1-\xi-\eta$.

Node	Shape function
Node 1	$N_1 = \sigma^2(10\sigma - 15\sigma^2 + 6\sigma^3 + 30\xi\eta(\xi+\eta))$ $N_2 = \xi\sigma^2(3 - 2\sigma - 3\xi^2 + 6\xi\eta)$ $N_3 = \eta\sigma^2(3 - 2\sigma - 3\eta^2 + 6\xi\eta)$ $N_4 = \xi^2\sigma^2(1 - \xi + 2\eta)/2$ $N_5 = \xi\eta\sigma^2$ $N_6 = \eta^2\sigma^2(1 + 2\xi - \eta)/2$
Node 2	$N_7 = \xi^2(10\xi - 15\xi^2 + 6\xi^3 + 15\eta^2\sigma)$ $N_8 = \xi^2(-8\xi + 14\xi^2 - 6\xi^3 - 15\eta^2\sigma)/2$ $N_9 = \xi^2\eta(6 - 4\xi - 3\eta - 3\eta^2 + 3\xi\eta)/2$ $N_{10} = \xi^2(2\xi(1-\xi)^2 + 5\eta^2\sigma)/4$ $N_{11} = \xi^2\eta(-2 + 2\xi + \eta + \eta^2 - \xi\eta)/2$ $N_{12} = \frac{1}{4}\xi^2\eta^2\sigma + \frac{1}{2}\xi^3\eta^2$
Node 3	$N_{13} = \eta^2(10\eta - 15\eta^2 + 6\eta^3 + 15\xi^2\sigma)$ $N_{14} = \eta^2\xi(6 - 4\eta - 3\xi - 3\xi^2 + 3\eta\xi)/2$ $N_{15} = \eta^2(-8\eta + 14\eta^2 - 6\eta^3 - 15\xi^2\sigma)/2$ $N_{16} = \frac{1}{4}\eta^2\xi^2\sigma + \frac{1}{2}\eta^3\xi^2$ $N_{17} = \eta^2\xi(-2 + 2\eta + \xi + \xi^2 - \eta\xi)/2$ $N_{18} = \eta^2(2\eta(1-\eta)^2 + 5\xi^2\sigma)/4$

functions in which the shape functions are zero at the three vertex nodes and have polynomial variation along a given edge, and also (ii) bubble functions which are again zero at the vertex nodes and are zero also along the edges. One way to think of these additional degrees of freedom is to view them as "Fourier-like" terms in the representation of a function in terms of these shape functions. As we know, Fourier coefficients are not tied to any nodal degrees of freedom.

9.4 Defining curved edges

We have already mentioned that by representing the global coordinates in terms of finite element shape functions we can obtain the isoparametric representation of curves and surfaces. This concept is commonly used in geometric modeling and computational geometry. A recent review, with historical comments, is given by Farin and Hamann.[10] We follow the presentation given by Mortenson,[11] Pratt,[12] and Farin.[13] We first develop the methodology for the piecewise representation of curves that is in general use, and employ this for the definition of planar elements.

9.4.1 An element on a parametric curve

Let us represent a segment of a curve using a cubic Hermite polynomial in u, where $0 \le u \le 1$ is a parameter. As in Fig. 9.5, let the points on the curve $\mathbf{r}(u)$ have the coordinates $x = x(u)$ and $y = y(u)$. The cubic parameterization is given by

$$\mathbf{r}(u) = \mathbf{a}_0 + \mathbf{a}_1 u + \mathbf{a}_2 u^2 + \mathbf{a}_3 u^3, \qquad (9.18)$$

where the coefficients $\mathbf{a}_i$ are vectors. For Hermite interpolation, we need the coordinates and the slopes of the curve of the end points. In matrix notation, we can write

$$\begin{pmatrix} \mathbf{r}(0) \\ d\mathbf{r}(0)/du \\ \mathbf{r}(1) \\ d\mathbf{r}(1)/du \end{pmatrix} = \begin{pmatrix} 1\,0\,0\,0 \\ 0\,1\,0\,0 \\ 1\,1\,1\,1 \\ 0\,1\,2\,3 \end{pmatrix} \cdot \begin{pmatrix} \mathbf{a}_0 \\ \mathbf{a}_1 \\ \mathbf{a}_2 \\ \mathbf{a}_3 \end{pmatrix}. \qquad (9.19)$$

Letting

$$\mathbf{Q} = (\mathbf{r}(0),\ d\mathbf{r}(0)/du,\ \mathbf{r}(1),\ d\mathbf{r}(1)/du)^T$$

and $\mathbf{A} = (\mathbf{a}_0, \mathbf{a}_1, \mathbf{a}_2, \mathbf{a}_3)^T$, we write the above equation in the form

$$\mathbf{Q} \equiv \mathbf{M}^{-1} \cdot \mathbf{A}. \qquad (9.20)$$

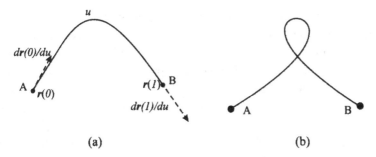

Fig. 9.5 (a) The parametric representation of a curve passing through the points A and B in terms of Hermite functions. (b) The appearance of a loop becomes possible as the tangent vectors are magnified.

Given the end coordinates of the curve and the tangents at those two points, we can solve for $\mathbf{a}_i$. We have

$$\mathbf{r}(u) = (2u^3 - 3u^2 + 1)\,\mathbf{r}(0) + (u^3 - 2u^2 + u)\frac{d}{du}\mathbf{r}(0)$$

$$+ (3u^2 - 2u^3)\mathbf{r}(1) + (u^3 - u^2)\frac{d}{du}\mathbf{r}(1)$$

$$\equiv F_1(u)\mathbf{r}(0) + F_2(u)\frac{d}{du}\mathbf{r}(0) + F_3(u)\mathbf{r}(1) + F_4(u)\frac{d}{du}\mathbf{r}(1)$$

$$= \mathbf{F}^T(u) \cdot \mathbf{Q}. \qquad (9.21)$$

Here the components of $\mathbf{Q}$ are themselves vectors. The polynomial coefficients $F_i(u)$ are Hermite interpolation polynomials defined over the interval $[0, 1]$. If we let

$$\mathbf{U}^T = (1,\ u,\ u^2,\ u^3), \qquad (9.22)$$

the vector of polynomials $\mathbf{F}$ is given by

$$\mathbf{F} = \mathbf{U}^T \cdot \mathbf{M}, \qquad (9.23)$$

and the coordinate vector $\mathbf{r}(u)$ is given by

$$\mathbf{r}(u) = \mathbf{U}^T \cdot \mathbf{M} \cdot \mathbf{Q}. \qquad (9.24)$$

If the magnitudes of the tangents $dr(0)/du$ and $dr(1)/du$ are increased steadily, the curve connecting $\mathbf{r}(0)$ and $\mathbf{r}(1)$ will eventually cross itself forming a loop. This is shown in Fig. 9.5b. Empirically, it has been determined that the loop does not form if the tangent vectors are less than three times the length of the chord AB.

The advantage of Hermite interpolation is that we will have a very smooth transition from one segment of a curve to another since we have the continuity of the function and its tangential derivative specified here. By using the Hermite parameterization it is possible to represent segments of straight lines, conic sections, and also arcs of circles subtending up to 30° to 45° at their centers with a high degree of accuracy.

From a practical point of view, it is more convenient to specify coordinates rather than derivatives in the input. In that case, the coordinates $\mathbf{r}_i$ are specified at the two end points, $u_1 = 0$ and $u_4 = 1$, and at two interior points, $u_2 = 1/3$ and $u_3 = 2/3$, for example. Here, in principle, u_2, u_3 have no restriction other than $0 < u_2 < u_3 < 1$. Using equation (9.21) we solve for the slopes at the two ends. This four-point interpolation does not provide interelement derivative continuity unless we explicitly constrain the slopes to be equal across the nodal boundary between elements.

The cubic Bezier[14] form of a segment of curve uses the so-called Bernstein polynomials. The curve is represented in the range $0 \leq u \leq 1$ by

$$\mathbf{r}(u) = (1-u)^3\mathbf{p}_0 + 3u(1-u)^2\mathbf{p}_1 + 3u^2(1-u)\mathbf{p}_2 + u^3\mathbf{p}_3. \quad (9.25)$$

Here the coefficients $\mathbf{p}_i$ are position vectors of points defining an "open polygon" such that the sides $(\mathbf{p}_1 - \mathbf{p}_0)$ and $(\mathbf{p}_3 - \mathbf{p}_2)$ are tangents to the parametric curve at $u = 0$ and $u = 1$, respectively. The points are specified by the relations

$$
\begin{aligned}
\mathbf{r}(0) &= \mathbf{p}_0, \\
d\mathbf{r}(0)/du &= 3(\mathbf{p}_1 - \mathbf{p}_0), \\
\mathbf{r}_1 &= \mathbf{p}_3, \\
d\mathbf{r}(1)/du &= 3(\mathbf{p}_3 - \mathbf{p}_2).
\end{aligned}
\quad (9.26)
$$

By shifting the so-called control points away from the curve itself we are able to define the form of the curve and specify the end point derivatives. In Fig. 9.6, the controlling vertices AP_1P_2B define the first portion of the curve, while the second and third portions are shown to have the curve segments crossing their open polygons. The tangents have been assigned directions in each segment consistent with the solid curve being drawn from A to D. When connected to

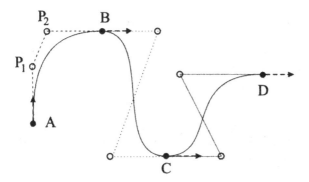

Fig. 9.6 Parametric representation of a curve using Bezier control points that lie away from the curve. Three segments are shown, demonstrating the derivative continuity of Bezier parameterization. Open circles are control points not on the curve. Points A,B,C,D are also control points.

the next segment, we have to ensure that the adjacent open polygons have a common point and also have collinear straight segments on either side of the common point.

Other parameterizations, employing B-spline interpolation, are also available[11–13] for the parametric representation of curves. With such functions we can represent element edges that are curved, and with a family of these functions we can construct the complete area elements with curved edges. These considerations can be very readily extended to curves defined in 3D space, once the control points are given three coordinates.

9.4.2   Parametric form of 2D surfaces

For a 2D curved rectangular element on a plane, we define two families of curves crossing each other. We again use Hermite interpolation functions for this parameterization. We specify four degrees of freedom at the four corner nodes of the curved "rectangular" element corresponding to

$$\mathbf{r}(u, v), \ \frac{\partial}{\partial u}\mathbf{r}(u, v), \ \frac{\partial}{\partial v}\mathbf{r}(u, v), \ \frac{\partial^2}{\partial u \partial v}\mathbf{r}(u, v),$$

at $(u, v) = (0, 0), (0, 1), (1, 0), (1, 1)$. Here the parameters u, v vary over $[0, 1]$. In effect, we are mapping the standard square element, of unit area, on the u, v plane into the curved rectangular element with global coordinates $\mathbf{r}(u = 0, v = 0)$, etc. In each of the "directions" u

or v, using equation (9.21), we can write

$$\mathbf{r}(u,0) = F_1(u)\mathbf{r}(0,0) + F_2(u)\mathbf{r}'_u(0,0) + F_3(u)\mathbf{r}(1,0) + F_4(u)\mathbf{r}'_u(1,0)$$
$$= \mathbf{F}^T(u) \cdot \mathbf{Q}(u) \qquad (9.27)$$

and

$$\mathbf{r}(0,v) = F_1(v)\mathbf{r}(0,0) + F_2(v)\mathbf{r}'_u(0,0) + F_3(v)\mathbf{r}(0,1) + F_4(v)\mathbf{r}'_v(0,1)$$
$$= \mathbf{F}^T(v) \cdot \mathbf{Q}(v). \qquad (9.28)$$

Here $\mathbf{Q}_u$ and $\mathbf{Q}_v$ are arrays with coordinate vectors and tangent vectors as elements. We construct the matrix of vectors

$$\mathcal{B} = \begin{pmatrix} \mathbf{r}(0,0) & \mathbf{r}'_v(0,0) & \mathbf{r}(0,1) & \mathbf{r}'_v(0,1) \\ \mathbf{r}'_u(0,0) & \mathbf{r}''_{uv}(0,0) & \mathbf{r}'_u(0,1) & \mathbf{r}''_{uv}(0,1) \\ \mathbf{r}(1,0) & \mathbf{r}'_v(1,0) & \mathbf{r}(1,1) & \mathbf{r}'_v(1,1) \\ \mathbf{r}'_u(1,0) & \mathbf{r}''_{uv}(1,0) & \mathbf{r}'_u(1,1) & \mathbf{r}''_{uv}(1,1) \end{pmatrix}. \qquad (9.29)$$

Now the interpolated coordinate vector in the curved element is given by

$$\mathbf{r}(u,v) = \mathbf{F}^T(u) \cdot \mathcal{B} \cdot \mathbf{F}(v). \qquad (9.30)$$

This representation for a rectangular element is preferable over a similar element using cubic Lagrange interpolation polynomials with 4×4 specified nodal coordinates. This is because we can guarantee interelement continuity with the Hermite polynomials.

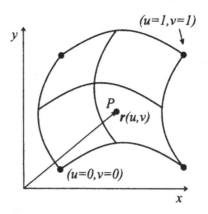

Fig. 9.7 Parametric representation of a rectangular element with curved sides. The coordinate interpolation is done with products of Hermite interpolation polynomials.

The 2D element with curved sides has been expressed in terms of two parameters u and v, and this is illustrated in Fig. 9.7. The coordinates x and y are also given in terms of these variables. We can therefore define the Jacobian of the transformation using the partial derivatives of x and y in terms of u and v so that the integration procedure used in the FEM can be mapped from the curved element to the standard square element in the u, v variables.

9.5 The action in 2D problems

It is useful to consider first the action integral for the Poisson problem. Suppose we have a distribution of charges with a surface charge density $\rho(x, y)$ in a region R confined by the boundary Γ. Let the potential function $\phi(x, y)$ be specified on the boundary Γ by the relation $\phi(s) = V_0(s)$. We have at hand a Dirichlet boundary value problem with an action integral given by

$$\mathcal{A} = \iint_R dx dy \left[\frac{1}{2} \nabla \phi(x, y) \cdot \nabla \phi(x, y) - \phi(x, y) \rho(x, y) \right]. \quad (9.31)$$

We can verify that the variation of the action with respect to the potential function ϕ leads to the Poisson equation, since

$$\delta_\phi \mathcal{A} = - \iint_R dx \, dy \, \delta\phi(x, y) \left(\nabla^2 \phi(x, y) + \rho(x, y) \right)$$

$$+ \oint_\Gamma ds \, \delta\phi(s) \frac{\partial}{\partial n} \phi(s). \quad (9.32)$$

Here an integration by parts leads to the "surface" integral along Γ of the normal derivative of ϕ at the boundary. Use has been made of Gauss's theorem in two dimensions. Since the boundary value of ϕ is fixed the surface term vanishes and we obtain

$$\nabla^2 \phi(x, y) = -\rho(x, y). \quad (9.33)$$

In the finite element approach we discretize the action, given in equation (9.31), directly by breaking up the region R into rectangular or triangular elements. A variation of the nodal values is invoked to implement the principle of least action (we are concerned here with the time-independent Schrödinger equation, so that the partial differential equation is of the elliptic kind).

In the case of Neumann, or more generally, mixed boundary value problems, we have the BCs

$$\frac{\partial \phi(s)}{\partial n} = h(s), \qquad \text{Neumann BC,} \qquad (9.34)$$

$$\frac{\partial \phi(s)}{\partial n} + g(s)\,\phi(s) = h(s), \qquad \text{mixed BC.} \qquad (9.35)$$

We start with the modified action integral

$$\mathcal{A} = \iint_R dx\,dy\left[\frac{1}{2}\nabla\phi(x,y)\cdot\nabla\phi(x,y) - \phi(x,y)\,\rho(x,y)\right]$$

$$+ \oint_\Gamma ds\left[\frac{1}{2}g(s)\,\phi^2(s) - h(s)\,\phi(s)\right], \qquad (9.36)$$

in which a line integral ("surface term") is added. We will see below that this term allows us to incorporate the BCs into the action integral.

We wish to find that function, ϕ_0, which minimizes the action. We use the standard method here of letting $\phi(x,y) = \phi_0(x,y) + \alpha\eta(x,y)$, where α is a parameter and η is an arbitrary function. We evaluate the minimum of the action

$$\mathcal{A}[\phi_0 + \alpha\eta] = \iint_R dx\,dy\left[\frac{1}{2}\nabla(\phi_0 + \alpha\eta)\cdot\nabla(\phi_0 + \alpha\eta) - (\phi_0 + \alpha\eta)\rho\right]$$

$$+ \oint_\Gamma ds\left(\frac{1}{2}g(s)[\phi_0(s) + \alpha\eta(s)]^2 - h(s)[\phi_0(s) + \alpha\eta(s)]\right), \qquad (9.37)$$

with respect to α. After an integration by parts, we have

$$\left.\frac{\delta\mathcal{A}}{\delta\alpha}\right|_{\alpha=0} = 0 = \iint_R dx\,dy\,\eta\left(-\nabla^2\phi_0 - \rho\right)$$

$$+ \oint_\Gamma \eta\left(g(s)\phi_0(s) - h(s) + \frac{\partial\phi_0}{\partial n}\right). \qquad (9.38)$$

Since η is arbitrary, we first consider those η that vanish on the boundary Γ. This leads to the condition

$$\nabla^2\phi_0(x,y) = -\rho(x,y), \qquad (9.39)$$

which is the Poisson equation. Next, we consider those η which do not vanish on the boundary. The condition for the minimum of the action, equation (9.38), together with the relation equation (9.39), leads to

$$\frac{\partial \phi_0(s)}{\partial n} + g(s)\, \phi_0(s) = h(s), \qquad (9.40)$$

which is the mixed BC. If the term proportional to $g(s)$ is absent, we have the Neumann BC. Finally, if both $h(s)$ and $g(s)$ are zero, we need add no additional line integral term to the action integral; now ϕ_0 is arbitrary on the boundary, while $\partial \phi_0 / \partial n$ vanishes there. This corresponds to what is termed the *natural* BC. The original action, equation (9.36), is discretized and evaluated directly using shape functions. Then the variation of the action with respect to the nodal coefficients is set to zero in order to obtain the minimized action. The BCs are inserted into the set of simultaneous equations that are given by the minimization. Terms arising from $h(s)$ in the action generate the nonhomogeneous terms, or the driving terms, which are transferred to the right side of the set of simultaneous equations before solving for the unknown nodal values of ϕ.

For bound state problems in quantum mechanics, with the wave-function falling off to zero at the boundary Γ of a physical region R, we begin with the action

$$\mathcal{A} = \iint_R dx\, dy \left(\frac{\hbar^2}{2m} \nabla \Psi^\dagger(x,y) \cdot \nabla \Psi(x,y) \right.$$
$$\left. + \Psi^\dagger(x,y)(V(x,y) - E)\Psi(x,y) \right). \qquad (9.41)$$

Here the wavefunction Ψ is continuous and has second derivatives in R. A variation of $\mathcal{A}$ with respect to the function $\Psi^\dagger$ leads to Schrödinger's equation, on noting again that the surface term, generated by the integration by parts, vanishes. It is therefore straightforward to discretize the action and integrate out the spatial dependence by employing interpolation polynomials. The resulting action is bilinear in the nodal variables $\Psi_\mu^\dagger$ and Ψ_μ. The variation of $\mathcal{A}$ with respect to $\Psi_\mu^\dagger$ gives the finite element matrix form of Schrödinger's equation. The final task is to include BCs at the periphery (*à la* benediction) and solve the resulting generalized eigenvalue problem.

We return to the question of BCs for 2D quantum scattering problems in Chapter 13.

Table 9.4 The Gauss coordinates and weights for the standard triangle are tabulated with the power p of the polynomial in ξ, η, which can be integrated exactly.[17] The numbers for $p = 2, 3, 4,$ and 5 are shown.

Order	ξ	η	Weight
		$p=2$	
3	0.666666666666667	0.166666666666667	0.333333333333333
	0.166666666666667	0.666666666666667	0.333333333333333
	0.166666666666667	0.166666666666667	0.333333333333333
		$p=3$	
4	0.333333333333333	0.333333333333333	-0.562500000000000
	0.600000000000000	0.200000000000000	0.520833333333333
	0.200000000000000	0.600000000000000	0.520833333333333
	0.200000000000000	0.200000000000000	0.520833333333333
		$p=4$	
6	0.108103018168070	0.445948490915965	0.223381589678011
	0.445948490915965	0.445948490915965	0.223381589678011
	0.445948490915965	0.108103018168070	0.223381589678011
	0.816847572980459	0.091576213509771	0.109951743655322
	0.091576213509771	0.816847572980459	0.109951743655322
	0.091576213509771	0.091576213509771	0.109951743655322
		$p=5$	
7	0.333333333333333	0.333333333333333	0.225000000000000
	0.059715871789770	0.470142064105115	0.132394152788506
	0.470142064105115	0.059715871789770	0.132394152788506
	0.470142064105115	0.470142064105115	0.132394152788506
	0.797426985353087	0.101286507323456	0.125939180544827
	0.101286507323456	0.797426985353087	0.125939180544827
	0.101286507323456	0.101286507323456	0.125939180544827

9.6   Gauss integration in two dimensions

The evaluation of the action integral over the area of an element is easily performed when the element is a (curved) rectangular element. The global element defined on the x, y plane is mapped into the standard square ranging over $[-1, 1]$ in the local variables (η, ξ).

A double sum over 1D Gauss sums is performed to obtain an approximate estimate of the integral. The 2D integral over a function $f(\xi, \eta)$ is given by

$$\iint_{-1}^{1} d\eta \, d\xi f(\xi, \eta) \simeq \sum_{i}^{N_g} \sum_{j}^{N_g} f(\xi_i, \eta_j) w_i w_j, \qquad (9.42)$$

where the sums extend over the N_g Gauss points and weights[15] $\{\xi_i, w_i\}$ and $\{\eta_j, w_j\}$, respectively. In the case of triangular elements the integration over the standard triangle entails evaluating the double integral

$$I = \int_0^1 d\eta \int_0^{1-\eta} d\xi \ f(\xi, \eta). \qquad (9.43)$$

This integral can be evaluated exactly if $f(\xi, \eta)$ is a polynomial in ξ and η.[16] For example,

$$I = \int_0^1 d\eta \int_0^{1-\eta} d\xi \ \xi^\alpha \eta^\beta (1 - \xi - \eta)^\gamma$$

$$= \frac{\alpha! \beta! \gamma!}{(\alpha + \beta + \gamma + 2)!}. \qquad (9.44)$$

For an arbitrary function, we resort to special Gauss weights and points for the triangle. It is preferable to use special weights and points designed to provide a single direct summation. The integral, equation (9.43), can then be written as a single sum

$$I \simeq \sum_{i}^{N_g} f(\xi_i, \eta_i) w_i, \qquad (9.45)$$

where ξ_i, η_i, and w_i are specific to the standard triangle. An extensive table of coordinates of Gauss points over the triangle and the corresponding weights is given by Dunavant.[17] The low order Gauss–Dunavant coordinates and weights are reproduced in Tables 9.4 and 9.5. A similar single-sum formula for integration over square regions is also given by Dunavant.[18].

With the shape functions and quadrature rules at hand, we investigate in the next chapter the procedures for breaking up a 2D region into elements through mesh generation.

Table 9.5 The Gauss coordinates and weights for the standard triangle are tabulated, with the power p of the polynomial in ξ, η, which can be integrated exactly.[17] The numbers for $p = 6$ and 7 are shown.

Order	ξ	η	Weight
		$p = 6$	
12	0.501426509658179	0.249286745170910	0.116786275726379
	0.249286745170910	0.501426509658179	0.116786275726379
	0.249286745170910	0.249286745170910	0.116786275726379
	0.873821971016996	0.063089014491502	0.050844906370207
	0.063089014491502	0.873821971016996	0.050844906370207
	0.063089014491502	0.063089014491502	0.050844906370207
	0.053145049844817	0.310352451033784	0.082851075618374
	0.053145049844817	0.636502499121399	0.082851075618374
	0.310352451033784	0.053145049844817	0.082851075618374
	0.310352451033784	0.636502499121399	0.082851075618374
	0.636502499121399	0.053145049844817	0.082851075618374
	0.636502499121399	0.310352451033784	0.082851075618374
		$p = 7$	
13	0.333333333333333	0.333333333333333	-0.149570044467682
	0.479308067841920	0.260345966079040	0.175615257433208
	0.260345966079040	0.479308067841920	0.175615257433208
	0.260345966079040	0.260345966079040	0.175615257433208
	0.869739794195568	0.065130102902216	0.053347235608838
	0.065130102902216	0.869739794195568	0.053347235608838
	0.065130102902216	0.065130102902216	0.053347235608838
	0.048690315425316	0.312865496004874	0.077113760890257
	0.048690315425316	0.638444188569810	0.077113760890257
	0.638444188569810	0.048690315425316	0.077113760890257
	0.638444188569810	0.312865496004874	0.077113760890257
	0.312865496004874	0.048690315425316	0.077113760890257
	0.312865496004874	0.638444188569810	0.077113760890257

References

[1] O. C. Zienkiewicz and R. L. Taylor, *The Finite Element Method*, Vol. 1, 4th ed. (McGraw-Hill, New York, 1994).

[2] G. Dhatt and G. Touzot, *The Finite Element Displayed* (Wiley,

New York, 1984).

[3] M. Zlamal, *Int. J. Num. Methods Eng.* **7**, 98 (1973).

[4] A. A. Ball, *Int. J. Num. Methods Eng.* **15**, 773 (1980).

[5] H. R. Schwarz, *Finite Element Methods* (Academic Press, New York, 1988).

[6] A. R. Mitchell and R. Wait, *The Finite Element Method in Partial Differential Equations* (Wiley, New York, 1977).

[7] L. Lapidus and G. F. Pinder, *Numerical Solution of Partial Differential Equations in Science and Engineering* (Wiley, New York, 1982).

[8] R. Courant, *Bull. Am. Math. Soc.* **49**, 1 (1943).

[9] J. P. Webb and R. Aboucharcra, *Int. J. Num. Methods Eng.* **38**, 245 (1995).

[10] G. Farin and B. Hamann, *J. Comput. Phys.* **138**, 1 (1997).

[11] M. E. Mortenson, *Geometric Modeling* (Wiley, New York, 1985).

[12] M. J. Pratt, in *The Mathematics of Surfaces*, ed. J. A. Gregory (Clarendon Press, Oxford, 1986).

[13] G. Farin, *Curves and Surfaces For Computer Aided Geometric Design*, 3rd ed. (Academic Pres, San Diego, 1993).

[14] P. Bezier, *Numerical Control: Mathematics and Applications* (Wiley, New York, 1972).

[15] A. H. Stroud and D. Secrest, *Gaussian Quadrature Formulas* (Prentice Hall, Englewood Cliffs, NJ, 1966).

[16] A. H. Stroud, *Approximate Calculation of Multiple Integrals* (Prentice Hall, Englewood Cliffs, NJ, 1971).

[17] D. A. Dunavant, *Int. J. Numer. Methods Eng.* **21**, 1129 (1985).

[18] D. A. Dunavant, *Int. J. Numer. Methods Eng.* **21**, 1777 (1985).

10
Mesh generation

10.1 Meshing simple regions

In the finite element approach to problems in greater than one dimension we are immediately faced with the complex issue of discretization of the physical domain. In this chapter we shall consider the essentials of 2D mesh generation. It is evident that for simple rectangular regions a straightforward breakup into rectangles or triangles can be performed without any difficulty. As seen in Fig. 10.1, the region is divided into rectangles which are then split up into a total of 12 triangular elements. We employ the vector right hand rule to define the direction of the area vector of each triangle to be out of the page. We label the three *local* node numbers $1, 2$, and 3 to correspond to this rule for each triangle. The *global* node numbers for a given triangle in the figure could have any value from 1 to the maximum number of nodes, and is determined by the mesh generator. While connecting the points in the desired manner to form the mesh, we define the connectivity relation between the elements and the global node numbers. For example, in Fig. 10.1, the triangular corner element with global node index 12 has the three global nodes $(12,10,11)$, in this order. We require two sets of data, one to specify the global coordinates (x_i, y_i) for each node i, and the other to specify the global node numbers for the nodes in each element, stated in the order consistent with the positive-area convention. This connectivity relation holds the key to placing the element matrices calculated in the finite element method (FEM) into a global matrix. The connectivity data should be generated and verified first, before any finite element calculation is performed.

When the nodes are connected together in a pattern and the connectivity is essentially independent of the location of the nodes, we have a *structured mesh*, as in Fig. 10.1. For physical regions with more complex boundaries we require the creation of what are called *unstructured meshes*. Any region with curved boundaries can be

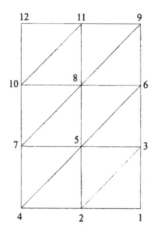

Fig. 10.1 A regular grid of triangles for a rectangular region.

meshed using either quadrilateral or triangular elements. Typically, we have to select the location of nodal points on the boundaries and in the interior. These points are connected to their neighbors in a way that varies from point to point leading to a lack of structure or pattern. Unstructured meshes have the advantage that they allow us to place a greater density of elements in regions where the solutions can be expected to vary rapidly. In this manner we can select a mesh such that it reflects the complexity in the solutions. In such unstructured meshes, setting up the connectivity relation between elements and the global numbering of their vertex nodes can be time consuming. As in structured meshes, the connectivity is again of primary importance. Automatic mesh generation algorithms have been devised that can use boundary nodes and generate a mesh in the interior with triangles that satisfy criteria suitable for finite element modeling. In the following, we consider various strategies for mesh generation.

10.1.1 Distortion of regular regions

We begin by meshing regions that are simple enough to be reproduced through a distortion of a regular region. Consider a circular region for meshing in polar coordinates. We begin with a rectangular grid in r and θ shown in Fig. 10.2. The circular symmetry is imposed by first overlaying the nodes from side IV on the corresponding nodes from side II. The nodes 1 through 7 and nodes 29 through 35 are now reduced to just one set. The operation is akin to "zippering" the two sides II and IV together. The connectivity rela-

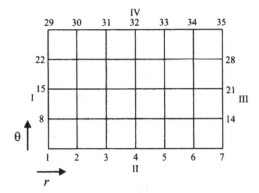

Fig. 10.2 The initial grid for a circular region. Sides I, II, III, and IV are labeled for reference in Figs. 10.3 and 10.4.

tion needs to be altered accordingly. The resulting annular region is shown in Fig. 10.3. The last step is to pull together nodes 1, 8, 15, and 22, so that the value of the angle at $r = 0$ is made indeterminate. This operation, which can be thought of as pulling the strings of a duffel-bag to close its top, is shown in Fig. 10.4. The resulting connectivity condition requires that the element matrices are suitably overlaid to form the final global matrix. One way to achieve this is to alter the connectivity conditions before calculating the global matrix so that they reflect the above node overlays. Alternatively, we can form the global matrix corresponding to the grid in Fig. 10.2. The overlay operation of Fig. 10.3 corresponds now to adding the rows with indices 29–35 to the corresponding rows with indices 1–7;

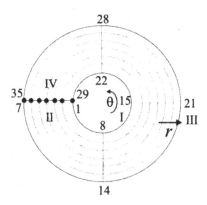

Fig. 10.3 The nodes along sides II and IV from Fig. 10.2 are overlaid so as to define an annular region, as an intermediate step towards gridding a circular region. Also see Figs. 10.2 and 10.4.

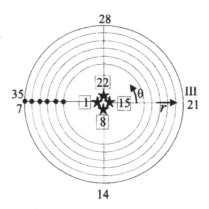

Fig. 10.4 The nodes on the innermost circle, indicated by stars, are "pulled together" and overlaid to form the final circular region.

the columns are treated similarly since the action integral is bilinear in the nodal variables for the wavefunction. The rows and columns 29–35 are then discarded. Next, rows 22, 15, 8 are added to row 1, and the corresponding columns added to column 1 of the global matrix. Now the rows 8, 15, 22, and the corresponding columns are removed from the matrix. This again achieves the desired global matrix for a circular region. In this procedure, the element matrices for elements with nodes 1, 8, 15, 22, and 29 are evaluated as if they are rectangular and Gauss quadrature is performed using a double loop over the 1D Gauss points and weights for integration over r and θ.

We have to exercise additional care if each node has derivative degrees of freedom. For Hermite interpolation, the degrees of freedom at each node correspond to

$$\left\{ f(r,\theta), \ \frac{\partial f}{\partial r}, \ \frac{\partial f}{\partial \theta}, \ \frac{\partial^2 f}{\partial r \partial \theta} \right\}. \tag{10.1}$$

The overlay of nodes along sides II and IV is straightforward since the derivative degrees of freedom should have the same values for any function periodic with angular period 2π. However, along the side $r = 0$, each of the derivative degrees of freedom has a directionality that has to be accounted for. This is illustrated in Fig. 10.5, where the first derivatives along the directions $\hat{r}$ and $\hat{\theta}$ are shown at a distance r from the origin for the sake of clarity. We must transform these derivatives into linear combinations of derivatives on the first

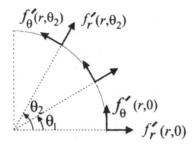

Fig. 10.5 The derivatives with respect to r, θ are shown at nodal points along a circular arc. In order to map the nodal derivative degrees of freedom at $\theta_i \neq 0$ onto the ones at $\theta = 0$ we employ a rotation, as discussed in the text.

node located at $(r = 0, \theta = 0)$. For a coordinate rotation given by

$$\begin{pmatrix} x \\ y \end{pmatrix} = \begin{pmatrix} \cos\theta_i & -\sin\theta_i \\ \sin\theta_i & \cos\theta_i \end{pmatrix} \cdot \begin{pmatrix} x' \\ y' \end{pmatrix}, \tag{10.2}$$

we have the relation

$$\begin{pmatrix} \dfrac{\partial}{\partial r_i} \\[2ex] \dfrac{\partial}{\partial \theta_i} \end{pmatrix} = \begin{pmatrix} \cos\theta_i & \sin\theta_i \\ -\sin\theta_i & \cos\theta_i \end{pmatrix} \cdot \begin{pmatrix} \dfrac{\partial}{\partial r_0} \\[2ex] \dfrac{\partial}{\partial \theta_0} \end{pmatrix}. \tag{10.3}$$

We therefore replace the partial derivatives at nodes at $\theta \neq 0$ in terms of the partial derivatives corresponding to node 1 at $\theta = 0$. This can be performed using matrix notation as follows. In order to be specific, consider the example provided by Fig. 10.2. If each degree of freedom for a given node appears in the order shown in equation (10.1), then the global matrix in this case is four times larger than the initial 35×35 global matrix for a single degree of freedom. The first row of nodes has four degrees of freedom associated with it. Let us suppose that the corresponding matrix elements occupy the first 28 rows and columns of the global matrix. Now, for node 8 we have row 29 to correspond to the function degree of freedom, and the $\partial_r f$ and $\partial_\theta f$ degrees of freedom to correspond to rows and columns 30 and 31 in the global matrix $\mathcal{M}$. We therefore transform these two rows using the transformation matrix appearing in equation (10.3). In the action integral, due to the bilinearity of the Lagrangian in the field variables, we have to apply the conjugate transformation to the columns as well. Once these transformations are performed (using temporary arrays to multiply into the old rows or columns and then restore the new rows or columns in the global matrix in the place of the old rows and columns) we can once again overlay these nodal variables onto the corresponding ones at $(r = 0, \theta = 0)$. This leads to a unique solution at the origin. A rotation of the second-order cross-derivative degree of freedom $\partial_{r\theta}^2$ would lead to terms corresponding to ∂_{rr}^2 and $\partial_{\theta\theta}^2$ also. This is undesirable since in 2D Hermite interpolation we have not allowed for these degrees of freedom. We therefore permit these variables to "float" and find their natural values without requiring additional constraints on them. The global matrices are now passed to the equation solver or the diagonalization subroutines, as the case may be.

Another example[1] of altering a regular grid to obtain a mesh for a region with a different symmetry is shown in Fig. 10.6. We wish to grid the quarter-circle shown in Fig. 10.6b using rectangular elements. We first construct a rectangle and split it into four regions. The entire rectangle is meshed with a regular grid of rectangular elements. (Figure 10.6a does not display this finer mesh.) Once the connectivity data is generated, we remove from it the elements of region IV of Fig. 10.6a. Next the coordinates of nodes along the line 5–6 and along the line 5–8 are made to coincide and their global node numbers are made the same. The last step is to smooth the sides 7–8 and 6–3 to conform to the arc of a circle by adjusting

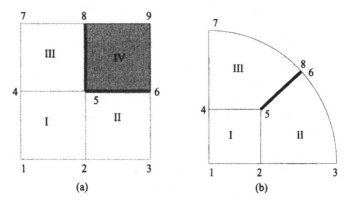

Fig. 10.6 Region IV of a regular grid shown in (a) is removed, and the nodes along the lines 5–6 and 5–8 are overlaid. This is shown in (b), leading to a mesh for a pie-shaped region. (From Ref. 1).

the boundary nodal coordinates and by using, say, isoparametric quadratic elements. We see that, with some ingenuity, it is possible to mesh fairly complex areas in two dimensions by initially starting with a regular grid.[2]

10.1.2   Using orthogonal curved coordinates

A physical way of generating orthogonal curvilinear coordinates to define finite elements is to use the theory of complex variables and perform a conformal mapping.[3,4] A numerical way of achieving a similar result is to specify all internal and external boundaries to be equipotential curves, and to solve Laplace's equation to obtain the potential function and the electrostatic field lines. This leads to gradients of equipotential contours and the potential contours which are locally orthogonal curves. The natural crowding together of field lines at surfaces with small radii of curvature would insert more elements there in a natural manner. The numerical work can be performed using the method of relaxation and finite differences, or by an application of the boundary element method. A finite element approach using linear elements with simple or approximate meshing could be used to initiate the grid generation process, followed by more refinements using higher order interpolation functions.

10.2   Regions of arbitrary shape

A number of algorithms exist for the automatic meshing of an arbitrary 2D region with a minimal amount of input for defining the shape of the region.

10.2.1   Delaunay meshing

A methodology for the breakup of a region into triangles was given by Dirichlet in 1850. A set of points distributed throughout the region is defined. With each point we construct a convex polygon by using the bisectors of the lines joining the point with all its neighbors. This is analogous to the construction of the Wigner–Seitz cell in configuration space or the Brillouin zone in reciprocal space that is used in solid state physics.[5] This Dirichlet tessellation leads to filling the region with convex polygons called Voronoi regions.[6] The edges of the polygons bisect the line joining two points, each being inside the Voronoi polygon on either side of the edge. By connecting only such points we end up with a triangulation of the region that is called

Delaunay triangulation.[7] Not all the triangles generated by the Delaunay triangulation will be optimal triangles from the point of mesh generation and we attempt to improve the mesh by either inserting or removing some of the points. Bowyer[8] and also Watson[9] devised an algorithm that is widely used in this context. They require that each triangle not have any other point within its circumcircle. Any such triangle is removed from the grid. Bowyer and Watson show that if a new point which violates this rule is inserted, all triangles for which this point is within their circumcircles are contiguous and their removal creates a polygonal void around the point. Now the new point is used to reconstruct triangles by connecting the new point to the neighboring vertices. This local reallocation of the triangles allows one to sequentially consider the insertion of additional points. One starts with an initial grid of conveniently placed points. Using linked lists in C programming it is possible to determine the neighboring triangles to be removed for each additional inserted point. At its best performance, this algorithm requires $\mathcal{O}(N \ln N)$ operations for obtaining an acceptable mesh. The reader is referred to the original papers quoted above and to Rebay's article[10] for further details and for approaches to improve the efficiency of the algorithm. Other approaches are discussed in Cheung *et al.*[11]

10.2.2   Advancing front algorithms

In the advancing front approach,[12,13] the boundaries are discretized into linear elements. The outermost boundary and the internal material boundaries are considered to be the initial location of the ad-

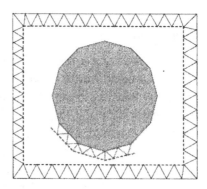

Fig. 10.7 The advancing front approach to meshing a physical region. The boundary and the internal material boundaries are incorporated into the mesh right from the beginning.

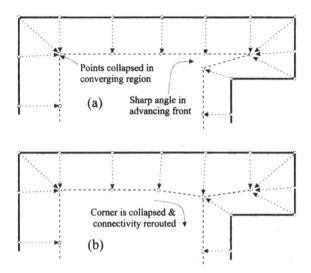

Fig. 10.8 The advancing front approach of Johnston and Sullivan using a normal offsetting method. Nodes leading to sharp angles or the cross-over and overlap of two advancing fronts are collapsed as in the figure. (After Ref. 13.)

vancing front which moves inwards systematically. First, a segment of the front is chosen and a point is inserted further into the interior so that an optimal (equilateral) triangle is formed with the linear element on the front as the base of the triangle. Nearby points that are inserted in this manner are now connected so as to form inverted triangles. Figure 10.7 illustrates the procedure. This process is continued until the advancing fronts cross and pinch off isolated regions from the rest of the unmeshed regions. Mavriplis uses the Delaunay triangulation with the Bowyer–Watson[8,9] algorithm to control the insertion of new points to the interior of the front. Johnston and Sullivan[13] use an offset vector that is normal to the advancing front in order to insert additional points. Again, selection criteria are used in order to move nodes, or remove them, when two adjacent offset segments cross. If the new front contains sharp angles, the nearby nodes are coalesced into one node in order to smooth out the advancing front. These aspects are illustrated in Fig. 10.8. This also minimizes the crossing of the offset segments used in the next cycle of input points. A mesh density function is defined in order to increase or decrease the density of point insertions and thereby generate a nonuniform mesh.

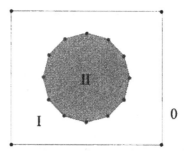

Fig. 10.9 The physical region is defined using linear boundary elements with specification of the few nodal coordinates, their connectivity, and the material index on either side of the linear elements. This information is supplied in the input file via a graphical user interface.

10.2.3 The algebraic integer method

We now consider in detail an efficient method for the generation of adaptive triangular meshes for arbitrarily shaped regions with multiple material regions.[14] The boundaries of the material regions are discretized using linear boundary elements. The coordinates of the boundary nodes are entered in an input file. These are treated as specialty nodes and are not displaced in the following algorithm. The nodes can be numbered in any order, and, correspondingly, we provide the boundary element connectivity and the material index for the materials occurring on either side of each boundary element. For example, in Fig. 10.9, there are two material regions labeled I and II, with the exterior labeled as material 0. Four boundary elements at the periphery of the square have materials 0 and I on either side of them. The materials I and II are separated by the 12 boundary elements defining the circular region. As we will see below, it does not matter which side is labeled as material 0, I, or II about the linear elements in the input file.

The algebraic integer method (AIM)[†] begins by considering a region slightly larger than the rectangle defined by (x_{min}, y_{min}) and (x_{max}, y_{max}). A flowchart of the steps in the algorithm is displayed in Fig. 10.10. This region is filled with regular hexagons split into equilateral triangles. The grid points in each row are offset with respect to the previous row, as shown in Fig. 10.11a. The size of the triangles is determined by the input file as some fraction of the average length of the boundary elements. This procedure already

---

[†]This nomenclature is used by J. Sullivan.

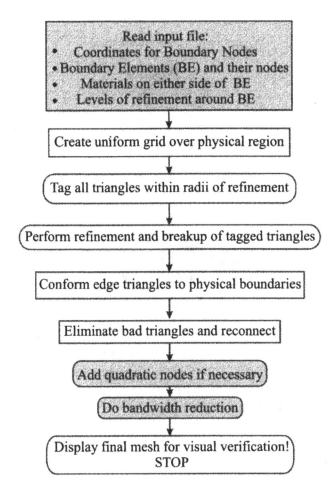

Fig. 10.10 A flowchart of the programming for the algebraic integer method of generating adaptive meshing. Adding mid-side nodes for quadratic elements and using bandwidth reduction immediately after mesh generation are optional steps.

ensures that the physical region will be filled mostly by equilateral triangles.

A smaller mesh size may be desired in certain regions. We specify the zone of refinement to be about particular boundary elements with the size of the zone determined by a radius vector, as shown in Fig. 10.11b. Any triangle, from the initial meshing, which has its centroid within the zone of refinement is tagged and then split up into four smaller triangles by connecting the midpoints of the sides of the triangle. Since the element tagging precedes the breakup of

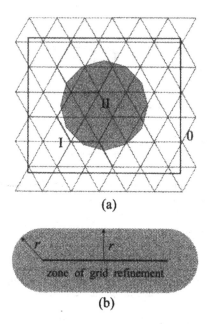

(a)

(b)

Fig. 10.11 In (a), the physical region is overlaid with a grid of equilateral triangles covering a region somewhat larger than the rectangle defined by (x_{min}, y_{min}) and (x_{max}, y_{max}). Part (b) of the figure illustrates the zone of refinement specified about each linear boundary element or any additional linear element specified by the input file.

the element, elements common to more than one zone of refinement are split up into four elements only once for each level of refinement. Multiple levels of refinement can be set up, each with its own radius of refinement about particular boundary elements. Additional regions can be refined by providing special linear elements of any length (including just a point). In this manner we have most of the mesh set up with equilateral triangles of various sizes. This procedure requires the determination of whether a particular triangle falls within a zone of refinement or not. Here, a fast algorithm is used to test the distance of the centroid of the triangle from a linear boundary element.[15]

The breakup of a given triangle into four smaller triangles leads to nodes being generated at the midpoints of the sides of the triangle. In order to obtain a conforming mesh in which adjacent triangles fully share the common side, we break up the neighboring triangles as well. This is shown in Fig. 10.12. Neighboring triangles are classified as having one, two, or three additional mid-side nodes due to the re-

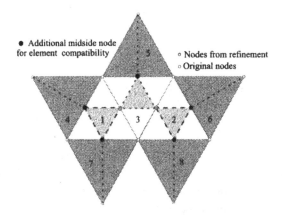

Fig. 10.12 The breakup of the triangles during the refinement process is shown. The gray circles refer to the nodes generated by the refinement and the black circle is included for element compatibility. The triangles with just one mid-side node are split into two, thereby limiting the range of refinement.

finements. A triangle with three additional mid-side nodes generated by the refinement is tagged for breakup into four smaller triangles. Triangles with two mid-side nodes are also split into four by adding a third mid-side node. Next, those triangles with just one additional mid-side node are made conforming by joining the mid-side node to the opposite vertex. These situations are shown in Fig. 10.12. Triangles labeled 1 and 3 were refined first. This led to triangle 2 having two mid-side nodes. We then introduce the third node (black circle) in triangle 2 and divide it into four triangles. Now triangles 4–8 all have just one mid-side node and these are split into two triangles.

The next step is to ensure that there are no elements outside the physical region, and that the triangles lying across boundary elements have their nodes shifted onto the boundary. The procedure is to loop over all the boundary elements and to test whether sides of triangles intersect them. This is done using the two line-segment intersection algorithm.[16,17] All intersecting triangles have some of their vertices shifted onto the nearest boundary such that no edge of any triangle crosses a physical boundary. This ensures that the mesh is faithful to the original boundaries. Some of the triangles near the boundaries will now be distorted. A test is performed to break up obtuse-angled triangles into two, and to eliminate those triangles that have very small angles with one side much smaller than the others.

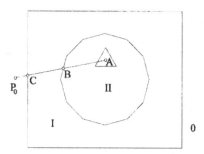

Fig. 10.13 The material assigned to a typical triangle is determined by the number of material boundaries crossed by a line joining the centroid A of the triangle to an external point P_0.

Now a smoothing procedure is applied and small shifts are given to nodes so that almost all triangles are nearly equilateral. This is done by Laplacian smoothing, as described by Hermann.[18] Each interior node is shifted to the position corresponding to the average of the x- and y-coordinates of the nearest nodes. Within a few passes through the subroutine the mesh converges to its optimal form.

We have not yet associated a material index with each triangle. This is done by connecting the centroid of a given triangle with a point P_0 outside the physical region, as in Fig. 10.13. Using the line-segment crossing algorithm, we note that the line AP_0 crosses the boundary of the material region I twice and the boundary of region II just once. This allows us to tag the triangle as being within region II. In other words, if the line from a triangle inside any material region to the exterior region crosses a material boundary only once, then the triangle lies inside that particular material.[‡] This procedure is carried out for every triangle in the physical region so that the finite element calculations can be performed with the appropriate material properties for each triangle.

At this stage, the mesh is passed through a bandwidth reduction program so that the node numbers of the various triangles generate a global matrix that is tightly banded about its diagonal. This concerns only a rearrangement and an exchange of the nodal indices. All the indices are arranged within each element to correspond to the anticyclic numbering such that the area vector for each triangle is directed outward in a manner consistent with the positive area right hand rule.

[‡]This is a simple extension of the well-known Jordan curve theorem.

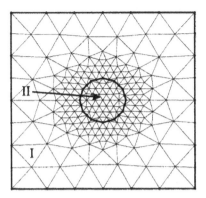

Fig. 10.14 The final mesh with two levels of refinement around the perimeter of the circular region.

Examples of mesh generation are given in Figs. 10.14 and 10.15. In Fig. 10.15, after the initial hexagonal gridding there were 81 nodes and 128 elements. After the first round of refinement there were 243 nodes and 420 elements. After the second round there were 381 nodes and 696 elements. After the third level of refinement there were 595 nodes and 1124 elements. After eliminating bad elements we had 553 nodes and 1044 elements. The bandwidth reduction program reduced the half bandwidth from 509 to 59. The radii of refinement were 0.23, 0.25, and 0.1 units.

With recent improvements in the CPU speed of computers and substantially larger computer memory available today, the finite element aspects of the computation are fairly rapid. The real bottleneck for large-scale computations arises in the process of mesh generation.

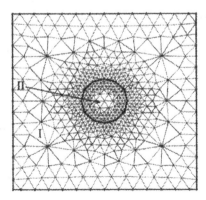

Fig. 10.15 The final mesh with two levels of refinement around the perimeter of the circular region.

For 3D problems, with solution requirements of, say, 5–10 hours on a workstation, the process of automatic mesh generation for general shapes could take many days. For this reason, investigations of different and novel algorithms for mesh generation continue to be pursued vigorously.

References

[1] T. R. Chandrupatla and A. D. Belegundu, *Introduction to Finite Elements in Engineering* (Prentice Hall, Englewood Cliffs, NJ, 1991).

[2] O. C. Zienkiewicz and R. L. Taylor, *The Finite Element Method* (McGraw-Hill, New York, 1994).

[3] J. F. Thompson, Z. U. A. Warsi, and C. W. Mastin, *Numerical Grid Generation* (North-Holland, New York, 1985).

[4] I. S. Kang and L. G. Leal, *J. Comput. Phys.* **102**, 78 (1992).

[5] C. Kittel, *Introduction to Solid State Physics*, 6th ed. (Wiley, New York, 1986).

[6] G. Voronoi, *J. Angew. Math.* **134**, 167 (1908).

[7] B. Delaunay, *Izv. Akad. Nauk SSR Otdelenie Matemat. Estestvennyka Nauk* **7**, 793 (1934).

[8] A. Bowyer, *Comput. J.* **24**, 162 (1981).

[9] D. F. Watson, *Comput. J.* **24**, 167 (1981).

[10] S. Rebay, *J. Comput. Phys.* **106**, 125 (1993).

[11] S. H. Lo, *Int. J. Numer. Methods Eng.* **21**, 1403 (1985). Also see: Y. K. Cheung, S. H. Lo, and A. Y. T. Leung, *Finite Element Implementation* (Blackwell Science, Oxford, 1996).

[12] D. J. Mavriplis, *J. Comput. Phys.* **117**, 90 (1995).

[13] B. P. Johnston and J. M. Sullivan, *Int. J. Numer. Methods Eng.* **33**, 425 (1992).

[14] J. M. Sullivan, in *CAD/CAM Robotics and Factories of the Future*, ed. Birendra Prasad, Vol. I (Springer-Verlag, Berlin, 1988). The author wishes to thank Dr. Sullivan for numerous discussions on mesh generation, for constructive comments on this chapter, and for generously providing his computer codes for mesh generation.

[15] *Graphic Gems I*, ed. A. S. Glassner (Academic Press, San Diego, 1990), p49.

[16] *Graphic Gems III*, ed. D. Kirk (Academic Press, San Diego, 1992), p199.

[17] M. E. Mortenson, *Geometric Modeling* (Wiley, New York, 1985).

[18] L. R. Hermann, *J. Eng. Mech. Div. ASCE* **102**, 749 (1976).

11
Applications in atomic physics

11.1 The H atom in a magnetic field

An externally applied perturbation can provide information about the structure of energy levels in a quantum mechanical system. The investigation of the effect of an external magnetic field on the energy levels of the hydrogen atom, the simplest of atoms, has fascinated each generation of theoretical physicists in the last century. There is a long history in the literature of attempts at solving this problem by perturbative and variational methods.[1-8] The problem is of such interest because of its relevance to astronomy and astrophysics, atomic physics, solid state spectroscopy, and the more recent exploration of quantum chaos exhibited by hydrogenic states with large Rydberg number n.

The spherical symmetry of the Coulomb potential is broken by the presence of the external magnetic field. As a consequence, the Hamiltonian is no longer separable in spherical or cylindrical coordinates. Using variational and perturbative methods, numerically accurate solutions have been obtained in the two domains corresponding to very low or very high magnetic fields. In the former, the magnetic field, being small, can be treated as a perturbation, while for high magnetic fields the Coulomb potential can be treated as a perturbation on the energy levels of the electron moving in a magnetic field. Unfortunately, most of these methods fail when the Coulomb energy and magnetic energy are comparable, i.e., in the transition from a hydrogen-like wavefunction to a Gaussian-type wavefunction in the plane perpendicular to the field.[5,7] For weak fields, the wavefunction is similar to that for hydrogen in the absence of a field. In the strong field limit, the wavefunction behaves like a Gaussian in the plane perpendicular to the magnetic field; parallel to the field the wavefunction is still basically "hydrogen-like." Hence this problem displays features of both the simple harmonic oscillator and the hydrogen atom with no external field.

It is shown below that very accurate results[9-11] are obtainable by the finite element method (FEM) in the intermediate regime in the applied field, where the Coulomb energy is comparable with the cyclotron energy of the electron in the magnetic field. Energies and wavefunctions for the $1s_0$, $2s_0$, $2p_0$, and $2p_{-1}$ states in fields from 10^5 to 10^{12} G are determined with an accuracy of 1 part in 10^6. Using only 100 elements and without excessively fine-tuning the grid for each calculation, the FEM gives highly accurate lower bounds for the binding energy of the excited states. This is a direct consequence of using local interpolation functions rather than a global basis set of hydrogenic states or Gaussian states. Since this calculation was reported in 1990, adaptive FEMs and also semianalytic methods, using a series solution,[12] have extended the accuracy of the eigenvalues to 1 part in 10^{12}. We draw attention to the early work of Kaschiev[13] who had independently investigated the same problem a decade earlier using the FEM.

11.1.1 Schrödinger's equation and the action

Schrödinger's equation for the hydrogenic electron in a magnetic field is obtained from the equation without the field by using the minimal gauge substitution $\mathbf{p} \rightarrow \mathbf{p} + |e|\mathbf{A}/c$ in the Hamiltonian. Appendix C contains a brief treatment of gauge invariance. Let the z-axis be parallel to the magnetic field $\mathbf{B}$, and choose the symmetric gauge $\mathbf{A} = (-By/2,\ Bx/2,\ 0)$. With only cylindrical symmetry surviving the application of the magnetic field, the acceptable wavefunctions in cylindrical coordinates ρ, ϕ, and z are of the form

$$\Psi(\rho, \phi, z) = \sum_m c_m \mathcal{F}_m(\rho, z)\, e^{im\phi}. \tag{11.1}$$

Schrödinger's equation for the wavefunction $\mathcal{F}$ is given by

$$-\frac{\hbar^2}{2m_e}\left(\frac{\partial^2}{\partial\rho^2} + \frac{1}{\rho}\frac{\partial}{\partial\rho} + \frac{\partial^2}{\partial z^2} - \frac{m^2}{\rho^2}\right)\mathcal{F}(\rho, z) - \frac{e^2}{\sqrt{\rho^2 + z^2}}\,\mathcal{F}(\rho, z)$$

$$+ \frac{|e|B}{2m_e c}(xp_y - yp_x)\,\mathcal{F}(\rho, z) + \frac{e^2 B^2}{8m_e c^2}\,\rho^2\,\mathcal{F}(\rho, z)$$

$$\pm \mu_B B\,\mathcal{F}(\rho, z) \;=\; E\,\mathcal{F}(\rho, z), \tag{11.2}$$

where m is the magnetic quantum number, B is the magnetic field strength, and μ_B is the Bohr magneton. In the following, only the positive sign is considered in the dipole interaction term between the electron's dipole moment and the external field. This corresponds to the electron spin being oriented antiparallel to the magnetic field. The coordinates are expressed in units of the Bohr radius a_0, the energy in units of the Rydberg R_y, and the energy due to the magnetic field in units of $\hbar\omega_B/2$, where $\omega_B = |e|B/m_e c$ is the cyclotron frequency of the electron in the magnetic field. The corresponding action integral is given by

$$\mathcal{A} = \int dz\,\rho d\rho \left[\frac{\partial \mathcal{F}^*}{\partial \rho} \frac{\partial \mathcal{F}}{\partial \rho} + \frac{\partial \mathcal{F}^*}{\partial z} \frac{\partial \mathcal{F}}{\partial z} + \mathcal{F}^* \left(\frac{m^2}{\rho^2} - \frac{2}{\sqrt{\rho^2 + z^2}} \right) \mathcal{F} \right.$$
$$\left. + \mathcal{F}^* \left(\gamma m + \gamma + \frac{\gamma^2 \rho^2}{4} - \epsilon \right) \mathcal{F} \right]. \quad (11.3)$$

Here ϵ is the energy in Rydbergs. The ratio γ of the unit of magnetic energy to the Rydberg is given by

$$\gamma = \frac{\hbar\omega_B}{2R_y} = \frac{a_0^2}{\mathcal{R}_0^2}. \quad (11.4)$$

The parameter γ is also seen to be the square of the ratio of the Bohr radius a_0 to the Landau radius $\mathcal{R}_0$ of the cyclotron orbit. In these units, $\gamma = 1$ corresponds to a magnetic field of $B = 2.35 \times 10^9$ G. The unknown wavefunction $\mathcal{F}(\rho, z)$ is required to vanish as $\rho \to \infty$ and $\pm z \to \infty$.

11.1.2   Applying the FEM

The action integral is discretized into 2D elements in the ρ,z plane. The infinite range of the action integral in ρ and in z is truncated by choosing cutoff values ρ_c and $\pm z_c$ where it is appropriate to set the wavefunction and its derivatives equal to zero. We apply the criterion that the actual wavefunction should be about six orders of magnitude smaller near the cutoff values than at its maximum value.

Within each 2D element we assign nine nodes: one at each of the corners, at midpoints of the sides, and at the center of the element. Hermite interpolation appropriate to nine nodes per element and four degrees of freedom per node is used to represent the wavefunction within each element. These polynomial interpolates have the special property that the 36 expansion coefficients are the values of $\mathcal{F}_i$, $\partial \mathcal{F}_i/\partial \rho$, $\partial \mathcal{F}_i/\partial z$, and $\partial^2 \mathcal{F}_i/\partial \rho \partial z$ at each of the nine nodes i in the element. The 2D interpolation functions are simply the products of

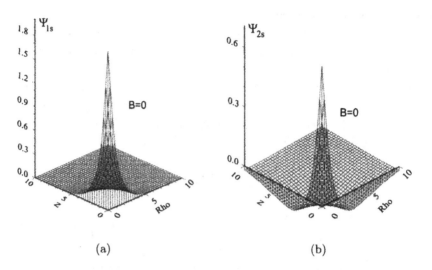

(a)   (b)

Fig. 11.1 The normalized s-state wavefunctions of the hydrogen atom at zero magnetic field: (a) the $1s_0$ state, and (b) the $2s_0$ state.

quintic 1D Hermite interpolation polynomials in ρ and in z. We require that $\mathcal{F}$, $\partial\mathcal{F}/\partial\rho$, $\partial\mathcal{F}/\partial z$, and $\partial^2\mathcal{F}/\partial\rho\partial z$ be continuous across the boundaries of the elements. A local coordinate system (ξ, η) which has the range $[-1, 1]$ in both directions is introduced for each element iel. In terms of local coordinates the nodes are located at the points where ξ and η have the value 0 or ± 1. If the size of the element in global coordinates is $h_\rho^{iel} \times h_z^{iel}$, it is simple to show that the local coordinates ξ and η are related to their respective global coordinates ρ and z by $\rho = \rho_0^{(iel)} + h_\rho^{(iel)}(1 + \xi)/2$ and $z = z_0^{(iel)} + h_z^{(iel)}(1 + \eta)/2$, respectively. Here $\rho_0^{(iel)}$ and $z_0^{(iel)}$ are the global coordinates at the corner $\xi = -1$, $\eta = -1$ of element iel. In the z-direction, parallel to the field, the wavefunction with principal quantum number n decays asymptotically as $z^{n-1} e^{(-|z|/n)}$ in the presence of a weak field. The effect of a strong magnetic field ($\gamma > 1$) is subtle and depends on the quantum numbers and z-parity of the state. Rösner[5] has shown that the tightly bound states are compressed in the z-direction, while some states with higher quantum numbers are slightly elongated in the z-direction. In the finite element calculation, the value of z_c was adjusted to accommodate the compression and elongation of the wavefunction where appropriate. The asymptotic behavior of the wavefunction in the ρ-direction is well represented by the Gaussian

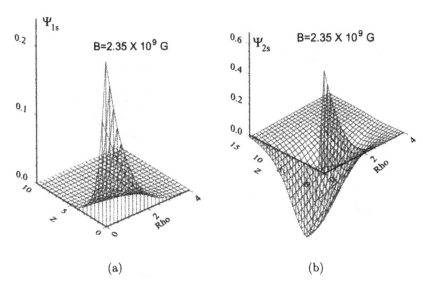

Fig. 11.2 The normalized s-states of the hydrogen atom in a magnetic field of $B = B_0$ ($B_0 = 2.35 \times 10^9\,\mathrm{G}$) for (a) the $1s_0$-state and (b) the $2s_0$-state.

form $\exp(-\gamma\rho^2/4)$ in fields $\gamma > 1$. Hence the cutoff value ρ_c was adjusted to reflect this behavior with increasing magnetic field. Since the $1s_0$ wavefunction bunches up sharply at the origin with increasing B, as seen in Figs. 11.2 and 11.3, we allocate more elements for the region $0 < \rho < 4a_0$. The p-states are found to be extremely insensitive to the location of the nodes. Values for the discretization of the grid are given in Table 11.1. Identical grids were used for all calculations in low magnetic fields.

For fields $\gamma > 1$, the wavefunction in the ρ-direction is approximately Gaussian; the grid was discretized into even intervals along ρ for both s- and p-states. In the z-direction, the p-states are slightly compressed towards the origin, and less extended in space. As z_c was decreased, additional nodes were placed at small values of z, retaining the overall number of elements. For the s-states, the situation is slightly more complicated. The $1s_0$-state is strongly compressed in the z-direction, and rapidly falls off to zero. Although the $2s_0$-state is also compressed near the origin, the exponential tail of the wavefunction is actually elongated slightly. Hence it is difficult to construct a grid with 100 elements that has an adequate number of nodes near $z=0$ and a sufficiently large value of z_c. This problem becomes more pronounced as the field strength increases. As a result, the value ob-

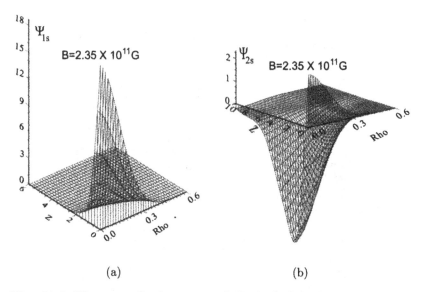

Fig. 11.3 The normalized s-states of the hydrogen atom in a magnetic field of $B = 100\,B_0$ ($B_0 = 2.35 \times 10^9\,\text{G}$) for (a) the $1s_0$-state and (b) the $2s_0$-state.

tained for the binding energy of the $2s_0$-state at $\gamma = 1000$ is slightly less accurate than the other results. Calculations were first carried out (in cylindrical coordinates) for the case $\gamma = B = 0$ to determine the accuracy of the finite element results. The integrations over the element area in the discretized action integral, equation (11.3), can be done exactly with the exception of the Coulomb potential term; all integrations were done numerically using 16-point Gauss quadrature, which is exact for polynomials up to degree 31. This may appear to be excessive, since all of the integrals with the exception of the Coulomb term are polynomials of degree 13 or less. However, the binding energy is extremely sensitive to the error in evaluating the Coulomb term, particularly near the origin. Using eight-point Gauss quadrature, for example, may result in an error in the final energy as large as 1 part in 10^5 for the $1s_0$-state, but significantly less for the excited states. Since numerical error in integration can raise or lower the energy, it is critical that this numerical integration error does not lead to false "rigorous bounds." At $B = 0$, it was verified that the error in the energy due to numerical integration of the Coulomb term with 16-point Gauss quadrature is less than 1 part in 10^6 for the $1s_0$-state and even less for the excited states.

Table 11.1 Finite element discretization: coordinates of nodes.

		$\gamma=0$,			**0.0001,**	**0.001,**	**0.01,**	**0.1**				
s	ρ,z	0	0.25	0.5	1	2	4	8	12	16	20	24
p	ρ,z	0	1	2	4	8	12	16	20	24	28	32
						$\gamma=1$						
s	ρ	0	0.5	1	1.5	2	2.5	3	4	5	6	7
s	z	0	0.25	0.5	1	2	4	8	12	16	20	24
p	z	0	1	2	4	8	12	16	20	24	28	32
						$\gamma=10$						
s	ρ	0	0.25	0.5	0.75	1.0	1.25	1.5	1.75	2	2.25	2.5
s	z	0	0.25	0.5	1	2	4	8	12	16	20	24
p	z	0	0.5	1	2	4	8	12	16	20	24	28
						$\gamma=100$						
s	ρ	0	0.07	0.14	0.21	0.28	0.35	0.42	0.49	0.56	0.63	0.7
s	z	0	0.125	0.25	0.5	1	2	4	8	12	18	24
p	z	0	0.25	0.5	1	2	4	8	12	16	20	24
						$\gamma=1000$						
s	ρ	0	0.025	0.05	0.075	0.1	0.125	0.15	0.175	0.2	0.225	0.25
s	z	0	0.0625	0.125	0.25	0.5	1	2	4	8	16	24
p	z	0	0.125	0.25	0.5	1	2	4	8	12	16	20

Table 11.2 Binding energy of the $1s_0$-state.

γ	Liu and Starace lower bound upper bound	Rösner et al. lower bound upper bound	Finite element lower bound
0.0			0.499 999
0.0001			0.500 049
0.001		0.500 500	0.500 499
0.01		0.504 975	0.504 974
0.1	0.5474	0.547 526	0.547 525
1.0	0.8167	0.831 169	0.831 168
	0.8418		
10		1.747 797	1.747 79
100	3.7360	3.789 1	3.789 78
	3.8027	3.790 3	
1000		7.662 1	7.662 36
		7.662 7	

Table 11.3 Binding energy of the $2s_0$-state.

γ	Liu and Starace lower bound upper bound	Rösner et al. lower bound upper bound	Finite element lower bound
0.0			0.125 000
0.0001			0.125 050
0.001		0.125 496 5	0.125 496
0.01		0.129 651 6	0.125 651
0.1	0.0873	0.148 089 2	0.148 089
	0.1503		
1.0	0.1357	0.160 468 9	0.160 469
	0.1594		
10		0.208 89	0.208 951
		0.208 99	
100	0.2555	0.256 170	0.256 179
	0.2565	0.256 189	
1000		0.295 855	0.295 85
		0.295 859	

Table 11.4 Binding energy of the $2p_0$-state.

γ	Liu and Starace lower bound upper bound	Rösner et al. lower bound upper bound	Finite element lower bound
0			0.125 000 0
0.0001		-	0.125 050 0
0.001		0.125 498 5	0.125 498 5
0.01		0.129 850 4	0.129 850 4
0.1		0.162 410 4	0.162 410 0
1	0.2577 0.2622	0.260 006 6	0.260 006 5
10		0.382 648 7 0.382 651 8	0.382 649 8
100	0.4637 0.4638	0.463 617 7	0.463 617 7
1000		0.492 495 0	0.492 495 0

Table 11.5 Binding energy of the $2p_{-1}$-state.

γ	Liu and Starace lower bound upper bound	Rösner et al. lower bound upper bound	Finite element lower bound
0.0			0.125 000 0
0.0001			0.125 100 0
0.001		0.125 997 0	0.125 997 0
0.01		0.134 701 2	0.134 701 1
0.1	0.1882 0.2013	0.200 845 7	0.200 845 6
1	0.4524 0.4595	0.456 597 1	0.456 596 9
10		1.125 422	1.125 422
100		2.634 74 2.634 80	2.634 758
1000		5.638 41 5.638 44	5.638 416

The local elements generate 36×36 matrices, whose matrix elements are overlaid to assemble the global matrices. This is done keeping in mind that the wavefunction $\mathcal{F}$ and its derivatives are continuous at the element boundaries.

The action integral equation (11.3) takes the form

$$\mathcal{A} = \mathcal{F}_{\mu}^{*} \, \mathcal{H}_{\mu\nu} \, \mathcal{F}_{\nu} - \epsilon \mathcal{F}_{\mu}^{*} \, \mathcal{U}_{\mu\nu} \, \mathcal{F}_{\nu}. \qquad (11.5)$$

The global indices have been indicated by μ, ν, and the global matrix $\mathcal{U}$, also called the overlap matrix, is identified as the coefficient of the energy ϵ. The expansion polynomials are nonorthogonal in each element and hence the overlap matrix is only block diagonal, and not the unit matrix. The variational principle requires $\delta \mathcal{A}/\delta \mathcal{F}_{\mu}^{*} = 0$, leading to

$$\mathcal{H}_{\mu\nu} \, \mathcal{F}_{\nu} - \epsilon \mathcal{U}_{\mu\nu} \, \mathcal{F}_{\nu} = 0. \qquad (11.6)$$

The global matrices $\mathcal{H}$ and $\mathcal{U}$ are symmetric and banded.

11.1.3 Magnetic fields

Equation (11.6) has the form of a standard generalized eigenvalue problem, where the components of the vector $\mathcal{F}_{\mu}$ are the values of the wavefunction and its derivatives at all the nodes in the grid. In order to satisfy the boundary conditions (BCs), components of the vector $\mathcal{F}_{\mu}$ which correspond to $r = r_c$ or $z = z_c$ are set equal to zero. This guarantees that the wavefunction and its derivatives vanish asymptotically.

Because the wavefunctions are known to be of even or odd parity with respect to z, the range $[-z_c, z_c]$ can be replaced by the half-range $[0, z_c]$, with the appropriate BCs at $z = 0$. Although two calculations, one for even and another for odd z-parity, must now be performed for each value of B and the orbital quantum number m, this is still more efficient than using the full range $[-z_c, z_c]$ and obtaining both even and odd states simultaneously. When we are working on the half-range $[0, z_c]$, BCs must also be imposed at $z = 0$, depending on the z-parity of the state. For the odd z-parity state $2p_0$, the wavefunction (and $\partial \mathcal{F}/\partial \rho$) must vanish at $z = 0$; for the even z-parity states, the derivative $\partial \mathcal{F}/\partial z$ (and $\partial^2 \mathcal{F}/\partial z \partial \rho$) are set equal to zero. These BCs are quite trivial to implement; if the nodal variable $\mathcal{F}_{\mu_0}$ is set to zero, then the corresponding μ_0th row and column are eliminated from the global matrices $\mathcal{H}$ and $\mathcal{U}$. These constraints result in a further reduction of the size of the global matrices, without destroying symmetry or bandedness.

The reduced global matrices can now be solved by standard algorithms. Subspace iteration,[14] an algorithm developed for finite element calculations, is particularly efficient, exploiting both the symmetry and banded nature of the matrices. It is more efficient than vector iteration, because it does not require orthogonalization of the eigenvectors until the approximate subspace has converged to the subspace spanned by the lowest eigenvectors.[†]

Equation (11.6) was solved for the four low-lying states: (i) $m = 0$, even z-parity: $1s$, $2s$; (ii) $m = 0$, odd z-parity: $2p_0$; and (iii) $m = -1$, even z-parity: $2p_{-1}$. The range of γ considered is $\gamma = 0 - 1000$, corresponding to zero field to $10^5 - 10^{12}$ G. The energies reported in Tables 11.2–11.5 are rigorous lower bounds to the binding energy, to within the last digit reported in the tables. The tables compare the results obtained using finite elements to the calculation of Liu and Starace[7] and Rösner et al.[5] The results of Larsen,[15] Santos and Brandi,[16] and Baldereschi and Bassani[3] are not included in the tables because they are significantly less accurate than the recent works and cover only a small range of magnetic fields. Nevertheless, they are noteworthy calculations because they mark the first attempts to calculate rigorous bounds for the excited states. The results clearly show that the lower bounds obtained via the FEM are far superior to those obtained using a conventional variational approach. (Note that the upper bound obtained by Liu and Starace[7] for the $2p_0$-state at $\gamma = 100$ is incorrect.) Although the modified Hartree–Fock approach of Rösner et al.[5] does not guarantee rigorous bounds to the true energy, their values agree extremely well with the finite element results for fields up to $\gamma = 1$. For higher fields, the finite element results lie between the upper and lower limits of the Hartree–Fock energy.[5] It is apparent that the results of the Rösner group are accurate as they have claimed.

The wavefunctions for the $1s_0$- and $2s_0$-states for $\gamma = 0$, $\gamma = 1$, and $\gamma = 100$ are shown in Figs. 11.1, 11.2, and 11.3, respectively. It is interesting to note that the wavefunctions at high fields, shown in Fig. 11.3, have nonzero slopes at $\rho = 0$ in the ρ-direction. This is due to the fact that the Coulomb potential energy term dominates over the magnetic terms in the action integral, equation (11.3), very close to the origin, even at high magnetic fields. This leads to the deviation from a Gaussian shape for small radial distances. The scales along

---

[†]Methods for diagonalizing large matrices are treated in Chapter 15.

the various axes should be noted in order to appreciate the changes in the wavefunction of the electron produced by the applied magnetic field at these representative values of the B-field.

It is of interest to emphasize some of the special features of the FEM for this problem. It may be noted that while the FEM too is a variational method it does not require making accurate guesses at the form of the trial variational solution. We have shown that the use of piecewise interpolation functions for solving Schrödinger's equation is clearly superior to the standard variational approach when the wavefunction cannot be readily approximated by simple global basis functions. In other words, the FEM succeeds in cases where standard variational and perturbative calculations fail. Schrödinger's equation for the hydrogen atom in an external magnetic field is a nonseparable partial differential equation; however, we have employed locally separable interpolation polynomials which are products of quintic Hermite interpolation polynomials in ρ and z. By limiting the product form to each element we are able to substantially improve on the usual global variational functions. The integration over the finite elements smoothes out the numerical differentiation errors which are expected to occur with the finite difference method. The singularities in the potential $V(r) \sim 1/r^{\alpha}$, for $0 < \alpha \leq 2$, do not present any special difficulty for the FEM since $r^2 r^{-\alpha} dr$ is bounded, whereas such singularities are problematic within the finite difference computational schemes for the solution of differential equations. The FEM leads to banded matrices and involves integrations over finite regions, unlike the Rayleigh–Ritz procedures which involve fully occupied matrices requiring $n(n+1)/2$ integrations for the matrix elements. The banded matrices generated in the FEM require less computer memory and also less computational time for obtaining the eigenvalues and eigenfunctions. This also introduces less round-off errors in the results since far fewer numerical operations are performed in the algorithm for determining eigenvalues. The global matrices $\mathcal{H}$ and $\mathcal{U}$ for 10×10 elements have dimensions of 1764. Such large matrices require sparse matrix diagonalization techniques. We used the subspace iteration method[14] to obtain the lowest few eigenvalues. A re-evaluation on an AlphaStation 500 running at 400 MHz required elapsed times of 9–25 s, depending on the number of elements and whether the symmetry in z is exploited or not.[‡] This suggests that the FEM can

---

[‡]Computer run-times are given merely to indicate approximate times. Clearly,

be extended today to larger global matrix sizes, and to include adaptive gridding for obtaining superior estimates for eigenvalues without excessive computational times on a computer workstation.

A related problem is to calculate the energies of a 2D hydrogen atom in the presence of an external magnetic field perpendicular to the 2D plane. A realization of this system occurs as shallow donors in deep quantum wells made up of semiconductor heterostructures. Shinada and Tanaka[17] investigated the 2D hydrogen atom in an external magnetic field and obtained the ground state energy of the electron by integrating the radial equation. MacDonald and Ritchie[18] have obtained analytical expressions for the low-lying energy levels for interpolation between the high field and the low field regimes. It would be of interest to apply the FEM to this problem in order to determine the eigenvalues more accurately and to provide interpolation formulas as functions of the applied magnetic field.

11.2   Ground state energy in helium

The calculation of the ground state energy of the two-electron helium atom represents a theoretical challenge which has been addressed by variational and numerical approaches over the years.[19-22] No exact analytical solution is possible in this case. The aim of all global variational schemes is to provide a happy choice of the global wavefunction which will minimize the Hamiltonian. In all bound state problems the kinetic and potential energy contributions come with opposite signs so that large errors in both could cancel out, leaving an acceptable value for the energy. However, the same wavefunctions could yield large errors in the expectation values of other operators such as $\langle r^2 \rangle$, $\langle (1/r) \rangle$, $\langle 1/r_{12} \rangle$, etc. For this reason it is advantageous to turn to a local basis set that can provide improved results for both the ground state energy and the non-Hamiltonian expectation values. The first application of the FEM to He was done in the classic paper by Levin and Shertzer.[23] We follow the details given in this article.

Schrödinger's equation for the two-electron atom is given by

$$\left[-\frac{1}{2}\nabla_1^2 - \frac{1}{2}\nabla_2^2 - \frac{2}{r_1} - \frac{2}{r_2} + \frac{1}{r_{12}} - E \right] \psi(\mathbf{r}_1, \mathbf{r}_2) = 0. \qquad (11.7)$$

with the faster machines of today, these times can be improved substantially. Also, more recent algorithms for diagonalization of sparse matrices can provide faster results.

Here, the interelectron separation is given by

$$r_{12} = (r_1^2 + r_2^2 - 2r_1 r_2 \cos \theta_{12})^{1/2}. \tag{11.8}$$

The Hamiltonian involves six coordinate variables for the two electrons and is nonseparable. It has been shown by Breit[20] that Schrödinger's equation is separable if the coordinate system is chosen such that the z-axis is aligned with one of the radius vectors. In this coordinate system the wavefunction is expressed as a product of two functions

$$\psi(r_1, r_2) = \sum_i f_i(r_1, r_2, \theta_{12}) u_i(\phi, \theta', \phi'), \tag{11.9}$$

where u_i satisfies the equation for a symmetric top. For the ground state the u_i are constant, and f_i satisfies the equation

$$\begin{aligned} \bigg[&-\frac{1}{2r_1^2} \frac{\partial}{\partial r_1} \left(r_1^2 \frac{\partial f}{\partial r_1} \right) - \frac{1}{2r_2^2} \frac{\partial}{\partial r_2} \left(r_2^2 \frac{\partial f}{\partial r_2} \right) \\ &- \frac{1}{2 \sin^2 \theta_{12}} \left(\frac{1}{r_1^2} + \frac{1}{r_1^2} \right) \frac{\partial}{\partial \theta_{12}} \left(\sin \theta_{12} \frac{\partial f}{\partial \theta_{12}} \right) \\ &+ \left(\frac{1}{r_{12}} - \frac{2}{r_1} - \frac{2}{r_2} - E \right) f \bigg] = 0. \end{aligned} \tag{11.10}$$

This equation was solved using the symmetry under the interchange of the two electrons. Working in the three coordinates r_1, r_2, and $\cos \theta_{12}$, Levin and Shertzer used the FEM with $9 \times 9 \times 4$ brick elements with cubic Hermite interpolation in each direction. The resulting eigenvalue problem, involving global matrices of size ~ 2000, was solved using the subspace iteration method. The ground state energy obtained in this calculation was -2.9032 a.u. The solution was then used to obtain the expectation values of $\langle r^n \rangle$, for $n = -2$, -1, 1, and 2. This calculation showed how the FEM could be used to locally improve the wavefunctions and obtain better results for the expectation values for all operators. Shertzer and Levin have revisited the problem to include mass polarization.[24]

Since 1985, when the above work was reported, there have been significant advances in the application of the FEM. In particular, Ackermann and Roitzsch[25] have developed an adaptive finite element approach§ in which the 3D discretization is carried out by inserting additional elements where they are needed. The physical domain

---

§For a simple introduction to issues in adaptive FEM, see Chapter 4.

is divided using a coarse grid of tetrahedra. Initially, linear interpolation functions are used in each element in order to obtain an
approximate solution. This approximate solution is used to determine which of the tetrahedra need refinement. The degree p of the
polynomials used for the interpolation of the solution is increased
systematically from 1 to 5 (p-refinement), in order to reduce the error in the approximation. If the error is still larger than a set value
for the tolerance, the tetrahedron is subdivided (h-refinement) into
more tetrahedra. This procedure leads to a new grid, and the new
solution for the global problem is obtained by the inverse vector iteration method. This is used to further refine the results. This hybrid
hp-adaptive method of refining the mesh and improving the results
has led to remarkable results.[26] The dimension of the sparse matrices in the final generalized eigenvalue problem is about 130 000. The
ground state energy of He is obtained as $-2.903\,724\,377\,021$ a.u.

The FEM has also been used for calculations on the higher excited
states in He. Braun et al.[27] have reported results for the 1^1S to the
4^1S state energies; matrix elements of $1/r$, r, r^2, and $\pi\delta(r)$ have also
been calculated.

Clearly, the FEM has given us a flexible and highly accurate
variational procedure for obtaining results of high accuracy.

11.3   Other results

A very cursory review of other results based on the FEM in atomic
physics and in solid state electronic bandstructure is presented here.

- Ackermann[26] has calculated the ground state energy of the H^-
  ion using the FEM to obtain $-0.527\,751\,016\,532$ a.u.

- Levin and Shertzer[28] have used the FEM to calculate accurate
  s-wave phase shifts in e^+–H scattering, without employing the
  Hylleraas functions as is typical of other approaches.

- Low energy e^-–H scattering was investigated by Shertzer and
  Botero[29] to obtain accurate phase shifts for $0 \leq L \leq 3$ below
  the $n = 2$ threshold. The presence of resonances due to qua-
  sistable doubly excited states of H^-, below the $n = 2$ threshold,
  is accounted for by computing the phase shift and monitoring
  the abrupt changes of π in it. The finite element approach was
  also shown to be successful in describing multichannel scattering.

- In the early work by Friedman *et al.*, the FEM was used to investigate the energy levels of the H atom in confined regions in the presence of external electric fields.[30,31] These articles together with the pioneering work of Askar[32] helped popularize the FEM in the atomic physics community.

- The two-body finite element solution of the Dirac equation was derived by Scott *et al.*[33] The radial equations were solved for the low-lying bound states with a noteworthy accuracy of 1 part in 10^{18}. The negative energy solutions do not cause any problem whatsoever.

- The bound state of an electron–positron pair, the positronium, has a lifetime of 1.25×10^{-10} s. In the presence of crossed electric and magnetic fields the positronium has been shown to have long-lived states. This is due to a spatial separation of several thousand angstroms in delocalized but stable configurations of the system.[34] The lack of wavefunction overlap has been shown to lead to lifetimes of up to a year. Such states could be used to stabilize other particle–antiparticle systems.

- Babb and Shertzer[35] have investigated Schrödinger's equation for the ground state of the hydrogen molecular ion, H_2^+. Expectation values for the bond length, the rotational constant, the permanent quadrupole moment, and other operators are calculated.

- Further applications of the FEM to scattering problems are given, for example, in Refs. 36 and 37. Laser–atom interactions investigated using the FEM are discussed in Refs. 38 and 39.

- Recently, the application of the FEM to solid state electronic structure is becoming a focus of research, as an alternative to the plane-wave expansion methods used earlier. Significant advantages of this approach have been presented by Klein and his collaborators,[40] and by Tsuchida and Tsukada,[41] and their papers may be consulted for relevant details.

There is a rapidly increasing interest in the application of the FEM to atomic and molecular systems, and the next few years can be expected to generate novel results based on new physical applications of advances in the computational capabilities of high speed computers.

References

[1] For a review up to 1977 see R. H. Garstang, *Rep. Prog. Phys.* **40**, 105 (1977). For a recent reference see: Y. P. Kravchenko, M. A. Liberman, and B. Johansson, *Phys. Rev. Lett.* **77**, 619 (1996), and references therein.

[2] A. R. P. Rau and L. Spruch, *Astrophys. J.* **207**, 671 (1976).

[3] A. Baldareschi and F. Bassani, in *Proceedings of the 10th International Conference on Physics of Semiconductors*, ed. S. P. Keller, J. C. Hensel, and F. Stern (USAEC Division of Technical Information, Washington, DC, 1970), p191.

[4] C. Aldrich and R. L. Greene, *Phys. Status Solidi* (b) **93**, 343 (1979).

[5] W. Rösner, G. Wunner, H. Herold, and H. Ruder, *J. Phys.* B **17**, 29 (1984).

[6] J. C. Le Guillou and J. Zinn-Justin, *Ann. Phys. (NY)* **147**, 57 (1983).

[7] C. Liu and A. F. Starace, *Phys. Rev.* A **35**, 647 (1987).

[8] C. R. Handy, D. Bessis, G. Sigismondi, and T. D. Morley, *Phys. Rev. Lett.* **60**, 253 (1988).

[9] J. Shertzer, *Phys. Rev.* A **39**, 3833 (1989).

[10] J. Shertzer, L. R. Ram-Mohan, and D. Dossa, *Phys. Rev.* A **40**, 4777 (1989).

[11] L. R. Ram-Mohan, S. Saigal, D. Dossa, and J. Shertzer, *Comput. Phys.* **4**, 50 (1990).

[12] Y. P. Kravchenko *et al.*, Ref. 1 above.

[13] M. Kaschiev, *Phys. Rev.* A **22**, 557 (1980).

[14] K.-J. Bathe and E. Wilson, *Numerical Methods in Finite Element Analysis* (Prentice Hall, Englewood Cliffs, NJ, 1976); K.-J. Bathe, *Finite Element Procedures in Engineering Analysis* (Prentice Hall, Englewood Cliffs, NJ, 1982); K.-J. Bathe, *Finite Element Procedures* (Prentice Hall, Englewood Cliffs, NJ, 1996).

[15] D. M. Larsen, *J. Phys. Chem. Solids* **29**, 271 (1968).

[16] R. R. dos Santos and H. S. Brandi, *Phys. Rev.* A **13**, 1970 (1976).

[17] M. Shinada and K. Tanaka, *J. Phys. Soc. Jpn* **29**, 1258 (1970).

[18] A. H. MacDonald and D. S. Ritchie, *Phys. Rev.* B **33**, 8336 (1986).

[19] E. A. Hylleraas, *Z. Phys.* **54**, 347 (1929).

[20] G. Breit, *Phys. Rev.* **35**, 569 (1930).

[21] H. A. Bethe and E. E. Salpeter, *Quantum Mechanics of One- and Two-Electron Atoms* (Plenum Press, New York, 1977).

[22] D. E. Freund, B. D. Huxtable, and J. D. Morgan, *Phys. Rev.* **29**, 980 (1984); J. Baker, D. E. Freund, R. N. Hill, and J. D. Morgan, *Phys. Rev.* A **41**, 1247 (1990).

[23] F. S. Levin and J. Shertzer, *Phys. Rev.* A **32**, 3285 (1985).

[24] J. Shertzer and F. S. Levin, *Phys. Rev.* A **43**, 2531 (1991).

[25] J. Ackermann and R. Roitzsch, *Chem. Phys. Lett.* **214**, 109 (1993); J. Ackermann, B. Erdmann, and R. Roitzsch, *J. Chem. Phys.* **101**, 7642 (1994).

[26] J. Ackermann, *Phys. Rev.* A **52**, 1968 (1995).

[27] M. Braun, W. Schweizer, and H. Herold, *Phys. Rev.* A **48**, 1916 (1993).

[28] F. S. Levin and J. Shertzer, *Phys. Rev. Lett.* **61**, 1089 (1988).

[29] J. Shertzer and J. Botero, *Phys. Rev.* B **49**, 3673 (1994).

[30] M. Friedman, A. Rabinovitch, and R. Thieberger, *J. Comput. Phys.* **33**, 359 (1979).

[31] S. Goshen, M. Friedman, R. Thieberger, and J. A. Weil, *J. Chem. Phys.* **79**, 4363 (1983); M. Friedman, A. Rabinovitch, and R. Thieberger, *Z. Phys. A – At. Nucl.* **316**, 1 (1984).

[32] A. Askar, *J. Chem. Phys.* **62**, 732 (1975).

[33] T. C. Scott, J. Shertzer, and R. A. Moore, *Phys. Rev.* A **45**, 4393 (1992).

[34] J. Ackermann, J. Shertzer, and P. Schmelcher, *Phys. Rev. Lett.* **78**, 199 (1997).

[35] J. F. Babb and J. Shertzer, *Chem. Phys. Lett.* **189**, 287 (1992).

[36] J. Botero and J. Shertzer, *Phys. Rev.* A **46**, R1155 (1992).

[37] F. Abdolsalami, M. Abdolsalami, and P. Gomez, *Phys. Rev.* A **50**, 360 (1994).

[38] J. Shertzer, A. Chandler, and M. Gavrila, *Phys. Rev. Lett.* **73**, 2039 (1994).

[39] H. Yu and A. D. Bandrauk, *J. Chem. Phys.* **102**, 1257 (1995).

[40] J. E. Pask, B. M. Klein, P. A. Sterne, and C. Y. Fong, *Comput. Phys. Commun.* **135**, 1 (2001); P. A. Sterne, J. E. Pask, and B. M. Klein, *Appl. Surf. Sci.* **149**, 238 (1999); J. E. Pask, B. M. Klein, C. Y. Fong, and P. A. Sterne, *Phys. Rev.* B **59**, 12352 (1999).

[41] K. Tagami, E. Tsuchida, and M. Tsukada, *Surf. Sci.* **446**, L108 (2000); E. Tsuchida, Y. Kanada, and M. Tsukada, *Chem. Phys. Lett.* **311**, 236 (1999); E. Tsuchida and M. Tsukada, *J. Phys. Soc. Jpn* **67**, 3844 (1998); *Phys. Rev.* B **54**, 7602 (1996); *Phys. Rev.* B **52**, 5573 (1995); *Solid State Commun.* **94**, 5 (1995).

12

Quantum wires

12.1 Introduction

The ability to confine electrons spatially in a controlled way in layered semiconductor heterostructures has led to the observation of remarkable new optical and transport properties in such structures.[1] Consequences of 1D confinement in planar layered structures and the resultant quantized energy levels have been explored in great detail over the past two decades. In 1980, Sakaki[2] considered theoretically the consequences of growing heterostructures which would confine electrons in two dimensions, the 2D quantum well or the quantum well wire (QWW). Since then, control over growth at an atomic level has allowed the construction of such heterostructures.[3] As in 1D confinement, the principal effect is to profoundly change the energy spectra of electrons which in turn influence the optical and transport properties of the composite material.[4]

In this chapter, we obtain the energy levels of electrons and holes in a quantum wire of GaAs with a finite confining potential arising from the band offset of the surrounding AlGaAs medium. Initial theoretical treatments either assumed that the confining potential is infinite,[5] or used a periodic arrangement of QWs which converts the problem to the determination of narrow energy bands in a periodic structure using tight-binding models.[6] More recent calculations have included a more realistic offset potential as well as strain and multiband effects.[7]

We focus here on the symmetry aspects of energy levels and consider QWWs with square or rectangular cross-section. We describe a typical QWW calculation for an arbitrary cross-section in Section 12.6. The results shown below demonstrate that the energy levels for the finite potential are significantly lower than those obtained with the infinite barrier. More striking, however, is the lifting of the degeneracies of certain levels of a quantum wire with square cross-section when a finite barrier potential is used in calculating the energy spectrum. The "accidental degeneracies" of the wire with an infinite potential barrier are removed when the barrier height corresponding to the band offset of the surrounding medium is used in calculating the energy levels. Group theory considerations of the square are used to explain the removal of these accidental degeneracies.

In Section 12.4, we consider the electronic energy bands of stacked quantum wires. The periodic array of QWWs may be thought of as a checkerboard superlattice (CBSL). The energy bands and the optical nonlinearity of the CBSL are calculated to illustrate how the finite element method (FEM) permits the treatment of the Schrödinger equation with complex physical boundary conditions (BCs).

12.2   Quantum wires and the FEM

The FEM is used to solve Schrödinger's equation in the effective mass approximation. Results are presented for the energy levels of carriers in $GaAs/Al_{0.37}Ga_{0.63}As$ for quantum wires of typical dimensions.

In the envelope function approximation (EFA),[8,9] the differential

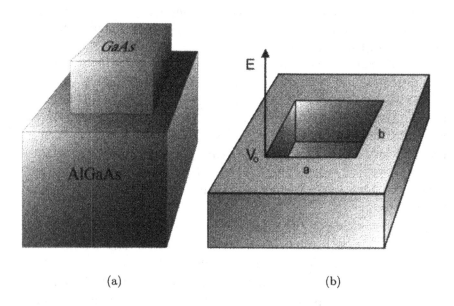

(a)   (b)

Fig. 12.1 (a) A quantum wire of GaAs of square cross-section embedded in AlGaAs. (b) The "kitchen-sink" 2D potential in the quantum wire. The corresponding potential function $V(x, y)$ is not separable in x and y.

equation for the electron's envelope function $f(x, y)$ contains a non-separable potential $V(x, y)$ corresponding to a finite barrier height,

$$-\frac{\hbar^2}{2m^*}\left(\frac{\partial^2}{\partial x^2} + \frac{\partial^2}{\partial y^2}\right)f(x, y)$$

$$+ V(x, y)\,f(x, y) = E\,f(x, y), \quad (12.1)$$

where m^* is the carrier effective mass m_w^* or m_b^* in the well or in the barrier, respectively. The potential $V(x, y) = 0$ for $|x| \leq a/2$ and $|y| \leq b/2$, and $V = V_0$ outside the well of dimension $a \times b$ centered at the origin (see Fig. 12.1). The input parameters for the effective masses of conduction electrons, light holes, and heavy holes in GaAs and in AlGaAs are obtained from the Landolt–Börnstein tables of semiconductor properties.[10] A more recent compilation of band parameters is given in Ref. 11. The variation of the energy levels with the barrier height V_0 is shown in Fig. 12.2. We return to a discussion of this below.

The FEM is used to solve equation (12.1) for the energy levels as well as the eigenfunctions $f(x, y)$. The efficacy of this method in solving quantum mechanical problems with nonseparable potentials has already been established[12] (see Chapter 11). The region of interest is partitioned into small elements and the unknown function in each element is approximated by products of local Hermite interpolation functions along the x- and y-directions. The global function $f(x, y)$ is constructed by joining the locally defined interpolation functions and requiring that $f(x, y)$ and its derivatives are continuous across the element boundaries. In the FEM, it is quite easy to implement the BC at the well–barrier interface, which requires the continuity of $f(x, y)$ and continuity of the effective mass derivative. The resultant eigenvalue problem can be solved for the energy spectra and the values of f, $\partial f/\partial x$, $\partial f/\partial y$, and $\partial^2 f/\partial x \partial y$ at the nodes.

The FEM values for the energy levels of conduction electrons, light holes, and heavy holes in rectangular quantum wires of GaAs-$Al_{0.37}Ga_{0.63}As$ of dimension $100\,\text{Å} \times 50\,\text{Å}$, and square quantum wires of size $50\,\text{Å} \times 50\,\text{Å}$, $100\,\text{Å} \times 100\,\text{Å}$, are given in Tables 12.1–12.4. A conduction band offset of $0.6\,\Delta E_g$ at the heterointerface was used in the calculations. Energy values reported here are accurate to within $0.1\,\text{meV}$. As in the case of 1D quantum well confinement, it is convenient to label the energy levels using the quantum numbers (n_x, n_y) associated with the infinite well, where n_x and n_y are the quantum

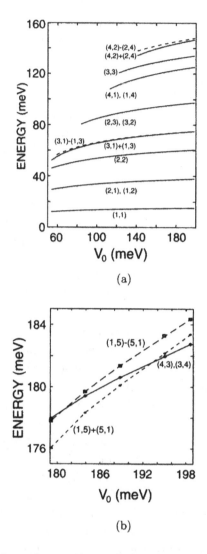

Fig. 12.2 The dependence of heavy-hole energy levels on the barrier height in a $100\,\text{Å} \times 100\,\text{Å}$ QWW for (a) the first 10 energy levels and (b) for levels 11–13.

numbers for the 1D infinite square well in the x- and y-directions. The energy levels obtained using the infinite barrier approximation are also included in Tables 12.1–12.4. Naturally, all of the energy levels are lowered from the values obtained with an infinite barrier. The effect of the finite potential is greater for small a and b and for energy levels approaching V_0. For the $50\,\text{Å} \times 50\,\text{Å}$ QWW, the single

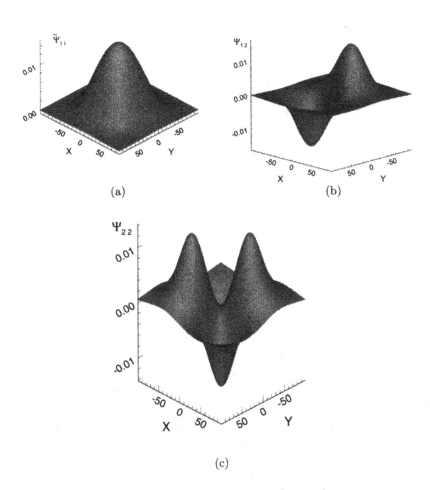

Fig. 12.3 Heavy-hole wavefunctions in a $100\,\text{Å} \times 100\,\text{Å}$ QWW for (a) the ground state (1,1); (b) the first excited state (2,1); and (c) the excited state (2,2).

bound state of the conduction electron and of the light hole are reduced in energy by a factor of 3 from the infinite barrier result.

In the infinite barrier situation any linear combination of the eigenstates with quantum numbers (m, n) and (n, m) are allowed states. In the finite barrier problem the symmetry of the square requires that the states be classified as $(m, n) \pm (n, m)$.

Wavefunctions for three low-lying states for heavy holes in a $100\,\text{Å} \times 100\,\text{Å}$ QWW are shown in Fig. 12.3. As expected, the FEM wavefunctions are similar to their infinite well analogs ex-

Table 12.1 Conduction electron energy levels in GaAs/Al$_{0.37}$Ga$_{0.63}$As QWWs, with $m_w^* = 0.0665\, m_0$ and $m_b^* = 0.0858\, m_0$. The quantum numbers (n_x, n_y) are indicated.

Cross-sectional area $a \times b\,(\text{Å}^2)$	Energy (meV) $V_0 = 276\,\text{meV}$		Energy (meV) $V_0 = \infty$
50 × 50	155.3	(1,1)	452.4
100 × 50	111.1	(1,1)	282.7
	197.6	(2,1)	452.4
100 × 100	63.5	(1,1)	113.1
	155.2	(1,2)	282.7
	155.2	(2,1)	282.7
	239.6	(2,2)	452.4
	274.2	(1,3)+(3,1)	565.5

cept that they leak out into the classically forbidden barrier region. The amount of barrier penetration increases as the energy level approaches the barrier height.

The degeneracy for the (2,2) and (4,1) states of the infinite rectangular well (100 Å × 50 Å) is removed. This degeneracy is present only in the infinite well because the energy is simply related to the dimension of the well and the quantum number, and the rectangular well was chosen to have commensurate sides. It is also interesting

Table 12.2 Light hole energy levels in GaAs/Al$_{0.37}$Ga$_{0.63}$As QWWs, with $m_w^* = 0.0905\, m_0$ and $m_b^* = 0.1107\, m_0$.

Cross-sectional area $a \times b\,(\text{Å}^2)$	Energy (meV) $V_0 = 184\,\text{meV}$		Energy (meV) $V_0 = \infty$
50×50	110.7	(1,1)	332.4
100× 50	79.8	(1,1)	207.8
	141.3	(2,1)	332.4
100×100	46.0	(1,1)	83.1
	111.7	(1,2)	207.8
	111.7	(2,1)	207.8
	171.0	(2,2)	332.4

to note that there is a mixing of the states (5,1) and (1,3) for the finite rectangular well. The state lower in energy is predominantly (1,3), and the state which is slightly higher in energy is predominantly (5,1). At an Al concentration of $x = 0.37$ only the lower state is bound, and its energy is reported in Tables 12.3 and 12.4. However, at an Al concentration $x = 0.4$, both states are bound and the FEM wavefunctions exhibit this mixing of states. Such mixing can be expected only if two states that are close in energy have the same symmetry in x and y. Of course, this mixing will depend on the dimensions of the rectangular well.

For wires of wider cross-section of the order of 500 Å, the effect of the finite barrier is less pronounced than in the examples shown in Tables 12.1–12.4. In a test calculation for the energy levels of conduction electrons in a square quantum wire of side 1000 Å, using a finite barrier of 276 meV, the first few levels are lowered by about 5% from the value obtained in the infinite barrier approximation. For higher levels, the effect is accentuated as the levels approach the barrier height V_0. In the following, we examine the removal of certain degeneracies of the infinite square-well spectra.

Table 12.3 Heavy hole energy levels in $GaAs/Al_{0.37}Ga_{0.63}As$ QWWs of dimensions 50×50 and $100 \times 50 \, Å^2$, with $m_w^* = 0.3774 \, m_0$ and $m_b^* = 0.3865 \, m_0$.

Cross-sectional area $a \times b$ (Å^2)	Energy (meV) $V_0 = 184$ meV		Energy (meV) $V_0 = \infty$
50×50	46.5	(1,1)	79.7
	112.2	(1,2)	199.3
	112.2	(2,1)	199.3
	172.7	(2,2)	318.8)
100×50	30.9	(1,1)	49.8
	53.1	(2,1)	79.7
	89.5	(3,1)	129.5
	97.4	(1,2)	169.4
	119.0	(2,2)	199.3
	138.2	(4,1)	199.3
	154.0	(3,2)	249.1
	182.8	(1,3)+(5,1)	288.9

12.3   Symmetry properties of the square wire

The symmetry properties of the envelope functions for the square quantum wire are governed by the symmetry of the potential, C_{4v}. The character table for C_{4v} is given in Ref. 13. The effect of the operators $\{E, C_2, 2C_4, 2\sigma_v, 2\sigma_d\}$ for this group on the function $f(x, y)$ are

$$
\begin{aligned}
E\, f(x, y) &= f(x, y), \\
C_2\, f(x, y) &= f(-x, -y), \\
C_4\, f(x, y) &= f(y, -x), \\
C_4^{-1}\, f(x, y) &= f(-y, x),
\end{aligned}
$$

together with the effect of reflection operations given by

Table 12.4 Heavy hole energy levels in GaAs/ AlGaAs QWW with Al concentration $x=0.37$ and cross-section $100 \times 100\,\text{Å}^2$, with $m_w^* = 0.3774\, m_0$ and $m_b^* = 0.3865\, m_0$.

Energy (meV) $V_0=184\,\text{meV}$		Energy (meV) $V_0 = \infty$
15.1	(1,1)	19.9
37.4	(1,2)	49.8
37.4	(2,1)	49.8
59.7	(2,2)	79.7
74.0	(1,3)+(3,1)	99.6
74.2	(1,3)−(3,1)	99.6
96.2	(2,3)	129.5
96.2	(3,2)	129.5
123.4	(1,4)	169.4
123.4	(4,1)	169.4
132.2	(3,3)	179.4
144.2	(2,4)+(4,2)	199.3
145.9	(2,4)−(4,2)	199.3
179.4	(3,4)	249.1
179.4	(4,3)	249.1
178.32	(1,5)+(5,1)	259.1
179.70	(1,5)−(5,1)	259.1

$$\begin{aligned}
\sigma_v \, f(x,y) &= f(-x,y), \\
\sigma_v^{-1} \, f(x,y) &= f(x,-y), \\
\sigma_d \, f(x,y) &= f(y,x), \\
\sigma_d^{-1} \, f(x,y) &= f(-y,-x).
\end{aligned}$$

It is straightforward to determine the representation corresponding to the infinite well eigenfunctions (n_x, n_y). The (odd, odd) singlet
states with $n_x = n_y$ belong to the A_1 representation and the (even,
even) singlet states belong to B_2. The degenerate states with $n_x \neq n_y$
can be classified as follows:

$$\begin{array}{cc}
\text{(even, odd) and (odd, even)} & E \\
\text{(even, even)} & A_2 + B_2 \\
\text{(odd, odd)} & A_1 + B_1.
\end{array}$$

Since the (even, even) and (odd, odd) degenerate states are combinations of two distinct irreducible representations, the degeneracy
of these levels is not a consequence of the symmetry group of the
square, but rather is due to the separability of the infinite square-
well potential. Using linear combinations of the standard (n_x, n_y)
eigenfunctions it is then possible to construct eigenfunctions which
correspond to one of the 1D irreducible representations A_1, A_2, B_1,
or B_2. For example, $(1,3)+(3,1)$ transforms as A_1 and $(1,3)-(3,1)$
transforms as B_1. As we shall see, these are a more natural choice for
the basis functions of this 2D subspace in that they are the $V_0 \to \infty$
limit of the finite barrier eigenfunctions.

For finite barriers, the potential is nonseparable and the accidental degeneracy which was present in the infinite barrier case for the
(even,even) and (odd,odd) levels is lifted (see Tables 12.3 and 12.4).
The state that is antisymmetric about the diagonal of the square
(B_1, A_2) is less bound than its symmetric counterpart (A_1, B_2). As
expected, the splitting of these particular energy levels decreases as
the barrier height increases, and in the limit $V_0 \to \infty$, the states
are truly degenerate. In Fig. 12.2, we show this dependence of
level splitting on barrier height for heavy holes in a $100\,\text{Å} \times 100\,\text{Å}$
GaAs/AlGaAs wire for the compositional range $0.1 > x > 0.4$. In some
cases, the splitting of the doublet results in one state being bound,
and the other free; for example, at $x = 0.1$, the state $(1,3)+(3,1)$ is
bound, but $(1,3)-(3,1)$ is unbound. Also note that the energy levels
for the states $(1,5) \pm (5,1)$ actually cross over the $(3,4)$ level; hence,
even the ordering of the energy levels is a function of V_0.

The FEM wavefunctions for the square well with finite barrier

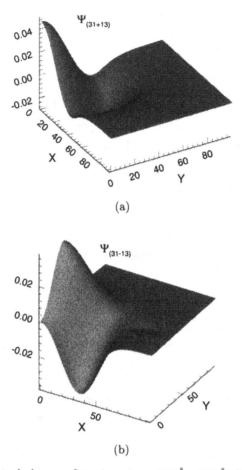

Fig. 12.4 Heavy-hole wavefunctions in a $100\,\text{Å} \times 100\,\text{Å}$ QWW for (a) the excited state $(1,3)+(3,1)$, and (b) the excited state $(1,3)-(3,1)$, in the first quadrant.

are similar to their infinite barrier analogs except that there is penetration of $f(x,y)$ into the barrier region. For the degenerate states of the E representation, any two orthogonal states which span the subspace are acceptable eigenstates. In the cases where the degeneracy is removed, the wavefunctions for the finite barrier must correspond to a single representation as required by the group properties; the wavefunctions are either even or odd with respect to reflection through the diagonals. The FEM wavefunctions for the states $(1,3)+(3,1)$ and $(1,3)-(3,1)$ are shown in Fig. 12.4 over a single quadrant $(x,y \geq 0)$ for clarity of presentation; the antisymmetric state vanishes at $x = y$ as expected.

The extended symmetry group of the square which includes the accidental degeneracy has been derived recently.[14] The dynamical symmetry operator

$$\hat{D}^{(B_1)} = \frac{\partial^2}{\partial x^2} - \frac{\partial^2}{\partial y^2} \tag{12.2}$$

connects states $(n, m) \pm (m, n)$ with $(n, m) \mp (m, n)$, and is analogous to the Runge–Lenz vector (belonging to the group $O(4)$) in the hydrogenic system. The full group is

$$\mathcal{G} = D(1) \wedge C_{4v}, \tag{12.3}$$

which is a semi-direct product of the set of one-parameter operators of the form $\exp(i\alpha \hat{D})$ and C_{4v}.

By studying the symmetry properties of the confining potential, it is straightforward to predict which of the degeneracies present in the infinite barrier approximation are due to the separability of the potential, and hence are accidental and will be removed in the presence of a finite barrier. One can apply this analysis to the more interesting case of the 3D quantum cubic dot, where the lowest 40 levels of infinite barrier spectrum contain at most three-, six-, nine-, and twelve-fold degeneracies. Degeneracies that arise from nonidentical quantum numbers (e.g., (2,2,5) and (4,4,1)) are automatically broken for a finite barrier since the energy is no longer simply related to the quantum numbers. The other degenerate states for the infinite barrier quantum cubic dot can be analyzed by looking at the behavior of the analytic solutions under the operations for the cubic group O_h. Since all of the representations of O_h are 1D, 2D, or 3D,[13] the energy spectra for the finite barrier quantum dot can have at most three-fold degeneracies. Consequently, the spectrum for the finite barrier will contain many doublets and triplets whose splitting depends on the barrier heights. Finite element calculations for the cubic quantum dot would provide visualizations of the wavefunctions and hence would be of interest.

12.4   The checkerboard superlattice

The recent advances in sub-micron epitaxy, which have allowed the growth of quantum wire heterostructures with electronic confinement in two dimensions,[3] motivate this study of a quantum wire structure consisting of rectangular semiconductor wires stacked alternately and having 2D periodicity. While such structures have not been grown in

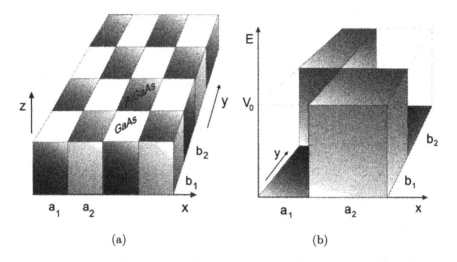

(a)   (b)

Fig. 12.5 (a) A checkerboard arrangement of quantum wires of GaAs and AlGaAs of rectangular cross-section stacked alternately to form a CBSL. (b) The 2D periodic potential in the CBSL is shown. The potential is not separable in x and y.

practice, it is useful to illustrate the use of the FEM for investigating electronic and optical properties of such a complex structure. These properties could have applications to optical switching, optical amplification in fibers, etc., and such applications are the motivation for these studies.

For concreteness, let us suppose that such a 2D CBSL is made of GaAs and AlGaAs. The effects of overlap of carrier wavefunctions across barriers in the directions parallel to the CBSL axes, and the free-carrier-induced nonlinear optical properties of such a structure with carriers in the lowest conduction miniband, are the topics of interest here.

Consider the structure shown in Fig. 12.5a. In the envelope function approximation,[8] we solve Schrödinger's equation

$$-\frac{\hbar^2}{2m^*}\left(\frac{\partial^2}{\partial x^2}+\frac{\partial^2}{\partial y^2}\right)f(x,y)+V(x,y)\,f(x,y) \;=\; E\,f(x,y) \quad (12.4)$$

for the energy E. Here m^* is the effective mass $m_w^* = 0.0665\,m_0$ in the GaAs wells, or $m_b^* = 0.0858\,m_0$ in the $Al_{0.3}Ga_{0.7}As$ barriers. The potential $V(x,y)$, shown over one period in Fig. 12.5b, is

not separable in the spatial coordinates for a finite barrier height
($V_0 = 0.274$ eV) in the AlGaAs region.

The well and barrier thicknesses along the x- and y-directions
are taken to be (a_1/a_2) and (b_1/b_2) respectively, with periodicities of
lengths $a = a_1 + a_2$ and $b = b_1 + b_2$ expressed in units of $d = 5.642$ Å
(the lattice parameter of GaAs). Here again we have a nonseparable
potential.

The BCs imposed on the solutions to equation (12.4) are the
continuity of the envelope functions $f(x,y)$ and the continuity of
$(1/m^*)f'(x,y)$ across all interfaces. In addition, Bloch's conditions

$$f(a,y) = e^{ik_x a} f(0,y); \quad f(x,b) = e^{ik_y b} f(x,0), \qquad (12.5)$$

are imposed to account for the super-translational symmetry of the
CBSL. Note that with derivative degrees of freedom we would have
to impose also the current continuity conditions, accounting for the
variations in the masses across interfaces.

Equation (12.4) is solved subject to these BCs using the FEM.
The basic "unit cell" of the CBSL of dimension $a \times b$ is split up into

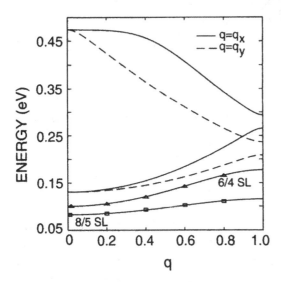

Fig. 12.6 The two lowest conduction minibands of a CBSL with well and
barrier thicknesses of (6/4) and (8/5) (in units of $d = 5.642$ Å) along x (solid
curves) and along y (dashed curves), respectively. For comparison, the
lowest minibands of 6/4 (solid curve with open triangles) and 8/5 (dashed
curve with open squares) planar superlattices are also shown.

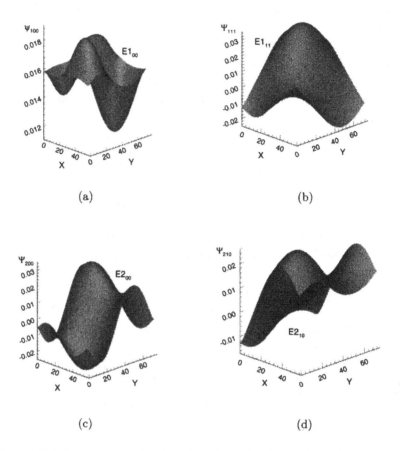

Fig. 12.7 Wavefunctions for the 6/4–8/5 CBSL (a) for $(q_x, q_y)=(0, 0)$, (b) for $(q_x, q_y)=(1, 1)$ for the first conduction band, (c) for $(q_x, q_y)=(1, 0)$, and (d) for $(q_x, q_y)=(1, 1)$ for the second conduction band.

a number of elements. The eigenvalue problem, equation (12.4), is set up in each element by assuming that the wavefunction is given by Hermite interpolation polynomials. The global wavefunctions $f(x, y)$ are constructed by joining the locally defined interpolation functions and matching the function and its derivatives across the element boundaries. The *heterointerface* BCs and Bloch's periodicity conditions mentioned above are easily incorporated in the FEM. The global matrix is constructed by element overlays as usual, and at the final stage the Bloch periodicity is implemented by overlaying the matrix elements associated with the last node along x or y onto the first node. This reduces the size of the global matrix correspondingly.

The resultant global eigenvalue problem is solved for the eigenenergies for each value of $(k_x, k_y) \equiv [(\pi/a)q_x, (\pi/b)q_y]$ to obtain the energy bands, as shown in Fig. 12.6 for a CBSL of dimensions (6/4) along x and (8/5) along y. For comparison, we have displayed the lowest energy minibands for planar superlattices of GaAs/AlGaAs. The bandedges at $q = 0$ are lower for these planar structures than in the CBSL as the carriers in the superlattice are "confined" only in one direction. Note that the bands in the CBSL do not arise from a simple additive effect from the planar superlattice bands. Some typical wavefunctions obtained using the FEM are shown in Fig. 12.7.

12.5 Optical nonlinearity in the CBSL

The nonparabolicity of the energy bands induces a nonvanishing third-order nonlinear optical susceptibility, $\chi^{(3)}$, in bulk semiconductors[15] and in superlattices.[16-18] We describe the origin of this optical nonlinearity in the following.

In small gap semiconductors the conduction band is given approximately by (the two-band model)

$$E(k) = \hbar^2 k^2 / 2m_0 + E_g/2 \pm [E_g^2/4 + 2k^2 P^2/3]^{1/2}. \quad (12.6)$$

In effect, $E(k)$ has "nonparabolic" terms in $\hbar k$. This means that the velocity $v = \partial E / \partial p$ has terms with powers of momentum: p, p^3, etc. The electronic current $J = n\,e\,v$ then contains powers of momentum p, p^3, etc.

In the presence of electromagnetic fields we have $\mathbf{p} \to (\mathbf{p} + |e|\mathbf{A}/c)$, leading to terms proportional to A^3, or equivalently to the third power in the electric field, E^3.

In four-wave mixing the carriers are driven by three photon beams of frequency ω_1, ω_2, ω_3 which induce a field, with a frequency ω_4, obeying the field equation

$$\frac{d^2}{dz^2} E_4(z) + \frac{\omega_4^2}{n\,c^2} E_4(z) = (4\pi/c^2)\,dJ/dt$$

$$= -\frac{4\pi \omega_4^2}{c^2} \chi^{(3)}\, E_1\, E_2\, E_3^*. \quad (12.7)$$

Esaki and Tsu[19] anticipated the fact that the induced nonparabolicity of the conduction bands due to Brillouin zone folding will lead to a free carrier nonlinearity in superlattices. This nonlinearity was calculated by Bloss and Friedman in a Kronig–Penney model and by

Xie *et al.* using the eight-band transfer-matrix algorithm for the calculation of the energy bandstructure of GaAs SLs. Experiments by Walrod et al.,[20] have confirmed this increase in optical nonlinearity for carrier excitation in the growth direction in superlattices.

We evaluate the free-carrier-induced optical nonlinearity due to the band nonparabolicity generated by the Brillouin zone folding of the conduction band in the CBSL in *two directions*. The analysis[18] of the optical nonlinearity in planar superlattices has revealed, contrary to expectations, that superlattices with wider wells and/or barriers have larger values of $\chi^{(3)}$. The following presentation reconfirms this result for the particular case investigated.

We now evaluate the optical nonlinearity $\chi^{(3)}$ due to the carrier band nonparabolicity.[15,16] The nonlinear susceptibility $\chi_i^{(3)}$ $(i = x, y)$ is given by

$$\chi_i^{(3)} = -\frac{e^4\, n\, \langle \partial^4 E/\partial k_i^4 \rangle}{24\, \hbar^4\, \omega_1\, \omega_2\, \omega_3\, (\omega_1 + \omega_2 - \omega_3)}, \qquad (12.8)$$

where ω_1, ω_2 correspond to incoming CO_2 laser beams in a four-wave mixing experiment with photon wavelengths of $\lambda = 10.6\,\mu$m, ω_3 corresponds to photons with $\lambda = 9.2\ \mu$m, and the outgoing photon has energy $\hbar(\omega_1 + \omega_2 - \omega_3)$. The index i refers to the electric field polarizations, in the x- or the y-direction. For further elaborations, see Ref. 21.

The nonlinearity $\chi_i^{(3)}$ is proportional to the fourth derivative $\partial^4 E/\partial k_i^4$ averaged over the Fermi distribution of the carriers:

$$n\, \langle \partial^4 E/\partial k_i^4 \rangle = [2/(2\pi)^3] \int d^3k\ \partial^4 E/\partial k_i^4\ f(E_F, E, T). \qquad (12.9)$$

At $T = 0\,$K, the Fermi function $f(E_F, E, T)$ reduces to a step-function $\theta(E_F - E(q_x, q_y) - E_z)$. The nonseparability of the potential does not allow the energy to be represented by a sum of terms dependent on q_x or q_y alone, so the integrals are performed numerically by evaluating $E(q_x, q_y)$ over a grid of values. The fourth derivatives are obtained using a nine-point difference formula on the energy dispersion. Figure 12.8 displays $\chi_i^{(3)}$ as a function of the number density n of carriers.

As seen from Fig. 12.8, the optical nonlinearity $\chi_y^{(3)}$ is larger than $\chi_x^{(3)}$, even though $\partial^4 E/\partial q_x^4$ is larger. A systematic investigation[6] of GaAs/AlGaAs superlattices, with 1D periodicity, over a wide range

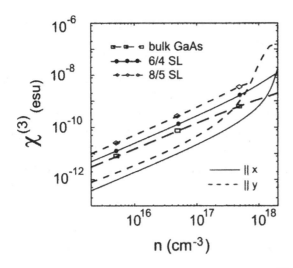

Fig. 12.8 The free carrier optical nonlinear susceptibility $\chi^{(3)}$ for the electric vector parallel to the x (solid curve) and to the y (dashed curve) directions. For reference, the values of $\chi^{(3)}$ for bulk GaAs (long-dashed curve with open squares), and for a (6/4) planar superlattice (solid curve with solid circles), and an (8/5) superlattice (dashed curve with open diamonds) are also shown as functions of carrier density.

of well and barrier thicknesses has shown that this can be understood in terms of the scale factors $(a/\pi)^4$ and $(b/\pi)^4$, the effects of band filling, and the details of the phase space integrations. The downturn in $\chi_y^{(3)}$ for number densities above $\sim 10^{18}\,\mathrm{cm}^{-3}$ is due to the form of $\partial^4 E/\partial q_y^4$, which has a broad minimum at $q_y \sim 0.9$ and which becomes positive near the zone edge. This leads to a partial cancellation in the contribution over the Brillouin zone as the number density pushes the Fermi level to the bandedge near the Brillouin zone boundary. The fact that for certain values of layer thicknesses $\chi^{(3)}$ can be lower than that in bulk GaAs has been noted by Chang.[17] This is in fact the case for $\chi_{x,y}^{(3)}$ (see Fig. 12.8) in our particular calculation with the specific choice of layer thicknesses for carrier concentrations below $5 \times 10^{17}\,\mathrm{cm}^{-3}$, whereas $\chi_y^{(3)}$ is about one–two orders of magnitude larger than $\chi_{GaAs}^{(3)}$ for larger carrier concentrations.

The envelope function approximation has been used extensively to solve for the energy levels and bands in quantum semiconductor heterostructures involving 1D confinement.[8] The FEM permits the use of this approximation in studying structures with 2D electronic

confinement; an extension to the investigation of periodic structures
with nonrectangular geometry is particularly straightforward in the
FEM.

We note from Fig. 12.8 that control over the well and barrier
thicknesses (a_1, a_2) and (b_1, b_2) in the two directions can allow for a
much broader range of choices for the layer thicknesses in order to
increase $\chi^{(3)}$. (The results of a systematic study of the dependence of
$\chi^{(3)}$ on layer thicknesses in planar GaAs/AlGaAs superlattices have
been reported in Ref. 18.) This is of importance in applications such
as intensity-dependent optical switching and optical signal process-
ing.

12.6 Quantum wires of any cross-section

In the above examples, for illustrative purposes, we used quantum
wires of square or rectangular geometry. However, the FEM comes
into its own when we start considering the physics of quantum wires
of an arbitrary cross-section. We limit ourselves to just one example
here and discuss how an FEM calculation for the energy eigenvalues can be performed for a quantum wire with a "bread-box" cross-
section. Such a cross-section can arise for wires grown by the process
of ion etching a substrate (denoted by region I with material labeled
I in Figs. 12.9 and 12.10), followed by filling the "notch" on the
surface with another material (material II occupying region II), and
overlaying with a capping layer of material III.

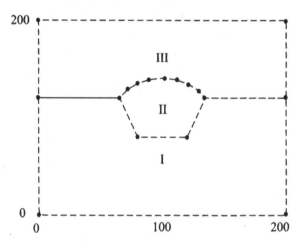

Fig. 12.9 The boundary elements defining the quantum wire of arbitrary
cross-section.

The most time-consuming part of the finite element programming used to be the generation of a physically desirable mesh, with higher mesh density in regions of interest where the solution may be expected to vary rapidly. With the advent of automatic adaptive meshing this issue has now been resolved for over a decade. We begin the FEM calculation by delineating the three regions using boundary nodes, as in Fig. 12.9. The boundary is constructed using linear elements which separate the regions. Every boundary line is assigned a material on either side of it. The region outside the box is labeled as material 0. We considered triangular meshes for our calculations.

The triangular elements are easier to break up for adaptive meshing and can be made to conform to the physical boundary as closely as desired. There are several approaches to triangulating the physical region for further finite element computations. These are discussed in Chapter 10. In the Delaunay triangulation, a number of internal node points are supplied which then form the vertices of the triangles. We have used the "algebraic integer method," developed by Sullivan,[22] that is faster than Delaunay meshing and requires only the outline boundary elements to be specified. The physical region is discretized by assigning a uniform grid of interior points which are used for triangulation of the entire region. Now the grid points near the boundary are adjusted to conform to the boundaries of the materials. Every linear boundary element is also assigned a radius

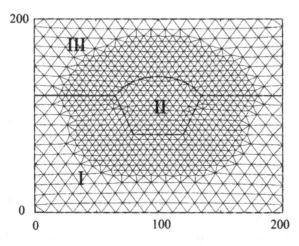

Fig. 12.10 The result of adaptive meshing with zones of refinement around boundary elements defining the periphery of the quantum wire.

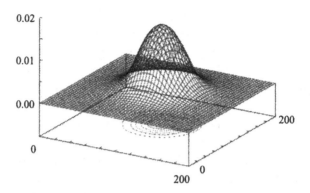

Fig. 12.11 The wavefunction of the conduction electron in the quantum
wire of Fig. 12.9.

of refinement around it so that the mesh generated by the interconnecting lines can be refined in specific regions of interest where we
might expect to see a rapidly varying solution. The result of this
adaptive meshing is shown in Fig. 12.10. The boundary elements of
the wire itself were assigned two levels of refinement; in each level
of refinement, the radii of refinement were chosen such that the regions associated with the boundary elements overlapped to provide
a finer mesh over the entire cross-section of the wire. Once the mesh
is created, it is checked automatically for the shape of each triangle
to ensure that the mesh is compatible with the FEM calculations,
and that it will represent faithfully the physical region. Each triangular element is tagged for its material properties. The nodes of
the elements are now renumbered in order that the global matrices
generated in the finite element procedure are highly banded. This
bandwidth reduction is essential since our adaptive procedure introduces additional nodes for more triangles in regions of higher mesh
density. We employ the reverse Cuthill–McGee bandwidth reduction
algorithm. This node renumbering precedes any actual FEM calculation. Further details parallel the presentation in Chapter 10.   In
the present calculation, we chose materials I and III to be AlGaAs
with 30% Al concentration, and material II to be GaAs. We employed linear triangular interpolation functions in the application of
the FEM. There were 117 nodes at the initial phase of triangulation.
With the first refinement this increased to 277, and with the second
level of refinement there were 515 nodes and 982 elements generated.
After verification of the acceptance level of the triangles we had 501

nodes and 958 elements. The half-bandwidth prior to bandwidth reduction was 459, while after the band reduction this was reduced to 34. The global matrices depend on the model employed. For each energy band we expect to have a matrix dimension of about 501. The discretized action integral associated with Schrödinger's equation was evaluated over each element. An empirical two-band model[23] was used to represent the conduction band states. The variation of the nodal wavefunction parameters in the action integral leads as usual to a generalized eigenvalue problem. The global matrix is stored in a sparse matrix format which retains in computer memory only those elements that are not zero. The diagonalization itself is carried out using the Lanczos, the subspace, or the Davidson algorithm modified to account for the complex-number arithmetic required in the multiband calculations, and for the sparse matrix format. These sparse matrix considerations permit the direct in-core solution of the eigenvalue problem for fairly large matrix sizes. In this calculation the matrix dimension was 1002. The six eigenvalues for the bound states in the conduction band were determined for the present complex-matrix generalized eigenvalue problem using the method of subspace iteration.

Figure 12.11 shows the conduction band component of the electron's wavefunction. The resulting wavefunctions are then used to compute optical matrix elements, overlap integrals, etc.

The methodology outlined here is still in its infancy. The FEM can be extended to triangular Hermite interpolation functions for much higher accuracy at the cost of having to work with larger matrices. We can clearly anticipate immediate extensions of the above considerations to the eight-band model and to much more complex 2D heterostructure geometries, with profound consequences for wavefunction engineering. We can expect the interplay between the geometry of the heterostructure and the optical and electronic properties of the system to lead to the discovery of new electronic processes and mechanisms that will be amenable to the design of new devices.

References

[1] R. Dingle, in *Festkörperprobleme* (Advances in Solid State Physics), ed. H. J. Queisser (Pergamon, Braunschweig, 1975), Vol. 15, p21; R. Dingle, *Semiconductors and Semimetals*, ed. R. K.

Willardson and A. C. Beer, Vol. 24 (Academic Press, San Diego, 1987).

[2] H. Sakaki, *Jpn. J. Appl. Phys.* **19**, L735 (1980).

[3] P. M. Petroff, A. C. Gossard, R. A. Logan, and W. Wiegmann, *Appl. Phys. Lett.* **41**, 635 (1982); J. Cibert, P. M. Petroff, G. J. Dolan, S. J. Pearton, A. C. Gossard, and J. H. English, *Appl. Phys. Lett.* **49**, 1275 (1986).

[4] H. Temkin, G. J. Dolan, M. B. Panish, and S.-N. G. Chu, *Appl. Phys. Lett.* **50**, 413 (1987); E. Kapon, D. M. Hwang, and R. Bhat, *Phys. Rev. Lett.* **63**, 430 (1989).

[5] J. Lee, *J. Appl. Phys.* **54**, 5482 (1983); H. H. Hassan and H. N. Spector, *J. Vac. Sci. Technol.* A3, 22 (1985); H. S. Cho and P. R. Prucnal, *Phys. Rev.* B39, 11150 (1989); D. A. B. Miller, D. S. Chemla and S. Schmitt-Rink, *Appl. Phys. Lett.* **52**, 2154 (1988).

[6] K. B. Wong, M. Jaros, and J. P. Hagon, *Phys. Rev.* B **35**, 2463 (1987); K. S. Dy and S.-Y. Wu, *Phys. Rev.* B **38**, 5709 (1988).

[7] M. Grundmann, O. Stier, and D. Bimberg, *Phys. Rev.* B **50** 14187 (1994); D. Stier and D. Bimberg, *Phys. Rev.* B **55** 7726 (1997); V. Turck, O. Stier, F. Heinrichsdorff, M. Grundmann, and D. Bimberg, *Phys. Rev.* B **55** 7733 (1997).

[8] G. Bastard, *Phys. Rev.* B **24**, 5693 (1981); *ibid.*, **25**, 7584 (1982). Also see L. R. Ram-Mohan, K. H. Yoo, and R. L. Aggarwal, *Phys. Rev.* B **38**, 6151 (1988), and references therein.

[9] G. Bastard, *Wave Mechanics Applied to Semiconductor Heterostructures* (Les Editions de Physique, Les Ulis, France, 1988).

[10] *Landolt-Börnstein Numerical Data and Functional Relationships in Science and Technology*, ed. O. Madelung Group III (Springer-Verlag, Berlin, 1982), Vol. 17.

[11] I. Vurgaftman, J. R. Meyer, and L. R. Ram-Mohan, *J. Appl. Phys.* **89**, 5815 (2001).

[12] J. Shertzer, *Phys. Rev.* A **39**, 3833 (1989); J. Shertzer, L. R. Ram-Mohan, and D. Dossa, *Phys. Rev.* A **40**, 4777 (1989); L. R. Ram-Mohan, S. Saigal, D. Dossa, and J. Shertzer, *Comput. Phys.* **4**, 50 (1990).

[13] M. Tinkham, *Group Theory and Quantum Mechanics* (McGraw-Hill, New York, 1964), p325, p329.

[14] F. Leyvarz, A. Frank, R. Lemus, and M. V. Andres, *Am. J. Phys.* **65**, 1087 (1997).

[15] C. K. N. Patel, R. E. Slusher, and P. A. Fleury, *Phys. Rev. Lett.* **17**, 1011 (1966); P. A. Wolff and G. A. Pearson, *Phys. Rev. Lett.*

17, 1015 (1966).

[16] W. L. Bloss and L. R. Friedman, *Appl. Phys. Lett.* **41**, 1023 (1982); G. Cooperman, L. R. Friedman, and W. L. Bloss, *Appl. Phys. Lett.* **44**, 977 (1984).

[17] Y. C. Chang, *J. Appl. Phys.* **58**, 499 (1985).

[18] H. Xie, L. Friedman, and L. R. Ram-Mohan, *Phys. Rev.* B **42**, 7124 (1990).

[19] L. Esaki and R. Tsu, *IBM J. Res. Dev.* **14**, 61 (1970).

[20] D. Walrod, S. Y. Auyang, P. A. Wolff, and M. Sugimoto, *Appl. Phys. Lett.* **59**, 2932 (1991); also see: D. Walrod, S. Y. Auyang, P. A. Wolff, and Won Tsang, *Appl. Phys. Lett.* **56**, 218 (1990).

[21] J. R. Meyer, F. J. Bartoli, E. R. Youngdale, and C. A. Hoffman, *J. Appl. Phys.* **70**, 4317 (1991).

[22] J. M. Sullivan, in *CAD/CAM Robotics and Factories of the Future*, Vol. 1, ed. Birendra Prasad (Springer-Verlag, Berlin, 1989) pp60-64.

[23] K. H. Yoo, L. R. Ram-Mohan, and D. F. Nelson, *Phys. Rev.* B **39**, 12808 (1989).

13

Quantum waveguides

When the dimensions of semiconductor heterostructure devices are reduced to about 50 nm, electrons in the device start behaving more like waves than particles. Heterostructures of such dimensions are called "mesoscopic" devices. With the mean free path becoming comparable to the dimensions of the electronic device, an electron is transmitted through the device ballistically with no scattering. The electron waves will display phase coherence, however, and undergo quantum interference effects as defined by the geometry of the meso-scopic device. These features of electron transport change the usual macroscopic nature of resistance, leading to a quantization of resistance. In this chapter, we give a qualitative derivation of the Landauer relation[1] defining the quantized conductance in terms of the transmission coefficient of electron waves through the device. The electron waves are guided through the device in a manner similar to optical waveguides.[2]

Here, we wish to study the propagation of electrons through narrow channels. With the geometry playing a central role in the behavior of electron waves, it is very natural to employ the finite element method (FEM). We consider various strategies for calculating the transmission coefficient through 2D waveguides. The boundary conditions (BCs) at the various input and output ports correspond to incoming or outgoing currents and we discuss the implementation of the mixed BCs for such waveguide problems.

The prediction that *any* bend or variation in cross-section in a waveguide gives rise to a bound state and to a peak in the transmission has led to novel ideas for device applications. We review one example of such a device.

The computational methods presented here should be useful in the context of 2D and 3D quantum mechanical scattering in open and confining (waveguide) regions. One of the major difficulties in numerically evaluating scattering amplitudes and cross-sections is the need to discretize the physical region all the way to the asymptotic region, where the BCs guaranteeing outgoing scattered waves are easy to implement. Such an approach is computationally intensive, prone to accumulation of error in the numerical analysis, and, in some sense, represents an excessive reliance on computational resources. The key aspect of any such numerical calculation should be to truncate the physical region close enough to the scattering region. We discuss some strategies for such calculations. A parallel, and essentially independent, presentation using the method of boundary elements is given in Chapter 18.

13.1 Quantization of resistance

A 2D distribution of electrons can be created at the interface between a thin layer of GaAs and n-doped AlGaAs. By placing contacts on the surface of the GaAs layer, as in Fig. 13.1, it is possible to investigate the electrical resistance R of the 2D electron gas in its flow through the split-gate channel. When the channel is constricted by the application of a gate voltage the resistance of the device is essentially the resistance of the channel. In the absence of any scattering one might anticipate that the carriers will go through ballistically with no resistance at all; however, this is not so, and the resistance of the device remains finite.

One of the results of Landauer's theory[1] of conductance in mesoscopic devices is that the device may be thought of as two contact reservoirs of carriers maintained at constant electrochemical potentials μ_1 and μ_2 with the "quantum device" in between, as shown in Fig. 13.2. The current I through the device is proportional to

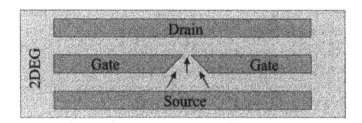

Fig. 13.1 A schematic of the experimental setup showing a 2D electron gas (2DEG) with a split-gate arrangement. The arrows indicate the direction of current flow. The narrow portion of the channel is about 250 nm.

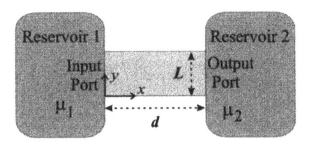

Fig. 13.2 The 2D electron waveguide equivalent to that in Fig. 13.1 is shown, with the angular sides of the constriction made parallel. The electron reservoirs are connected by the waveguide of dimensions $L \times d$.

$\mu_1 - \mu_2$, and its conductance $G = 1/R$ is given by

$$G = \left(\frac{2e^2}{\hbar}\right) \sum_{\mu\nu} T_{\mu\nu}. \tag{13.1}$$

The device is treated as a 2D waveguide with input and output modes labeled by μ and ν, with the transmission coefficients $T_{\mu\nu}$ at the Fermi energy being labeled accordingly.

Experiments performed independently by Wharam et al.,[3] and by van Wees et al.,[4] confirm this remarkable quantization of resistance. Here we follow the arguments given by Wharam et al.[3] and Sharvin[5] to derive this result.

The waveguide of Fig. 13.2, which is an idealization of Fig 13.1, has transverse modes, labeled by an index μ, having energy

$$E_y^\mu = \frac{\hbar^2 (k_y^\mu)^2}{2\,m^*} = \frac{\hbar^2 \pi^2 \mu^2}{2\,m^* L^2}. \tag{13.2}$$

Here L is the transverse width of the waveguide. The electron's mass in the medium, the effective mass, is denoted by m^*. The current in a given subband μ associated with the energy of motion along the lateral direction of the waveguide is given by

$$I_\mu = n_\mu\, e\, \delta v_\mu, \tag{13.3}$$

where n_μ is half of the number of carriers per unit length in the subband. (Only half the electrons have wavevectors in the $+x$-direction.) In the above equation, e is the electron charge and δv_μ is the increase in velocity. The number of carriers n_μ contributing to the current, including a spin degeneracy factor of 2, is

$$n_\mu = \frac{2}{2\pi} \int_0^{k_{max}} dk = 2 \int_{E_y^\mu}^{E_F} \frac{dE}{\hbar} \sqrt{\frac{m^*}{2E}}$$

$$= \frac{2m^* v_{\mu F}}{2\pi\hbar}. \qquad (13.4)$$

The change in kinetic energy is given in terms of the applied voltage as

$$eV = \mu_1 - \mu_2 = \frac{1}{2}m^*(v_{\mu F} + \delta v_{\mu F})^2 - \frac{1}{2}m^* v_{\mu F}^2, \qquad (13.5)$$

so that for $\delta v_\mu \ll v_{\mu F}$ we have $\delta v_\mu = eV/m^* v_{\mu F}$. Now the current in the channel is given by

$$I = \frac{e^2 V}{\pi\hbar}, \qquad (13.6)$$

and the resistance corresponding to carriers in one subband is

$$R_\mu = \frac{h}{2e^2} = \frac{2\pi\hbar c}{2e^2 c}$$

$$= \frac{\pi}{\alpha c} \simeq 1.44 \times 10^{-8} \text{ s cm}^{-1} \quad (\text{cgs units})$$

$$\simeq 13 \text{ k}\Omega. \qquad (\text{mksa units}) \quad (13.7)$$

Here, $\alpha = e^2/\hbar c \simeq 1/137$ is the Sommerfeld fine-structure constant. This resistance is dependent only on fundamental physical constants and is the same for any transmission mode. For transmission in more than one transverse mode of the waveguide we can estimate the total resistance assuming that the transport of carriers in the various modes is independent of each other. In this case the double sum over incoming and outgoing modes reduces to just one. We then have

$$\frac{1}{R_{tot}} = \sum_\mu \frac{1}{R_\mu}. \qquad (13.8)$$

This derivation leads to a resistance quantization essentially because there is a cancellation of the velocity dependence of the number density and the change in the velocity, in the expression for the current.

The resistance of the system may be thought of as arising from the contacts between the waveguide and the reservoirs at the two ends.

For more elaborate treatments for the calculation of the quantized resistance the reader is referred to the original articles of Landauer,[1] and the review articles by Beenakkar and van Houten,[6] and by Imry.[7] A treatment using quantum kinetic equations is given by Datta and co-workers.[8] A derivation using the usual Kubo formalism for linear response theory is available.[9,10]

The behavior of the conductance as a function of the Fermi level and gate voltage can be complex for short quantum channels. The above semiclassical result is a first approximation to a more elaborate theory. Kirczenow[11] has calculated the exact transmission through a short ballistic channel connecting two reservoirs by treating the electrons as free carriers in the reservoirs and using any potential in the transverse direction in the channel in between. The transmission is predicted to have discernible oscillations at the edges of the steps in the quantized conductance, due to resonant conduction at low temperatures. While we have assumed that there is no scattering in the waveguide, Bagwell[12] has investigated the effect of a few scattering centers in the channel and included the scattering in traveling modes and also evanescent modes. An evanescent mode begins to influence the conductance if the exponential tail of the mode overlaps with other modes from other scatterers. Also, if the Fermi energy approaches a sub-bandedge, so that the decay length of the evanescent mode becomes very large, the transmission properties are affected. It is well known from the theory of scattering[13] that other effects, such as the occurrence of bound states below the lowest sub-bandedge and quasibound states below the higher sub-bandedges, can occur if the scattering centers have attractive potentials.[12]

13.2 The straight waveguide

It is evident from the above discussion that the conductance of a general mesoscopic device depends on the transmission coefficients through output ports of a multiport device. In this section, we consider the calculation for the transmission through the simplest example, that of a 2D straight waveguide. We are concerned here with setting up the formalism for the calculation of transmission and reflection coefficients for electron waves in two dimensions in the waveguide shown in Fig. 13.3. Schrödinger's equation in two dimensions is given by

$$-\frac{\hbar^2}{2m^*}\left(\frac{\partial^2}{\partial x^2} + \frac{\partial^2}{\partial y^2}\right)\psi(x,y) + (E - V(x,y))\,\psi(x,y) = 0. \quad (13.9)$$

As is the case in semiconductors, the electron's effective mass m^* differs from the free-electron mass m_0. We shall assume that the potential energy $V(x,y)$ of the electron is zero within the waveguide and infinite outside the side walls. We transform the coordinates using $x = \tilde{x}\ell_0$ and $y = \tilde{y}\ell_0$, where $\ell_0 = 1$ Å. For convenience, in the following, we drop the tildes on the coordinate once the scaling is performed. Defining $\mathcal{C} = (\hbar^2/2m_0\ell_0^2)$ and letting $k_0^2 = (m^*E)/(m_0\mathcal{C})$, we obtain Helmholtz's equation for the electron propagation

$$\left(\frac{\partial^2}{\partial x^2} + \frac{\partial^2}{\partial y^2}\right)\psi(x,y) + k_0^2\,\psi(x,y) = 0. \quad (13.10)$$

Recall that in 1D tunneling there are just two propagating modes corresponding to wavefunctions of the form $\exp(\pm ik_x x)$. The BCs for ensuring that the scattered waves are outgoing waves in one dimension are simply given by $\psi'(x) = \pm ik_x\psi(x)$ for the transmitted and reflected waves, respectively. In the case of the 2D waveguide shown in Fig. 13.3, we have two types of BCs. First, the wavefunction vanishes along the side walls of the waveguide. Next, at the input and

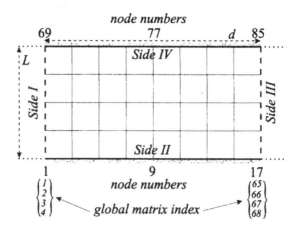

Fig. 13.3 The rectangular waveguide, showing the region meshed with 2D Hermite elements with nine nodes per element. Examples of node numbers and the global numbers for the four degrees of freedom at each node are shown.

output ports, the BCs are not as simple as in the 1D case. Here, the confinement in the y-direction gives rise to quantization of the transverse energy and the formation of eigenmodes of specific transverse mode numbers, together with propagation in the x-direction. In order to ensure that we have outward-propagating transmitted and reflected waves, we first have to perform a "modal decomposition" of these scattered waves. Then we can impose the condition that each mode satisfies a mixed BC analogous to the 1D case. In other words, the BCs for extruding currents cannot be stated simply in terms of (mixed) BCs on the *total* wavefunctions: we require an initial modal decomposition of the outgoing amplitudes, in terms of which we can conform to BCs demanding outgoing currents.

For a given electron energy, the wavevector k_x^μ is specified by the relation

$$k_x^\mu = \sqrt{\left(k_0^2 - \frac{\mu^2\pi^2}{L^2}\right)}. \tag{13.11}$$

When k_x^μ becomes imaginary we have evanescent waves at the boundaries. The wavefunction at the input port, located at $x = 0$, is a combination of incoming and reflected waves, and it is given by

$$\psi_i(0,y) = \sqrt{\frac{2}{L}}\sum_{\mu=1}\left(a_\mu\, e^{ik_x^\mu x}\sin(\mu\pi y/L)\right)\Big|_{x=0}$$
$$+\sqrt{\frac{2}{L}}\sum_{\nu=1}\left(r_\nu\, e^{-ik_x^\nu x}\sin(\nu\pi y/L)\right)\Big|_{x=0}$$
$$= \sum_\mu \psi_{\mu,Inc}(0,y) + \sum_\nu \psi_{\nu,Refl}(0,y). \tag{13.12}$$

Here we have allowed for different numbers of incoming (μ) and reflected (ν) modes that constitute the wavefunction. At the output port, the outgoing wavefunction is again expanded in terms of the normal modes in the form

$$\psi(L,y) = \sqrt{\frac{2}{L}}\sum_{\lambda=1}\left(t_\lambda\, e^{ik_x^\lambda x}\sin(\lambda\pi y/L)\right)\Big|_{x=d}$$
$$= \sum_\lambda \psi_{\lambda,Trans}(d,y). \tag{13.13}$$

Fig. 13.4 The nine-noded element showing the local node numbers.

The BCs are imposed on the individual modes at the two ports in the form

$$\psi'_{\mu,Inc}(0,y) = i\,k_\mu\,\psi_{\mu,Inc}(0,y),$$
$$\psi'_{\nu,Refl}(0,y) = -i\,k_\nu\,\psi_{\nu,Refl}(0,y), \qquad (13.14)$$
$$\psi'_{\lambda,Trans}(d,y) = i\,k_\lambda\,\psi_{\lambda,Trans}(d,y).$$

We solve equation (13.10) using the Galerkin formulation of the FEM. The waveguide is discretized using nine-noded rectangular elements in which the convention for the local node numbering is chosen as in Fig. 13.4. We approximate the wavefunction in equation (13.10) using 2D quintic Hermite interpolation polynomials, which are obtained as the product of the 1D quintic polynomials along x and y. This approximate wavefunction uses parameters corresponding to $\{\psi,\ \psi'_{,x},\ \psi'_{,y},\ \psi''_{,xy}\}$ at each node. We label the $4 \times 9 = 36$ interpolation functions by a single index which goes over the four degrees of freedom at each of the nine nodes. We write

$$\psi(x,y) = \sum_{\beta=1}^{36} N_\beta(x,y)\,\psi_\beta. \qquad (13.15)$$

The unknown parameters are determined by setting the projections of the "residual" for each interpolation polynomial to zero. We have

$$\int_{x_i}^{x_{i+1}} dx \int_{y_i}^{y_{i+1}} dy\ N_\alpha(x,y)\left(\frac{\partial^2}{\partial x^2} + \frac{\partial^2}{\partial y^2}\right)\psi(x,y)$$
$$+ \int_{x_i}^{x_{i+1}} dx \int_{y_i}^{y_{i+1}} dy\ N_\alpha(x,y)\,k_0^2\,\psi(x,y) = 0. \qquad (13.16)$$

The integrations here are over each finite element i because the interpolation polynomials are zero outside a given element. The substitution of equation (13.15) into equation (13.16) leads to the equations for the ith element

$$\int_{x_i}^{x_{i+1}} dx \int_{y_i}^{y_{i+1}} dy \ N_\alpha(x,y) \left(\frac{\partial^2}{\partial x^2} + \frac{\partial^2}{\partial y^2} \right) N_\beta(x,y) \, \psi_\beta$$
$$+ \int_{x_i}^{x_{i+1}} dx \int_{y_i}^{y_{i+1}} dy \ N_\alpha(x,y) \, k_0^2 \, N_\beta(x,y) \, \psi_\beta = 0.$$

We combine equations containing the same nodal variables from neighboring elements. This amounts to the overlay of element matrices in order to obtain the global matrix, as has been mentioned in earlier chapters. The resulting equations, of which there are four times the total number of nodes, are expressed compactly in the form

$$\mathcal{M} \cdot \Psi = 0, \tag{13.17}$$

where the right side is a null vector.

For interpolation polynomials of lower degree than the product of quintics suggested here, it is useful to integrate by parts the Laplacian term. A gradient operator then acts on each of the interpolation polynomials leading to a form symmetrical in the derivatives for the kinetic energy term. This is at the price of generating a "surface" term which corresponds to a contour integral along the exterior boundary of the region. The line integral is evaluated by taking account of those interpolation functions and their normal derivatives that do not vanish at the periphery. These terms are added to the global matrix appropriately. The corresponding book-keeping is not too involved.

We first apply BCs at nodes along the side walls where the wavefunction is zero. The nodal variables ψ_i at the nodes along the sides are set to zero, while the three derivative degrees of freedom at these nodes have no conditions on them. This Dirichlet BC on the ψ_i is implemented by setting to zero the corresponding row and column of the global matrix and inserting unity on the diagonal element while leaving a zero in the corresponding row on the vector on the right side of equation (13.17). We have referred to this procedure earlier as "benediction." In this manner the function values at such nodes on sides II and IV in Fig. 13.3 solve to zero.

Next, we consider the BCs at the input and output ports. At this stage, it is useful to view the global matrix equation, equation (13.17), explicitly as a set of simultaneous equations. The variables ψ_i and the derivative degrees of freedom at each node on the input and output ports can be expressed in terms of the input, transmitted, and reflected amplitudes for each mode. We substitute for ψ_i and the derivative degrees of freedom at each of the nodes in the ports in terms of equations (13.12), (13.13), and their partial derivatives. For example, in the case of the port on side III, suppose there are five nodes ($i = 1, \ldots, 5$), with the corresponding coordinates (L, y_i). Let the number of modes under consideration be $\lambda = 1, 2$. For each mode λ in the output port we have the relations

$$\psi_1 = \sqrt{\frac{2}{L}} \left(t^{(1)} e^{ik_x^{(1)}d} \sin(\pi y_1/L) + t^{(2)} e^{ik_x^{(2)}d} \sin(2\pi y_1/L) \right),$$

$$\psi'_{1,x} = \sqrt{\frac{2}{L}} \left(ik_x^{(1)} t^{(1)} e^{ik_x^{(1)}d} \sin(\pi y_i/L) \right.$$
$$\left. + ik_x^{(2)} t^{(2)} e^{ik_x^{(2)}d} \sin(2\pi y_i/L) \right),$$

$$\psi'_{1,y} = \sqrt{\frac{2}{L}} \left((\pi y_i/L) t^{(1)} e^{ik_x^{(1)}d} \cos(\pi y_i/L) \right. \tag{13.18}$$
$$\left. + (2\pi y_i/d) t^{(2)} e^{ik_x^{(2)}d} \cos(\pi y_i/d) \right),$$

$$\psi''_{1,xy} = \sqrt{\frac{2}{L}} \left(ik_x^{(1)} (\pi y_i/L) t^{(1)} e^{ik_x^{(1)}d} \cos(\pi y_i/L) \right.$$
$$\left. + ik_x^{(2)} (2\pi y_i/L) t^{(2)} e^{ik_x^{(2)}d} \cos(\pi y_i/L) \right),$$

and so on for nodal variables at other nodes i on the output port. Again, the same modal analysis is performed for the input port. Regrouping terms we obtain equations involving a_μ, r_ν, and t_λ only, and the usual nodal variables at interior nodes. These simultaneous equations are solved using LU-decomposition and Gaussian elimination or with the help of iterative conjugate gradient methods.

The total outgoing probability current is given by the integral over the output port of the surface current density in each propagat-

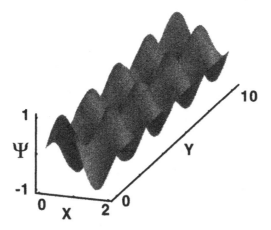

Fig. 13.5 The real part of the wavefunction in a waveguide with transverse dimension of 2 units and a length of 10 units. The incident wave is in the second transverse mode.

ing mode. We have

$$
\begin{aligned}
\mathcal{J}_t &= \frac{\hbar}{2m^*i} \sum_\lambda \int_0^L dy \left(\psi^*_{\lambda,Trans} \partial_x \psi_{\lambda,Trans} - \partial_x \psi^*_{\lambda,Trans} \psi_{\lambda,Trans} \right) \\
&= \frac{\hbar}{m^*} \sum_\lambda k_x^{(\lambda)} |t_\lambda|^2 .
\end{aligned}
\tag{13.19}
$$

The transmission coefficient is obtained as the outgoing probability current divided by the incident current, and is given by

$$
T = \frac{\sum_\lambda k_x^{(\lambda)} |t_\lambda|^2}{\sum_\mu k_x^{(\mu)} |a_\mu|^2} .
\tag{13.20}
$$

This is the total transmission that enters the conductance formula.

Table 13.1 The transmission and reflection coefficients in two modes for a simple waveguide. The incident wave is in the second transverse mode.

Mode number n	Transmission coefficient	Reflection coefficient
1	2.80×10^{-23}	2.178×10^{-23}
2	$0.999\,999\,92$	8.2078×10^{-8}
Total:	$0.999\,999\,92$	8.21×10^{-8}

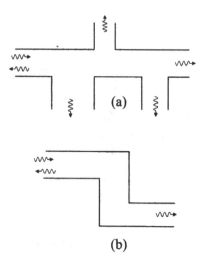

Fig. 13.6 (a) A multiport 2D waveguide. (b) A double-bend waveguide.

Figure 13.5 shows the solution for the straight waveguide in which the transverse dimension is 2 units and the length is 10 units. The input wave of unit amplitude is in the second transverse mode, and the total wavevector is $k = \sqrt{k_x^2 + k_y^2} = 4$. The wavefunction is shown over a propagation length of four wavelengths. A total of 10×10 quintic Hermite elements were used in the calculation. After the BCs were implemented there were 1764 equations. A standard LU decomposition method was used to solve for the nodal amplitudes. The condition number for the matrix was determined to be 10^{-7} which corresponds in the LINPACK matrix software package to a loss of accuracy of about 7 digits out of 15. The wavefunction shown in Fig. 13.5 has been reconstructed from these values on an x–y grid of 30×30 points for the graphical display. The transmission and reflection coefficients are shown in Table 13.1 for each mode. We should mention that similar levels of accuracy can be achieved in the case of quadratic Lagrange interpolation polynomials or cubic Hermite interpolation polynomials by judiciously employing more elements.

We have illustrated the method for calculating the transmission coefficient for a simple waveguide. This method lends itself to the calculation of multiport rectangular waveguides such as the ones shown in Fig. 13.6. Baranger[14] has investigated multi-port waveguides (Fig. 13.6a) and shown that the junctions act as filters and redistribute the electrons among the different modes, and also give rise to large longitudinal resistance and bend resistances. The quantum mechanical calculations are compared with classical (Boltzmann equation) results in order to show the differences between the two.

Double-bend waveguides, such as the one shown in Fig. 13.6b, were fabricated to form a 2D electron gas in a GaAs layer and the split-gate technique was used to form the double bend.[15,16] At low temperature, resonant peaks were seen in the lowest quantized conductance plateau. The number of peaks depended on the length of the double-bend cavity which sets up standing waves in the cavity. A similar dependence of the cavity geometry has also been theoretically investigated by Yuan and Gu.[17] With cavities of different shapes the internal scattering from the walls begin to influence the transport through the device. Transmission peaks just below the threshold for the opening of a new modal channel are ascribed to resonant states in the cavity. In all of these reported calculations the mode matching method was used to solve Schrödinger's equation in the waveguides. Kawamura and Leburton[18] employ the recursive Green's function technique to achieve the same purposes.

The transmission through circular bends has been calculated by Sols and Macucci,[19] and by Ji.[20] Finite element calculations by Lent on circular bends[21] and other structures[22,23] have provided a detailed picture of the current distribution in quantum waveguides showing current circulation and flow through the devices. We note here the calculation of Sprung et al., who approximate the current propagation through circular bends by a 1D problem with a square-well potential replacing the bend.[24] The approximate method is able to obtain analytic expressions for transmission in simple cases.

More general shapes for 2D waveguides are meshed using triangular elements. Again, a modal analysis can be carried out at ports so that the direction of the incoming and outgoing currents can be specified clearly.

13.3   Quantum bound states in waveguides

In a remarkable calculation, Schult et al.[25] showed that there exist bound states at the intersection of two open narrow channels perpendicular to each other, as in Fig. 13.7. The potential in the cross-channel is zero, while the walls are impenetrable. Classically, there is no confinement and hence there are no bound states. The threshold for propagation in any one arm is given by the energy for the

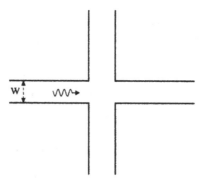

Fig. 13.7 Two waveguides crossing at right angles.[25] The four corners can bind electronic states in the central region.

lowest mode, $E_t = \hbar^2\pi^2/2m^*w^2$, where w is the width of the channel. The mode matching method uses $\sin(n\pi x/w)\,\exp(\pm K_n y)$ and $\sin(n\pi y/w)\,\exp(\pm K_n x)$ as the expansion functions in the open channels in order to locate bound states. (The problem is also amenable to the FEM.) With exponentially falling solutions the BCs are that the wavefunction falls off to zero at large distances from the crossing and that the wavefunction vanishes along the walls. It is found that there are two bound states, the symmetric ground state with energy $E_1 = 0.66\,E_t$ which is below the threshold E_t for propagation, and the antisymmetric bound state with energy $E_2 = 3.72\,E_t$ which is below the corresponding threshold $4\,E_t$ for propagating odd states.

These bound states are formed and are localized because of the presence of corners. The odd-parity bound state lies in the continuum of the even-parity channel and shows up as a resonance in the scattering just below the opening of the $n = 2$ channel. The odd-parity case is equivalent to the case of a single wire of half the width, bent at right angles with wavenumber $k = 0.96\pi/w$. An interesting question is whether a second electron of the opposite spin can also be bound to the single electron already bound. The reader is referred to the article already cited for the answer.

Another example of a bound state occurring in open channels is the case of the T-geometry. For all three arms open and of equal width the single bound state occurs at $k = 0.90\pi/w$.

It is evident that bound states in quantum channels are a universal phenomenon. The uncertainty principle provides an explanation. Consider an infinite waveguide with a local bulge in the middle. The particle propagating in this waveguide is able to have a lower transverse momentum and hence a lower transverse energy within the bulge. A bound state results because the quantum particle then has to squeeze back into the narrower portion of the waveguide in order to continue its propagation. The straight 3D tube or 2D waveguide with a slowly varying cross-section can be mapped into a 1D problem with an attractive potential corresponding to the bulge. In one dimension, an attraction, no matter how weak, leads to binding. It has been shown by Exner *et al.* that bound states occur in 2D channels of fixed width and small and slowly varying curvature.[26] Goldstone and Jaffe[27] have extended this work by showing that a waveguide of constant cross-section always has a bound state if it bends. A bend provides a region in which the quantum particle can relax, lowering its momentum in comparison with the straight regions; a general variational proof of this result is given.[27,28] These bound states affect the quantized steps in the conductance by modifying the shape of the lowest step in the conductance and by generating a resonance at the threshold of each successive mode opening up energetically. An effective Lagrangian approach for propagation in quantum waveguides has been studied by Simanek,[29] who uses the Feynman path integral approach and finds a curvature-induced attractive potential together with a quartic term in the particle velocity so that the effective mass becomes velocity dependent and also dependent on the curvature of the waveguide. Similar issues can be expected in optical waveguides, and further research in this area can be expected to produce general results applicable to other areas of physics.

13.4   The quantum interference transistor

In a layered double-barrier heterostructure the presence of transmission peaks in the 1D tunneling leads to a negative differential resistance in the tunneling current, and a consequent transistor action: if the device is on the upper potential side in an I–V curve, a small reduction in voltage gives rise to a large amplification of the current. In the case of quantum waveguides too the transmission peaks can be used in device applications. The sharp peaks in the conductance are due to the presence of bound states associated with the double bends. These states participate in the resonant transmission across the double bend. In fact, each bend gives rise to a bound state so that in the double bend we have two localized states that couple to each other and give rise to symmetric and antisymmetric bound

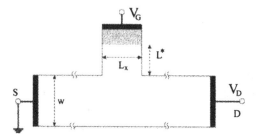

Fig. 13.8 The quantum interference transistor. (After Ref. 31.)

states. These states are displayed in Ref. 30.

It has been proposed by Sols *et al.*[31,32] that quantum interference in a T-shaped waveguide can be used to have transistor action. Consider the structure shown in Fig. 13.8, in which just the 2D waveguide is depicted. The actual structure consists of a layer of GaAs between two layers of doped AlGaAs so that there are free carriers in the GaAs layer. The layers have to be etched to form a T-shaped device with contacts attached on them for the source, drain, and gate as in Fig. 13.8. For a given $w \simeq 100\,\text{Å}$, the transmission T depends on the length L^* and the width of the T-stub and the energy of the carrier. For a gate voltage V_G the conductance can be shown to be proportional to

$$G \propto \frac{\partial T}{\partial L^*} \frac{\partial L^*}{\partial V_G} \tag{13.21}$$

and can be finite when $\partial T/\partial L^*$ is not zero. For a given L^*, as the gate voltage is varied the transmittance goes through broad maxima and sharp minima. As L^* is increased, the number of minima and maxima increase. This is due to the fact that the electron can propagate into the stub with more than one wavelength. It is clear from the earlier discussion that we can anticipate the formation of quasibound states in the stub and hence the peaks and valleys in the transmission can be attributed, as expected, to resonant transmission. Finite element calculations for the transmittance for resonant cavities and rings attached to a straight waveguide have been performed by Lent,[23] who demonstrates the effects of charging the cavity and displays circulation of the current within the cavity.

With increasing control over the epitaxial growth of such structures we can expect reductions in the dimensions of such quantum waveguides, and an increase in the complexity of waveguide structures, with a consequent development of very novel electronic devices operating at low power and high speed.

13.5   "Stealth" elements and absorbing BC

The BCs for scattering problems in general are difficult to implement numerically because of the need to ensure that only outgoing waves are present in the scattered waves. In analytical calculations the mixed BC that ensures this nature of scattered waves is built into the solution by starting directly with outgoing waves. In 2D scattering, Schrödinger's equation reduces to Helmholtz's equation outside the localized scattering center. For an open boundary the general solution for the total wavefunction away from the scattering potential is

$$\psi_{total}(\rho, \phi) = \psi_{inc}(\rho, \phi) + \psi_{sc}(\rho, \phi), \qquad (13.22)$$

with

$$\psi_{inc}(\rho, \phi) = e^{ik_x x},$$
$$\psi_{sc}(\rho, \phi) = \sum_{m=0}^{\infty} C_m H_m^{(1)}(k\rho) e^{im\phi}, \qquad k\rho \gg 1. \qquad (13.23)$$

The outgoing nature of the wavefunctions is ensured by the fact that the mth-order Hankel functions of the first kind, used in equation (13.23) above, all have the asymptotic form given by $e^{ik\rho}/\sqrt{\rho}$. The first p asymptotic amplitudes

$$\psi_{sc}^{asymp}(\rho, \phi) \sim \sqrt{\frac{1}{\pi k \rho}} e^{ik\rho - i\pi/2} \sum_{\ell=0}^{\infty} \frac{g_\ell(\phi)}{\rho^\ell}; \qquad k\rho \gg 1, \qquad (13.24)$$

satisfy the Bayliss–Turkel[33] radiation BCs

$$\Pi_{\ell=1}^{p} \left(\frac{\partial}{\partial \rho} + \frac{2\ell - 3/2}{\rho} - ik \right) \psi_{sc}^{asymp}(\rho, \phi) = 0. \qquad (13.25)$$

Earlier attempts at considering local BCs were made by Engquist and Majda.[34]

In an actual scattering experiment, a partial wave analysis is usually performed in order to determine the coefficients C_m by measuring the asymptotic amplitudes and their angular dependencies. In a numerical calculation it is inefficient to discretize the open domain. A truncation of the scattering region inevitably leads to inaccuracies, and more importantly to a possible nonzero reflected amplitude from the external surface (a curve in two dimensions) defining the truncation.

We employ, with simple modifications, the strategy presented earlier in this chapter by requiring the modes present in the scattering amplitude to have outward-propagating amplitudes. The finite number of nodal variables, associated with the outer boundary at a finite value of ρ, is mapped into a few coefficients C_m, depending on the number of partial waves we wish to retain. This modal analysis is used to write the mixed BC for outgoing currents for each term involving Hankel functions in equation (13.23). The relation

$$2\frac{\partial}{\partial z}H_{n-1}^{(1)}(z) = H_{n-1}^{(1)}(z) - H_{n+1}^{(1)}(z) \tag{13.26}$$

is used to provide conditions on the nodal amplitudes. This is the approach of Li and Cendes.[35] Other variations on this theme are given by Jin.[36]

More recently, Berenger[37,38] provided an unusual solution to the issue of applying BCs for scattered electromagnetic waves. He noted that one can clad the physical region away from the scattering center with an artificial layer, called the perfectly matched layer (PML), which is totally absorbing. Berenger modified Maxwell's equations in the PML in order to mathematically ascertain that there is no reflection from the PML. Rappaport[39] and also Chew and Weedon[40] showed that this is equivalent to making the coordinates complex in the PML. Lee and co-workers noted that the PML can be defined through the modification of the dielectric permittivity and the magnetic permeability to be totally absorbing for incident waves at any angle of incidence.[41] The major advantage of Lee *et al.*'s method is that we can retain the usual formulation of the FEM and evaluate element matrices for the stealth elements for which we have merely modified the material properties appropriately. These mathematical modifications are understood to be purely for the simplification of the BCs.

Here we wish to illustrate that the similar considerations permit us to define properties of a region such that it is a perfect absorber (PML), i.e., a *stealth* material, for Schrödinger waves. Consider a planar interface between the scattering region (outside the localized scattering center) and a stealth region, as shown in Fig. 13.9. Let Schrödinger's equation in the two regions I and II be

$$\frac{\partial}{\partial x}\left(\frac{1}{a}\frac{\partial}{\partial x}\psi(x,y)\right) + \frac{\partial}{\partial y}\left(b\frac{\partial}{\partial y}\psi(x,y)\right) + c\,k_0^2\,\psi(x,y) = 0. \quad (13.27)$$

Here the parameters a, b, and c are considered to be complex constants in each region, with the choice $a_1 = b_1 = c_1 = 1$ in region I. The wavevector k_0^2 is given by

$$k_0^2 = \frac{m^*}{m_0}\frac{E}{C}. \quad (13.28)$$

Consider an electron wave incident on the interface from the left, and let it be reflected and transmitted as shown in the figure. With solutions of the form

$$\psi_{inc}(x,y) = A\,e^{i(k_{0x}x+k_{0y}y)},$$
$$\psi_{refl}(x,y) = R\,e^{i(k_{0x}x-k_{0y}y)}, \quad (13.29)$$
$$\psi_{trans}(x,y) = T\,e^{i(k_x x+k_y y)},$$

the dispersion formulas in the two regions are obtained in the form

$$k_{0x}^2 + k_{0y}^2 = k_0^2 \qquad \text{in I;}$$
$$\frac{1}{a_2}k_x^2 + b_2\,k_y^2 = c_2\,k_0^2, \qquad \text{in II.} \quad (13.30)$$

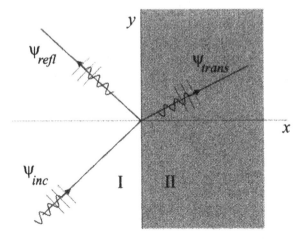

Fig. 13.9 Transmission and reflection at a stealth material interface. Conditions are determined for making the surface reflection zero.

The BCs at the interface are the continuity of the wavefunction and the continuity of the derivative of the wavefunction weighted by the coefficient a. We have

$$\psi_{inc}(0,y) + \psi_{refl}(0,y) = \psi_{trans}(0,y)$$

$$\frac{d}{dx}\psi_{inc}(0,y) + \frac{d}{dx}\psi_{refl}(0,y) = \frac{1}{a_2}\frac{d}{dx}\psi_{trans}(0,y). \quad (13.31)$$

Expressed in terms of the plane-wave forms of the wavefunctions, equations (13.30), we first obtain

$$(A + R)\,e^{ik_{0y}y} = T\,e^{ik_y y}, \quad (13.32)$$

showing that for the waves to match at all points at the interface we must have

$$k_{0y} = k_y. \quad (13.33)$$

The derivative condition now leads to

$$ik_{0x}(A - R) = i\frac{k_x}{a_2}\,T. \quad (13.34)$$

Equations (13.32) and (13.34) are used to solve for R and T in terms of A, and we obtain

$$\frac{R}{A} = \frac{k_{0x} - k_x/a_2}{k_{0x} + k_x/a_2}. \quad (13.35)$$

We can thus insist on there being no reflection ($R = 0$) *for all angles of incidence* by requiring $a_2\,k_{0x} = k_x$. Furthermore, from the dispersion relations, equations (13.30), we find that with $b_2 = a_2$ and $c_2 = a_2$ we have

$$\frac{1}{a_2^2}k_x^2 + k_y^2 = k_0^2, \quad (13.36)$$

or, what is the same,

$$k_{0x}^2 + k_{0y}^2 = k_0^2. \quad (13.37)$$

While the dispersion formulas are equivalent on both sides we have

$$\psi_{trans}(x,y) = T\,e^{ia_2 k_{0x}x}\,e^{ik_{0y}y}. \quad (13.38)$$

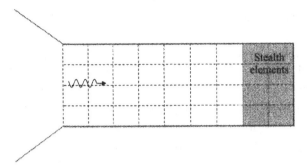

Fig. 13.10 The use of totally absorbing stealth elements at the end of a waveguide changes the scattering BCs to the Dirichlet BCs.

The parameter a_2 is now chosen to be the complex quantity $a = 1 + i\alpha$. We thus have a reflectionless surface, with a transmitted wavefunction that attenuates in the stealth region as given by

$$\psi_{trans}(x, y) = T\, e^{-\alpha k_{0x} x} e^{ik_{0x} x}\, e^{ik_{0y} y}. \tag{13.39}$$

For example, with $\alpha = 100/k_{0x}\lambda = 100/2\pi$, the attenuation is a factor of $\sim 10^{-7}$ for the amplitude and the intensity is reduced to essentially zero within a wavelength. By making the parameter α spatially dependent and smoothly increasing from zero at the interface we can further minimize reflection arising from numerical round-off. In summary, we have shown that we can define a stealth material through the use of complex coefficients in the differential equation such that we can have a planar reflectionless interface between the stealth material and the physical region. By using a complex parameter $a = 1 + i\alpha(x)$ we can attenuate the transmitted amplitude so the BC at the outer surface of the stealth layer can be reduced to the Dirichlet BC $\psi_{sc} = 0$. This substantially reduces our numerical programming problems in implementing the scattering BCs.

As an illustration, consider the waveguide of Fig. 13.10. We have inserted a stealth region at the output port. With a coordinate-dependent parameter $\alpha(x)$ we can ensure that the wavefunction dies out to essentially zero at the outer side of the stealth layer. This means that we can impose simple Dirichlet BCs along the entire periphery of the waveguide including all the ports. At the input port, however, we must ensure that the incoming wave does not encounter the stealth region. This known amplitude gives us a known contribution at the nodes of the input port. (We can completely surround the scattering region and yet circumvent the issue of how to bring in the incident waves by including source terms inside the scattering region. This is discussed below.) By transferring these known contributions to the right side of the global matrix equation we provide a "driving term" for the system of equations and obtain the solution everywhere in the interior of the waveguide. The transmitted and reflected amplitudes are obtained at the interfaces of the physical region in the interior and the stealth regions. Given the solutions for the nodal values of the scattered wavefunctions on these interfaces we can now perform a modal analysis and determine the transmittance in each mode and the total transmission coefficient associated with any given port.

If the scattering center is completely surrounded by the stealth elements, how do we allow the incident waves to enter the scattering region? We simply add a line source (an antenna) at $x = x_0$ inside the scattering region (in Fig. 13.11 the antenna is shown thicker than it need be). In Schrödinger's equation we add a source term to the right side such that it will generate a plane wave

$$\psi_{src}(\mathbf{r}) = Ae^{ik_x|x-x_0|}, \tag{13.40}$$

propagating away on both sides. For $x > x_0$ we have the incoming wave given by the above form; on the other hand, for $x < x_0$ we have an additional outgoing wave that will eventually be totally absorbed in the stealth region and is therefore of no consequence. The choice

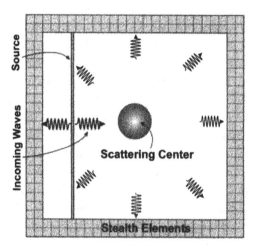

Fig. 13.11 2D scattering in an open domain, showing waves incident on a scattering center. Stealth elements are used to provide totally absorbing BCs only for the outgoing waves.

of ψ_{src} was based on ensuring that we have wavefunction continuity at x_0 for both the incident and the scattered waves.[†] So we add a term $\mathcal{S}(x, y)\delta(x - x_0)$ to the right side of equation 13.9. Integrating this equation over a small range $x_0 - \delta$ to $x_0 + \delta$ we obtain

$$\lim_{\delta \to 0} \left\{ [\psi'(x)]_{x_0-\delta}^{x_0+\delta} + 2\delta k^2 \psi(x_0) \right\} = \mathcal{S}. \qquad (13.41)$$

Here k^2 represents the effective wavevector at $x = x_0$; in any case, the k^2 term vanishes in the limit $\delta \to 0$. The jump in the derivative at $x = x_0$ is given by taking the derivative of ψ_{src} and we obtain

$$\psi'(x_0 - \delta) = -ik_x A e^{-ik_x(x-x_0)},$$
$$\psi'(x_0 + \delta) = +ik_x A e^{ik_x(x-x_0)},$$

so that the form of $\mathcal{S}$ is determined to be

$$\mathcal{S} = 2i\,kA\,e^{ik_x|x-x_0|}. \qquad (13.42)$$

We now solve Schrödinger's equation with the δ-function source term within the framework of the FEM. The element geometry is set up such that the source at x_0 falls on an edge of the 2D elements. The nodal values of elements along the source are allowed to have a jump in the values of their x-derivative. In the Galerkin calculation we now have to evaluate an explicit contribution on the right side from a line integral along the y-direction of the form $\int dx dy\, S(x, y)\delta(x - x_0)N_i(x, y)$ to account for the line source.

In 2D scattering in waveguides, the source term is modified to have contributions from various transverse modes. The results for the scattering amplitudes, the transmission coefficients, have been shown by Moussa[42] to be identical to those obtained from a modal analysis at the ports on the waveguides discussed earlier in this chapter.

We should emphasize that the absence of reflection at the interface can be guaranteed not only for the planar interface. The issues that arise at curved surfaces have been studied by Kuzuoglu and Mittra.[43] Further work on this problem has been done to show that we can indeed define perfectly matching layers for cylindrical geometry.[44]

[†]Note that the left-propagating portion can be eliminated if we allow for a discontinuity in the wavefunction at the antenna; however, this leads to more book-keeping issues in the FEM calculations.

The use of PMLs, or stealth regions as we have called them, should provide the means of applying the scattering, or radiation, BCs to fairly complex quantum mechanical scattering problems. The geometry that can be used in the finite element calculations is illustrated in Fig. 13.11. The approach suggested here is currently being used by the author in a systematic study of the scattering amplitudes both in the vicinity of scattering centers and farther away. Applications to quantum heterostructures and to other quantum systems are also being pursued at present.

13.6   The Ginzburg–Landau equation

The Ginzburg–Landau form of the nonlinear Schrödinger equation is given by

$$-i\hbar\frac{\partial}{\partial t}\,\psi(\mathbf{r},t) = -\frac{\hbar^2}{2m}\nabla^2\,\psi(\mathbf{r},t) + V(\mathbf{r})\psi(\mathbf{r},t)$$
$$+ \lambda|\psi(\mathbf{r},t)|^2\,\psi(\mathbf{r},t). \tag{13.43}$$

This form of the differential equation occurs in the Ginzburg–Landau theory of superconductivity and superfluidity, where the wavefunction $\psi(\mathbf{r},t)$ is interpreted as the order parameter which determines the coherence in the condensed or superfluid state.[45–47] A numerical solution of the Ginzburg–Landau Schrödinger (GLS) equation was obtained by Griffiths et al.[48] and Argyris and Haase,[49] who showed the formation of solitons or nonlinear coherent waves. More recently, Wang and Wang[50] have investigated the time-dependent d-wave GLS equation using the finite element solutions. The general approach for solving nonlinear equations is to first solve the linear Schrödinger equation

$$-i\hbar\frac{\partial}{\partial t}\,\psi(\mathbf{r},t) = -\frac{\hbar^2}{2m}\nabla^2\,\psi(\mathbf{r},t) + V(\mathbf{r})\psi(\mathbf{r},t), \tag{13.44}$$

in order to obtain a starting solution $\psi_{(0)}(\mathbf{r},t)$, and then to iterate the equation

$$-i\hbar\frac{\partial}{\partial t}\,\psi_{(k)}(\mathbf{r},t) = -\frac{\hbar^2}{2m}\nabla^2\,\psi_{(k)}(\mathbf{r},t) + V(\mathbf{r})\psi_{(k)}(\mathbf{r},t)$$
$$+ \lambda|\psi_{(k-1)}(\mathbf{r},t)|^2\,\psi_{(k)}(\mathbf{r},t), \tag{13.45}$$

in which the nonlinear term proportional to λ has $|\psi_{(k-1)}(\mathbf{r},t)|^2$ given by the previous iteration. A new solution $\psi_{(k)}(\mathbf{r},t)$ is now sought which satisfies equation (13.45), and is such that the wavefunction is normalized. This normalization condition also determines the strength of the effective potential arising from the nonlinear coupling. Invariably, just a few iterations lead to a self-consistent solution as determined by the difference in the wavefunctions, $|\psi_{(k)} - \psi_{(k-1)}|$ from any two successive iterations.

We conclude this chapter by noting that waves in confined regions can display the effects of the geometry of the physical region on the nonlinear behavior. An investigation of the time-independent GL equation, for traveling waves within a waveguide,

$$-\frac{\hbar^2}{2m}\nabla^2\,\psi_{(k)}(\mathbf{r},t) + \alpha\psi_{(k)}(\mathbf{r},t)$$
$$+ \beta|\psi_{(k-1)}(\mathbf{r},t)|^2\,\psi_{(k)}(\mathbf{r},t) = 0, \quad (13.46)$$

was performed by Rosenberg.[51] Interestingly, it was found that the transmission coefficient changes very rapidly with the strength of the nonlinear coupling. With $\alpha = -6$ and β varying over $0-10^{-4}$ the transmission coefficient was essentially one. Over the range $\lambda = 10^{-3}$ to $\lambda = 8 \times 10^{-3}$ the transmission drops to zero, and again recovers to unity. A number of ranges in β were found over which there is no transmission, interspersed with windows of large nonzero transmission.

It is clear that considerable work needs to be done in order to investigate solutions of nonlinear GLS equations in finite regions in order to fully understand the effect of the geometry or the shape of the regions as well as the effect of variation of the nonlinear coupling. This issue is of interest in the context of nonlinear effects in other areas of physics such as nonlinear optical waveguides,[52] soliton theories of elementary particles, nonlinear plasma stabilities and confinement, and so on. An example of a nonlinear problem of interest is the one described by Feynman,[53] who suggests that vortex rings should be formed when liquid helium flowing in a narrow tube is effluent into a large liquid helium reservoir. In two dimensions, the superfluid current should show circulation patterns.

References

[1] R. Landauer, *IBM J. Res. Dev.* **1**, 223 (1957); *Philos. Mag.* **21**, 863 (1970).

[2] M. Büttiker, Y. Imry, R. Landauer, and S. Pinhas, *Phys. Rev.* B **31**, 6207 (1985).

[3] D. A. Wharam, T. J. Thronton, R. Newbury, M. Pepper, H. Ahmed, J. E. F. Frost, D. G. Hasko, D. C. Peacock, D. A. Ritchie, and G. A. C. Jones, *J. Phys. C: Solid State Phys.* **21**, L209 (1988).

[4] B. J. van Wees, H. van Houten, C. W. J. Beenakker, J. G. Williamson, L. P. Kouwenhoven, D. van der Marel, and C. T. Foxon, *Phys. Rev. Lett.* **60**, 848 (1988); B. J. van Wees *et al.*, *Phys. Rev.* B **43**, 12431 (1991).

[5] Y. V. Sharvin, *JETP Lett.* **1**, 152 (1965).

[6] C. W. J. Beenakker and H. van Houten, "Quantum Transport in Semiconductor Nanostructures," in *Solid State Physics*, Vol. 44, ed. H. Ehrenreich and D. Turnbull (Academic Press, New York, 1991). Also see: H. van Houten, C. W. J. Beenakker, and B. J. van Wees, in *Semiconductors and Semimetals*, Vol. 35, ed. M. Reed (Academic Press, New York, 1992).

[7] Y. Imry, "Physics of Mesoscopic Systems," in *Directions in Condensed Matter Physics*, ed. G. Grinstein and G. Mazenko (World Scientific, Singapore, 1986).

[8] M. J. Mclennan, Y. Lee, and S. Datta, *Phys. Rev.* B **43**, 13846.

[9] A. D. Stone and A. Szafer, *IBM J. Res. Dev.* **32**, 384 (1988).

[10] H. U. Baranger and A. D. Stone, *Phys. Rev.* B **40**, 8169 (1989).

[11] G. Kirczenow, *Phys. Rev.* B **39**, 10452 (1989).

[12] P. F. Bagwell, *Phys. Rev.* B **41**, 10354 (1990); A. Kumar and P. F. Bagwell, *Phys. Rev.* B **43**, 9012 (1991).

[13] See, for example, M. L. Goldberger and K. M. Watson, *Collision Theory* (Wiley, New York, 1964); S. K. Adhikari, *Variational Principles and the Numerical Solution of Scattering Problems* (Wiley, New York, 1998).

[14] H. U. Baranger, *Phys. Rev.* B **42**, 11479 (1990).

[15] J. C. Wu, M. N. Wybourne, W. Yindeepol, A. Weisshaar, and S. M. Goodnick, *Appl. Phys. Lett.* **59**, 102 (1991).

[16] J. C. Wu, M. N. Wybourne, A. Weisshaar, and S. M. Goodnick, *J. Appl. Phys.* **74**, 4590 (1993).

[17] S. Q. Yuan and B. Y. Gu, *J. Appl. Phys.* **73**, 7496 (1993).

[18] T. Kawamura and J. P. Leburton, *J. Appl. Phys.* **73**, 3577 (1993).

[19] F. Sols and M. Macucci, *Phys. Rev.* B **41**, 11887 (1990).

[20] Z. L. Ji, *J. Appl. Phys.* **73**, 4468 (1993).

[21] C. S. Lent, *Appl. Phys. Lett.* **56**, 2554 (1990).

[22] C. S. Lent, S. Sivaprakasam, and D. J. Kirkner, in *Nanostructure*

Physics and Fabrication, ed. M. A. Reed and W. P. Kirk (Academic Press, New York, 1989).

[23] C. S. Lent and D. J. Kirkner, *J. Appl. Phys.* **67**, 6353 (1990).

[24] D. W. L. Sprung, H. Wu, and J. Martorell, *J. Appl. Phys.* **71**, 515 (1992).

[25] R. L. Schult, D. G. Ravenhall, and H. W. Wyld, *Phys. Rev.* B **39**, 5476 (1989); *Phys. Rev. Lett.* **62**, 1780 (1989).

[26] P. Exner, *Phys. Lett.* A **141**, 213 (1989); P. Exner and P. Seba, *J. Math. Phys.* **30**, 2574 (1989); M. S. Ashbough and P. Exner, *Phys. Lett.* A **150**, 183 (1990); P. Exner, P. Seba, and P. Stovicek, *Phys. Lett.* A **150**, 179 (1990).

[27] J. Goldstone and R. L. Jaffe, *Phys. Rev.* B **45**, 14100 (1992).

[28] K. Lin and R. L. Jaffe, *Phys. Rev.* B **54**, 5750 (1996).

[29] E. Simanek, *Phys. Rev.* B. **57**, 14634 (1998). For vorticity aspects in waveguides see: E. Simanek, *Phys. Rev.* B. **59**, 10152 (1999).

[30] C. K. Wang, K. F. Berggren, and Z. L. Ji, *J. Appl. Phys.* **77**, 9565 (1995).

[31] F. Sols, M. Macucci, U. Ravaioli, and K. Hess, *Appl. Phys. Lett.* **54**, 350 (1989); *J. Appl. Phys.* **66**, 3892 (1989).

[32] K. Hess and G. J. Iafrate, "Approaching the quantum limit," *IEEE Spectrum*, p44, July 1992.

[33] A. Bayliss, M. Gunzburger, and E. Turkel, *SIAM J. Appl. Math.* **42**, 430 (1982).

[34] B. Engquist and A. Majda, *Math. Comput.* **31**, 629 (1977).

[35] Y. Li and Z. J. Cendes, *IEEE Trans. Magn.* **29**, 1835 (1993).

[36] J. Jin, *The Finite Element Method in Electromagnetics* (Wiley, New York, 1993).

[37] J. P. Berenger, *J. Comput. Phys.* **114**, 185 (1994).

[38] J. P. Berenger, *J. Comput. Phys.* **127**, 363 (1996).

[39] C. M. Rappaport, *IEEE Trans. Magn.* **32**, 968 (1996).

[40] W. C. Chew and W. H. Weedon, *Microwave Opt. Technol. Lett.* **7**, 599 (1994).

[41] Z. S. Sacks, D. M. Kingsland, R. Lee, and J. F. Lee, *IEEE Trans. Antennas Propag.* **43**, 1460 (1995).

[42] J. E. Moussa, "Modeling the Aharonov Bohm effect: comparison with experiment," Senior Thesis, Worcester Polytechnic Institute, 2000.

[43] M. Kuzuoglu and R. Mittra, *IEEE Trans. Antennas Propag.* **45**, 474 (1997).

[44] J. Maloney, M. Kesler, and G. Smith, Proceedings of the

USNC/URSI Meeting, Baltimore, MD, July 1996, p365.

[45] D. Saint-James, E. J. Thomas, and G. Sarma, *Type II Superconductivity* (Pergamon, New York, 1969).

[46] A. D. Grassie, *The Superconducting State* (Sussex University Press, Brighton, 1975).

[47] L. P. Gorkov, *Sov. Phys.–JETP* **10**, 998 (1960).

[48] D. F. Griffiths, A. R. Mitchell, and J. L. Morris, *Comput. Methods Appl. Mech. Eng.* **45**, 177 (1984). A. R. Mitchell and S. W. Schoombie, *Numerical Methods in Coupled Systems*, ed. R. W. Lewis, P. Bettess, and E. E. Hinton (Wiley, New York, 1984).

[49] J. H. Argyris and M. Haase, *Comput. Methods Appl. Mech. Eng.* **61**, 71 (1987).

[50] Q. Wang and Z. D. Wang, *Phys. Rev.* B **54**, R15645 (1996).

[51] M. Rosenberg, "Finite element analysis of nonlinear Ginzburg–Landau type Schrödinger equations: applications to supercurrents in Bose condensates," Senior Thesis, Worcester Polytechnic Institute,1995.

[52] X. H. Wang, *Finite Element Methods for Nonlinear Optical Waveguides* (Gordon and Breach, Australia, 1995).

[53] R. P. Feynman, *Statistical Mechanics: a course of lectures* (W. A. Benjamin, Reading, MA, 1972), p341.

14

Time-dependent problems

14.1 Introduction

In this final chapter on the finite element method (FEM) we consider the essential features of modeling the time evolution of a quantum mechanical system. This problem has been investigated theoretically and numerically with great intensity over the past several decades, and deserves, at the very least, a separate volume to itself. However, of necessity, our treatment of the subject will provide no more than an introduction to the literature and describe some of the general methods of solving the initial value problem.

By including the time degree of freedom we introduce a fresh perspective on the subject of quantum mechanics. For example, the computer studies of Goldberg et al.,[1] of the scattering of a wave packet at a potential barrier and also at a potential well, showed the complex and beautiful formation of interference oscillations and the eventual build-up of the reflected and transmitted waves. Such calculations of scattering of wave packets are becoming more and more relevant in the understanding of quantum phenomena occurring, for example, in nanoscale devices and in the modeling of chemical reactions.

It is useful to compare techniques used for time development of the wave equation with the ones used for Schrödinger's equation, and it is in this context that we include some examples for the wave equation.

14.2 Standard approaches to time evolution

14.2.1 Schrödinger's equation and the method of finite differences

The time evolution of a quantum mechanical system that is described by its wavefunction, $\psi(x,t)$, is given by the time-dependent

Schrödinger equation

$$i\hbar \frac{\partial}{\partial t}\psi(x,t) = -\frac{\hbar^2}{2m}\frac{\partial^2}{\partial x^2}\psi(x,t) + V(x,t)\,\psi(x,t). \qquad (14.1)$$

In order to solve this parabolic partial differential equation[†] we require one boundary condition (BC) in the time variable t, and, being a second-order differential equation in the spatial variable x, we need two BCs on the spatial part. For example, for a simple particle-in-a-box problem with a time-independent potential $V(x)$ we may use the conditions that the particle is confined in a 1D box over the range $0 \le x \le L$ to write

$$\psi(x=0,t) = 0 = \psi(x=L,t), \qquad \text{for} \quad t_i \le t \le t_f, \qquad (14.2)$$

and

$$\psi(x,t=t_i) = \phi_0(x), \qquad \text{for} \quad 0 \le x \le L. \qquad (14.3)$$

Here $\phi_0(x)$ is some given initial configuration of the wavefunction at time $t = t_i$. The traditional way of treating this problem is to expand the wavefunction in terms of a complete set of states that satisfy the spatial BCs

$$\psi(x,t) = \sum_{n=1}^{\infty} C_n\, e^{i\omega_n t}\, \sin(n\pi x/L), \qquad (14.4)$$

where

$$\hbar w_n = \frac{\hbar^2 n^2 \pi^2}{2mL^2}. \qquad (14.5)$$

We then solve for the Fourier coefficients C_n at time $t = t_i$ using the known form of $\phi_0(x)$ as given in equation (14.3).

For more complex BCs, especially in higher spatial dimensions, we require the machinery developed in the earlier chapters. Unfortunately, a straightforward implementation of the usual variational approach to the space + time dimensions would be incomplete; *we do not have any BC at the final time $t = t_f$, resulting in the solution being not defined at t_f.*[‡]

[†]The classification of second-order differential equations in terms related to the classification of binomial forms is discussed, for example, in Courant and Hilbert.[2]

[‡]We return to this issue in Section 14.6 below.

In such a situation, the first inclination is to invoke the "shooting method." Given the value of the solution at $t = t_i$, we discretize the first derivative by setting $t_n + \Delta t = t_{n+1}$, and write

$$\frac{\partial \phi(x,t)}{\partial t}\bigg|_{t=t_i} = \frac{\phi(x, t_{i+1}) - \phi(x, t_i)}{\Delta t}. \tag{14.6}$$

For Schrödinger's equation, equation (14.1), with $V(x)$ independent of time, the formal solution is

$$\psi(x,t) = e^{-iH(t_f - t_i)/\hbar}\, \psi(x, t_i). \tag{14.7}$$

With the discretized derivative we would have

$$\psi(x)_{j+1} = (1 - iH\, \Delta t/\hbar)\, \psi(x)_j, \tag{14.8}$$

the index j referring to the discretized time. This is an unstable formulation for an iterative approach, and an implicit, stable, approach using

$$\psi(x)_{j+1} = (1 + iH\, \Delta t/\hbar)^{-1}\, \psi(x)_j \tag{14.9}$$

is recommended in *Numerical Recipes*.[3] In fact, the two forms of the discretized evolution operators can be combined to obtain a unitary finite difference form of the operator $e^{-iHt/\hbar}$

$$e^{-iHt/\hbar} \simeq \frac{(1 - \frac{1}{2}iH\, \Delta t/\hbar)}{(1 + \frac{1}{2}iH\, \Delta t/\hbar)}, \tag{14.10}$$

and *Numerical Recipes* restates the difference equation in the form

$$\left(1 + \frac{1}{2}iH\Delta t/\hbar\right)\psi(x)_{j+1} = \left(1 - \frac{1}{2}iH\Delta t/\hbar\right)\psi(x)_j. \tag{14.11}$$

The time evolution using finite differences can be reformulated to make use of fast Fourier transforms (FFTs). If we write the Hamiltonian as the sum of the kinetic and potential energy terms, $H = K + V(r)$, we can express the exponentiated time evolution factor in the "split-operator" form

$$e^{-iH\Delta t/\hbar} \simeq e^{-iK\Delta t/2\hbar} e^{-iV(r,t)\Delta t/\hbar} e^{-iK\Delta t/2\hbar} + \mathcal{O}(\Delta t^3) \tag{14.12}$$

that is accurate to order Δt^3. The Baker–Hausdorf theorem

$$e^A e^B = e^{A + B + \frac{1}{2}[A,B]} \tag{14.13}$$

is used in expressing the exponentiated operators in product form, and it then permits an estimation of the error. The initial state in configuration space is Fourier transformed using FFTs into a momentum state representation in which the factor $\exp(-iK\Delta t/2\hbar)$ is evaluated immediately, since the operator is given in momentum space. Then the state is Fourier transformed back to configuration space to evaluate the exponential factor with the potential energy. By further use of FFTs the final state is obtained at time $t+\Delta t$. This method has been used to study chemical reactions and scattering,[4-7] and the reader is directed to the book by Bayfield[8] for further details.

14.2.2 The finite difference method for the wave equation

We note that the same issues come up for the wave equation, which is a hyperbolic partial differential equation of second order in space (we limit ourselves to one spatial dimension) and time

$$\frac{n^2}{c^2} \frac{\partial^2}{\partial t^2} \phi(x,t) - \frac{\partial^2}{\partial x^2} \phi(x,t) = 0, \qquad (14.14)$$

when both the BCs on time are given at the initial time t_i. Here $v = c/n$ is the velocity of the wave, $n(x)$ being the refractive index. This is the usual physical circumstance where we attempt to follow the time evolution of an electromagnetic wave given initial conditions for it. In this case, with equation (14.14), we further substitute the discretized form for the second derivative at $t = t_n$

$$\left. \frac{\partial^2}{\partial t^2} \phi(x,t) \right|_{t=t_n} = \frac{\phi(x,t_{n+1}) - 2\,\phi(x,t_n) + \phi(x,t_{n-1})}{(\Delta t)^2}, \qquad (14.15)$$

in the differential equation. This leads to a three-term recursion relation for the time development; since the value of the function is known at the first two points on a discretized spatial/temporal grid, it can be solved for at the next point. This approach has been pursued vigorously since the early papers of Yee[9] and of Taflove and Brodwin,[10] and the method is known as the finite difference time domain (FDTD) approach. This is elaborated on in recent books by Taflove.[11,12]

An interesting way of improving the solution of the wave equation is to use the "eikonal approximation." We pull out a factor corresponding to the leading solution of the wave equation, $e^{ikx-i\omega t}$. Substituting

$$\phi(x,t) = A(x)\,e^{ikx-i\omega t} \qquad (14.16)$$

into the wave equation we obtain the reduced differential equation for the as-yet-unknown space-dependent coefficient $A(x)$:

$$-\frac{d^2}{dx^2}A(x) - 2\,i\,k\,\frac{d}{dx}A(x) + \left(k^2 - \frac{\omega^2 n^2(x)}{c^2}\right)A(x) = 0. \quad (14.17)$$

It is usual to assume that the amplitude $A(x)$ of the wave does not vary much over a few wavelengths and that $A''(x) \simeq 0$. Again we have at hand an initial value problem for $A(x)$.

We refer the reader to the papers by Feit and Fleck[13] for a more detailed discussion of the general methods of solution in two and three dimensions with applications to lasers, self-focusing, filament formation, and other effects in nonlinear media.

As with all discretized time-stepping schemes, Schrödinger's equation and the wave equation accumulate errors as the solution evolves in time. While various methods have been developed in terms of forward–backward differencing for improving the calculations, the net result is that errors *will* accumulate. It is fairly easy to see that this will be the case since the exact solution is expected to have an exponential form for the first-order (and also for the second-order) differential equations. The time discretization attempts to approximate the exponential form by a series of linear steps. Typically, the accumulated errors inevitably force the solution into runaway exponentials.

Recently, there have been proposals to use *nonstandard* finite difference schemes. A careful, but simple, redefinition of the discretized derivatives in two and three dimensions can be shown to lead to higher levels of accuracy. We refer the reader to articles by Cole and colleagues.[14-16]

14.3   A transfer matrix for time evolution

We can generate an *implicit* time development scheme as follows. Within a finite element framework, as discussed in the earlier chapters, the spatial part of the Schrödinger differential equation is discretized by setting up a variational or a Galerkin approach. The unknown wavefunction is expanded over each element in terms of local interpolation polynomials or shape functions. In the Galerkin approach, the projection onto each of the shape functions for the left side of equation (14.1) is obtained, generating the discretized form of Schrödinger's equation in the form of simultaneous equations for the nodal variables. The equations are rearranged, in a manner analogous to element matrix overlays, so that we finally arrive at a set of simultaneous differential equations in the time variable. This leads to a vector differential equation of the general form

$$i\hbar \, \mathbf{B} \, \frac{d}{dt} \Psi(t) = \mathbf{A} \, \Psi(t). \tag{14.18}$$

Here $\mathbf{A}$ and $\mathbf{B}$ are the global matrices obtained after the spatial integration is performed, and Ψ is the vector of unknown coefficients in the spatial interpolation scheme. The matrix $\mathbf{A}$ includes the kinetic and potential energy terms, and we again assume here that the potential $V(x)$ is independent of time. By working out the details it is seen that the matrix $\mathbf{B}$ is made up of overlap integrals (we are using local nonorthogonal basis functions as our shape functions in each element) and is invertible. The formal solution of equation (14.18) is

$$\Psi(t) = e^{-i(\mathbf{B}^{-1}\mathbf{A})\,(t-t_i)/\hbar} \, \Psi(t_i). \tag{14.19}$$

For convenience, let us define

$$\mathbf{C} = \mathbf{B}^{-1}\mathbf{A}. \tag{14.20}$$

As is usual, the exponential function of the matrix $\mathbf{C}$ is given a meaning by representing it as the infinite series expansion in terms of powers of the matrix $\mathbf{C}$,

$$\exp[-i\mathbf{C}(t-t_i)/\hbar] = 1 - \frac{i}{\hbar}(t-t_i)\,\mathbf{C} - \frac{(t-t_i)^2}{2!\hbar^2}\mathbf{C}^2 + \dots. \tag{14.21}$$

The exponentiated matrix, the left side of the above equation, is called a transfer matrix,[17] in that it propagates the solution from time $t = t_i$ to another time t. The exponentiation of an $N \times N$ matrix can be performed using the Cayley–Hamilton theorem,[18] and the function can be expressed as a terminating series in powers of the matrix $\mathbf{C}$ from $0, \dots, (N-1)$. In effect, we have provided a discretized version of the time evolution operator that is developed in sections on time-dependent perturbation theory in books on quantum mechanics.

More directly, we can obtain the exponential function of a matrix by first diagonalizing $\mathbf{C}$. Consider the eigenvalue problem

$$\mathbf{C}\,\xi^{(j)} = (\hbar\omega_j)\,\xi^{(j)}, \tag{14.22}$$

where $\hbar\omega_j$ and $\xi^{(j)}$ are eigenvalue and eigenfunction pairs with $j = 1, \dots, N$. We place the vectors $\xi^{(j)}$ in N columns to form a new $N \times N$

matrix, $\mathbf{P}$, of eigenvectors. We also arrange the eigenvalues along the diagonal of an $N \times N$ matrix $\mathbf{\Omega}$. Now the relation, equation (14.22), can be written as

$$\mathbf{C} \cdot \mathbf{P} = \mathbf{P} \cdot \mathbf{\Omega}. \tag{14.23}$$

In other words,

$$\mathbf{C} = \mathbf{P} \cdot \mathbf{\Omega} \cdot \mathbf{P}^{-1}. \tag{14.24}$$

This states that the unitary transformation matrix $\mathbf{P}$ which diagonalizes the matrix $\mathbf{C}$ is made up of its eigenfunctions. Once $\mathbf{P}$ is determined, we can express the exponential matrix in the form

$$\exp(-i\,\mathbf{C}\,t/\hbar) = \mathbf{P} \cdot \left(1 - i\mathbf{\Omega} t/\hbar - \frac{t^2}{2}\mathbf{\Omega}^2/\hbar^2 + \cdots \right) \cdot \mathbf{P}^{-1}.$$

$$= \mathbf{P} \operatorname{diag}(e^{-i\omega_1 t}, e^{-i\omega_2 t}, \ldots, e^{-i\omega_N t})\,\mathbf{P}^{-1}. \tag{14.25}$$

In practice, since $\mathbf{B}^{-1}\mathbf{A}$ is not symmetric we rearrange the eigenvalue problem by performing a Cholesky decomposition $\mathbf{B} = \mathbf{L}\mathbf{L}^{\mathbf{T}}$, and restoring the symmetry by writing

$$\mathbf{A}\,\psi = \lambda\,\mathbf{B}\,\psi = \lambda\mathbf{L}\left\{\mathbf{L}^{\mathbf{T}}\psi\right\},$$

$$\left[\mathbf{L}^{-1} \cdot \mathbf{A} \cdot (\mathbf{L}^{T})^{-1}\right]\left\{\mathbf{L}^{T}\psi\right\} = \lambda\left\{\mathbf{L}^{T}\psi\right\}. \tag{14.26}$$

This reduces the equation to a standard eigenvalue problem. The eigenfunctions $\mathbf{L}^{\mathbf{T}}\psi$ are then transformed back to the original basis.

Given this exponential matrix, the solution $\Psi(t)$ is now determined as in equation (14.19). We can deduce from the above discussion that the behavior of the solution depends in an essential manner on the eigenvalues of the $\mathbf{C}$ matrix; these eigenvalues correspond to the ω_j of equation (14.22). If some of the eigenvalues have imaginary parts to them, we would have exponentially rising or falling solutions. This development allows us to quantify the error accumulation mentioned in the previous section. This is the basic idea of stability analysis of solutions of differential equations.[19-21]

A corresponding treatment of the wave equation by integrating out the spatial variables leads to a time-dependent vector $\mathcal{F}(t)$ of interpolation coefficients[§] that satisfies the equation:

$$\mathbf{B} \cdot \frac{d^2\mathcal{F}(t)}{dt^2} + \mathbf{A} \cdot \mathcal{F}(t) = 0. \tag{14.27}$$

[§]If there is damping in the problem we could have a first-order derivative term as well. This causes very minor changes in the formalism. The solutions, however, could reflect this as a substantial change.

Now the second-order differential equation is written in terms of two coupled first-order differential equations

$$\frac{d}{dt}\begin{pmatrix} \mathcal{F}(t) \\ d\mathcal{F}/dt \end{pmatrix} = \begin{pmatrix} \mathbf{0} & \mathbf{1} \\ -\mathbf{B}^{-1}\mathbf{A} & \mathbf{0} \end{pmatrix} \cdot \begin{pmatrix} \mathcal{F}(t) \\ d\mathcal{F}/dt \end{pmatrix}. \qquad (14.28)$$

Once again, stability analysis can be performed by diagonalizing the $2N \times 2N$ matrix on the right side, and examining its eigenvalues. The formally exact solution is a transfer matrix relating the initial solution at $t = t_i$ to the solution at $t = t_f$.

What are the limitations of the time-stepping approach and of the transfer matrix approach? In the time-stepping, or the time-slicing scheme as it is sometimes referred to, we are approximating the solution by a series of linear, quadratic, or other interpolation schemes in the time variable. As mentioned earlier, this accumulates errors such that over time scales of the order $[\Im(\omega_j) \times (t_f - t_i)] \geq 1$ we can anticipate deviations from the exact answer.

On the other hand, the transfer matrix approach has its own problems. The diagonalization of the coefficient matrix $\mathbf{C}$ will in general have round-off errors that increase with increasing matrix size. The eigenfunctions are obtained with less precision. The transformation matrix $\mathbf{P}$ constructed from the eigenfunctions will carry these errors forward. The eigenfunctions have to be orthogonal to each other, and this requires a Gram–Schmidt orthogonalization. As the orthogonalization is carried out, one accumulates error very rapidly. Despite using the so-called reverse Gram–Schmidt orthogonalization[¶] the accumulated errors in setting up the transformation matrix becomes prohibitively large already for matrices of order $N \simeq 100$–500. We also have to contend with calculating the inverse of the transformation matrix $\mathbf{P}$ and this further compounds the errors. The condition number of the $\mathbf{P}$ matrix provides a measure of the loss of accuracy during the inversion. The condition number may be such that for particular problems the numerical coefficients could lead to considerable roundoff in the inversion. Thus the implementation of the transfer matrix is fraught with numerical round-off error arising from these steps in the calculation for matrices larger than, say, 100×100. All these issues are important for determining if a solution is an acceptable one.

[¶]See Ref. 22 and other references given in Chapter 15.

14.4   Lanczos reduction of transfer matrices

A practical compromise has been developed by Park and Light[23] to overcome in part the difficulties mentioned above with large transfer matrices. They suggest that the transfer matrix can be evaluated using the Lanczos method[24-26] for diagonalizing the $\mathbf{C}$ matrix. The Lanczos method is discussed in Chapter 15. The Lanczos method naturally converges towards the highest eigenvalues. Here we follow their treatment *except that we develop the algorithm by seeking the lowest eigenvalues ω_i of the matrix $\mathbf{C}$ rather than the highest eigenvalues.* We anticipate that the states of very high frequency will not dominate the time evolution.$^{\parallel}$

Consider the $N \times N$ transfer matrix $\exp[-i\mathbf{C}(t - t_i)]$ and its expansion in a power series as in equation (14.21). We can obtain an approximate solution at time t by truncating the power series for the transfer matrix after p terms. This truncation will lead to a solution that will be within an error tolerance for times $t \le \tau$ provided certain criteria are met. We will elaborate on this below.

We generate a sequence of independent vectors $\{r_0,\, \mathbf{C}\, r_0,\, \mathbf{C}^2\, r_0,$ ..., $\mathbf{C}^{p-1}\, r_0\}$, where r_0 is an arbitrary starting vector of dimension N, and the number of vectors r_i is p. The solution $\Psi(t)$ is then expanded in terms of *mutually orthogonal* Lanczos vectors q_i which are actually linear combinations of the r_i. The vectors q_i are chosen such that they satisfy B-orthogonality

$$q_i^\dagger\, \mathbf{B}\, q_j = \delta_{ij}. \tag{14.29}$$

They are defined by the method of Lanczos discussed in Chapter 15, and are given by the relation

$$(\mathbf{A}^{-1}\mathbf{B})\,[\mathbf{q}_1, \mathbf{q}_2, \ldots, \mathbf{q}_p] = [\mathbf{q}_1, \mathbf{q}_2, \ldots, \mathbf{q}_p]\,\mathbf{T}, \tag{14.30}$$

where the vectors $\mathbf{q}_i$ are arranged column by column into an $N \times p$ matrix which we will label by $\mathbf{Q}$, and

$$\mathbf{T} = \begin{bmatrix} \alpha_1 & \beta_1 & 0 & 0 & & & \\ \beta_1 & \alpha_2 & \beta_2 & 0 & & & \\ 0 & \beta_2 & \alpha_3 & \beta_3 & & & \\ & & \ddots & \ddots & \ddots & & \\ & & & & & \beta_{p-1} & \\ & & & & \beta_{p-1} & \alpha_p \end{bmatrix}. \tag{14.31}$$

$^{\parallel}$In fact, we can focus our search for the frequency spectrum around any given value σ by seeking the eigenvalues $1/(\omega_i - \sigma)$ of $(\mathbf{A} - \sigma\mathbf{B})^{-1}\mathbf{B}$.

The transformation of the original matrix $\mathbf{C}$ into the tridiagonal matrix $\mathbf{T}$ is given by

$$[\mathbf{Q}^\dagger \, \mathbf{B} \, \mathbf{A}^{-1} \, \mathbf{B} \, \mathbf{Q}]_{\alpha\beta} = \mathbf{T}_{\alpha\beta}, \qquad (14.32)$$

where the indices α, β range over $1, \ldots, p$. Let the eigenfunctions of $\mathbf{T}$ be given by

$$\mathbf{T}_{\alpha\beta} \, s_\beta = \gamma_\alpha \, s_\alpha. \qquad (14.33)$$

Let the $p \times p$ matrix $\mathbf{s}$ be formed from the s_α by placing them column by column into the matrix. When written in matrix notation we have

$$\mathbf{s}^{-1} \cdot \mathbf{T} \cdot \mathbf{s} = \Gamma. \qquad (14.34)$$

The matrix $\mathbf{s}$ is unitary and we obtain

$$\mathbf{s}^\dagger \, \mathbf{Q}^\dagger \, (\mathbf{A}^{-1} \, \mathbf{B}) \, \mathbf{Q} \, \mathbf{s} = \Gamma. \qquad (14.35)$$

The eigenfunctions of $\mathbf{A}^{-1}\mathbf{B}$ are the same as those of $\mathbf{B}^{-1}\mathbf{A}$ and the first p eigenfunctions are given by $\mathbf{Q} \cdot \mathbf{s}$. The approximate eigenvalues γ_α can be shown[22] to approach the eigenvalues $1/\omega_\alpha$. This convergence is remarkably rapid – each addition to the dimension p gives an additional approximate eigenvalue while noticeably improving the earlier ones towards convergence to their true values.

In brief, the above procedure solves the eigenvalue problem

$$(\mathbf{A}^{-1} \, \mathbf{B}) \, \mathbf{Q} \, \mathbf{s} = \mathbf{Q} \, \mathbf{s} \, \Gamma, \qquad (14.36)$$

and thereby determines the lower end of the eigenspectrum together with the corresponding eigenfunctions. If we pursue the iterative method using the relation

$$(\mathbf{B}^{-1} \, \mathbf{A}) \cdot \mathbf{Q} \, \mathbf{s} = \mathbf{Q} \, \mathbf{s} \, \Omega, \qquad (14.37)$$

we would obtain the eigenvalues at the higher end of the spectrum and the corresponding eigenfunctions.

How does one pick the starting vector? In this case we have a ready-made answer: we use the known initial vector $\Psi(t_i)$ as the starting vector q_1 in the Lanczos development of the solution $\Psi(t)$ in the space of q vectors. The initial vector $\Psi(t_i)$ then corresponds to the coefficients $\tilde{s}_1^\dagger = (1, 0, 0, \ldots)$ of the q_i vectors. Note that $\tilde{s}_1$ is in general not an eigenvector of the matrix $\mathbf{T}$.

Now, the initial differential equation, equation (14.18), is recast as a differential equation for the coefficient vector $\tilde{s}_1(t)$

$$i\hbar\,\mathbf{B}\cdot\mathbf{Q}\,\frac{d}{dt}\tilde{s}_1(t) = \mathbf{A}\cdot\mathbf{Q}\,\tilde{s}_1(t), \qquad (14.38)$$

in the $p \times p$ space. We now multiply both sides by $\mathbf{Q}^{\dagger}$, use the B-orthonormality of the q vectors, and substitute $\mathbf{A}\,\mathbf{Q} = \mathbf{B}\,\mathbf{Q}\,\mathbf{T}^{-1}$ using equation (14.30). This leads to the equation

$$\tilde{s}_1(t) = e^{-i\mathbf{T}^{-1}(t-t_i)/\hbar}\,\tilde{s}_1(t_i). \qquad (14.39)$$

As time increases, this initial p-dimensional coefficient vector evolves into a vector that is still a linear combination of the same p eigenvectors of $\mathbf{T}$, or equivalently of $\mathbf{T}^{-1}$. The vector $\tilde{s}_1(t)$ represents the coefficients of q_i at time t that will give $\Psi(t)$ such that

$$\Psi(t) = \mathbf{Q}\,\tilde{s}_1(t). \qquad (14.40)$$

Problems with this approximation arise when $\Psi(t)$ requires additional q vectors that are outside the p-dimensional space to represent it, and the expansion of the exponential in equation (14.39) to just p terms is inadequate. The time τ over which an expansion to p terms will be sufficient may be characterized by the requirement that the pth component of $\tilde{s}_1(t)$ has a probability less than a tolerance ϵ

$$|\tilde{s}_1(t)_p| \le \epsilon. \qquad (14.41)$$

In other words, the contribution of the last Lanczos N-vector q_p is small. The time at which this limit is reached will provide the temporal range of accuracy of the truncated expansion. Following the discussion of Park and Light, given a value of ϵ we can determine τ by noting that

$$|\tilde{s}_1(t)_p| \simeq \epsilon = \left[\left(\sum_{j=0}^{p-1}\frac{(-i)^j(\tau - t_i))^j}{j!\hbar^j}\mathbf{T}^{-j}\right)\tilde{s}_1(t_i)\right]_p. \qquad (14.42)$$

The estimation here seems harder to extract than in their procedure because they focus on the larger eigenvalues and work with $\mathbf{T}$ rather than its inverse. Finally, if the final wavefunction is accurate enough, it may be reused as the initial guess for the next step in time development along these lines. Examples of the use of this method are given by them.

14.5   Instability with initial conditions

If we insist on employing the FEM for the calculation of the initial
value problem we end up facing numerical instability. The best way
to see this is to use a simple illustrative 1D problem to bring out the
difficulty. We discuss two examples here to show that the problem is
in the global matrix which becomes progressively badly conditioned
as we *increase* the number of elements. Surprisingly, there is a way
to recover from this bad conditioning of the matrix. We show how
this can be done in the following.

14.5.1   Comparing IVBC and two-point BCs

Here we wish to explicitly distinguish between the two-point boundary conditions (2PBC) and the initial value boundary conditions,
which we may label as IVBC, problems. In order to illustrate the
issue let us consider the simple 1D problem over the range $[0, 1]$ with
a unit inhomogeneous term

$$\frac{d^2}{dt^2}u(t) = f(t) = 1. \tag{14.43}$$

This is similar to the equation describing projectile motion in the
presence of a constant acceleration, $u''(t) = -g$. Let the initial
boundary conditions (IVBC) be

$$u(t = 0) = -0.5, \qquad u'(t = 0) = 0. \tag{14.44}$$

The two-point boundary value (2PBC) problem that we can compare
with the above is

$$u(t = 0) = -0.5, \qquad u'(t = 1) = 1. \tag{14.45}$$

The analytical solution given by

$$u(t) = \frac{(t^2 - 1)}{2} \tag{14.46}$$

is shown in Fig. 14.1.

Only the 2PBC problem can be expressed as an action integral
in the *traditional* way, whereas the Galerkin formulation can be applied to both 2PBC and IVBC. Here we make use of the Galerkin
method and show that the Galerkin approach with the IVBC has
an instability. In the following section we will show how to overcome this limitation by fundamentally altering the action integral to
accommodate the IVBC.

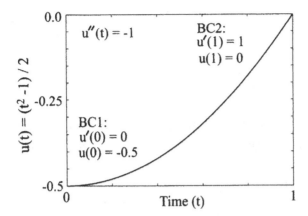

Fig. 14.1 Solution of the differential equation $u''(t) = -1$ over the range $[0, 1]$ with "shooting" BCs only at one end, which is analytically given by $u(t) = (t^2 - 1)/2$.

As an aside, note that the action integral will have to be expressed as

$$\mathcal{A} = \int_0^1 dt \left[\frac{1}{2} \left(\frac{du(t)}{dt} \right) \left(\frac{du(t)}{dt} \right) + u(t)f(t) \right]. \qquad (14.47)$$

We have the "source term" $f(t) = 1$ in our present example. With the usual variational method we obtain surface terms through integration by parts of the first term in the action:

$$[\delta\, u(t)\, u'(t)]_{t=0}^{t=1}. \qquad (14.48)$$

At $t = 0$, the value of $u(0)$ is fixed, so that $\delta\, u(0)$ is zero. However, the surface term at $t = 1$ does not vanish. We might think that subtracting a term $u(t = 1)\, u'(t = 1)$ from the action would do the trick. However, these quantities are unknown, and under the variations of the modified action our correction term would generate two terms

$$- \left[\delta\, u(t = 1)\, u'(t = 1) + u(t = 1)\, \delta\, u'(t = 1) \right].$$

Even if the first term cancels a term generated from the action given above, we do not know how to evaluate the second term since both $u(t)$ and $u'(t)$ are unknown there.

Let us return to the Galerkin approach to the FEM. We break up the region $[0,1]$ into a number of elements, and then represent the function $u(t)$ in terms of quintic Hermite interpolation polynomials in each element,

$$u(t) = \sum_{i=1}^{3} u_i N_i(t) + \sum_{j=1}^{3} u_j' \overline{N}_j(t), \qquad (14.49)$$

where the N_i, $\overline{N}_i$ are Hermite interpolation or shape functions. While the choice of the quintic Hermite interpolation polynomial in this instance is "overkill" it allows us to accommodate the IVBC in a convenient manner. The left side of equation (14.43) is multiplied by each of the shape functions and integrated over each element. This projection of $(u'' - 1)$ onto each shape function is set to zero. We obtain

$$\sum_{iel} \int_{t_{iel}}^{t_{iel+1}} dt N_\mu N_\nu'' u_\nu = \sum_{iel} \int_{t_{iel}}^{t_{iel+1}} dt\, N_\mu, \qquad \mu = \{i, j\}. \, (14.50)$$

On integrating by parts the second-derivative term we have

$$-\sum_{iel} \int_{t_{iel}}^{t_{iel+1}} dt N_\mu' N_\nu' u_\nu + \sum_{iel} \left[N_\mu N_\nu' u_\nu \right]_{t_{iel}}^{t_{iel+1}}$$

$$= \sum_{iel} \int_{t_{iel}}^{t_{iel+1}} dt\, N_\mu. \qquad (14.51)$$

Such equations with projections on shape functions that are associated with the same global nodal variable from adjacent elements iel and $(iel +1)$ are added together in a manner analogous to the overlay of element matrices in the action formulation (see Chapters 1 and 2). The "surface terms" reduce to just two. With this operation completed, we can write the global set of equations in matrix form

$$M_{\alpha\beta} U_\beta + \left[N_\mu N_\nu' u_\nu \right]_{t_i}^{t_f} = R_\alpha, \qquad (14.52)$$

where the matrix $M_{\alpha\beta}$ is obtained after integration from the first term in equation (14.51), and the indices α, β refer to the ordered set of indices for the function and its derivative at each of the nodes in the global system. The components of the vector R_α correspond to the integration of interpolation polynomials in adjacent elements that have the coefficients U_α.

For the sake of clarity, it is now best to limit ourselves to one quintic element while considering the implementation of BCs. For

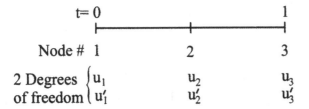

Fig. 14.2 BCs are discussed in the text using a single element for the physical region in order to simplify the book-keeping. The element has three nodes with two degrees of freedom at each node. The shape functions are quintic Hermite interpolation polynomials.

just one element with three nodes each having two degrees of freedom, as in Fig. 14.2, we have

$$
\begin{pmatrix}
M_{11} & M_{12} & M_{13} & M_{14} & M_{15} & M_{16} \\
M_{21} & M_{22} & M_{23} & M_{24} & M_{25} & M_{26} \\
M_{31} & M_{32} & M_{33} & M_{34} & M_{35} & M_{36} \\
M_{41} & M_{42} & M_{43} & M_{44} & M_{45} & M_{46} \\
M_{51} & M_{52} & M_{53} & M_{54} & M_{55} & M_{56} \\
M_{61} & M_{62} & M_{63} & M_{64} & M_{65} & M_{66}
\end{pmatrix}
\cdot
\begin{pmatrix}
u_1 \\
u_1' \\
u_2 \\
u_2' \\
u_3 \\
u_3'
\end{pmatrix}
$$

$$
+ \delta_{\{5,6\}} u_3' - \delta_{\{1,2\}} u_1' = -
\begin{pmatrix}
R_1 \\
R_2 \\
R_3 \\
R_4 \\
R_5 \\
R_6
\end{pmatrix} .
\tag{14.53}
$$

The so-called surface terms at the two ends of the physical range for time are evaluated using the properties of the Hermite interpolation polynomials. They can be included into the matrix $\mathbf{M}$ by adding unity to M_{56} and subtracting 1 from M_{12}, and this takes care of surface terms in equation (14.53).

If we were to apply BCs at the two end points, we would insert the known values of u_1 and u_3' into equation (14.53) and transfer the known terms from the left side to the right side of the equation with a change of sign. We would be left with six equations for four unknown variables, i.e., two equations more than the number of unknowns. In a larger set of equations, arising from employing a larger number of elements, we could safely eliminate two equations corresponding to the two variables associated with the BCs, namely the first and the sixth in the above one-element example, to obtain

$$
\begin{pmatrix} M_{22} & M_{23} & M_{24} & M_{25} \\ M_{32} & M_{33} & M_{34} & M_{35} \\ M_{42} & M_{43} & M_{44} & M_{45} \\ M_{52} & M_{53} & M_{54} & M_{55} \end{pmatrix} \cdot \begin{pmatrix} u_1' \\ u_2 \\ u_2' \\ u_3 \end{pmatrix} = - \begin{pmatrix} R_2 - M_{21}\,u_1 - M_{26}\,u_6 \\ R_3 - M_{31}\,u_1 - M_{36}\,u_6 \\ R_4 - M_{41}\,u_1 - M_{46}\,u_6 \\ R_5 - M_{51}\,u_1 - M_{56}\,u_6 \end{pmatrix}.
$$

$$(14.54)$$

We can show that this is the same set of simultaneous equations we would obtain from the action integral formulation of the FEM for the (two-point) BCs. In fact, the elimination of the first and sixth rows in some sense is encouraged by the way the action formulation works. The global test functions must satisfy the BCs of the problem. In this case, the numerical solution converges to the analytically obtained solution fairly rapidly as the number of elements is increased. Just one element would more than suffice to reproduce the simple solution function here. What is important is to note that adding more elements does not destabilize the solution in this case.

We now return to the IVBC. In the Galerkin approach, we insert the known values for u_1 and u_1' into equation (14.53) and transfer the known terms to the right side. Again, this leaves us with four unknowns and six simultaneous equations. Eliminating the first two equations to reduce the number of equations to the number of unknowns, we again obtain a matrix equation of the general form $\mathbf{A} \cdot \mathbf{x} = \mathbf{b}$.

Goloskie and Ram-Mohan[27] noted a decade ago that *the larger the number of finite elements used in the calculation of initial value problems the more ill-conditioned the coefficient matrix becomes* in this case.

We can understand this issue by considering the introduction of more elements providing more flexibility in the solution with no constraints at the upper limit, $t = 1$. This may be likened to determining the shape of a bicycle chain with just the first link positioned properly with its location and slope being specified. The opposite end having no constraint on it has the freedom to be located anywhere.

With just 10 elements, the condition number, as defined by the LINPACK subroutines, indicates that we have lost all accuracy in a double-precision calculation working with 15 significant figures in intermediate steps in the calculations! Moreover, the idea of convergence to the exact solution by using more and more elements fails dramatically in this case.

A practical solution to this problem was suggested in 1992.[27] Note that the application of the two BCs reduces the number of unknowns by two, while keeping the number of equations the same. Let us continue to use just one quintic element in order to present a clearer picture during this discussion. At this stage, there is no physically compelling reason to select two particular equations to be discarded. A better criterion would be to drop the first equation, the one associated with u_1, since the Galerkin weight functions must satisfy the BCs, and to investigate the selection of the second equation that would improve the condition number of the coefficient matrix. When the condition number is charted against the equation number to be eliminated, one finds that the most favorable choice occurs when either one of the last two equations connected with the variable u_3 or u_3' is dropped. Furthermore, the problem with the condition number is ameliorated substantially by this procedure, as is seen in Fig. 14.3 in which we have shown only two graphs with

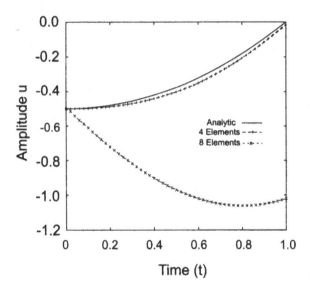

Fig. 14.3 Solution of the differential equation $u''(t) = 1$ over the range $[0, 1]$ with "shooting" BCs at one end, $u(0) = 0.5$, $u'(0) = 0$. The analytic solution is given by $u(t) = (t^2 - 1)/2$. The one-sided BCs lead to instability that increases as the number of elements is increased.

Table 14.1 The condition number of the global matrix is tabulated for increasing number of quintic Hermite elements for the example problem $u(t)'' = 1$, $0 \leq t \leq 1$, with initial conditions $u(0) = -0.5$ and $u'(0) = 0$. The condition numbers when rows 1, 2 are eliminated and when rows $1, N - 1$ are removed are given. The matrix dimension is $N = 4 \times nelem + 2$.

Number of elements	Rows (1,2) removed Condition number	Rows $(1, N - 1)$ removed Condition number
1	3.72632 E-04	2.91999 E-03
2	3.67164 E-06	1.92855 E-04
3	1.10841 E-07	6.38362 E-05
4	4.77013 E-09	2.74292 E-05
5	2.48053 E-10	1.74835 E-05
10	3.29683 E-16	2.55450 E-06
15	3.24357 E-20	8.13854 E-07
20	8.47680 E-21	2.90739 E-07
30	1.00259 E-20	8.87835 E-08
40	1.18717 E-21	3.81221 E-08
50	2.22587 E-21	2.06716 E-08
100	3.95541 E-22	2.48256 E-09

increasing elements. In Table 14.1, we have tabulated the condition number as given by LINPACK matrix inversion subroutines where the logarithm of the condition number is an approximate measure of the number of digits of accuracy lost in the matrix inversion. What is remarkable is that the procedure of eliminating the equations with the largest "separation" in the row index leads to a recovery of accuracy. The corresponding solutions are virtually coincident with the analytic solution and have not been displayed in Fig. 14.3. All calculations were performed in double-precision arithmetic, so that loss of accuracy of nine significant figures would still allow us to extract the result accurate to about five or six significant figures.

Another simple example we consider is the solution of the initial value differential equation

$$\frac{d^2}{dt^2}u(t) + 100\ u(t) = 0, \qquad 0 \leq x \leq 1;$$
$$u(0) = 1,$$
$$u'(0) = 0. \tag{14.55}$$

Table 14.2 The condition number of the global matrix is tabulated for increasing number of quintic Hermite elements for the example problem $u(t)'' + 100u(t) = 1$, $0 \leq t \leq 1$, with initial conditions $u(0) = 1$ and $u'(0) = 0$. The condition numbers when rows 1, 2 are eliminated and when rows $1, N - 1$ are removed are given. The matrix dimension is $N = 4 \times nelem + 2$.

Number of elements	Rows $(1, 2)$ removed Condition number	Rows $(1, N - 1)$ removed Condition number
1	4.40914 E-04	4.90690 E-04
2	2.21134 E-05	4.90460 E-04
3	1.24922 E-06	2.67398 E-04
4	9.49368 E-08	1.97238 E-04
5	1.32501 E-08	5.93843 E-05
10	2.14778 E-12	7.57716 E-06
15	8.35554 E-13	1.96659 E-06
20	1.31966 E-13	1.00107 E-06
30	7.03054 E-15	3.71075 E-07
40	7.96849 E-16	1.31356 E-07
50	1.42842 E-16	6.39990 E-08
100	6.27832 E-19	1.27110 E-08

The analytic solution is $u(x) = \cos(10x)$. The numerical simulations using the FEM are performed with quintic Hermite elements. The consequence of employing the usual method of applying BCs on the condition number of the global coefficient matrix is shown in Table 14.2 for increasing number of elements. In each case, the elimination of the first equation associated with the starting coordinate $t = 0$ and the second-to-last equation corresponding to the final time coordinate $t = 1$ are shown to lead to a substantial increase in the expected accuracy of the solution. The results of implementing the BCs using the standard choice for the eliminated equations is shown in Fig. 14.4. With increasing number of elements the instability grows. On employing the procedure outlined above we substantially recover accuracy in the resulting curves, and for all practical purposes the curves are indistinguishable from the analytic solution.

These numerical experiments have been extended to space–time calculations by Kieweg.[28] He considers the solution of the wave equation in 1+1 dimensions and solves the boundary value problem for the vibrations of a string held fast at $x = 0$ and at $x = L$ at all

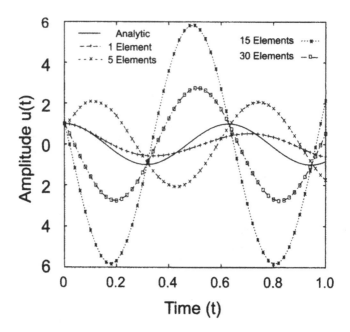

Fig. 14.4 Solution of the differential equation $u''(t) + 100u(t) = 0$ over the range $[0, 1]$ with "shooting" BCs at one end, which is analytically given by $u(t) = \cos(10t)$. The one-sided BCs lead to instability that increases as the number of elements is increased.

times. The initial configuration of the string $u(x, t = 0)$ and its velocity $u'(x, t = 0)$ are specified and the time evolution of the wave is solved using rectangular Hermite space+time finite elements. Here again he found improvement in the condition number of the global matrix and the consequent improvement in the accuracy of the solution does occur.

In summary, in this section we considered the solution of a second-order differential equation in the FEM with two initial conditions. When we make the "natural" choice of eliminating equations 1 and 2 in the finite element development with Hermite interpolation functions we obtain global matrices that are ill-conditioned. From a variational point of view, the initial value problem does not pin down the solution at both ends, as mentioned earlier in a mechanical analogy. When we pick equations numbered 1 and $N - 1$, associated with the first and the last nodes, for elimination the conditioning improves substantially. There are two ways of understanding this improvement in the condition number. First, given N equations for

$N - 2$ variables it is not always clear that the natural choices are the most appropriate. Perhaps all we have shown is that empirically it is almost always better to choose the first and the last equation in the set of global equations! Secondly, in keeping the second equation and eliminating the last we are effectively replacing the initial value derivative boundary condition by a condition $u(t = 1) = h$, where h is unknown, but is determined through $u'(t = 0) = f(0)$, in our examples. As proposed by Noble[29] (also see Ref. 30), we solve the problem as if h is implicitly known and at the very end use the derivative condition to express $u(t = 1)$ in terms of its derivative value at the initial point.

Note that this section focused on a second-order differential equation instead of a first-order equation; this was just to consider the more difficult example.

14.6   The variational approach

For concreteness, let us consider the Schrödinger action integral for a particle in an infinite potential well with integrals over space and time of the Lagrangian density. Let the potential $V(x,t)$ be a function of space and time inside the well $0 \le x \le L$. The BCs are given in equation (14.3), with the initial configuration of the wavefunction $\psi(x,t)$ being $\phi_0(x)$.

14.6.1   A variational difficulty

We begin by considering the action

$$\mathcal{A}_0[\psi^*, \psi] = \int_0^L dx \int_0^T dt \left[-\frac{i\hbar}{2} \left(\psi^* \frac{\partial \psi}{\partial t} - \frac{\partial \psi^*}{\partial t} \psi \right) \right.$$

$$\left. + \frac{\hbar^2}{2m} \frac{\partial \psi^*}{\partial x} \frac{\partial \psi}{\partial x} + \psi^* V(x,t) \psi \right] \tag{14.56}$$

where the space–time dependence of $\psi(x,t)$ has not been explicitly displayed. Based on the principle of stationary action, a functional variation of the action with respect to ψ^*, together with the usual integrations by parts, should be set to zero:

$$\delta_{\psi^*} \mathcal{A}_0[\psi^*, \psi] = 0. \tag{14.57}$$

We obtain

$$\int_0^L dx \int_0^T dt \, \delta\psi^* \left[-i\hbar \frac{\partial\psi}{\partial t} - \frac{\hbar^2}{2m} \frac{\partial^2\psi}{\partial x^2} + V(x,t)\,\psi \right]$$

$$+ \frac{i\hbar}{2} \int_0^L dx \, [\delta\psi^*\psi]_{t=0}^{t=T} + \frac{\hbar^2}{2m} \int_0^T dt \left[\delta\psi^* \frac{\partial\psi}{\partial x} \right]_{x=0}^{x=L} = 0. \quad (14.58)$$

Now, $\delta\psi^*$ is completely arbitrary within the domain $0 < x < L$ and $0 < t < T$, and hence the argument within square brackets in the double integral term in equation (14.58) is zero. We therefore recover the time-dependent Schrödinger equation. Now consider the terms in equation (14.58) with single integrals. If we note that the spatial dependence of $\psi^*(x,t)$ is the same as the complex conjugate of $\psi(x,t)$ then $\delta\psi^*(0,t)=\delta\psi^*(L,t)=0$, since $\psi(x,t)$ is specified there. Thus the term with the single integral over time drops out. The term with the integration over x is of the form

$$\frac{i\hbar}{2} \int_0^L dx \, [\delta\psi^*(x,T)\psi(x,T) - \delta\psi^*(x,0)\psi(x,0)], \quad (14.59)$$

where the replacement $\psi(x,0) = \phi_0(x)$ can be made. With the initial function at $t = 0$ specified, we have $\delta\psi^*(x,0) = 0$. However, we are left with a term

$$\frac{i\hbar}{2} \int_0^L dx \, \delta\psi^*(x,T) \, \psi(x,T) \quad (14.60)$$

that depends on the value of the solution at $t = T$ and cannot be set to zero as required by the principle of least action. This is the difficulty with the standard implementation of the action principle: we are unable to define the behavior of the solution at $t = T$.

14.6.2   Variations using adjoint functions

Variational principles for linear initial value problems were developed by Gurtin[31] who considered their application to elastodynamics. Noble[29] has recast this work making use of adjoint functions in the action in order to circumvent the problem in the conventional variational approach revealed above. Consider the action

$$\mathcal{A}_0[\psi_{(-)}^*, \psi_{(+)}] = \int_0^L dx \int_0^T dt \left[\frac{-i\hbar}{2} \left(\psi_{(-)}^* \frac{\partial\psi_{(+)}}{\partial t} - \frac{\partial\psi_{(-)}^*}{\partial t} \psi_{(+)} \right) \right.$$

$$\left. + \frac{\hbar^2}{2m} \frac{\partial\psi_{(-)}^*}{\partial x} \frac{\partial\psi_{(+)}}{\partial x} + \psi_{(-)}^* V(x,t)\psi_{(+)} \right]. \quad (14.61)$$

Here the function $\psi_{(+)}(x,t)$ is our usual Schrödinger function, while $\psi_{(-)}(x,t)$, called the adjoint function, is defined to be $\psi_{(+)}$ with a time dependence of $T - t$. We have

$$\psi_{(-)}(x,t) = \psi_{(+)}(x, T - t). \tag{14.62}$$

In effect, the adjoint wavefunction evolves backward in time! Following Noble, we invoke the variational principle and set the variation of the action with respect to the adjoint function $\psi_{(-)}^*$ to obtain

$$\delta_{\psi_{(-)}^*} \mathcal{A} = 0. \tag{14.63}$$

This leads to

$$\int_0^L dx \int_0^T dt \, \delta\psi_{(-)}^* \left(-i\hbar \frac{\partial \psi_{(+)}}{\partial t} + \frac{\hbar^2}{2m} \frac{\partial^2 \psi_{(+)}}{\partial x^2} + V(x,t)\psi_{(+)} \right)$$

$$+ \frac{i\hbar}{2} \int_0^L dx \left[\delta\psi_{(-)}^* \psi_{(+)} \right]_0^T + \frac{\hbar^2}{2m} \int_0^T dt \left[\delta\psi_{(-)}^* \frac{\partial}{\partial x} \psi_{(+)} \right]_0^L (14.64)$$

As before, Schrödinger's equation obtains from the term with the space–time integrals: with an arbitrary $\delta\psi_{(-)}^*$ within the interior of the space–time domain the integrand within the large parentheses must be zero. The "surface term" with an integral over time vanishes because

$$\delta\psi_{(-)}(0,t) = \delta\psi_{(+)}(0, T - t) = 0,$$
$$\delta\psi_{(-)}(L,t) = \delta\psi_{(+)}(L, T - t) = 0, \tag{14.65}$$

and these relations hold due to the given BCs on $\psi_{(+)}(x,t)$ at the spatial boundary. Hence $\psi_{(-)}$ cannot be varied on the spatial boundary. We now consider the surface term with an integral over the coordinate

$$\frac{i\hbar}{2} \int_0^L dx \left[\delta\psi_{(-)}^*(x,T)\psi_{(+)}(x,T) - \delta\psi_{(-)}^*(x,0)\psi_{(+)}(x,0) \right]. \tag{14.66}$$

Here $\delta\psi_{(-)}^*(x,T)$ is $\delta\psi_{(+)}^*(x,0)$ and the known initial condition on the wavefunction $\psi(x,0) = \phi_0(x)$ ensures that its variation $\delta\phi_0^*(x) = 0$. We therefore have a residual term in the variation of the action

$$-\frac{i\hbar}{2} \int_0^L dx \, \delta\psi_{(-)}^*(x,0)\psi_{(+)}(x,0)$$

$$= -\frac{i\hbar}{2} \int_0^L dx \, \delta\psi_{(+)}^*(x,T)\phi_0(x), \tag{14.67}$$

which can be eliminated by adding a compensating term to the action in equation (14.61). The final form of an acceptable action integral is then given by

$$A[\psi_{(-)}^*, \psi_{(+)}] = A_0[\psi_{(-)}^*, \psi_{(+)}] + \frac{i\hbar}{2} \int_0^L dx \, \psi_{(-)}^*(x, 0)\phi_0(x). \quad (14.68)$$

The above formulation of the action for initial value problems can be combined with the FEM for a frontal numerical attack on quantum time-evolution issues. An important advantage this method provides in treating low-dimensional (1D and 2D + time) problems is that, by directly including time as one of the coordinates, we can cope with potentials that are functions of time. Computational limitations at the present time would make a 3D+time calculation prohibitive; however, computational resources are improving dramatically year by year. Typical textbook treatments of time-dependent potentials involve the development of a perturbative expansion with time-ordering issues arising in the order-by-order calculation of the wavefunction in perturbation theory. Here, we would be able to numerically evaluate the time evolution of the wavefunction for time-dependent potentials.

14.6.3   Adjoint variations for the wave equation

For the sake of completeness, we note that the variational approach of Noble can be applied to the wave equation with initial conditions as well. Consider the 1D wave equation, equation (14.14), with BCs given by

$$
\begin{aligned}
\phi(0, t) &= 0 = \phi(L, t) & &\text{(a string tied at two ends),} \\
\phi(x, 0) &= f(x) & &\text{(initial configuration),} \qquad (14.69) \\
\partial_t \phi(x, 0) &= g(x) & &\text{(initial velocity).}
\end{aligned}
$$

We again write

$$A_0 = \int_0^T dt \int_0^L dx \left[\frac{\partial \phi_{(-)}}{\partial x} \frac{\partial \phi_{(+)}}{\partial x} - \frac{1}{c^2} \frac{\partial \phi_{(-)}}{\partial t} \frac{\partial \phi_{(+)}}{\partial t} \right], \quad (14.70)$$

where $\phi_{(-)}(x, t) = \phi_{(+)}(x, T-t)$. Finding the extremum of the action by a variation with respect to the adjoint function leads to

$$\delta_{\phi_{(-)}} A_0 = 0, \quad (14.71)$$

so that

$$\int_0^T dt \int_0^L dx \left[-\delta\phi_{(-)} \frac{\partial^2 \phi_{(+)}}{\partial x^2} + \frac{1}{c^2}\delta\phi_{(-)}\frac{\partial^2 \phi_{(+)}}{\partial t^2} \right]$$
$$+ \int_0^T dt \left[\delta\phi_{(-)} \frac{\partial \phi_{(+)}}{\partial x} \right]_0^L - \frac{1}{c^2}\int_0^L dx \left[\delta\phi_{(-)} \frac{\partial \phi_{(+)}}{\partial t} \right]_0^T = 0. \quad (14.72)$$

With $\delta\phi_{(-)}$ arbitrary, the integrand within square brackets in the double integral term must be zero, so that we reproduce the wave equation, equation (14.14). The surface term with an integral over time is zero since

$$\delta\phi_{(-)}(0,t) = \delta\phi_{(+)}(0, T - t) = 0,$$
$$\delta\phi_{(-)}(L,t) = \delta\phi_{(+)}(L, T - t) = 0, \quad (14.73)$$

because the spatial BCs specify $\phi_{(+)}$ at the spatial boundaries at all times. The final surface term with an integral over space is of the form

$$-\frac{1}{c^2} \int_0^L dx\ \delta\phi_{(-)}(x,T) \left. \frac{\partial \phi_{(+)}(x,t)}{\partial t} \right|_{t=T}$$
$$+ \frac{1}{c^2} \int_0^L dx \delta\phi_{(-)}(x,0) \left. \frac{\partial \phi_{(+)}(x,t)}{\partial t} \right|_{t=0}. \quad (14.74)$$

Using the fact that $\phi_{(-)}(x,T) = \phi_{(+)}(x,0) = f(x)$ is fixed, we see that the first term vanishes. The second term can be written as $(1/c^2) \int_0^L dx \phi_{(-)}(x,0) g(x)$. It does not vanish but can be compensated for through the addition of a term to the action. We write

$$\mathcal{A} = \mathcal{A}_0 - \frac{1}{c^2} \int_0^L dx\ \phi_{(-)}(x,0)\ g(x). \quad (14.75)$$

This variational approach is a field ripe for detailed numerical investigations of time-dependent phenomena, with applications to a number of areas of research, including diffusion phenomena, wave propagation, and quantum time evolution in mesoscopic devices. A unified formulation of the construction of variational principles is discussed in the review article by Gerjuoy et al.,[32] which also provides an excellent bibliography through 1983 of earlier work.

14.6.4 Connection with quantum field theory

In the above, we have developed the action in the Schrödinger representation in which physical operators for observables are time independent while the wavefunctions contain the time dependence. In this so-called Schrödinger picture, we have

$$H_0|\psi_S(x,t)\rangle = i\hbar\partial_t|\psi_S(x,t)\rangle. \tag{14.76}$$

If H_0 is not a function of time, the solution is

$$|\psi_S(x,t)\rangle = e^{-iH_0(t-t_0)/\hbar}|\psi_S(x,t_0)\rangle. \tag{14.77}$$

Throughout this book we have not considered the Heisenberg representation, in which the burden of time evolution is passed on to the operators, and the wavefunctions (state vectors) are time independent. In the Heisenberg representation we have

$$\begin{aligned}|\psi_H(x)\rangle &= |\psi_S(x,t_0)\rangle \\ &= e^{-iH_0(t-t_0)/\hbar}|\psi(x,t)\rangle,\end{aligned} \tag{14.78}$$

and an observable Ω has the expectation value

$$\Omega_H(t) = \langle\psi_S(x,t_0)|e^{iH_0(t-t_0)/\hbar}\Omega_S e^{-iH_0(t-t_0)/\hbar}|\psi_S(x,t_0)\rangle. \tag{14.79}$$

This expression allows us to derive the result that the observable Ω obeys the Heisenberg equation of motion

$$\begin{aligned}\frac{\partial\Omega_H(t)}{\partial t} &= \frac{\partial}{\partial t}\left(e^{iH_0(t-t_0)/\hbar}\Omega_S e^{-iH_0(t-t_0)/\hbar}\right) \\ &= \frac{-i}{\hbar}[\Omega_H(t),H_0]. \end{aligned}\tag{14.80}$$

In quantum field theory, the time evolution is usually determined in terms of the *interaction picture*. Suppose we are given the Hamiltonian $H = H_0 + H_1(t)$. In the interaction picture, the state vectors evolve with the time dependence given by the time-independent Hamiltonian H_0. The advantages of this interaction picture were first appreciated by Schwinger and Tomonaga while developing their theory of quantum electrodynamics. It is assumed that we can solve for the eigenstates and their energies for the time-independent Hamiltonian. Now we have

$$i\hbar\,\partial_t\,|\psi_S(x,t)\rangle = (H_0 + H_1(t))|\psi_S(x,t)\rangle. \qquad (14.81)$$

Expressed in terms of the interaction picture wavefunctions

$$|\psi_I(x,t)\rangle = e^{iH_0t/\hbar}|\psi_S(x,t)\rangle \qquad (14.82)$$

the above Schrödinger equation can be rewritten as

$$\begin{aligned} i\hbar\partial_t|\psi_I(x,t)\rangle &= e^{iH_0t/\hbar}H_1(t)e^{-iH_0t/\hbar}|\psi_I(x,t)\rangle \\ &= H_1^I(t)|\psi_I(x,t)\rangle. \end{aligned} \qquad (14.83)$$

In other words, $|\psi_I(x,t)\rangle$ satisfies a Schrödinger equation with the interaction Hamiltonian $H_1^I(t)$, expressed in the interaction picture. This leads to the definition of a time-evolution operator, $U(t,t_0)$, with the properties

$$\begin{aligned} |\psi_I(x,t)\rangle &= U(t,t_0)|\psi_I(x,t_0)\rangle; \\ U(t,t) &= 1; \\ U^\dagger(t,t_0) &= U^{-1}(t,t_0) = U(t_0,t); \\ U(t,t_1)U(t_1,t_0) &= U(t,t_0) \\ |\psi_I(x,t)\rangle &= e^{iH_0t/\hbar}e^{-iH_0(t-t_0)/\hbar}e^{-iH_0t/\hbar}|\psi_I(x,t_0)\rangle. \end{aligned} \qquad (14.84)$$

We can show from these properties of the time-evolution operator that

$$i\hbar\,\partial_t\,|\psi_I(x,t)\rangle = H_1^I(t)|\psi_I(x,t)\rangle, \qquad (14.85)$$

where $H_1^I(t)$ is in the interaction picture representation. This equation is used to obtain a differential equation satisfied by the time-evolution operator, and we have

$$i\hbar\,\partial_t\,(U(t,t_0)|\psi_I(x,t_0)\rangle) = H_1^I(t)\,U(t,t_0)|\psi_I(x,t_0)\rangle. \qquad (14.86)$$

Since this is true for every state vector, we obtain

$$i\hbar\,\partial_t\,U(t,t_0) = H_1^I(t)\,U(t,t_0). \qquad (14.87)$$

The solution of this differential equation can be written as a Volterra integral equation

$$U(t,t_0) = 1 - \frac{i}{\hbar}\int_{t_0}^t dt\,H_1^I(t)\,U(t,t_0). \qquad (14.88)$$

As discussed in books on field theory,[33-35] an iteration of the integral equation leads to the time-ordered form of the time-evolution operator

$$
\begin{aligned}
U(t, t_0) &= 1 - \frac{i}{\hbar} \int_{t_0}^{t} dt H_1^I(t) \\
&+ \frac{(-i)^2}{2! \, \hbar^2} \int_{t_0}^{t} \int_{t_0}^{t} dt_1 dt_2 \, T\left(H_1^I(t_1) \, H_1^I(t_2)\right) + \dots \\
&= T e^{-\frac{i}{\hbar} \int_{t_0}^{t} dt' H_1^I(t')}.
\end{aligned} \tag{14.89}
$$

This is the traditional way of calculating the time-evolution order by order in perturbation theory by taking each term in the expansion of the exponential function, while keeping the time ordering in place. We will not go into the details of this perturbative development here. The variational approach presented in the subsection above attempts to bypass this entire issue of a perturbative evaluation of the time evolution by proposing a space–time finite element calculation!

The adjoint variational approach suggested here has been arrived at in a very simple way by providing arguments that are based on numerical considerations of stability. Earlier, Dirac developed a time-dependent variational method for Schrödinger's equation.[36] This method is referred to as the Dirac–Frenkel[37] variational method in the literature, and has been used extensively by chemists and nuclear physicists in the application of time-dependent Hartree–Fock perturbation theory to the variational calculation of energy levels in many-particle systems.[38,39] These ideas have been extended by Balian and Veneroni[40] to density matrix theory. The density matrix ρ satisfies the equation of motion given by

$$
i\hbar \frac{d\rho}{dt} = [H, \rho]. \tag{14.90}
$$

The Schrödinger picture was used earlier by Schwinger[41] in constructing a time contour method for the time evolution within quantum mechanics of systems far from thermal equilibrium. (The time contour method is very reminiscent of the adjoint variational approach discussed here.) This idea of using a time contour was developed by Keldysh[42] into a Green's function (Feynman diagrammatic) technique for a field-theoretic description of a system including thermal nonequilibrium.

Nearly all books in quantum field theory are concerned with calculating properties that do not depend on specified initial values for wavefunctions at finite times. Rather, they direct their efforts towards obtaining scattering cross-sections, lifetimes for particle decays, and the like. However, over the past two decades, the issue of time evolution in the Schrödinger picture in quantum field theory has been revisited by field theorists. Jackiw and Kerman[43] show that the time-dependent action

$$\Gamma = \int dt \langle \psi_{(-)}(x,t) | i\hbar \partial_t - H | \psi_{(+)}(x,t) \rangle \qquad (14.91)$$

for the time-dependent states $\psi_{(\pm)}(x,t)$ must be stationary against variation of $\psi_{(\pm)}(x,t)$, subject to the BC

$$\langle \psi_{(-)}(x,t) | \psi_{(+)}(x,t) \rangle = 1, \qquad (14.92)$$

at all times. Cooper and collaborators have investigated applications of this method.[44]

Rajagopal has extended the time-dependent variational formulation to construct an effective action for density functional theory;[45] he has shown how this approach can allow one to extract Berry's phase, a geometric phase associated with evolving quantum states. An extension of this formulation to nonequilibrium statistical dynamics based on the Lindblad density matrix evolution has also been demonstrated.[46]

14.7 Concluding remarks

Throughout the treatment of the FEM in this book, we have emphasized that the FEM is intimately related to the principle of stationary action. Once this connection is made, need one say more? It is clear from the recent literature that the FEM is making a definite impact on research in all fields of physics, and the number of articles reported is increasing very rapidly.

In many respects, the presentation here has been made less mathematical: the issues of existence of the integrals, uniqueness of solutions, convergence, etc., have not been treated. These were dealt with by mathematicians in the 1960s. The reader may consult the corresponding references.[47–50]

We have also not reviewed the work on the application of the FEM to operator quantum mechanics.[51–57] In this approach, the

Heisenberg equations of motion are solved for the quantum mechanical operators using a finite element framework for the extrapolation from the initial time $t = 0$. This is a promising line of development that will provide new insights into lattice field theory.

Time evolution in quantum mechanics is a fairly complex and a very active area of research. It is the author's view that the method of finite elements and variational methods will play a fundamental role in elucidating dynamical aspects of basic quantum mechanics. The application of this experience and the FEM in designing and modeling a new generation of quantum devices will define the field of quantum wavefunction engineering.

References

[1] A. Goldberg, H. M. Schey, and J. L. Schwartz, *Am. J. Phys.* **35**, 177 (1967).

[2] R. Courant and D. Hilbert, *Methods of Mathematical Physics*, Vol. I (Interscience, New York, 1953).

[3] W. H. Press, S. A. Teukolsky, W. T. Vetterling, and B. R. Flannery, *Numerical Recipes* (Cambridge University Press, Cambridge, 1992).

[4] C. Leforestier, R. H. Bisseling, C. Cerjan, M. D. Feit, E. Friesner, A. Guldberg, A. Hammerich, G. Jolicard, W. Karrlein, H. D. Meyer, N. Lipkin, O. Roncero, and R. Kosloff, *J. Comput. Phys.* **94**, 59 (1991).

[5] D. Neuhauser, M. Baer, R. S. Judson, and D. J. Kouri, *Comput. Phys. Commun.* **63**, 460 (1991).

[6] R. Kosloff, *J. Phys. Chem.* **92**, 2087 (1988); *Annu. Rev. Phys. Chem.* **45**, 145 (1994).

[7] D. E. Weeks and D. J. Tannor, *Chem. Phys. Lett.* **207**, 301 (1993). M. J. MacLachlan and D. E. Weeks, *J. Phys. Chem.* A **102**, 9489 (1998).

[8] J. E. Bayfield, *Quantum Evolution: An Introduction to Time-dependent Quantum Mechanics* (Wiley-Interscience, New York, 1999).

[9] K. S. Yee, *IEEE Trans. Antennas Propag.* **AP-14**, 302 (1966).

[10] A. Taflove and M. E. Brodwin, *IEEE Trans. Microwave Theory Tech.*, **MTT-23**, 623 (1975).

[11] A. Taflove, *Computational Electrodynamics: The Finite-Difference Time-Domain Method* (Artec House, Norwood, MA, 1995).

[12] A. Taflove, *Advances in Computational Electrodynamics: The Finite-Difference Time-Domain Method* (Artec House, Norwood, MA, 1998).

[13] M. D. Feit and J. A. Fleck, *J. Opt. Soc. Am.* B **5**, 633 (1988); R. P. Ratowsky, J. A. Fleck, and M. D. Feit, *Opt. Lett.* **17**, 10 (1992); R. P. Ratowsky, J. A. Fleck, and M. D. Feit, *J. Opt. Soc. Am.* A Lett. **17**, 10 (1992); R. P. Ratowsky, J. A. Fleck, and M. D. Feit, *Opt. Lett.* **19**, 1284 (1994).

[14] J. B. Cole, *Comput. Phys.* **8**, 730 (1994).

[15] J. B. Cole, R. A. Krutar, S. K. Numrich, and D. B. Creamer, *Comput. Phys.* **9**, 235 (1995).

[16] J. B. Cole, *Comput. Phys.* **12**, 82 (1998).

[17] The transfer matrix approach has been presented in another context by exponentiating 16×16 matrices in: L. R. Ram-Mohan, K. H. Yoo, and R. L. Aggarwal, *Phys. Rev.* B **38**, 6151 (1988). Also see: B. Chen, M. Lazzouni, and L. R. Ram-Mohan, *Phys. Rev.* B **45**, 1204 (1992).

[18] L. A. Pipes and L. R. Harvill, *Applied Mathematics for Engineers and Physicists* (McGraw-Hill, New York, 1971).

[19] T. J. R. Hughes, "Analysis of Transient Algorithms with Particular Reference to Stability Behavior," in *Computational Methods for Transient Analysis*, ed. T. Belytschko and T. J. R. Hughes (Elsevier Science, New York, 1983).

[20] W. L. Wood, "Some Transient and Coupled Problems -- A State-of-the-Art Review," in *Numerical Methods for Transient and Coupled Problems*, ed. R. W. Lewis, E. E. Hinton, P. Bettess, and B. A. Schreffer (Wiley, New York, 1987).

[21] M. A. Dokainish and K. Subbaraj, *Comput. Struct.* **32**, 1371, 1387 (1989).

[22] G. H. Golub and C. F. van Loan, *Matrix Computations* (Johns Hopkins University Press, Baltimore, MD, 1983).

[23] T. J. Park and J. C. Light, *J. Chem. Phys.* **85**, 5870 (1986).

[24] C. Lanczos, *J. Res. NBS* **45**, 255 (1950).

[25] B. N. Parlett, *The Symmetric Eigenvalue Problem* (Prentice Hall, Englewood Cliffs, NJ, 1980).

[26] T. J. R. Hughes, *The Finite Element Method* (Prentice Hall, Englewood Cliffs, NJ, 1987).

[27] R. Goloskie and L. R. Ram-Mohan, unpublished notes (1992).

[28] D. Kieweg, "Boundary Conditions for the 2D Wave Equation in FEM," Senior Thesis, Worcester Polytechnic Institute, 1996.

[29] B. Noble, "Variational FEM for Initial Value Problems," in *Mathematics of Finite Elements and Applications*, Vol. I, ed. J. Whiteman (Academic Press, New York, 1973).

[30] A. J. Davies, *The Finite Element Method* (Clarendon Press, Oxford, 1980).

[31] M. E. Gurtin, *Q. Appl. Math.* **22**, 252 (1964).

[32] E. Gerjuoy, A. R. P. Rau, and L. Spruch, *Rev. Mod. Phys.* **55**, 725 (1983).

[33] J. D. Bjorken and S. D. Drell, *Relativistic Quantum Fields* (McGraw-Hill, New York, 1965).

[34] J. J. Sakurai, *Advanced Quantum Mechanics* (Addison-Wesley, Reading, MA, 1967).

[35] A. L. Fetter and J. D. Walecka, *Theoretical Mechanics of Particles and Continua* (McGraw-Hill, New York, 1980).

[36] P. A. M. Dirac, *Proc. Cambridge Philos. Soc.* **26**, 376 (1930).

[37] J. Frenkel, *Wave Mechanics: Advanced General Theory* (Clarendon, Oxford, 1934), pp253 and 435. In a footnote, Frenkel refers to an Appendix to the Russian edition of Dirac's book, *Principles of Quantum Mechanics*, which apparently contains Dirac's development of his time-dependent theory.

[38] P. Langhoff, S. Epstein, and M. Karplus, *Rev. Mod. Phys.* **44**, 602 (1972).

[39] A. Kerman and S. Koonin, *Ann. Phys. (NY)* **100**, 332 (1976).

[40] R. Balian and M. Veneroni, *Ann. Phys. (NY)* **164**, 334 (1985).

[41] J. Schwinger, *J. Math. Phys.* **2**, 407 (1961).

[42] L. V. Keldysh, *Sov. Phys.–JETP* **20**, 1018 (1965).

[43] R. Jackiw and A. Kerman, *Phys. Lett.* **71A**, 158 (1979); R. Jackiw, *Int. J. Quantum Chem.* **17**, 41 (1980). Also see: J. P. Blaizot and G. Ripka, *Quantum Theory of Finite Sytems* (MIT Press, Cambridge, MA, 1986) Chapter 9, *Time Dependent Self-consistent Fields*.

[44] F. Cooper, S.-Y. Pi, and P. N. Stancioff, *Phys. Rev.* D **34**, 3831 (1986); also see further references in this article.

[45] A. K. Rajagopal, *Phys. Rev.* B **54**, 3916 (1996).

[46] A. K. Rajagopal, *Phys. Lett.* A **228**, 66 (1997).

[47] D. Braess, *Finite Elements: Theory, Fast Solvers, and Applications in Solid Mechanics*, trans. L. L. Schumaker (Springer-Verlag, Berlin, 1997).

[48] G. Strang and G. J. Fix, *An Analysis of the Finite Element Method* (Prentice Hall, Englewood Cliffs, NJ, 1973).

[49] P. G. Ciarlet, *The Finite Element Method for Elliptic Problems* (Elsevier North-Holland, New York, 1978).

[50] P. G. Ciarlet and J. L. Lions, general editors, *Handbook of Numerical Analysis* (Elsevier North-Holland, New York, 1991).

[51] C. M. Bender, G. S. Guralnik, and D. H. Sharp, *Nucl. Phys.* B **207**, 54 (1982).

[52] C. M. Bender and D. H. Sharp, *Phys. Rev. Lett.* **50**, 1535 (1983).

[53] C. M. Bender, *Physica* **124A**, 91 (1984).

[54] C. M. Bender, K. A. Milton, D. H. Sharp, L. M. Simmons, and R. Strong, *Phys. Rev.* D **32**, 1476 (1985).

[55] F. Cooper, K. A. Milton, and L. M. Simmons, *Phys. Rev.* D **32**, 2056 (1985).

[56] C. M. Bender and K. A. Milton, *Phys. Rev.* D **34**, 3149 (1986).

[57] D. Miller, K. A. Milton, and S. Siegemund-Broka, *Phys. Rev.* D **46**, 806 (1992).

Part IV

Sparse matrix applications

15

Matrix solvers and related issues

15.1 Introduction

The aim of this chapter is to review methods for the solution of
linear matrix equations and for the solution of complex Hermitian
generalized eigenvalue problems when the matrices are of large dimension. For example, within the framework of the finite element
method (FEM) the Galerkin approach to scattering problems can
generate a matrix equation for the scattering amplitude with matrix dimensions $N \geq 10^3$. Similarly, bound state problems in two or
three dimensions give rise to matrix eigenvalue equations of comparable dimensions. The matrices generated in the FEM are banded
and sparse. As a very rough criterion for sparseness we can suppose
that matrices with element occupancy of say, N^α, $\alpha \simeq 1.2$, may be
considered to be sparse matrices.

We assume that the standard methods for the solution of these
problems for small matrices are accessible through the widely available library of matrix subroutines such as LAPACK,[1] LINPACK,[2]
and EISPACK.[3] The algorithms typically used for smaller matrices
are described in the collection *Numerical Recipes*,[4] and in the books
by Wilkinson,[5] Parlett,[6] Jennings,[7] and Golub and van Loan.[8] Here
we focus on the iterative methods for the solution of such matrix
problems.

15.2 Bandwidth reduction

The bandwidth of a matrix $\mathbf{M}$ can be defined as the maximum value
of $|i - j|$ for occupied matrix elements M_{ij}, plus one to account for
the main diagonal. In general, the global matrices generated in the
FEM are banded and are typically symmetric or Hermitian. For
1D problems that are solved by the FEM using linear interpolation,
the matrices are tridiagonal and hence have a bandwidth of 2. In 2D
problems, the matrices tend to be less tightly banded and have supra-
and sub-diagonals which are sparsely occupied. We may refer to

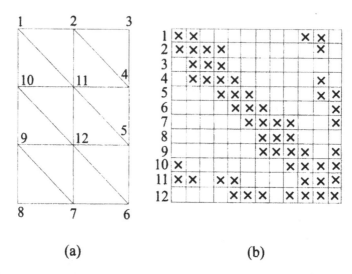

(a)   (b)

Fig. 15.1 An example of node numbering resulting in a large occupancy bandwidth of 9 in the global matrix. In (a) we see a finite element mesh with node numbering that leads to a global matrix, shown in (b) which is poorly banded with the occupied matrix elements denoted by crosses. (Adapted from Collins, Ref. 9.)

these as poorly banded matrices. The banded nature of the matrices depends on the connectivity of the nodes and the node numbering used.

If a matrix is not tightly banded, with a minimal number of zeros within its bandwidth, there are two disadvantages. First, the storage requirements in computer memory for such a matrix are higher. Secondly, the procedure of Gauss elimination in the solution of the FEM equations, or the diagonalization procedures in eigenvalue problems, will tend to insert "fills" in the unoccupied elements within the bandwidth, leading to increasing round-off error. A simple rearrangement of node numbers could alleviate both problems simultaneously. A new connectivity relation is then used to keep track of the nodes in each element and to assign coordinates to each of the nodes. Such a bandwidth reduction can be performed even before the finite element integrations are performed through a proper permutation of the node numbers. This is also equivalent to the permutation of the rows and columns of the global matrices after they are constructed.

For concreteness, consider the regular triangular mesh on a rectangular region shown in Fig. 15.1. It is clear that the choice of node numbering has led to a bandwidth of 9. We can expect many of the

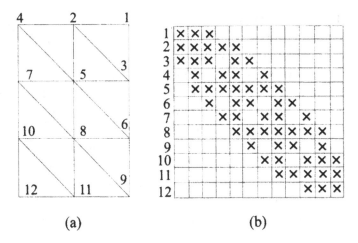

(a)   (b)

Fig. 15.2 An example of bandwidth reduction by node renumbering of the matrix in Fig. 15.1. The bandwidth associated with the node numbering of a finite element mesh shown in (a) is now 4, and the corresponding occupancy in the global matrix is shown in (b). (Adapted from Collins, Ref. 9.)

unoccupied matrix elements to be filled during intermediate stages of Gauss elimination or LU-decomposition. We see from Fig. 15.2 that a simple reassignment of the node numbers leads to a reduction of the bandwidth to 4, and it is evident that the LU-decomposition of the matrix will lead to a very small increase in the occupancy.

The theory of bandwidth reduction is fairly well developed and employs graph theory algorithms. Automatic node renumbering algorithms are available which will minimize the bandwidth. Once the vector solution to the system of simultaneous equations, or an eigenvector in an eigenvalue problem, is obtained, the components of the vector are re-permuted back to the original order in case the bandwidth is reduced after the global matrix has been formed.

In an automatic node renumbering scheme one begins with a classification of all nodes in terms of their "degree of connectivity," which specifies how many nodes are connected to a particular node. The general idea is to start from an extremity and label the node with the smallest degree of connectivity as node number 1. Nodes connected to it are sequentially labeled depending on their degree of connectivity. Each of these nodes is considered in turn, and their connected neighbors are numbered according to their degree of connectivity. This procedure is continued until all the nodes are numbered. The

Cuthill–McGee algorithm[10] formalizes this procedure. A Fortran version of this algorithm has been developed by Collins.[9] After the Cuthill–McGee algorithm was published it was shown by George and Liu[11] that reversing the ordering of the node numbering can improve the bandwidth. The two algorithms have been compared by Liu and Sherman.[12] One other algorithm of note is the one by Gibbs *et al.*,[13] which produces a bandwidth reduction comparable to the one by the reverse Cuthill–McGee scheme, while requiring less computational time. Issues related to the application of graph theory to bandwidth reduction are somewhat further afield, and the reader is directed to these references and to Duff *et al.*,[14] for elaboration. In the following discussion, it will be assumed that a bandwidth reduction has already been performed on the global matrices.

15.3   Solution of linear equations

The FEM applied to scattering problems naturally leads to a set of simultaneous equations for the nodal variables. The global matrix equation,

$$\mathbf{A} \cdot \boldsymbol{\psi} = \mathbf{b}, \tag{15.1}$$

is solved using methods related to Gauss elimination or by iterative methods, such as the conjugate gradient method. Given the known vector $\mathbf{b}$, the aim is to solve for the nodal values $\boldsymbol{\psi}$ without directly inverting the coefficient matrix $\mathbf{A}$. In fact, inverting large matrices should be avoided as far as possible. Once the nodal connectivity has been arranged so as to produce a global matrix with a minimal bandwidth, we apply one of these methods.

15.3.1   Gauss elimination

A thorough discussion of Gauss elimination and related matrix solvers is available in *Numerical Recipes*. We shall defer to their treatment of the LU-decomposition, $\mathbf{A} = \mathbf{LU}$, and the Cholesky decomposition, $\mathbf{A} = \mathbf{LL}^T$, approaches. The decomposition represents the matrix $\mathbf{A}$ as the product of a lower triangular matrix and an upper triangular matrix. Consider the case of Cholesky decomposition of a symmetric positive definite $\mathbf{A}$ matrix into the lower triangular $\mathbf{L}$ and its transpose (or Hermitian conjugate). We can write equation (15.1) in the form

$$\mathbf{L} \cdot \mathbf{y} = \mathbf{b},$$
$$\mathbf{L}^T \cdot \boldsymbol{\psi} = \mathbf{y}. \tag{15.2}$$

Forward substitutions in the first equation to obtain the vector $\mathbf{y}$, and back-substitutions in the second equation, lead to the solution vector ψ. The brief algorithms associated with the decomposition and equation solving are given in *Numerical Recipes*.

The LU-decomposition or the Cholesky decomposition of large sparse matrices can be computationally costly, and iterative methods become competitive with these methods as the matrix dimensions become on the order of 10^4 or larger. We describe here only the conjugate gradient method for the solution of simultaneous equations. Other iterative methods, such as the Gauss–Seidel and successive over-relaxation (SOR) methods are discussed by Golub and van Loan.[8] The multigrid method is another approach that has been explored extensively. The reader is referred to the literature.[15,16]

15.3.2   The conjugate gradient method

The problem of solving N linear simultaneous equations,

$$\mathbf{A} \cdot \mathbf{x} = \mathbf{b}, \tag{15.3}$$

is equivalent to minimizing the quadratic function

$$\mathcal{F}(\mathbf{x}) = \frac{1}{2}\mathbf{x}^T \cdot \mathbf{A} \cdot \mathbf{x} - \mathbf{x}^T \cdot \mathbf{b}$$
$$= \frac{1}{2}x_i A_{ij} x_j - x_i b_i. \tag{15.4}$$

Minimizing $\mathcal{F}$ by differentiation leads to

$$\frac{\partial \mathcal{F}}{\partial x_i} = A_{ij} x_j - b_i = 0, \tag{15.5}$$

which is just equation (15.3). The minimum value is given by

$$\mathcal{F}(\mathbf{x}) = -\frac{1}{2}\mathbf{b}^T \cdot \mathbf{A}^{-1} \cdot \mathbf{b}. \tag{15.6}$$

For the following, we assume that $\mathbf{A}$ is a symmetric positive definite matrix, a constraint that can be overcome, as shown in references given below.

Consider an arbitrary starting point $\mathbf{x}_0$. The gradient vector $\nabla \mathcal{F}$ at this arbitrary point gives the direction of maximum increase of

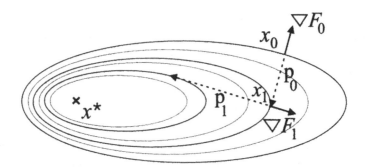

Fig. 15.3 Contours of the function $\mathcal{F}(\mathbf{x})$ are shown. The gradient vectors $\nabla\mathcal{F}$ and their negatives are used to indicate the direction of relaxation from the starting point $\mathbf{x}_0$ towards the location of the minimum at $\mathbf{x}^*$. The figure illustrates the method of steepest descent.

the contours, so that we should consider moving along a vector $\mathbf{p}_0$ that is negative to this. The residual $\mathbf{r}_0$ at $\mathbf{x}_0$ is defined as

$$\mathbf{r}_0 = (\mathbf{A} \cdot \mathbf{x}_0 - \mathbf{b}) = \nabla\mathcal{F}(\mathbf{x}_0). \tag{15.7}$$

We search from $\mathbf{x}_0$ along the line $\mathbf{p}_0 = -\mathbf{r}_0$ for the next point $\mathbf{x}_1$, where $\mathcal{F}$ is at a local minimum. Let

$$\mathbf{x}_1 = \mathbf{x}_0 + \lambda_0\,\mathbf{p}_0; \tag{15.8}$$

we determine this point by minimizing

$$\mathcal{F}(\mathbf{x}_1) = \mathcal{F}(\mathbf{x}_0) + \lambda_0\,\mathbf{p}_0^T \cdot (\mathbf{A} \cdot \mathbf{x}_0 - \mathbf{b}) + \frac{1}{2}\lambda_0^2\,\mathbf{p}_0^T \cdot \mathbf{A} \cdot \mathbf{p}_0 \tag{15.9}$$

with respect to λ_0. We obtain

$$\lambda_0 = -\frac{\mathbf{p}_0^T \cdot \mathbf{r}_0}{\mathbf{p}_0^T \cdot \mathbf{A} \cdot \mathbf{p}_0}. \tag{15.10}$$

We multiply both sides of equation (15.8) by $\mathbf{A}$, subtract $\mathbf{b}$, and use (15.10) to obtain

$$\mathbf{r}_1 = \mathbf{r}_0 - \frac{\mathbf{r}_0^T \cdot \mathbf{r}_0}{\mathbf{r}_0^T \cdot \mathbf{A} \cdot \mathbf{r}_0}\,\mathbf{A} \cdot \mathbf{r}_0. \tag{15.11}$$

This equation and its subsequent iterations show that the successive residuals are orthogonal to the previous search direction. If we continue to select $\mathbf{p}_i$ to be along $-\mathbf{r}_i$ we have at hand the "method of steepest descent." Figure 15.3 illustrates the procedure in two dimensions. For N-dimensional hyperellipsoids with large eccentricities, this choice could lead to a zig-zag traversing of the "valley," resulting in a painfully slow convergence to the global minimum of $\mathcal{F}$.

To speed up the convergence we must choose $\mathbf{p}_i$, $i = 1, 2, \ldots$, in a more general way, keeping in mind that the new search directions should not negate the line minimizations performed earlier. We let

$$\mathbf{p}_i = -\mathbf{r}_i + \epsilon_{i-1}\,\mathbf{p}_{i-1}, \qquad i = 1, 2, 3, \ldots; \qquad (15.12)$$

with the condition that the $\mathbf{p}_i$ are A-orthogonal to $\mathbf{p}_{i-1}$. In other words, we require

$$\mathbf{p}_i^T \cdot \mathbf{A} \cdot \mathbf{p}_{i-1} = 0. \qquad (15.13)$$

As shown below, this orthogonality is preserved through each iteration such that $\mathbf{p}_i$ are orthogonal to all previous ones. This A-orthogonality is referred to as *conjugacy*, and the procedure is called the "method of conjugate gradients." [17]

We determine ϵ_{i-1} by using the A-orthogonality or conjugacy condition, equation (15.13). We have

$$(-\mathbf{r}_i + \epsilon_{i-1}\mathbf{p}_{i-1})^T \cdot \mathbf{A} \cdot \mathbf{p}_{i-1} = 0, \qquad (15.14)$$

leading to

$$\epsilon_{i-1} = \frac{\mathbf{r}_i^T \cdot \mathbf{A} \cdot \mathbf{p}_{i-1}}{\mathbf{p}_{i-1}^T \cdot \mathbf{A} \cdot \mathbf{p}_{i-1}}. \qquad (15.15)$$

The vectors $\mathbf{p}_i$ constitute a mutually conjugate basis set, and the set of vectors $\{\mathbf{p}_0, \mathbf{p}_1, \ldots\}$ is said to define a Krylov subspace. Since

$$\begin{aligned}
\mathbf{x}_i &= \mathbf{x}_{i-1} + \lambda_{i-1}\,\mathbf{p}_{i-1} \\
&= \mathbf{x}_0 + \lambda_0\,\mathbf{p}_0 + \lambda_1\,\mathbf{p}_1 + \lambda_2\,\mathbf{p}_2 + \cdots + \lambda_{i-1}\,\mathbf{p}_{i-1}, \quad (15.16)
\end{aligned}$$

we see that through each iteration the vector $\mathbf{x}_i - \mathbf{x}_0$ is built up as a linear combination of the conjugate vectors $\mathbf{p}_k$. If we represent the linear combination of $\mathbf{p}_k$ in a matrix form, we have

$$\mathbf{x}_i - \mathbf{x}_0 = \mathbf{P}_{i-1} \cdot \mathbf{\Lambda}_{i-1} = \mathbf{P}_{i-2} \cdot \mathbf{\Lambda}_{i-2} + \lambda_{i-1}\mathbf{p}_{i-1}, \qquad (15.17)$$

where $\mathbf{\Lambda}_{i-1}$ is a diagonal matrix with the λ_k as the diagonal elements, and $\mathbf{P}_{i-1}$ contains the vectors $\mathbf{p}_k$, column by column. Let us suppose for the moment that $\mathbf{x}_0 = 0$. At $\mathbf{x}_i$, we have

$$
\begin{aligned}
\mathcal{F}(\mathbf{x}_i) &= \mathcal{F}(\mathbf{P}_{i-2} \cdot \mathbf{\Lambda}_{i-2} + \lambda_{i-1}\mathbf{p}_{i-1}) \\
&= \mathcal{F}(\mathbf{x}_{i-1}) + [\mathbf{\Lambda}_{i-2}^T \cdot \mathbf{P}_{i-2}^T \cdot \mathbf{A} \cdot \mathbf{p}_{i-1}\lambda_{i-1}] \\
&\quad + \left(\frac{1}{2}\lambda_{i-1}^2\mathbf{p}_{i-1}^T \cdot \mathbf{A} \cdot \mathbf{p}_{i-1} - \lambda_{i-1}\mathbf{p}_{i-1}^T \cdot \mathbf{b}\right). \quad (15.18)
\end{aligned}
$$

Here, if we select $\mathbf{p}_{i-1}$ to be A-conjugate to all the earlier $\mathbf{p}$ vectors, we will have

$$
\mathbf{p}_{i-1}^T \cdot A \cdot \mathbf{P}_{i-2} = 0. \quad (15.19)
$$

This constraint eliminates the cross-term in $\mathcal{F}(\mathbf{x_i})$. The last two terms in parentheses in equation (15.18) can be minimized with respect to λ_{i-1} to obtain

$$
\begin{aligned}
\lambda_{i-1} &= -\frac{\mathbf{p}_{i-1}^T \cdot \mathbf{b}}{\mathbf{p}_{i-1}^T \cdot \mathbf{A} \cdot \mathbf{p}_{i-1}} \\
&= -\frac{\mathbf{p}_{i-1}^T \cdot (\mathbf{A}\mathbf{x}_{i-1})}{\mathbf{p}_{i-1}^T \cdot \mathbf{A} \cdot \mathbf{p}_{i-1}} + \frac{\mathbf{p}_{i-1}^T \cdot \mathbf{r}_{i-1}}{\mathbf{p}_{i-1}^T \cdot \mathbf{A} \cdot \mathbf{p}_{i-1}}. \quad (15.20)
\end{aligned}
$$

The first term vanishes due to the A-conjugacy, and the second term reproduces the result, equation (15.10), obtained originally for $i = 0$, that is valid for all i.

With only N independent vectors in the N-dimensional space, the conjugate gradient method is expected to converge to the minimum in at most N steps. Usually, numerical round-off leads to a breakdown of the A-orthogonality of the $\mathbf{p}_i$; this is not an issue if the iteration is carried out more than N times, except that more computing is called for. With large sparse matrices, the method comes into its own and compares very favorably with methods based on Gauss elimination.

The procedure of solving the system of equations (15.3) begins by selecting a random starting vector $\mathbf{x}_0$ and $\mathbf{p}_0 = -\mathbf{r}_0$. Next, we determine sequentially

$$
\{\lambda_0, \mathbf{x}_1, \mathbf{r}_1, \epsilon_0, \mathbf{p}_1\},
$$
$$
\{\lambda_1, \mathbf{x}_2, \mathbf{r}_2, \epsilon_1, \mathbf{p}_2\},
$$

and so on, in that order. At the ith iteration we test for convergence by comparing the norm of the difference of the vectors $\mathbf{x}_i$ and $\mathbf{x}_{i-1}$

with the norm of $\mathbf{x}_i$. If $|\mathbf{x}_i \cdot \mathbf{x}_i - \mathbf{x}_{i-1} \cdot \mathbf{x}_{i-1}|/|\mathbf{x}_i \cdot \mathbf{x}_i|$ is less than a tolerance we stop the iteration.

If the matrix $\mathbf{A}$ is ill-conditioned, the rate of convergence of the conjugate gradient method is reduced. For a symmetric positive definite matrix the condition number is defined as the ratio of the largest eigenvalue to the smallest. If the eigenvalues of $\mathbf{A}$, in increasing order, are $\mu_1, \mu_2, \ldots, \mu_N$, then the condition number κ is

$$\kappa(\mathbf{A}) = \frac{|\mu_N|}{|\mu_1|}. \tag{15.21}$$

With an ill-conditioned matrix, the inverse of the matrix is numerically obtained with a large error that is representative of the condition number. This means that the solution vector $\mathbf{x}$ will correspondingly have large loss of accuracy. In an attempt to reduce this problem, and also to accelerate the convergence of the solution for well-conditioned matrices, equation (15.3) is multiplied by a preconditioner matrix $\mathbf{M}^{-1}$. The essential idea with the preconditioner is that $(\mathbf{M}^{-1} \cdot \mathbf{A})$ has a denser distribution of eigenvalues, thereby having a smaller condition number than the original matrix $\mathbf{A}$. The closer $\mathbf{M}$ is to $\mathbf{A}$, the lower is the condition number of $\mathbf{M}^{-1}\mathbf{A}$. However, we wish to use a preconditioner that is easier to invert than the original matrix! A simple choice for a preconditioner is to take a diagonal $\mathbf{M}$ with $M_{ii}^{-1} = A_{ii}^{-1}$. Further improvements in the convergence occur if we let $\mathbf{M}$ be the symmetric, banded, invertible matrix with a small bandwidth, having the same matrix elements as $\mathbf{A}$ over the occupied bandwidth. *In other words, using a tridiagonal matrix $\mathbf{M}$ whose elements are the same as the diagonal and the next-diagonal elements, we can improve the convergence substantially.* (Getting the inverse of the tridiagonal preconditioner matrix can be easily programmed into the algorithm. The price of using a tridiagonal form is therefore not prohibitive at all. As the bandwidth of the selected matrix $\mathbf{M}$ increases, the number of iterations needed continues to decrease, for a solution of a given accuracy. In the limit that the preconditioner is $\mathbf{A}^{-1}$ itself, naturally we have the answer in one iteration!) With a preconditioner present, we solve

$$\mathbf{M}^{-1} \cdot \mathbf{A} \cdot \mathbf{x} = \mathbf{M}^{-1} \cdot \mathbf{b}. \tag{15.22}$$

The method of solution is to use the Cholesky decomposition $\mathbf{M} = \mathbf{C} \cdot \mathbf{C}^T$, where $\mathbf{C}$ is a lower triangular matrix with unit diagonal elements. We have

$$\mathbf{C}^{-T} \cdot (\mathbf{C}^{-1} \cdot \mathbf{A} \cdot \mathbf{C}^{-T}) \cdot (\mathbf{C}^T \cdot \mathbf{x}) = \mathbf{C}^{-T} \cdot (\mathbf{C}^{-1} \cdot \mathbf{b}), \quad (15.23)$$

where an intermediate step and a cancellation of the matrix $\mathbf{C}^{-T}$ from both sides is indicated, and we solve for $\mathbf{y} = (\mathbf{C}^T \cdot \mathbf{x})$. A back-substitution then leads to the required solution $\mathbf{x}$.

Observe that $(\mathbf{C}^{-1} \cdot \mathbf{A} \cdot \mathbf{C}^{-T})$ equals $(\mathbf{M}^{-1} \cdot \mathbf{A})$ since

$$(\mathbf{C}^T \cdot \mathbf{C}^{-T}) \cdot (\mathbf{C}^{-1} \cdot \mathbf{A} \cdot \mathbf{C}^{-T}) = \mathbf{C}^T \cdot (\mathbf{M}^{-1} \cdot \mathbf{A}) \cdot \mathbf{C}^{-T}. \quad (15.24)$$

Thus the two forms of the matrix product have the same eigenspectrum, and hence the same condition number. We work with equation (15.23) since the matrix $(\mathbf{C}^{-1} \cdot \mathbf{A} \cdot \mathbf{C}^{-T})$ is symmetric and positive definite, and we can therefore apply the conjugate gradient method to equation (15.23). The algorithm presented in *Numerical Recipes*[4] avoids the direct inversion of $\mathbf{C}$ or, for that matter, of $\mathbf{M}$, and requires knowing only the effect of applying $\mathbf{M}^{-1}$ on any vector.

Other variants on the conjugate gradient method and its extension to an asymmetric real coefficient matrix, and to a complex coefficient matrix are available. The general minimum residual method (GMRES), based on the conjugate gradient method, is used to solve linear equations with such $\mathbf{A}$ matrices. The GMRES uses the quadratic function, $\mathcal{G} = \mathbf{r}^T \cdot \mathbf{r}$, of the residual $\mathbf{r} = \mathbf{A} \cdot \mathbf{x} - \mathbf{b}$. The reader is referred to Jennings,[7] Axelsson,[18] Saad and Schultz,[19] and to *Numerical Recipes* cited earlier.

A final comment is that when we require solutions to the simultaneous equations for a number of right hand vectors $\mathbf{b}_i$, but with the same coefficient matrix $\mathbf{A}$, the iterative approach could become more costly in terms of computations that need to be performed for each case. On the other hand, a single LU-decomposition can be used again and again to obtain the solution vectors for the different right hand vectors. Such considerations have to be taken into account in large-scale matrix computations.

15.4 The standard eigenvalue problem

The standard eigenvalue problem is represented by the equation

$$\mathbf{A} \cdot \psi = \lambda \mathbf{I} \cdot \psi, \quad (15.25)$$

where $\mathbf{I}$ is the unit matrix.[†]

[†]In the following, we reserve boldface symbols for matrices only. This should cause no confusion.

A number of methods are available for the solution of this equation for the eigenvalues λ_i and the corresponding eigenfunctions ψ_i for small matrices. The most straightforward method applicable to very small matrices is to find the roots of the characteristic equation, $\det(\mathbf{A} - \lambda\,\mathbf{I}) = 0$. Once the eigenvalue λ_i is determined, the corresponding eigenvector is developed by assuming, for example, that the first component of the array ψ_i is unity,

$$\mathbf{\Psi}_i^T = \{1, \psi_2^{(i)}, \psi_3^{(i)}, \ldots\}, \tag{15.26}$$

and explicitly solving for the other components using the eigenvalue equation (15.25) with $\lambda = \lambda_i$. However, as the matrix dimension increases towards $\mathcal{O}(10^2)$ it is convenient to turn to transformation methods. The typical method for complex matrices is to employ a similarity transformation of the form

$$\mathbf{P}^{-1} \cdot \mathbf{A} \cdot \mathbf{P} = \mathbf{\Lambda}. \tag{15.27}$$

This can also be done using a series of unitary transformation matrices $\mathbf{P}_i$ to systematically evolve the matrix $\mathbf{A}$ towards the diagonal form $\mathbf{\Lambda}$. The Jacobi and Householder methods are representative of this approach. We note that when the eigenfunctions are arranged in columnar form in a matrix $\mathbf{\Psi}$ we can write the eigenvalue problem in the form

$$\mathbf{A} \cdot \mathbf{\Psi} = \mathbf{\Psi} \cdot \mathbf{\Lambda}. \tag{15.28}$$

Multiplying both sides by the inverse of the matrix $\mathbf{\Psi}$ we obtain

$$\mathbf{\Psi}^{-1} \cdot \mathbf{A} \cdot \mathbf{\Psi} = \mathbf{\Lambda}. \tag{15.29}$$

We see that the similarity transformation performed by the matrix $\mathbf{\Psi}$ is the one that diagonalizes the matrix $\mathbf{A}$.

When the dimension of the matrix becomes larger than, say, 100 we have to invoke iterative methods to obtain the eigenvalues of interest. These are essentially similar to the ones for the generalized eigenvalue problem considered below.

15.5   The generalized eigenvalue problem

We have seen in earlier chapters that the variation of the discretized action integral for Schödinger's equation leads to a generalized eigenvalue problem

$$\mathbf{A} \cdot \psi = \lambda\,\mathbf{B} \cdot \psi. \tag{15.30}$$

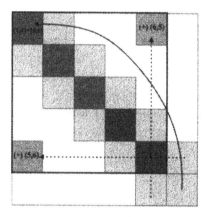

Fig. 15.4 The reduction in dimensionality of the usual global matrix due to Bloch periodicity is shown. The arrows indicate the repositioning of the matrix elements. Elements (6,5) and (5,6) are added to element (1,5) and (5,1), respectively, just as element (6,6) is added to (1,1). In finite element applications the extremely off-diagonal elements are typically absent before this Bloch folding is performed. Further details are given in the text.

We wish to solve this eigenvalue problem for the eigenpairs (λ_i, ψ_i). Here $\mathbf{A}$ is an $N \times N$ Hermitian matrix and $\mathbf{B}$ is a Hermitian positive definite matrix.[‡] The presence of the so-called overlap matrix $\mathbf{B}$ on the right side is due to the nonorthogonal basis set used in the FEM which defines shape functions over each element. In most applications the matrix $\mathbf{B}$ is real and symmetric, as it is obtained from the integration of the finite element shape functions which are real polynomials. However, in periodic systems, with periodicity d, Bloch's theorem[20] applies, and the wavefunction has to satisfy the relation $\psi(d) = \psi(0) \exp(iqd)$. This boundary condition (BC) leads to complex Hermitian forms for $\mathbf{A}$ and also $\mathbf{B}$. In this case both matrices have a full bandwidth since the off-diagonal corners of the matrix are occupied.

One way to see that the matrices in Bloch-periodic systems are fully banded is to use matrix multiplication for inserting the factors of $\exp(\pm iqd)$. Consider a linear region discretized into, say, five elements with linear interpolation functions. The global matrices will be of dimension 6. The last node (6) will have its nodal value ψ_6 replaced by $\psi_1 \exp(iqd)$ in order to implement Bloch's periodicity condition. Before this condition is imposed the matrix $\mathbf{B}$ will be real

---

[‡]This means that the eigenvalues of $\mathbf{B}$ are greater than zero.

and made of the overlay of the usual 2×2 element matrices. Now let us define a 6×6 *diagonal* matrix $\mathbf{Q}$ with unit elements except for the last element, which is $\exp(iqd)$. Let us also write

$$\psi = \text{Transpose}\{\psi_1, \psi_2, \psi_3, \psi_4, \psi_5, \psi_6\},$$
$$\overline{\psi} = \text{Transpose}\{\psi_1, \psi_2, \psi_3, \psi_4, \psi_5, \psi_1\}, \qquad (15.31)$$

We then have

$$\psi^* \, \mathbf{B} \, \psi = \overline{\psi}^* \, \mathbf{Q}^\dagger \mathbf{B} \mathbf{Q} \, \overline{\psi}. \qquad (15.32)$$

Multiplying through the matrices leads to the elements in the last column of $\mathbf{B}$ acquiring a factor of $\exp(iqd)$ and the elements in the last row of the matrix being multiplied by $\exp(-iqd)$. The element in location (6,6) in our example continues to be real. We now implement the Bloch periodicity condition as follows. We relabel the node index $nelem + 1 = 6$ to be index 1. This effectively leads to the overlay of the last element matrix on to the first in our system of simultaneous equations. Next we note that the elements $(6, j) \rightarrow (1, j)$ and $(i, 6) \rightarrow (i, 1)$, where $(i, j = 1, \dots, 5)$. These elements are added to any previously nonzero elements present in these locations. With this reduction in dimension we in effect have

$$\mathbf{Q}^\dagger \mathbf{B} \mathbf{Q} = \begin{pmatrix} B_{11} & B_{12} & 0 & 0 & 0 & 0 \\ B_{21} & B_{22} & B_{23} & 0 & 0 & 0 \\ 0 & B_{32} & B_{33} & B_{34} & 0 & 0 \\ 0 & 0 & B_{43} & B_{44} & B_{45} & 0 \\ 0 & 0 & 0 & B_{54} & B_{55} & B_{56}e^{iqd} \\ 0 & 0 & 0 & 0 & e^{-iqd}B_{65} & e^{-iqd}B_{66}e^{iqd} \end{pmatrix}$$

$$\Rightarrow$$

$$\overline{\mathbf{B}} = \begin{pmatrix} B_{11} + B_{66} & B_{12} & 0 & 0 & e^{-iqd}B_{65} \\ B_{21} & B_{22} & B_{23} & 0 & 0 \\ 0 & B_{32} & B_{33} & B_{34} & 0 \\ 0 & 0 & B_{43} & B_{44} & B_{45} \\ B_{56}e^{iqd} & 0 & 0 & B_{54} & B_{55} \end{pmatrix}.$$

$$(15.33)$$

These rearrangements of the matrix elements are displayed in Fig. 15.4. Thus both the $\mathbf{A}$ and $\mathbf{B}$ matrices have maximally off-diagonal elements in periodic systems, and in particular $\mathbf{B}$ also becomes Hermitian when Bloch periodicity is required. On the brighter side of the situation, the matrices generated in the finite element approach are generally banded and sparse, and hence the rows and columns of such a matrix can be rearranged so as to substantially lower the bandwidth from $N - 1$ to a much smaller value using the methods discussed earlier. Once the wavefunctions are obtained, it is convenient to put back the final nodal value explicitly in the array for any graphical display of the solution.

When the dimension N of the matrix becomes larger than $\sim 10^2$, we are typically interested in a specific spectral range of eigenvalues $[\lambda_\ell, \lambda_h]$ and not the entire spectrum. This is particularly true with matrix dimensions increasing beyond, say, $N = 10^3$. Solving for all the eigenvalues and eigenfunctions in such a case would be a formidable task fraught with increasing numerical inaccuracies as the dimension of the matrix increases. Let $\sigma = (\lambda_\ell + \lambda_h)/2$ be the center of the spectral window. We then attempt a solution of

$$\mathbf{H} \cdot \psi = \mu \mathbf{B} \cdot \psi, \qquad (15.34)$$

where $\mathbf{H} = \mathbf{A} - \sigma \mathbf{B}$ and $\mu = (\lambda - \sigma)$. This shift of the spectrum does not alter the eigenvectors, while the eigenvalues of interest are located on either side of the center.

Usually, we can truncate the N-dimensional space to a subspace of dimension p, where $p \ll N$, while obtaining reasonably accurate values for μ_i. We select a set p of N-dimensional seed vectors in a manner such that this initial selection has some representation of the exact eigenvectors of the problem. Then through the iteration procedure, to be described below, we approach the exact eigenvectors ψ_i. This convergence will be exact if the dimension p approaches N and the number of iterations approaches infinity. In the following, we use the Sturm sequence determination in order to provide guidelines for selecting an appropriate dimension for the subspace, and show how to select seed vectors.

15.5.1   Sturm sequence check

We begin the computation by determining the number of eigenvalues in the range $[\lambda_\ell, \lambda_h]$ using the Sturm sequence check. This procedure requires the symmetric decomposition of $\mathbf{H}_\ell = \mathbf{A} - \lambda_\ell \mathbf{B}$, and of $\mathbf{H}_h$, into the factors $\mathbf{LDL}^T$. Here $\mathbf{L}$ is a lower triangular matrix with unit diagonals and zero matrix elements in the upper triangle, and $\mathbf{L}^T$ is its transpose. The matrix $\mathbf{D}$ is a diagonal matrix. We suppose that we have at hand either a real symmetric matrix or a complex Hermitian matrix $\mathbf{H}$. This decomposition essentially retains the banded nature of $\mathbf{H}$ and is performed using the algorithm given in *Numerical Recipes*.[4] §

The determinant of the matrix $\mathbf{H}_\ell$ is given by

$$\det(\mathbf{H}_\ell) = \det(\mathbf{L} \cdot \mathbf{D} \cdot \mathbf{L}^T)$$
$$= \det(\mathbf{L}) \cdot \det(\mathbf{D}) \cdot \det(\mathbf{L}^T). \qquad (15.35)$$

The determinants of the lower and upper triangular matrices are unity, and hence $\det(\mathbf{H}_\ell) = \det(\mathbf{D})$.

If we think of λ_ℓ as the variable in the characteristic equation for the initial diagonalization problem, then the characteristic polynomial is expressed here as $\det(\mathbf{D}) = \Pi D_{ii}$. This product, being already factored, provides the means of determining the number of eigenvalues below λ_ℓ. This number ν_ℓ is just the number of negative diagonal entries in $\mathbf{D}$. If one of the D_{ii} is zero, then we have already determined one of the eigenvalues to be λ_ℓ.

When $\mathbf{H}_\ell$ is complex we can either directly modify the **LDL**-algorithm, or separate the characteristic equation, $\mathbf{A} - \lambda_\ell \mathbf{B} = 0$, into real and imaginary parts, denoted by subscripts $\Re$ and $\Im$, and equate them separately to zero. This leads to the $2N$ equations

$$\begin{pmatrix} (\mathbf{H}_\ell)_\Re & -(\mathbf{H}_\ell)_\Im \\ (\mathbf{H}_\ell)_\Im & (\mathbf{H}_\ell)_\Re \end{pmatrix} \cdot \begin{pmatrix} \psi_\Re \\ \psi_\Im \end{pmatrix} = 0. \qquad (15.36)$$

We decompose the matrix on the left into its $\mathbf{LDL}^T$ form. The new diagonal matrix $\mathbf{D}$ contains twice the number, $2\nu_\ell$, of negative entries in this case.

This process is repeated with λ_ℓ replaced by λ_h to obtain the number ν_h of eigenvalues below λ_h. The difference, $\nu = \nu_h - \nu_\ell$, represents the number of eigenvalues in the interval $[\lambda_\ell, \lambda_h]$. This rather striking result is arrived at without going through any diagonalization procedure.

The number ν is used to decide on the working dimension for the subspace, and again in a final check to verify whether our iteration process has determined all the eigenvalues in the range. It has been empirically determined that selecting $p = \min\{2\nu, \nu + 8\}$ gives good results for the final eigenvalues.[21] It is preferable to err on the side

---

§Very minor, and obvious, changes in the algorithm for real matrices given there are needed for the complex Hermitian case.

of selecting a larger p, the limitation being computer memory or the time taken to diagonalize matrices in the subspace.

15.5.2 Inverse vector iteration

It is useful to consider just one vector in order to illustrate the nature of the block vector iteration procedure and its convergence. With just one vector, the procedure is called the method of inverse vector iterations. Suppose we have to determine a particular eigenvalue λ_0 and the corresponding eigenfunction ψ_0. We begin by using an approximate guess, λ_1, for the eigenvalue. The corresponding seed eigenvector X_1 is taken to be a vector array with unity for all the entries in the array.

Let us define $\lambda_0 = \lambda_1 + \Delta\lambda_1$, so that

$$(\mathbf{A} - \lambda_1\,\mathbf{B})\,\psi_0 = \Delta\lambda_1\,\mathbf{B}\,\psi_0. \tag{15.37}$$

The iteration is initiated by replacing ψ_0 on the right side by the known seed vector X_1 and the vector on the left side by the next iterate X_2 which has to be determined. Equation (15.37) is solved by performing an LU-decomposition of the matrix on the left and solving for X_2. While $\Delta\lambda_1$ is not known, we can imagine for the moment that this factor is absorbed in the unknown vector on the left side. The new iterate X_2 is used to determine a new approximation $\lambda_2 = \lambda_1 + \Delta\lambda_2$ using the Rayleigh quotient

$$\Delta\lambda_2 = \frac{\langle X_2|\mathbf{A} - \lambda_1\mathbf{B}|X_2\rangle}{\langle X_2|\mathbf{B}|X_2\rangle}. \tag{15.38}$$

The vector X_2 is normalized with respect to the matrix $\mathbf{B}$ in the sense that X_2 is scaled by the factor $\langle X_2|\mathbf{B}|X_2\rangle^{1/2}$. If this normalization is performed before evaluating the Rayleigh quotient, the denominator in the above equation reduces to unity.

This sequence of iterations converges to the appropriate eigenvalue λ_0 that is closest to λ_1, and to the corresponding eigenvector, as shown below. In other words, a good starting guess is a λ_1 that would have to be reasonably close to the particular eigenvalue that is being sought. This choice is usually dictated by physical considerations. The convergence can be accelerated if the shift of the eigenspectrum represented by λ_1 is updated to $\lambda_1 + \Delta\lambda_i$ at the end of $i = 5$ or 10 iterations. Then the LU-decomposition is performed every so often after several iterations. An eigenvector is assumed to have converged if the relative change $|\lambda_i - \lambda_{i+1}|/|\lambda_i|$ in the corresponding eigenvalue from iteration i to $i+1$ is less than a tolerance, which is typically set to 10^{-6}–10^{-10}. An alternate criterion is to evaluate $(|\psi_{i+1}^\dagger \mathbf{B}\psi_{i+1} - \psi_i^\dagger \mathbf{B}\psi_i|)/|\psi_{i+1}^\dagger \mathbf{B}\psi_{i+1}|$ and require that it be less than a similar tolerance. A very readable description of inverse vector iteration is presented by Kerner et al.,[22] with an extension to general non-Hermitian matrices as well.

We make very minor changes to the discussion in *Numerical Recipes*[4] to illustrate the reason why the inverse vector iteration converges rapidly. At the ith iteration, we have

$$(\mathbf{A} - \lambda_{\mathbf{i}} \mathbf{B}) \cdot X_{i+1} = \Delta\lambda_i \mathbf{B}X_i. \qquad (15.39)$$

If we expand X_{i+1} and X_i in terms of the exact eigenfunctions ψ_k as follows,

$$X_{i+1} = \sum_k \alpha_k \psi_k,$$

$$X_i = \sum_k \beta_k \psi_k,$$

then equation (15.39) is expressible as

$$\sum_k \alpha_k (\lambda_k^{(0)} - \lambda_i)\mathbf{B}\psi_k = \sum_k \beta_k \Delta\lambda_i \mathbf{B}\psi_k. \qquad (15.40)$$

We then have

$$\alpha_k = \beta_k/(\lambda_k^{(0)} - \lambda_i),$$

$$X_{i+1} = \sum_k \left(\frac{\beta_k \, \Delta\lambda_i}{\lambda_k^{(0)} - \lambda_i} \right) \psi_k. \qquad (15.41)$$

Here $\lambda_k^{(0)}$ are the exact eigenvalues. With every iteration, the above expansion picks up a power of the coefficient in front of each ψ_k. As the eigenvalue gets updated and the iteration cycle is completed, the denominator of one particular eigenvector ψ_j, of index j, tends towards zero, making it the dominant eigenvector. Thus, the eigenvector we are seeking, ψ_0, is identified as ψ_j. This argument assumes that the eigenvalues are nondegenerate.

If the next eigenvalue is sought, we start with a new seed vector Y_1, again with unit entries in its array. In order to converge to the next nearest eigenvalue, this vector is B-orthogonalized against the first solution vector ψ_0 (or the final form of X_1). This is done by a modification of the usual Gram–Schmidt orthogonalization by requiring that

$$\tilde{Y}_1 = Y_1 - |\psi_0\rangle \left(\frac{\langle\psi_0|\mathbf{B}|Y_1\rangle}{\langle\psi_0|\mathbf{B}|\psi_0\rangle} \right). \tag{15.42}$$

At each stage of the iteration the new approximation to the eigenvector is B-orthogonalized with respect to the eigenvectors already determined in order to ensure convergence to a new eigenvalue and the corresponding eigenvector.

15.5.3 The subspace vectors

In the method known as subspace iteration, or simultaneous vector iteration, a number of seed vectors are iterated together to obtain convergence to eigenvalues simultaneously. As mentioned earlier, the dimension of the subspace is empirically selected to be $p = \min(2\nu, \nu + 8)$, where ν is the number of eigenvalues within a spectral range as determined by using the Sturm sequence check.

As with any iteration process, we start with some wishful thinking, and hope for a good starting point. If the matrices $\mathbf{H}$ and $\mathbf{B}$ were actually diagonal matrices then the eigenvalues would be given by $\tilde{\mu}_i = H_{ii}/B_{ii}$ and the eigenvectors would be the unit vectors $\epsilon_i(k) = \delta_{ik}$. Recall that the eigenvectors of the problem with the spectral shift σ are the same as those of the original problem without the shift. The Rayleigh–Ritz variational method that will be the basis of our diagonalization method naturally selects the lowest eigenvalues. Since we are working in a p-dimensional subspace we pick the starting selection for the basis vectors X_i to be the unit vectors ϵ_i corresponding to the lowest $p - 1$ values of $\tilde{\mu}_i$ determined by sorting the $\tilde{\mu}_i$ by their magnitude. We now assume that the $\mathbf{H}$ and $\mathbf{B}$ matrices are "diagonally dominant" and use these unit vectors as our seed vectors. The last eigenvector is chosen such that all its components are unity. This is to ensure that our set of seed vectors is not accidentally orthogonal to one of the desired final eigenvectors.

The true eigenvectors ψ_i corresponding to the lowest p eigenvalues μ_i are approximated by a linear combination of these p seed vectors with coefficients q_{ji}. Inserting the column vectors X_i into the ith column of an $N \times p$ matrix we have

$$\psi_i(k) \simeq \tilde{\psi}_{ki} = \sum_{j=1}^{p} X_{kj} q_{ji}. \tag{15.43}$$

Using a linear combination of vectors, all of which are used in the iteration, will clearly lead to faster convergence than using individual seed vectors as was done in inverse vector iteration. We return to subspace iteration after considering the properties of the Rayleigh quotient.

15.5.4   The Rayleigh quotient

We derive the properties of the Rayleigh quotient:

$$\rho(\psi) = \frac{\langle \psi | \mathbf{H} | \psi \rangle}{\langle \psi | \mathbf{B} | \psi \rangle}. \tag{15.44}$$

Let the general vector ψ be expanded in terms of the exact orthogonal eigenfunctions ψ_i where the corresponding eigenvalues μ_i are arranged in ascending order. We write

$$\psi = \sum_{i}^{N} c_i \psi_i. \tag{15.45}$$

In terms of this expansion the Rayleigh quotient is given by

$$\rho(\psi) = \frac{\mu_1 |c_1|^2 + \mu_2 |c_2|^2 + \cdots + \mu_N |c_N|^2}{|c_1|^2 + |c_2|^2 + \cdots + |c_N|^2}. \tag{15.46}$$

Assuming that neither μ_1 nor μ_N is zero, we have

$$\rho(\psi) = \frac{\mu_1 (|c_1|^2 + (\mu_2/\mu_1)|c_2|^2 + \cdots + (\mu_N/\mu_1)|c_N|^2)}{|c_1|^2 + |c_2|^2 + \cdots + |c_N|^2}, \tag{15.47}$$

and

$$\rho(\psi) = \frac{\mu_N ((\mu_1/\mu_N)|c_1|^2 + (\mu_2/\mu_N)|c_2|^2 + \cdots + |c_N|^2)}{|c_1|^2 + |c_2|^2 + \cdots + |c_N|^2}. \tag{15.48}$$

We are assuming that the ψ_i are B-orthogonal, i.e., $\langle \psi_i | B | \psi_j \rangle = \delta_{ij}$. Since the μ_i are ordered, we see from equations (15.47) and (15.48) that $\rho(\psi)$ is bounded from above and from below:

$$\mu_1 \leq \rho(\psi) \leq \mu_N. \tag{15.49}$$

Secondly, when $\psi = \psi_i$ only one term survives in the numerator and the denominator of the Rayleigh quotient, so that

$$\rho(\psi_i) = \mu_i. \tag{15.50}$$

Thus the Rayleigh quotient obtained using any eigenvector equals the corresponding eigenvalue.

In our iterative process, we will be working with approximate eigenfunctions. If the approximate eigenfunction $\tilde{\psi}$ is close to an eigenfunction ψ_i, we can write

$$\tilde{\psi} = \psi_i + \sum_{\substack{j=1 \\ j \neq i}} \delta_j \psi_j, \tag{15.51}$$

where the δ_j are small coefficients. In this case, the Rayleigh quotient is

$$\rho(\psi) = \frac{\mu_i + \sum_{\substack{j=1 \\ j \neq i}} \mu_j |\delta_j|^2}{1 + \sum_{\substack{j=1 \\ j \neq i}} |\delta_j|^2}$$

$$= \mu_i + \mathcal{O}(|\delta_j|^2). \tag{15.52}$$

Thus the convergence to the exact eigenfunction will be quadratic in the error δ.

In our case, we have approximate vectors X_i representing our subspace, as in equation (15.43). We form a matrix with the vectors X_i arranged in columns, so that this matrix $\mathbf{X}$ will be of dimension $N \times p$. Using the Einstein notation in which it is implicit that repeated indices are summed over, the corresponding Rayleigh quotient is

$$\rho(\tilde{\psi}_i) = \frac{\langle \tilde{\psi}_i | \mathbf{H} | \tilde{\psi}_i \rangle}{\langle \tilde{\psi}_i | \mathbf{B} | \tilde{\psi}_i \rangle}$$

$$= \frac{q_{ij}^* X_{jk}^* H_{k\ell} X_{\ell m} q_{mi}}{q_{ij}^* X_{jk}^* B_{k\ell} X_{\ell m} q_{mi}}. \tag{15.53}$$

We consider the form $\tilde{\psi}$ to be a variational guess. Employing a variation with respect to the coefficients q_{ik}^* to determine the minimum of $\rho(\tilde{\psi})$ we obtain

$$\frac{\delta}{\delta q^*} \rho(q^*, q) = 0$$

$$= X_{jk}^* H_{k\ell} X_{\ell m} q_{mi} - X_{jk}^* B_{k\ell} X_{\ell m} q_{mi} \tilde{\mu}_i. \tag{15.54}$$

Here we have made use of the relation

$$\left(\frac{\langle \tilde{\psi}_i | \mathbf{H} | \tilde{\psi}_i \rangle}{\langle \tilde{\psi}_i | \mathbf{B} | \tilde{\psi}_i \rangle} \right)\bigg|_{min} = \tilde{\mu}_i. \tag{15.55}$$

If we define

$$\mathbf{h} = \mathbf{X}^\dagger \cdot \mathbf{H} \cdot \mathbf{X},$$
$$\mathbf{b} = \mathbf{X}^\dagger \cdot \mathbf{B} \cdot \mathbf{X}, \tag{15.56}$$

as shown in Fig. 15.5, and let $\tilde{\boldsymbol{\mu}}$ be a diagonal matrix with entries $(\tilde{\mu}_i)$ along the diagonal, we see that equation (15.54) is an eigenvalue equation

$$\mathbf{h} \cdot \mathbf{q} = \mathbf{b} \cdot \mathbf{q} \cdot \tilde{\boldsymbol{\mu}}, \tag{15.57}$$

in the p-dimensional subspace for the eigenpairs $\{\tilde{\mu}_i, q_i\}$.

15.5.5 Subspace iteration

We have folded the original $N \times N$ eigenvalue problem into a $p \times p$ matrix problem. This is at the price of expressing the original eigenfunction in the approximation of expanding it in terms of the basis vectors associated with the subspace. The hope, as with all variational schemes, is that a small number of the subspace basis vectors would be sufficient for the convergence to the exact eigenfunctions. The lowest eigenvalues converge rapidly to the exact values, whereas the higher ones will be more and more approximate. It is for this reason that we use a subspace dimension p which is somewhat larger than the desired number ν of eigenvalues. The resulting matrices $\mathbf{h}$ and $\mathbf{b}$ in general are full matrices.

With p of order $10-100$ we can afford to solve equation (15.57) using a full matrix solver. The eigenspectrum of $\mathbf{b}$ obtained by a projection of $\mathbf{B}$ onto the subspace should still be a positive definite, symmetric matrix. We should mention that if the vectors X_i are B-orthogonalized, then the matrix $\mathbf{b} = \mathbf{X}^\dagger \cdot \mathbf{B} \cdot \mathbf{X}$ is a unit matrix. In case this B-orthogonality is not included before the eigenvalue equation is set up, we then proceed as follows. We start by expressing $\mathbf{b}$ as the product of a lower triangular matrix $\boldsymbol{\ell}$ and an upper triangular matrix $\boldsymbol{\ell}^T$ obtained by the Cholesky decomposition. Once this decomposition is performed, we follow the algorithm described

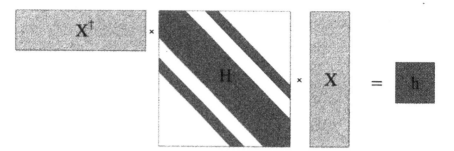

Fig. 15.5 Construction of the projection of global matrices into the subspace using subspace basis vectors. Matrix occupancy of **H** is schematically shown by the shading.

in *Numerical Recipes*[4] and rewrite the subspace eigenvalue problem in the standard form

$$(\ell^{-1}\, \mathbf{h}\, \ell^{-T}) \cdot (\ell^T q_i) = \mu_i\, (\ell^T q_i). \tag{15.58}$$

The matrix inverses appearing in the above equation will be filled matrices; however, with p being small, they will be of small enough dimension to allow us to perform this operation without excessive computer operation counts or loss of accuracy. At this stage, one of the usual schemes, such as the Householder method, for the diagonalization of a Hermitian matrix is used to obtain the eigenpairs of this standard eigenvalue problem. After $(\ell^T q_i)$ and $\tilde{\mu}_i$ are determined, we can construct the approximate eigenvectors for the global matrices using equation (15.43). An alternate approach is to use the QZ algorithm of Moler and Stewart,[23] in which elementary transformations are used to convert **h** into upper Hessenberg form and also **b** simultaneously.

The magnitudes of the eigenvalues are now sorted in ascending order and the corresponding vector components $X_i(k) = X_{kj}q_{ji}$ are placed into the matrix **X**, column by column, in order to provide an improved set of basis vectors for the next iteration. The new X_i are made B-orthogonal to each other using the Gram–Schmidt orthogonalization procedure mentioned earlier. The use of a reverse Gram–Schmidt procedure[7,8] is recommended here in order to minimize round-off error.

The $(i+1)$th iteration cycle consists of the following steps:[6,21,24]

1. Given vectors $\mathbf{X}^{(i)}$, obtain improved vectors $\mathbf{X}^{(i+1)}$

$$\mathbf{H} \cdot \mathbf{X}^{(i+1)} = \mathbf{B}\mathbf{X}^{(i)}. \tag{15.59}$$

This is done using LU-decomposition of $\mathbf{H}$ to solve the set of equations. The $\mathbf{LU}$ matrices are retained for the next iteration. An alternate way is to use the (bi-)conjugate gradient method to solve for the column vectors in $\mathbf{X}^{(i+1)}$. This has the advantage that it requires no additional matrices to be kept in computer memory, and the disadvantage that each vector solution requires iterative steps to be performed anew.

2. Form subspace matrices $\mathbf{h}$ and $\mathbf{b}$ and solve the eigenvalue problem

$$\mathbf{h}^{(i+1)} \cdot \mathbf{q}^{(i+1)} = \mathbf{b}^{(i+1)} \cdot \mathbf{q}^{(i+1)} \cdot \boldsymbol{\mu}^{(i+1)}. \tag{15.60}$$

3. Arrange $\mu_k^{(i+1)}$ in ascending order, and rearrange the columns of $\mathbf{q}^{(i+1)}$ accordingly.

4. Form the new eigenvectors as linear combinations of the old, with coefficients q_{ik}:

$$\mathbf{X}^{(i+2)} = \mathbf{X}^{(i+1)} \cdot \mathbf{q}^{(i+1)}. \tag{15.61}$$

5. Perform a B-orthogonalization of the vectors in $\mathbf{X}^{(i+2)}$ according to the reverse Gram–Schmidt scheme.

6. Stop the iteration if the eigenvalues have converged, as determined by some tolerance criterion, and if the eigenvectors have converged to within their acceptable tolerance. If the convergence criterion is not satisfied, return to Step 1 and continue the iteration.

The reader is referred to Parlett,[6] Rutishauser,[25] Yamamoto and Ohtsubo,[26] Jennings,[7] and Bathe and Ramaswamy[27] for a discussion of the Chebychev and other acceleration procedures for obtaining faster convergence in the subspace iteration scheme.

We note that the subspace iteration method for complex Hermitian matrices can be given a helping hand by identifying windows of eigenvalues containing four to eight eigenvalues each. This can be done using the Sturm sequence check and bisections of the initial window as necessary. The advantage of this step stems from the fact that it is faster to work with many smaller subspaces than a single large subspace.

15.5.6   The Davidson algorithm

The Davidson algorithm for the eigenvalue problem extends the subspace method by enlarging the subspace at each iteration cycle. In the following, we adapt the Davidson algorithm to the generalized eigenvalue problem.[28-30]

As in the subspace method, we begin by constructing guess vectors, X_k. We suppose that the guess eigenvalue for the vector X_k is λ_k. The vector is improved in the ith iteration to $X_k^{(i+1)} = X_k^{(i)} + \delta X_k^{(i)}$ by solving

$$\left(A - \lambda_k^{(i)} B\right) \delta X_k^{(i)} = -\left(A - \lambda_k^{(i)} B\right) X_k^{(i)}. \tag{15.62}$$

Here the eigenvalue is obtained from

$$\lambda_k^{(i)} = \frac{X_k^{(i)\dagger} \mathbf{A} X_k^{(i)}}{X_k^{(i)\dagger} \mathbf{B} X_k^{(i)}}. \tag{15.63}$$

While $\delta X_k^{(i)}$ obtained from above may be used to improve $X_k^{(i)}$ to form $X_k^{(i+1)}$, such an iteration cycle repeated with $X_k^{(i+1)}$ is found to have poor convergence characteristics after an initial rapid approach to acceptable eigenvalues, it being the basic subspace method. Instead, $\delta X_k^{(i)}$ is used to expand the subspace as follows.

In Davidson's approach, the solution of equation (15.62) is approximated by

$$\delta X_k^{(i)} = -(\mathbf{A}_D - \lambda_k^{(i)} \mathbf{B}_D)^{-1} \cdot (\mathbf{A} - \lambda_k^{(i)} \mathbf{B}) \cdot X_k^{(i)}. \tag{15.64}$$

Here A_D and $\mathbf{B}_D$ are diagonal matrices with just the diagonal elements of $\mathbf{A}$ and $\mathbf{B}$, respectively. This approximation attempts to bypass the need for an LU-decomposition of $(\mathbf{A} - \lambda_k^{(i)}\mathbf{B})$ for each eigenvalue. If any of the elements of $(\mathbf{A}_D - \lambda_k^{(i)} \mathbf{B}_D)$ is zero, the determination of $\delta X_k^{(i)}$ is abandoned for that value of k. Now the vector $\delta X_k^{(i)}$ is B-orthogonalized and added to the set of subspace vectors, as in Fig. 15.6, thereby enlarging the subspace. This enlargement of the subspace is done by considering each vector and determining its complement $\delta X_k^{(i)}$.[29] This improves the convergence of the subspace procedure and also improves the eigenvalue, according to the Hylleraas–Undheim–MacDonald variational principle which states that as the size of the basis set increases each eigenvalue tends to the true eigenvalue.[28,31]

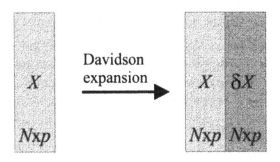

Fig. 15.6 The Davidson expansion of the subspace.

The iterative process is continued in this manner until the norm of the residual vector

$$R_i = (\mathbf{A} - \lambda_k^{(i)} \mathbf{B}) \cdot X_k^{(i)} \tag{15.65}$$

is less than a threshold value. Alternatively, we stop the iteration if the modulus of the coefficient $|q_{ik}|$ of the vector enlarging the subspace is small, so that it contributes very little in the expansion of ψ_k in terms of the subspace vectors $X_k^{(i)}$. The expansion of the subspace by one dimension each time is carried out for each of the initial eigenvectors.

In our implementation of the Davidson method, we have used subspace iteration three times followed by a doubling of the subspace using the residual vectors. (The original algorithm suggests alternating between subspace iteration and the Davidson expansion of the subspace.) We then repeat the subspace iteration three times and use the residual vectors of just the first p vectors to update the second set of vectors in columns $p + 1$, $p + 2$, $\ldots$, $2p$, in our expanded matrix with $2p$ columns. We do not expand the $\mathbf{X}$ matrix any further but keep updating the last p vectors after every three cycles of the subspace iteration. The subspace iterations improve the vectors and the eigenvalues. The updated vectors occupying the first p columns of $\mathbf{X}$ are the best approximation to the eigenvectors obtained by using the $2p$-dimensional subspace. The residuals inserted at each cycle bring in components of the vector space outside the original p-dimensional subspace and provide an accelerated convergence. We have found very rapid convergence to the exact values even with doubly degenerate eigenvalues present in the spectrum.

15.5.7 Least square residual minimization

In the Davidson method, we aim towards reducing the norm of the residual vector. This goal of minimizing the least square error in the eigenvector[32,33] can itself be cast in the form of a variational principle. Suppose we consider the quantity

$$\Omega(R_i) = \langle R_i | R_i \rangle$$

$$= \frac{\langle \psi_i | (\mathbf{A} - \lambda_i \mathbf{B})(\mathbf{A} - \lambda_i \mathbf{B}) | \psi_i \rangle}{\langle \psi_i | \mathbf{B} | \psi_i \rangle}$$

$$= \frac{q_{ik}^* X_{kj}^* (\mathbf{A} - \lambda_i \mathbf{B})_{j\ell} (\mathbf{A} - \lambda_i \mathbf{B})_{\ell m} X_{mn} q_{ni}}{q_{ik}^* X_{kj}^* \mathbf{B}_{j\ell} X_{\ell n} q_{ni}}. \quad (15.66)$$

The variation of $\Omega(R_i)$ with respect to q_{ik}^* leads to the eigenvalue equation in subspace matrices

$$X_{kj}^* (\mathbf{A} - \lambda_i \mathbf{B})_{j\ell} (\mathbf{A} - \lambda_i \mathbf{B})_{\ell m} X_{mn} q_{ni} = \rho_i X_{kj}^* \mathbf{B}_{j\ell} X_{\ell n} q_{ni}. \quad (15.67)$$

The coefficient vector q_{ni} associated with the smallest eigenvalue ρ_i is used to improve the linear combination of the X_k vectors employed in representing ψ_i. This improvement of the eigenvectors is done individually. This may be contrasted with the subspace method which uses simultaneous improvement of the p eigenvectors, but does not converge fast enough. The advantage of this minimum residual method is that its focus is on the improvement of the eigenvectors, while the subspace method converges to the eigenvalues reasonably quickly. With a small subspace, our hope is that these variants aimed at improving individual vectors do not lead to excessive computational overhead while providing faster overall convergence. It has been found by Murray et $al.$[30] that employing the method of minimum residual after every few cycles of the Davidson expansion of the subspace followed by subspace iteration provides a stable and convergent sequence of steps.

15.5.8 The Lanczos method

The Lanczos method[34] employs the idea of the subspace approach but differs somewhat in details. First consider the standard eigenvalue problem

$$\mathbf{A}\psi = \lambda\psi \quad (15.68)$$

with an $N \times N$ matrix $\mathbf{A}$. If we start with an arbitrary vector[¶] r_0 then the sequence of vectors $\{r_0, \mathbf{A}r_0, \ldots, \mathbf{A}^p r_0\}$ converges rapidly to the eigenvector corresponding to the largest eigenvalue. This can be understood simply by expanding the last vector in terms of the true eigenvectors v_i of $\mathbf{A}$ as follows. If the final vector is $\tilde{\psi}$ we have

$$\tilde{\psi} = \mathbf{A}^p r_0 = c_1 \lambda_1^p v_1 + c_2 \lambda_2^p v_2 + \ldots + c_n \lambda_n^p v_n. \qquad (15.69)$$

After scaling, we see that

$$\frac{\tilde{\psi}}{c_n \lambda_n^p} = \frac{c_1 \lambda_1^p}{c_n \lambda_n^p} v_1 + \frac{c_2 \lambda_2^p}{c_n \lambda_n^p} v_2 + \ldots + v_n, \qquad (15.70)$$

and as the iterative scheme proceeds the starting vector converges to the eigenvector corresponding to the largest eigenvalue.

Returning to the generalized eigenvalue problem

$$\mathbf{A}\psi = \lambda \mathbf{B}\psi, \qquad (15.71)$$

we note that we are typically interested either in the smallest eigenvalues, or in a few eigenvalues about a center, σ_0, and are not concerned about the largest eigenvalues. In the Lanczos algorithm we will then make use of the sequence of vectors

$$\{r_0, \mathbf{C}\, r_0, \mathbf{C}^2\, r_0, \ldots, \mathbf{C}^{p-1}\, r_0\}, \qquad (15.72)$$

where r_0 is an arbitrary starting vector of dimension N, and the number of vectors r_i is p. Here

$$\mathbf{C} = \mathbf{A}^{-1} \mathbf{B}, \qquad (15.73)$$

or

$$\mathbf{C} = (\mathbf{A} - \sigma_0 \mathbf{B})^{-1} \mathbf{B}; \qquad (15.74)$$

in the following we set $\sigma_0 = 0$ to simplify the discussion. We should reassure the reader that the method proceeds without explicitly inverting $\mathbf{A}$. A Cholesky decomposition or an LU-decomposition of $\mathbf{A}$ is employed in the intermediate steps below, and the matrix $\mathbf{A}$ (or the matrix $(\mathbf{A} - \sigma_0 \mathbf{B})$ is LU-factored only once. The convergence will be towards the smallest eigenvalues $\mu_i = 1/\lambda_i$ (or $1/(\lambda_i - \sigma_0)$ if an eigenvalue shift is used to "focus" the calculations into eigenvalues around σ_0).

[¶] Again, here we reserve boldface symbols for matrices only.

This sequence of independent vectors r_i defines a subspace called the *Krylov subspace*. The solution ψ will then be expanded in terms of these vectors as the basis vectors. Actually, in the Lanczos method we employ a linear combination of the r_i that form a mutually orthogonal set q_i and represent the solution ψ in this space. The aim of the method is to reduce $\mathbf{C}$ to a tridiagonal matrix $\mathbf{T}$ which is then diagonalized very easily. The matrix $\mathbf{T}$ is defined by

$$(\mathbf{A}^{-1}\mathbf{B}) \cdot [q_1, q_2, \ldots, q_p] = [q_1, q_2, \ldots, q_p] \cdot \mathbf{T}, \tag{15.75}$$

where the vectors q_i are arranged column by column into an $N \times p$ matrix, call it $\mathbf{Q}$, and

$$\mathbf{T} = \begin{bmatrix} \alpha_1 & \beta_1 & & & & \\ \beta_1 & \alpha_2 & \beta_2 & & & \\ & \beta_2 & \alpha_3 & \beta_3 & & \\ & & & \ddots & \ddots & \ddots & \\ & & & & & \beta_{p-1} \\ & & & & \beta_{p-1} & \alpha_p \end{bmatrix}. \tag{15.76}$$

Expanding out the first two equations in (15.75) we see that

$$\mathbf{A}^{-1}\mathbf{B}q_1 = \alpha_1 q_1 + \beta_1 q_2$$
$$\mathbf{A}^{-1}\mathbf{B}q_2 = \beta_1 q_1 + \alpha_2 q_2 + \beta_2 q_3. \tag{15.77}$$

We start with an arbitrary vector r_0 and normalize it with respect to the matrix $\mathbf{B}$ to obtain the starting vector q_1 for the Lanczos method. We define the B-normalization by

$$r_0^\dagger \mathbf{B} r_0 = c_1;$$
$$q_1 = r_0/\sqrt{c_1}, \tag{15.78}$$

and the vector q_1 is now B-normalized. It is understood that all the vectors q_i that are sequentially generated below are B-normalized.

In the first of equations (15.77) we multiply through by $\mathbf{B}$ and by $q_1^\dagger$. If we choose

$$\alpha_1 = q_1^\dagger \mathbf{B} \mathbf{A}^{-1} \mathbf{B} q_1, \tag{15.79}$$

then for $\beta_1 \neq 0$ we find that q_1 is B-orthogonal to q_2. The value of α_1 is determined by first forming the temporary vector

$$w_1 = \mathbf{B} q_1, \tag{15.80}$$

and then solving for the vector u_1 using the relation

$$\mathbf{A}u_1 = w \tag{15.81}$$

with an LU-decomposition for $\mathbf{A}$. Thus $u_1 = \mathbf{A}^{-1}\mathbf{B}q_1$.

We now construct q_2. Let us define a temporary vector

$$v_1 = \beta_1 q_2 = \mathbf{A}^{-1}\mathbf{B}q_1 - \alpha_1 q_1. \tag{15.82}$$

The vector v_1 is determined by the known terms on the right side of the above equation. Now

$$v_1^\dagger \mathbf{B}v_1 = \beta_1^* \beta_1 \cdot q_2^\dagger \mathbf{B}q_2. \tag{15.83}$$

Since q_2 is B-normalized we have

$$\beta_1 = \sqrt{v_1^\dagger \mathbf{B}v_1}, \tag{15.84}$$

where the positive sign may be used. We now have

$$q_2 = \frac{v_1}{|\beta_1|}. \tag{15.85}$$

The second of equations (15.77) is used to determine q_3 in a similar manner. We do this explicitly here. We multiply through by $\mathbf{B}$ and by $q_2^\dagger$ and use the B-orthonormality of the q_i to obtain

$$q_2^\dagger \mathbf{B}\mathbf{A}^{-1}\mathbf{B}\,q_2 = \alpha_2, \tag{15.86}$$

and with

$$v_2 = \beta_2 q_3 = \mathbf{A}^{-1}\mathbf{B}\,q_2 - \beta_1 q_1 - \alpha_2 q_2 \tag{15.87}$$

we obtain

$$q_3 = \frac{v_2}{|\beta_2|}, \tag{15.88}$$

with $|\beta_2|^2 = v_2^\dagger \mathbf{B}v_2$. In this manner we determine the other elements of the tridiagonal matrix $\mathbf{T}$.

It is possible to show that the eigenvalues ω_i of

$$\mathbf{T} = \mathbf{Q}^\dagger \mathbf{B}\mathbf{A}^{-1}\mathbf{B}\mathbf{Q} \tag{15.89}$$

approach the true eigenvalues $1/\lambda_i$ of the original generalized eigenvalue problem as $p \to N$. The convergence is fairly rapid so that accurate results can be extracted for the lowest eigenvalues even for a small number of vectors. The diagonalization of the tridiagonal matrix is a straightforward procedure, and with $p \ll N$, we can expect to obtain all the eigenvalues and eigenfunctions of the $\mathbf{T}$ matrix very rapidly on the computer. (Note that this is a standard eigenvalue problem and not a generalized eigenvalue problem.) With each increase in the dimension of $\mathbf{T}$ we obtain a new approximate eigenvalue while the earlier ones are systematically refined further.

Suppose that the eigenvectors of the $p \times p$ matrix $\mathbf{T}$ for the eigenvalue ω_i are given by s_i. These eigenvectors satisfy the relation

$$\mathbf{T} s_i = \omega_i s_i. \tag{15.90}$$

We can store the vectors s_i column by column in a matrix, call it $\mathbf{S}$. Then the above equation can be cast in a matrix form by employing a diagonal matrix Ω with elements ω_i along the diagonal. We have

$$\mathbf{T}\mathbf{S} = \mathbf{S}\Omega, \tag{15.91}$$

or

$$\mathbf{S}^{-1}\mathbf{T}\mathbf{S} = \Omega. \tag{15.92}$$

Here $\mathbf{S}$ is a unitary matrix with $\mathbf{S}^{-1} = \mathbf{S}^{\dagger}$.

The solution to the original problem $\mathbf{A}\psi = \lambda \mathbf{B}\psi$ is obtained by the variational minimization of the Rayleigh quotient

$$R = \frac{\mathbf{S}^{\dagger}\mathbf{Q}^{\dagger}\mathbf{B}\mathbf{A}^{-1}\mathbf{B}\mathbf{Q}\mathbf{S}}{\mathbf{S}^{\dagger}\mathbf{Q}^{\dagger}\mathbf{B}\mathbf{Q}\mathbf{S}} \tag{15.93}$$

with respect to the coefficients $\mathbf{S}$. This leads to the relation

$$\frac{\delta R}{\delta \mathbf{S}^{\dagger}} = 0 = \frac{\mathbf{Q}^{\dagger}\mathbf{B}\mathbf{A}^{-1}\mathbf{B}\mathbf{Q}\mathbf{S}}{\mathbf{S}^{\dagger}\mathbf{Q}^{\dagger}\mathbf{B}\mathbf{Q}\mathbf{S}} - \frac{\mathbf{Q}^{\dagger}\mathbf{B}\mathbf{Q}\mathbf{S}}{\mathbf{S}^{\dagger}\mathbf{Q}^{\dagger}\mathbf{B}\mathbf{Q}\mathbf{S}}\Omega. \tag{15.94}$$

This relation is analogous to equation (15.54). Since the $\mathbf{q}_i$ vectors are B-orthogonal, we have $\mathbf{Q}^{\dagger}\mathbf{B}\mathbf{Q}$ equal to a unit matrix. Therefore

$$(\mathbf{Q}^{\dagger}\mathbf{B})\,\mathbf{A}^{-1}\mathbf{B}\,\mathbf{Q}\mathbf{S} = (\mathbf{Q}^{\dagger}\mathbf{B})\,\mathbf{Q}\mathbf{S}\Omega. \tag{15.95}$$

We see that an approximation for the eigenvectors of the original problem, $\psi_{j(i)}$, expressed in component form with component index j and associated with the "nearly converged eigenvalues" ω_i are given by $Q_{jk}S_{k(i)}$. Note that these eigenvalues of $\mathbf{T}$ correspond to $1/\lambda_i$ of the original problem.

We have thus reduced the initial $N \times N$ generalized eigenvalue problem to the diagonalization of the $p \times p$ matrix $\mathbf{T}$, and the N-dimensional solution vectors are given by linear combinations of the Lanczos vector q_i, with coefficients determined by the eigenvectors of $\mathbf{T}$. In practical calculations, convergence to the actual eigenvalues of the large matrices is very rapid. However, the Lanczos vectors loose their orthogonality very quickly, and the method fails for the particular starting vector at hand. It is possible to recover from this by using a new starting vector. The procedure can be automated to make the algorithm reliable.

A thorough discussion of this method and further elaborations, including error analysis together with quantifying the error and extensions to the block-Lanczos approach, are given in the books by Parlett,[6] Hughes,[35] Cullum and Willoughby,[36] and by Saad.[37]

Obtaining eigenvalues of Hermitian (generalized) eigenvalue problems is a central issue in the computation of quantum states. It occurs in quantum chemistry, solid state bandstructure calculations, in atomic physics when nonorthogonal basis sets are used in the calculations, and in the finite element applications to quantum semiconductor heterostructures. It is therefore essential to have a reliable iterative diagonalizer that is capable of obtaining a few eigenvalues in a specified window in the spectrum. The solver should be capable of working with very large matrices of dimensions on the order of 10^6. The Davidson algorithm has been applied to such large matrices.

References

[1] E. Anderson, Z. Bai, C. Bischof, J. Demmel, J. Dongerra, J. Du Croz, A. Greenbaum, S. Hammarling, A. McKenney, S. Ostrouchov, and D. Sorensen, *LAPACK User's Guide* (Society for Industrial and Applied Mathematics, Philadelphia, 1992).

[2] J. J. Dongerra *et al.*, *LINPACK User's Guide* (Society for Industrial and Applied Mathematics, Philadelphia, 1979).

[3] B. T. Smith *et al.*, *Matrix Eigensystem Routines* - EISPACK *Guide*, 2nd ed., 'Lecture Notes in Computer Science', Vol. 6 (Springer-Verlag, New York, 1976).

[4] W. H. Press, S. A. Teukolsky, W. T. Vetterling, and B. P. Flannery, *Numerical Recipes* (Cambridge University Press, Cambridge, 1992).

[5] J. H. Wilkinson, *The Algebraic Eigenvalue Problem* (Oxford Universtity Press, Oxford, 1965).

[6] B. N. Parlett, *The Symmetric Eigenvalue Problem* (Prentice Hall, Englewood Cliffs, NJ, 1980).

[7] A. Jennings and J. J. McKeown, *Matrix Computation*, 2nd ed. (Wiley, New York, 1992).

[8] G. H. Golub and C. F. van Loan, *Matrix Computations* (Johns Hopkins University Press, Baltimore, MD, 1983).

[9] R. J. Collins, *Int. J. Numer. Methods Eng.* **6**, 345 (1973).

[10] E. Cuthill and J. McKee, *Proceedings of the 24th National Conference of the ACM*, ACM Publ P-69, pp157–172 (Association for Computing Machinery, New York, 1969).

[11] A. George and W. H. Liu, *SIAM Rev.* **31**, 1 (1989).

[12] W. H. Liu and A. H. Sherman, *SIAM J. Numer. Anal.* **13**, 198 (1976).

[13] N. E. Gibbs, W. G. Poole, and P. K. Stockmeyer, *SIAM J. Numer. Anal.* **13**, 236 (1976).

[14] I. S. Duff, A. M. Erisman, and J. K. Reid, *Direct Methods for Sparse Matrices* (Oxford University Press, Oxford, 1986).

[15] W. L. Briggs, V. E. Henson, and S. F. McCormick, *A Multigrid Tutorial* (Society for Industrial and Applied Mathematics, Philadelphia, 2000).

[16] U. Trottenberg, C. Oosterlee, and A. Schuller, *Multigrid* (Academic Press, New York, 2000).

[17] M. R. Hestenes and E. Stiefel, *J. Res. NBS* **49**, 409 (1952); M. R. Hestenes, *Optimization Theory – the finite dimensional case* (Wiley, New York, 1975); *Conjugate Direction Methods in Optimization* (Springer-Verlag, New York, 1980).

[18] O. Axelsson, *Iterative Solution Methods* (Cambridge University Press, Cambridge, 1994).

[19] Y. Saad and M. Schultz, *SIAM J. Sci. Stat. Comput.* **7**, 856 (1986). Also see: Y. Saad, *Iterative Methods for Sparse Linear Systems* (PWS, Boston, 1996).

[20] C. Kittel, *Introduction to Solid State Physics*, 6th ed. (Wiley, New York, 1986).

[21] K.-J. Bathe and E. Wilson, *Numerical Methods in Finite Element Analysis* (Prentice Hall, Englewood Cliffs, NJ, 1976); K.-J. Bathe, *Finite Element Procedures in Engineering Analysis* (Prentice Hall, Englewood Cliffs, NJ, 1982); K.-J. Bathe, *Finite Element Procedures* (Prentice Hall, Englewood Cliffs, NJ, 1996).

[22] W. Kerner, K. Lerbinger, and J. Steuerwald, *Comput. Phys. Commun.* **38**, 27 (1985).

[23] C. B. Moler and G. W. Stewart, *SIAM J. Numer. Anal.* **10**, 241 (1973). Also see: L. Kaufman, *ACM Trans. Math. Software* **1**, 271 (1975).

[24] N. S. Sehmi, *Large Order Structural Eigenanalysis Techniques* (Ellis Horwood, Wiley, New York, 1989).

[25] H. Rutishauser, *Numer. Math.* **13**, 4 (1969).

[26] Y. Yamamoto and H. Ohtsubo, *Int. J. Numer. Methods Eng.* **10**, 935 (1976).

[27] K. J. Bathe, and S. Ramaswamy, *Comput. Methods App. Mech. Eng.* **23**, 313 (1980).

[28] E. R. Davidson, *J. Comput. Phys.* **17**, 87 (1975); *Comput. Phys. Commun.* **53**, 49 (1989); *Comput. Phys.* **7**, 519 (1993).

[29] B. Liu, in *Numerical Algorithms in Chemistry: Algebraic Methods*, ed. C. B. Moler and I. L. Shavitt (Lawrence Berkeley Laboratory, Berkeley, CA, 1978).

[30] C. W. Murray, S. C. Racine, and E. R. Davidson, *J. Comput. Phys.* **103**, 382 (1992).

[31] A. Hylleraas and B. Undheim, *Z. Phys.* **65**, 759 (1930); J. K. L. MacDonald, *Phys. Rev.* **43**, 830 (1933); Also see: J. C. Slater, *Quantum Theory of Atomic Structure* (McGraw-Hill, New York, 1960), Vol. I, pp113–119.

[32] D. M. Wood and A. Zunger, *J. Phys. A* **18**, 1343 (1985).

[33] M. G. Feler, *J. Comput. Phys.* **14**, 341 (1974).

[34] C. Lanczos, *J. Res. NBS* **45**, 255 (1950).

[35] T. J. R. Hughes, *The Finite Element Method* (Prentice Hall, Englewood Cliffs, NJ, 1987).

[36] J. K. Cullum and R. A. Willoughby, *The Lanczos Algorithm for Large Symmetric Eigenvalue Computations*, Vols. 1 and 2 (Birkhäuser, Boston, 1985).

[37] Y. Saad, *Numerical Methods for Large Eigenvalue Problems* (Manchester University Press, Manchester, 1992).

Part V

Boundary elements

16

The boundary element method

16.1 Introduction

The boundary element method (BEM) is based on an integral equation formulation for the solution of differential equations. As we have seen in the context of the finite element method (FEM), analytical methods become cumbersome when the geometrical shape of the physical domain is complex or when the boundary conditions (BCs) for the physical property being calculated are complicated. If there are inhomogeneous terms in the differential equation, they lead to a "particular integral" which is difficult to evaluate in a direct analytical manner. In general, inhomogeneous terms in the differential equation are treated by the method of Green's function. A somewhat lengthy, but self-contained, exposition in Appendix C introduces the essential ideas of the Green's function technique. We show below that by using Green's theorem of vector calculus we can convert the differential equation under consideration into an integral equation on the boundary of the physical region. The discretization of the boundary integral leads to self-consistency conditions for the solution at the boundary. Once the solution at the boundary is obtained, the construction of the solution in the interior region proceeds in a straightforward manner. Thus the BEM employs the discretization of only the boundary, and not the interior of the physical region. This is a distinct advantage over the FEM, in which the entire physical region is discretized, which in turn leads to fairly large global matrices. The "reduction in dimensionality" associated with the BEM gives rise to global matrices of much fewer dimensions. The price we pay is the need for much greater care in ensuring that the BCs are properly implemented and that the boundary integrals are evaluated accurately. This is important because the discretized boundary integral equation is a self-consistency condition; in contrast, the FEM is based on determining a variational extremum for the action integral, a procedure that is more robust numerically. Another important difference is that the final global matrices are full matrices in the BEM, in contrast with global matrices generated by the FEM which are sparse.[†]

In the following, we restrict ourselves to 2D problems. In this chapter, we solve Laplace's equation for a simple rectangular domain, thus affording us an opportunity to discuss the method in specific terms. First, we derive series solutions for Green's functions in order to show how these problems are solved by the usual analytic methods.[1] *These analytical calculations are important for a fuller understanding of the discussion of the numerical method that follows.* For this reason they have been presented in some detail. We then consider the same examples using the infinite domain Green's function in the implementation of the BEM. We show the advantages of such an approach and discuss the computational issues involved. In the following chapters, we treat a physical system with inhomogeneous material properties due to the presence of regions occupied by different dielectric properties, and give examples of the application of the BEM to quantum mechanical problems.

16.2 The boundary integral

Let us consider the calculation of the electrostatic potential $\phi(\mathbf{r})$, with $\mathbf{r} = (x, y)$, in a rectangular region of dimension $a \times b$. The potential $\phi(\mathbf{r})$ satisfies Laplace's equation, $\nabla^2 \phi(\mathbf{r}) = 0$, in the interior region.

We now wish to set up an integral equation for the potential function. If we are given any two functions $\phi(\mathbf{r})$ and $g(\mathbf{r}, \mathbf{r}')$ then Green's theorem allows us to write

$$\int d^2 r' \left[g(\mathbf{r}, \mathbf{r}') \, \nabla'^2 \phi(\mathbf{r}') - \phi(\mathbf{r}') \, \nabla'^2 g(\mathbf{r}, \mathbf{r}') \right]$$

$$= \oint d\ell' \left(g(\mathbf{r}, \mathbf{r}') \frac{\partial \phi(\mathbf{r}')}{\partial n'} - \phi(\mathbf{r}') \frac{\partial g(\mathbf{r}, \mathbf{r}')}{\partial n'} \right). \quad (16.1)$$

The partial derivative with respect to n' denotes the gradient along the outward-drawn normal to the contour defining the physical region, and the prime on the gradient operator refers to the fact that it is a gradient with respect to the integration variable $\mathbf{r}'$. We now

---

[†]When we consider the multiregion BEM, we again encounter global matrices that are less than fully occupied.

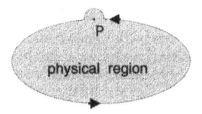

Fig. 16.1 The point of observation, P, on the boundary and the distortion of the physical region around it.

choose the function $g(\mathbf{r}, \mathbf{r}')$ to be the Green's function satisfying the equation

$$\nabla^2 g(\mathbf{r}, \mathbf{r}') = -4\pi\, \delta(\mathbf{r} - \mathbf{r}'), \tag{16.2}$$

with a δ-function "source term" on the right side.[†]

We substitute the condition $\nabla^2 \phi = 0$ and the above equation for g into equation (16.1) to obtain

$$\phi(\mathbf{r}) = \frac{1}{4\pi} \oint d\ell' \left(g\, \frac{\partial \phi}{\partial n'} - \phi\, \frac{\partial g}{\partial n'} \right). \tag{16.3}$$

Assuming that the form of $g(\mathbf{r}, \mathbf{r}')$ is known, the function ϕ can be determined everywhere in the physical region in terms of its given value or its normal derivative at the boundary. We now address the choices we have for the Green's function g.

Equation (16.3) is valid in the interior of the physical region. In other words, the potential $\phi(\mathbf{r})$ on the left side of equation (16.3) is specified by the integral relation at a point of observation, P, located at $\mathbf{r}$ away from the boundary. When the point P approaches the boundary from the interior region, it is necessary to distort the contour of integration so as to maintain the point of observation as an interior point. This distortion of the contour is shown in Fig. 16.1. The integral is divided into two parts, the original integral over the closed path that now has a small gap at P, and the semicircular path around the point P. We return to this issue once again when we develop and apply the BEM.

[†] In the boundary element literature the right side of the above equation is written with either sign, as well as with or without the factor of 4π. Choices other than the one made here lead to simple changes in the overall sign of the Green's function or its overall scale modulo 4π. Needless to say, this does not affect the final results.

16.3 An analytical approach

It is useful to consider simple examples initially in order to develop the BEM. In the following, we obtain analytical solutions to the Green's functions for two simple boundary value problems. This is followed by the discussion of the BEM using the infinite domain Green's function.

16.3.1 A Dirichlet problem

Let us consider a rectangular region $a \times b$ in which we wish to solve Laplace's equation. Suppose that the potential has the values at the boundary given by

$$\phi(0, y) = V_0 \sin \left(\frac{2\pi y}{b} \right), \qquad 0 \le y \le b,$$
$$\phi(a, y) = \quad 0, \qquad 0 \le y \le b,$$
$$\phi(x, 0) = \quad 0, \qquad 0 \le x \le a, \qquad (16.4)$$
$$\phi(x, b) = \quad 0, \qquad 0 \le x \le a,$$

specified along the boundary. Given the value of the potential, rather than its derivative, along the boundary we have a Dirichlet boundary value problem.

The analytical solution obtained by the method of separation of variables is

$$\phi(\mathbf{r}) = V_0 \sin \left(\frac{2\pi y}{b} \right) \frac{\sinh \left[2\pi (a - x)/b \right]}{\sinh \left(2\pi a/b \right)}. \qquad (16.5)$$

For comparison, we solve the above problem by explicitly constructing the Green's functions in two different ways to exemplify the variety of ways of proceeding in a traditional approach.

1. Since the value of ϕ is given on the boundary, it is natural to choose g to be zero there so that the first term in equation (16.3) vanishes. The potential everywhere is then given by the line integral over the normal derivative of g multiplied by the known boundary value of ϕ.

We can construct such a Green's function using eigenfunctions for a parallel problem

$$\nabla^2 \psi(x, y) + k^2 \psi(x, y) = 0, \qquad (16.6)$$

with solutions vanishing at the boundary. This can be recognized as the familiar problem of Schrödinger's equation for a particle in a 2D box with $0 \leq x \leq a$, and $0 \leq y \leq b$, if we identify k^2 with $(2mE/\hbar^2)$. The standard solution for the wavefunctions is

$$\psi_{nm}(x,y) = \frac{2}{\sqrt{ab}} \sin\left(\frac{n\pi x}{a}\right) \sin\left(\frac{m\pi y}{b}\right), \qquad (16.7)$$

and the "quantized" values for k^2 are $\pi^2(n^2/a^2 + m^2/b^2)$. We now let

$$g(x,y;x',y') = \sum_{n,m=1}^{\infty} C_{nm}(x',y') \frac{2}{\sqrt{ab}} \sin\left(\frac{n\pi x}{a}\right) \sin\left(\frac{m\pi y}{b}\right),$$

and substitute g into equation (16.2) to obtain

$$-\sum_{n,m=1}^{\infty}\left(\frac{\pi^2 n^2}{a^2} + \frac{\pi^2 m^2}{b^2}\right) C_{nm}(x',y')\psi_{nm}(x,y) =$$

$$-4\pi \sum_{n,m=1}^{\infty} \psi_{nm}(x,y)\psi_{nm}(x',y'). \qquad (16.8)$$

Here the δ-function has been expressed in terms of these eigenfunctions using the completeness of the eigenfunctions. Equating the coefficients of $\psi_{nm}(x,y)$ from both sides of the equation we identify

$$C_{nm}(x',y') = \frac{4\pi}{\left(\dfrac{\pi^2 n^2}{a^2} + \dfrac{\pi^2 m^2}{b^2}\right)}\psi_{nm}(x',y'), \qquad (16.9)$$

and hence

$$g(x,y;x',y') = 4\pi \sum_{n,m=1}^{\infty} \frac{\psi_{nm}(x,y)\,\psi_{nm}(x',y')}{k^2}$$

$$= \frac{16}{\pi ab} \sum_{n,m=1}^{\infty} \frac{\sin\left(\frac{n\pi x}{a}\right)\sin\left(\frac{m\pi y}{b}\right)\sin\left(\frac{n\pi x'}{a}\right)\sin\left(\frac{m\pi y'}{b}\right)}{\dfrac{n^2}{a^2} + \dfrac{m^2}{b^2}}. \qquad (16.10)$$

This function clearly vanishes at the boundary, and in equation (16.3) we are left with

$$\phi(\mathbf{r}) = \frac{-1}{4\pi} \oint d\ell' \left(\phi\frac{\partial g}{\partial n'}\right). \qquad (16.11)$$

The only nonzero segment of the line integral in the present example is over $0 \leq y' \leq b$ along which the outward-drawn normal is parallel to the unit vector $-\hat{x}$. A careful evaluation of the integral reproduces the desired result, equation (16.5). This presentation is an adaptation of Jackson's treatment[1] to two dimensions.

2. Let us take a slightly different approach to the determination of the Green's function. The Green's function must be symmetric under the interchange $\{x, y \leftrightarrow x', y'\}$. If we start with equation (16.2) and note that we can represent the y-dependence of the solution in terms of sine functions we can write

$$g(x, y; x', y') = \frac{2}{b} \sum_m f_m(x, x') \sin\left(\frac{m\pi y}{b}\right) \sin\left(\frac{m\pi y'}{b}\right). \quad (16.12)$$

We also have

$$\delta(y - y') = \frac{2}{b} \sum_m \sin\left(\frac{m\pi y}{b}\right) \sin\left(\frac{m\pi y'}{b}\right). \quad (16.13)$$

Substituting the above into equation (16.2) we have

$$\frac{\partial^2}{\partial x^2} f_m(x, x') - \frac{m^2 \pi^2}{b^2} f_m(x, x') = -4\pi\, \delta(x - x'). \quad (16.14)$$

In the two regions $x < x'$ and $x > x'$ the homogeneous form prevails, and we obtain the solutions vanishing at the boundaries to be of the form

$$f_m(x, x') \propto \begin{cases} \sinh\left(m\pi x_</b\right) & \text{for} \quad x < x', \\ \sinh\left(m\pi(a - x_>)/b\right) & \text{for} \quad x > x'. \end{cases} \quad (16.15)$$

Here $x_<$ ($x_>$) is the smaller (larger) of x, x'. We can now let $f_m(x, x') = A \sinh\left(m\pi x_</b\right) \sinh\left(m\pi(a - x_>)/b\right)$. The coefficient A is determined by integrating the differential equation for f_m to match the discontinuity in f'_m at $x = x'$. Substituting $f_m(x, x')$ in equation (16.14) and integrating the equation from $x = x' - \epsilon$ to $x = x' + \epsilon$ we obtain for the left side

$$\lim_{\epsilon \to 0} \frac{\partial f(x, x')}{\partial x}\Big|_{x' - \epsilon}^{x' + \epsilon} = -\left(\frac{Am\pi}{b}\right) \left\{ \cosh\left(\frac{m\pi}{b}(a - x)\right) \sinh\left(\frac{m\pi}{b}x\right) \right.$$
$$\left. + \sinh\left(\frac{m\pi}{b}(a - x)\right) \cosh\left(\frac{m\pi}{b}x\right) \right\}.$$

The integration of the right side gives -4π. Hence

$$A = \frac{4b}{m \sinh\left(\frac{m\pi a}{b}\right)},$$

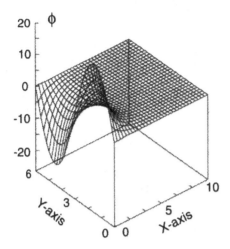

Fig. 16.2 The potential for the Dirichlet problem obtained using the BEM with $V_0 = 20\,\text{V}$. This agrees with the exact solution given in equation (16.5).

and g is given by

$$g(x, y; x', y') = \sum_m \left\{ \frac{2}{b} \sin\left(\frac{m\pi y}{b}\right) \sin\left(\frac{m\pi y'}{b}\right) \right\} \left(\frac{4b}{m \sinh\left(\frac{m\pi a}{b}\right)} \right)$$

$$\times \left\{ \sinh\left(\frac{m\pi x_<}{b}\right) \sinh\left(\frac{m\pi (a - x_>)}{b}\right) \right\}. \qquad (16.16)$$

The normal derivative of g with respect to x', along $-\hat{x}$, at $x' = 0$ is

$$-\frac{\partial}{\partial x'} g(x, y, x', y') \bigg|_{x'=0} = -\sum_m^{\infty} \left\{ \frac{2}{b} \sin\left(\frac{m\pi y}{b}\right) \sin\left(\frac{m\pi y'}{b}\right) \right\} \times$$

$$\left(\frac{4b}{m \sinh\left(\frac{m\pi a}{b}\right)} \right) \left\{ \frac{m\pi}{b} \cosh\left(\frac{m\pi x'}{b}\right) \sinh\left(\frac{m\pi (a - x)}{b}\right) \right\} \bigg|_{x'=0}, \qquad (16.17)$$

where we have used the definitions $x_< = x'$ and $x_> = x$. Substitution into equation (16.11) reproduces the solution given in equation (16.5).

We have reproduced the same result using the BEM, for illustrative purposes, in Fig. 16.2. It is evident from the two approaches that

$$\left(\frac{b}{m\pi \sinh\left(\frac{m\pi a}{b}\right)}\right) g \sinh\left(\frac{m\pi x_<}{b}\right)\sinh\left(\frac{m\pi(a-x_>)}{b}\right)$$

$$= \frac{2}{a}\sum_{n=1}^{\infty}\frac{\sin\left(\frac{n\pi x}{a}\right)\sin\left(\frac{n\pi x'}{a}\right)}{\frac{\pi^2 n^2}{a^2}+\frac{\pi^2 m^2}{b^2}}. \tag{16.18}$$

16.3.2 A Neumann problem

Now let us suppose that the normal derivative of the potential is specified at the boundary. Consider the example

$$\frac{\partial \phi}{\partial n} = \begin{cases} V_0\,x, & \text{along } y=0, \text{ and } 0\le x \le a, \\ -V_0\,x, & \text{along } y=b, \text{ and } 0\le x \le a, \\ V_0\,y, & \text{along } x=0, \text{ and } 0\le y \le b, \\ -V_0\,y, & \text{along } x=a, \text{ and } 0\le y \le b. \end{cases}$$

This is a Neumann BC. In this case it is natural to seek a Green's function whose normal derivatives vanish along the periphery so that again we have only one term surviving in the line integral in equation (16.3).

In analogy with the method described in the previous subsection, a simple choice for the Green's function is to cast it in terms of cosine functions which have zero derivatives along the periphery. Then g is given by

$$g(x,y;x',y') =$$

$$\frac{16}{\pi ab}\sum_{nm}\frac{\cos\left(\frac{n\pi x}{a}\right)\cos\left(\frac{m\pi y}{b}\right)\cos\left(\frac{n\pi x'}{a}\right)\cos\left(\frac{m\pi y'}{b}\right)}{\frac{n^2}{a^2}+\frac{m^2}{b^2}}, \tag{16.19}$$

and the solution is

$$\phi(\mathbf{r}) = \frac{1}{4\pi}\oint d\ell'\, g\,\frac{\partial \phi}{\partial n'}. \tag{16.20}$$

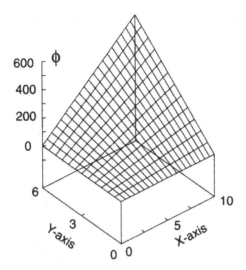

Fig. 16.3 The potential for the Neumann problem obtained using the BEM. The exact form is $\phi = -V_0 xy$, with $V_0 = 10\,\text{V}$.

An alternative form for the Green's function can be obtained by using an expansion in cosine functions in the y-direction and solving the differential equation obeyed by g. Following the technique delineated in example 2 in Section 16.3.1, we obtain

$$g(x,y;x',y') = \sum_m \left\{ \frac{2}{b} \cos\left(\frac{m\pi y}{b}\right) \cos\left(\frac{m\pi y'}{b}\right) \right\} \left(\frac{4b}{m \sinh\left(\frac{m\pi a}{b}\right)} \right)$$
$$\times \left\{ \cosh\left(\frac{m\pi x_<}{b}\right) \cosh\left(\frac{m\pi(a-x_>)}{b}\right) \right\}. \qquad (16.21)$$

Using this Green's function in equation (16.20) allows us to solve for the potential. With the derivative BCs the potential is determined up to an arbitrary constant. The reader may note that in this example the derivative conditions were constructed for the quadrupolar potential

$$\phi(x,y) = -V_0\, x\, y. \qquad (16.22)$$

The solution obtained from the BEM for $V_0 = 10\,\text{V}$ is shown in Fig. 16.3.

Further examples of the analytic approach to Green's functions and their applications are given in the classic text by Morse and Feshbach.[2] Integral equation methods in potential theory and in electrostatics have been well appreciated for a very long time.[3] The use of computational methods developed recently[4-7] has influenced the method of tackling such problems, especially when the complexity of the physical region puts them beyond the reach of analytical means.

16.4   Infinite domain Green's function

The Green's functions we have displayed in the above cases have been constructed to satisfy either the Dirichlet or the Neumann BCs at the boundary. They are such that one of the terms in the line integral, equation (16.3), vanishes.

It is important to recognize that Green's functions need not explicitly satisfy the BCs at the periphery of the finite domain. The essential aspects of the Green's function are (i) the singularity it exhibits at $\mathbf{r} = \mathbf{r}'$ and (ii) the property of reciprocity under the interchange of $\mathbf{r}$ and $\mathbf{r}'$. In a purely computational approach we need only consider the infinite domain Green's function and eschew the complications of first deriving the Green's function that satisfies the BCs. The task of constructing such a Green's function for a complex geometry would in itself be a formidable one, whereas Green's functions for most differential equations for the infinite domain are easy to derive.

The infinite domain Green's function for the Poisson problem is (see Appendix C for details)

$$g(\mathbf{r}, \mathbf{r}') = -2 \ln \left(\frac{|\mathbf{r} - \mathbf{r}'|}{R_0} \right). \tag{16.23}$$

Here R_0 is an arbitrary scale factor introduced in the logarithm to make the argument dimensionless. In the following, we will drop this factor since it contributes zero to both terms in equation (16.3) when integrated around the closed path. Now the solution, equation (16.3), has possible contributions from both terms in the line integral. The gradient of the Green's function with respect to $\mathbf{r}'$ is

$$\nabla' g(\mathbf{r}, \mathbf{r}') = -2 \frac{(\mathbf{r}' - \mathbf{r})}{|\mathbf{r} - \mathbf{r}'|^2}, \tag{16.24}$$

and equation (16.3) has the form

$$\phi(\mathbf{r}) = \frac{-1}{2\pi} \oint d\ell' \left(\ln(|\mathbf{r} - \mathbf{r}'|) \frac{\partial \phi(\mathbf{r}')}{\partial n'} - \frac{(\mathbf{r}' - \mathbf{r}) \cdot \mathbf{n}'}{|\mathbf{r} - \mathbf{r}'|^2} \phi(\mathbf{r}') \right). \tag{16.25}$$

We return later to the complications associated with allowing $\mathbf{r}$ to approach the boundary.

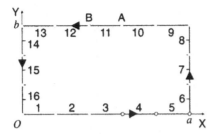

Fig. 16.4 The discretization into elements of the boundary of the rectangular physical region. The direction of the line integration is indicated by arrows.

In the boundary element approach, we imagine breaking up the line integral into a number of small segments, or "elements." For the rectangular region considered above, the elements are line segments along the periphery of the rectangle. Let us divide the sides parallel to the x-axis into five elements and the sides parallel to the y-axis into three elements. This is shown in Fig. 16.4. In each element, let us approximate the potential function by means of a linear interpolation between its values at the two end points (nodes) of the element.[§] We do the same for the normal derivative of the potential function. For example, the 11th element, shown in Fig. 16.4, will have the interpolation represented by

$$\phi(\mathbf{r}) = \phi_a^{(11)} \frac{(\mathbf{r}_b - \mathbf{r})}{(\mathbf{r}_b - \mathbf{r}_a)} + \phi_b^{(11)} \frac{(\mathbf{r} - \mathbf{r}_a)}{(\mathbf{r}_b - \mathbf{r}_a)},$$

$$\frac{\partial}{\partial n}\phi(\mathbf{r}) = \phi_a'^{(11)} \frac{(\mathbf{r}_b - \mathbf{r})}{(\mathbf{r}_b - \mathbf{r}_a)} + \phi_b'^{(11)} \frac{(\mathbf{r} - \mathbf{r}_a)}{(\mathbf{r}_b - \mathbf{r}_a)}.$$

$$(16.26)$$

Here $\phi_{a,b}^{(11)}$ refer to the values of the potential at end points A and B of element number 11, and $\mathbf{r}$ is on AB. With these explicit forms of ϕ and ϕ', and knowing the Green's function, we can evaluate spatial dependence of the line integral element by element.

In the BEM, we have the freedom to work with the Dirichlet and/or the Neumann BC *at each node* of the boundary. At first, we assume that we know either the value of the potential or its normal derivative. If there are n elements along the boundary we have n

[§]The reader would find it instructive to develop in parallel the theory with just a constant potential in each element.

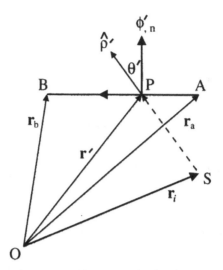

Fig. 16.5 The configuration of the element and the response point.

nodal parameters for the potential and the same number of nodal parameters for the normal derivative of the potential. Of the $2n$ nodal parameters in this example, n are known from the BCs. We evaluate equation (16.25) with the coordinate $\mathbf{r}$ at the nodes on the boundary, thereby setting the left side to $\phi(\mathbf{r}_i)$. With the given form of the interpolation, we can evaluate the line integrals and express the right side of equation (16.25) in terms of the $2n$ nodal parameters. This leads to a total of n equations for the unknown nodal values that could be either the potential or its normal derivative at each node.

We now consider several aspects of calculating the boundary integrals. We evaluate $\phi(\mathbf{r}_i)$ for each node $\mathbf{r}_i$ using equation (16.25). The line integral around the closed path is split into integrals over the total number, *nelem*, of elements, and therefore we have a total of *nelem* nodes. The coordinates of the end points of the element are $\mathbf{r}_a$ and $\mathbf{r}_b$, as in Fig. 16.5. Let us denote the length of the linear segment $\mathbf{r}_b - \mathbf{r}_a$ by ℓ_{iel}, where *iel* is a running index for the elements. In terms of a local variable ξ, ranging over $[0, 1]$, we can write $|\mathbf{r}' - \mathbf{r}_a| = \xi\, \ell_{iel}$ and $d\ell' = \ell_{iel}\, d\xi$. The expression for the interpolation also simplifies,

$$\phi^{(iel)}(\mathbf{r}') = \phi_a^{(iel)}\,(1 - \xi) + \phi_b^{(iel)}\,\xi,$$

and the normal derivative is written as

$$\frac{\partial}{\partial n}\phi^{(iel)}(\mathbf{r}') = \phi_a'^{(iel)}(1-\xi) + \phi_b'^{(iel)}\,\xi, \qquad (16.27)$$

with the integration ranging over $[0, 1]$ in the local variable ξ.

Let us suppose that O is the origin of the coordinate system and that $\mathbf{r}_i$ is the coordinate of the point S (see Fig. 16.5). Recall that Green's functions have two arguments, $\mathbf{r}'$ referring to the *source point* where the unit source is located, and $\mathbf{r}$ to the point of observation, or the *destination point*, of the response of the system to the source. Here, the nodal point S at $\mathbf{r}_i$ is the destination point for the response to the source at P with coordinates $\mathbf{r}'$. The Green's function depends on the distance $|\mathbf{r}_i - \mathbf{r}'|$ between any source point P to the point S.

The scalar product occurring in the second term of the line integral in equation (16.25) is evaluated as follows. The *rectangular* region is on the xy plane and the z-axis is perpendicular to it. We can define the unit vector $\mathbf{n}'$ as

$$\mathbf{n}' = \frac{\mathbf{r}' - \mathbf{r}_a}{|\mathbf{r}' - \mathbf{r}_a|} \times \hat{\mathbf{z}}. \qquad (16.28)$$

Let the unit vector along $(\mathbf{r}' - \mathbf{r}_i)$ be denoted by $\hat{\rho}'$. The calculation of the scalar product $\hat{\rho}' \cdot \mathbf{n}' = \cos\theta'$ can be automated in the computer program once the above position vectors are supplied. The angle θ' is indicated in Fig. 16.5. Other strategies can be used for other nonrectangular regions and for arcs.

With the above substitutions we have

$$\phi(\mathbf{r}_i) = \sum_{iel=1}^{nelem} \frac{\ell_{iel}}{2\pi} \int_0^1 d\xi \left[-\ln(|\mathbf{r}_i - \mathbf{r}'|)\left\{\phi_a'^{(iel)}(1-\xi) + \phi_b'^{(iel)}\,\xi\right\} \right.$$
$$\left. + \frac{\cos\theta'}{|\mathbf{r}_i - \mathbf{r}'|}\left\{\phi_a^{(iel)}(1-\xi) + \phi_b^{(iel)}\,\xi\right\} \right]. \qquad (16.29)$$

For a given $\mathbf{r}_i$, when the integral is calculated around the closed path we encounter two special elements for which the node i at $\mathbf{r}_i$ is at one end of the linear segment. As the variable $\mathbf{r}'$ ranges over such an element, the Green's function and its normal derivative become singular at the end point $\mathbf{r}_i = \mathbf{r}'$. This requires special treatment. We distort the contour integral to circumambulate around the point $\mathbf{r}_i$, along a semicircular path of infinitesimal radius ϵ, so as to make it a point interior to the region. This is shown in Figs. 16.6 and 16.7.

We have to separate the contributions of the straight segments of the line integral from the integration along the circular arcs. When

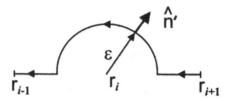

Fig. 16.6 The path of integration around the singular point r_i on a straight segment of path.

the destination point lies on straight segments of path, the normal derivative is perpendicular to the path of integration and to the line joining the destination and source points. Hence the singular integral is zero.

The integrals along the circular arcs, the unit vectors $\hat{\rho}'$, and the normal to the path n' are parallel to each other, while $|r_i - r'| = \epsilon$ and the increment $d\ell'$ is $\epsilon\, d\theta$. Thus, the integrals in equation (16.25) along the arcs become

$$\frac{1}{2\pi} \int d\theta \left(-\epsilon \ln(\epsilon) \frac{\partial}{\partial n'} \phi(r') \bigg|_{r'=r_i} + \phi(r')\big|_{r'=r_i} \right).$$

In the limit $\epsilon \to 0$ the first term vanishes since $\epsilon \ln \epsilon$ goes to zero, and the second term reduces to $(\Delta\theta/2\pi)\,\phi(r_i)$. The angle $\Delta\theta$ is measured to be positive in a counterclockwise direction as we move along the path around the point r_i. For a node on a straight segment of the contour, this contribution is then $\phi(r_i)/2$, whereas at a corner, as in Fig. 16.7, the contribution is $(3/4)\,\phi(r_i)$.

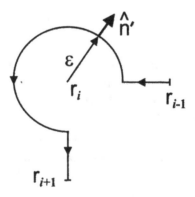

Fig. 16.7 The path of integration around the singular point r_i at a corner node.

On moving the contribution of the line integrals along the small circular arcs to the left side of equation (16.25), we have

$$\left(1 - \frac{\Delta\theta^i}{2\pi}\right)\phi(\mathbf{r}_i) = \sum_{iel=1}^{nelem}\left[\left\{\frac{\ell_{iel}}{2\pi}\int_0^1 d\xi\left(\frac{\cos\theta'}{|\mathbf{r}_i - \mathbf{r}'|}(1-\xi)\right)\right\}\phi_a^{(iel)}\right.$$
$$+ \left(\frac{\ell_{iel}}{2\pi}\int_0^1 d\xi\frac{\cos\theta'}{|\mathbf{r}_i - \mathbf{r}'|}\xi\right)\phi_b^{(iel)}$$
$$- \left(\frac{\ell_{iel}}{2\pi}\int_0^1 d\xi\ln(|\mathbf{r}_i - \mathbf{r}'|)(1-\xi)\right)\phi_a'^{(iel)}$$
$$\left. - \left(\frac{\ell_{iel}}{2\pi}\int_0^1 d\xi\ln(|\mathbf{r}_i - \mathbf{r}'|)\,\xi\right)\phi_b'^{(iel)}\right]. \qquad (16.30)$$

The above equation can be written more compactly as

$$\left(1 - \frac{\Delta\theta^i}{2\pi}\right)\phi_i = \sum_{iel=1}^{nelem}[\,\mathcal{H}_{ia}^{(iel)}\phi_a^{(iel)} + \mathcal{H}_{ib}^{iel}\phi_b^{(iel)}\,]$$
$$+ \sum_{iel=1}^{nelem}[\,\mathcal{G}_{ia}^{(iel)}\phi_a'^{(iel)} + \mathcal{G}_{ib}^{(iel)}\phi_b'^{(iel)}\,]. \qquad (16.31)$$

Here we have let $\phi(\mathbf{r}_i) \equiv \phi_i$.

It is clear from this notation that as the contribution of each element is calculated and included in the sum to obtain the complete line integral, we will have contributions from adjacent elements with the same coefficients $\phi_{a,b}^{(iel)}$ or $\phi_{a,b}'^{(iel)}$ but with different indices. If we define $j = iel$, we can label the parameter $\phi_{a,b}^{(iel)}$ by ϕ_j, ϕ_{j+1}, and the normal derivatives $\phi_{a,b}'^{(iel)}$ by ϕ_j', ϕ_{j+1}', respectively.

We perform the same notational change on $\mathcal{H}$ and $\mathcal{G}$ after combining terms to have common coefficients ϕ_j and ϕ_j'. It is also usual to lump the term on the left side of equation (16.31) with the corresponding ϕ_j term on the right side. Furthermore, the angle $\Delta\theta$ depends on the actual nature of the path around $\mathbf{r}_i$, and whether $\mathbf{r}_i$ is on a straight segment or at a corner. We can define $C_i = \Delta\theta/2\pi$, leading to

$$\mathcal{H}_{ij} = \mathcal{H}_{ia}^{(iel)} + \mathcal{H}_{ib}^{(iel-1)} - (1 - C_i),$$
$$\mathcal{G}_{ij} = \mathcal{G}_{ia}^{(iel)} + \mathcal{G}_{ib}^{(iel-1)}. \qquad (16.32)$$

The discretized integral equation may now be presented in the compact form

$$\sum_{j=1}^{nelem} [\mathcal{H}_{ij}\phi_j + \mathcal{G}_{ij}\phi_j'] = 0. \tag{16.33}$$

16.5   Numerical issues

We now consider the details of numerically evaluating the integrations and imposing the given BCs to obtain the unknown quantities in equation (16.33). The application of Neumann BCs for the normal derivatives of the potential at corner nodes requires some care. The simultaneous equations for the unknown quantities are solved using standard matrix methods.

16.5.1   Evaluation of the element integrals

We now discuss the types of problems we face in automating the computation of the integrals. The integrands are well behaved as long as the destination point is not at either of the two ends of the element that is being integrated. For linear elements these integrals can be carried out analytically. However, we can achieve reasonable accuracy in evaluating the two terms in the line integral by using an n-point Gauss quadrature, with an order $n \sim 8$–16, for integrating over (i) the logarithm multiplied by a polynomial and (ii) a polynomial multiplying the $1/r$ behavior of the derivative of the Green's function. We wish to reiterate that, when the destination point lies on a straight segment of the line integral, the normal derivative of the Green's function is orthogonal to the path and so this term vanishes anyway.

   We now consider the case when the destination point is at either end of an element. Let us suppose that in the local integration variable ξ ranging over $[0, 1]$ the destination point $\mathbf{r}_i$ is at $\xi = 0$. We split the integral into two parts, one over the range $[0, \epsilon]$, and the other over $[\epsilon, 1]$, with ϵ being a small positive quantity. For the second interval, we can use an adaptive Gauss integration procedure as described in Appendix A. In adaptive quadrature, the integral over the range $[\epsilon, 1]$ is first calculated using n-point Gauss integration. Next, the range is split into two, and the same quadrature procedure is employed over each of the subintervals. The sum of the two subintegrals is compared with the initial estimate, and the procedure is terminated if the difference is less than a relative tolerance value.

Otherwise, the bisection of each subinterval is continued recursively until the desired accuracy is obtained. In this method, breakup of the integration into further subintervals is done as needed. It is found that the convergence of the integral is fairly rapid, and one typically requires two to eight recursive bisections, depending on the value of ϵ (10^{-4}–10^{-8}). The integral over the range $[0, \epsilon]$ can be performed analytically and stored for use in every instance that the destination point is at one end of the element that is being integrated. Other alternatives to this approach are as follows. If ϵ is kept very small, we could even neglect the contribution of the integral over $[0, \epsilon]$ at the price of extending the level of recursion in the adaptive quadrature. A viable alternative is to use a special Gauss quadrature rule for the log function multiplied by polynomials.[8] This would require corresponding base points and weights, which are given in Appendix A, Table A.3.

If the path of integration is curved, the integration cannot be performed analytically. For a destination point on the curve, the term in the line integral with the normal derivative of the Green's function will not vanish in general. Here again we break the integral up into one portion close to the destination point and into one over the path that is further away. The former integral can be approximated as being over a straight line, and again it vanishes. The latter is finite and can be evaluated by adaptive Gauss quadrature.

16.5.2 Applying boundary conditions

We return to equation (16.33). For notational compactness, let $nelem \equiv n$. Writing out the first few equations, we have

$$\mathcal{H}_{11}\phi_1 + \mathcal{H}_{12}\phi_2 + \ldots + \mathcal{G}_{11}\phi'_1 + \mathcal{G}_{12}\phi'_2 + \cdots + \mathcal{G}_{1n}\phi'_n = 0$$
$$\mathcal{H}_{21}\phi_1 + \mathcal{H}_{22}\phi_2 + \cdots + \mathcal{G}_{21}\phi'_1 + \mathcal{G}_{22}\phi'_2 + \cdots + \mathcal{G}_{2n}\phi'_n = 0$$
$$\vdots \qquad \qquad \vdots \qquad \qquad \qquad \qquad (16.34)$$
$$\mathcal{H}_{n1}\phi_1 + \cdots + \mathcal{H}_{nn}\phi_n + \mathcal{G}_{n1}\phi'_1 + \mathcal{G}_{n2}\phi'_2 + \cdots + \mathcal{G}_{nn}\phi'_n = 0.$$

We now have to specify n values along the boundary *either* for the potential *or* for the corresponding normal derivative of the potential. Let us suppose that at the first m nodes ($m < n$) we are given the potentials $\overline{\phi_k}$, with $k = 1, \ldots, m$, and at nodes $m+1, \ldots, n$ we have the normal derivatives $\overline{\phi'_p}$, with $p = m+1, \ldots, n$. We then rearrange the above equations by leaving the terms with unknown coefficients on the left sides and moving the known terms to the right sides of the equations. This leaves us with n equations for the same number of unknown quantities.

To further clarify the situation, we can use a specific example such as a Dirichlet boundary value problem. We have a known potential $\bar{\phi}_j$ at every node, and we can use a matrix notation to write

$$\mathcal{G}_{ij}\,\phi'_j = -\mathcal{H}_{ij}\bar{\phi}_j, \qquad (16.35)$$

leaving all the unknowns on the left. In a purely Neumann boundary value problem the normal derivative of the potential $\bar{\phi}'_j$ is given at each node. Moving all the known terms to the right side we have

$$-\mathcal{H}_{ij}\,\phi_j = \mathcal{G}_{ij}\,\bar{\phi}'_j. \qquad (16.36)$$

In the more general case, we would have the set of equations given by

$$\mathcal{H}_{1,m+1}\phi_{m+1} + \mathcal{H}_{1,m+2}\phi_{m+2} + \cdots + \mathcal{G}_{1,1}\phi'_1 + \mathcal{G}_{1,2}\phi'_2 + \ldots =$$

$$-\mathcal{H}_{1,1}\bar{\phi}_1 - \mathcal{H}_{1,2}\bar{\phi}_2 - \cdots - \mathcal{G}_{1,m+1}\bar{\phi}'_{m+1} - \mathcal{G}_{1,m+2}\bar{\phi}'_{m+2} - \cdots ,$$

$$\mathcal{H}_{2,m+1}\phi_{m+1} + \mathcal{H}_{2,m+2}\phi_{m+2} + \cdots + \mathcal{G}_{2,1}\phi'_1 + \mathcal{G}_{2,2}\phi'_2 + \ldots =$$

$$-\mathcal{H}_{2,1}\bar{\phi}_1 - \mathcal{H}_{2,2}\bar{\phi}_2 - \cdots - \mathcal{G}_{2,m+1}\bar{\phi}'_{m+1} - \mathcal{G}_{2,m+2}\bar{\phi}'_{m+2} - \cdots ,$$

$$\vdots \qquad \vdots \qquad \vdots$$

$$(16.37)$$

$$\mathcal{H}_{n-1,m+1}\phi_{m+1} + \cdots + \mathcal{H}_{n-1,n}\phi_n + \mathcal{G}_{n-1,1}\phi'_1 + \cdots + \mathcal{G}_{n-1,m}\phi'_m =$$

$$-\mathcal{H}_{n,1}\bar{\phi}_1 - \cdots - \mathcal{G}_{n-1,m+1}\bar{\phi}'_{m+1} - \cdots - \mathcal{G}_{n-1,n}\bar{\phi}'_n ,$$

$$\mathcal{H}_{n,m+1}\phi_{m+1} + \cdots + \mathcal{H}_{n,n}\phi_n + \mathcal{G}_{n,1}\phi'_1 + \cdots + \mathcal{G}_{n,m}\phi'_m =$$

$$-\mathcal{H}_{n,1}\bar{\phi}_1 - \mathcal{H}_{n,2}\bar{\phi}_2 - \cdots - \mathcal{H}_{n,m}\bar{\phi}_m - \mathcal{G}_{n,m+1}\bar{\phi}'_{m+1}-$$

$$\mathcal{G}_{n,m+2}\bar{\phi}'_{m+2} - \cdots .$$

The subscripts on the matrix elements have been separated by a comma for clarity. These equations are solved by using standard matrix linear equation solvers.

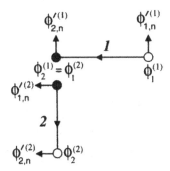

Fig. 16.8 The application of BCs at a corner node. Elements labeled 1 and 2 meet at a corner node. Node 2 of element 1 and node 1 of element 2 are at the same location. While $\phi_2^{(1)}$ and $\phi_1^{(2)}$ are the same, $\phi_2'^{(1)}$ and $\phi_1'^{(2)}$ are not.

16.5.3   Boundary condition at the corner node

There is ambiguity in the application of the BC at a corner node. In the two elements terminating at such a corner node, the normal derivatives of the potential at the node can be different. The angle between the two normals depends on the angle at which the elements meet; these are perpendicular to each other in Fig. 16.8. The corner node has three variables ϕ, ϕ_ℓ', and ϕ_r', where the subscripts ℓ and r refer to the elements on the left and right sides of the node. On the other hand the standard BCs would give either ϕ or a unique ϕ'. There are at least five ways to deal with this issue.

1. Pulling back the nodes: We can pull back the nodes from the two elements at the corner so that they no longer coincide. We continue to use the linear interpolation polynomials between the nodes, in each element. We now allow the same expressions to represent the potential and its normal derivative outside the element by extrapolating out to the corner. We evaluate the boundary integral along the contour up to the original corner. In effect, we have invoked the existence of an extra node by separating the node at the corner into two. The problem reduces to one in which we have two independent variables at each node. At the price of some numerical inaccuracy, this procedure provides a suitable solution. A variant on this option is to simply discard linear interpolation for the derivative of the potential in the elements next to the corner and to suppose that the normal derivatives do not vary at all over the element. Then the two

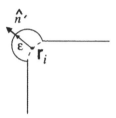

Fig. 16.9 The boundary node is pulled in to round out the corner. The contour is distorted outward, as usual, to make the node at $\mathbf{r}_i$ an interior point.

normal derivatives at the corner are dropped since they are determined at the alternate node in each element. Correspondingly, we also discard one of the equations from among the *nelem* equations.

2. Rounding out the corner: Another stratagem is to round out the corner. This gives a unique normal derivative at the corner node, as shown in Fig. 16.9. Again every node on the periphery has only two variables associated with it. The integration is performed along the curved path.

3. Redefining the normal derivatives: At the corner we let the normal derivative $\phi_1'^{(2)}$ be defined by $(\phi_2^{(1)} - \phi_1^{(1)})/h_1$ where the quantities are as in Fig. 16.8 and h_1 is the length of element 1. Similarly, suppose that $\phi_2'^{(1)}$ is defined by $(\phi_2^{(2)} - \phi_1^{(2)})/h_2$. We have then eliminated the two derivative values at the corner. When these definitions are substituted in the set of n simultaneous equations (16.35), we are left with $n - 1$ variables. We now discard one of the n equations to obtain as many equations as there are unknowns. The discarded equation should not correspond to the one associated with the corner node.

4. Inserting additional points: As seen above, the main difficulty with a corner is that there are one too many variables there. The system of equations is limited to the n boundary integrals corresponding to the n destination points (nodes). So what is needed is yet another equation! We can generate an extra equation by evaluating the potential at a convenient point on the boundary *where the value of the potential is known.*[9] Let the coordinates of this point be $\mathbf{r}^*$ where $\phi = \phi^*$. Following the calculations leading to equation (16.33), we obtain

$$\sum_{j=1}^{nelem} \left(\mathcal{H}_{ij}\phi_j + \mathcal{G}_{ij}\phi'_j \right) = (1 - C^*)\,\phi^*. \qquad (16.38)$$

The right side is completely known, where C^* is chosen in a manner similar to the C_i for the regular nodes. Just as in equations (16.33), we have to distort the contour to make point $\mathbf{r}^*$ an interior point. However, the interpolation polynomials for the regular nodes are still over the entire element and the line integrals are now split into two parts in the element in which the extra point is placed. The integrable singularity is treated as described above, with the distorted semicircular contour contributing in the usual manner. For each corner we pick an additional point on the contour, a convenient choice being at the midpoint of an element, and add a corresponding equation.

5. *Use of the Hermite interpolation polynomials:* A more sophisticated solution is to use the Hermite interpolation polynomials for the potential as well as for its normal derivative.[10] The improved continuity properties of the interpolated functions also enhance the accuracy of the results. Let the two sets of interpolation polynomials used in the Hermite interpolation be $N_i(\xi)$ and $\overline{N}_i(\xi)$, with

$$N_i(\xi_j) = \delta_{ij}, \frac{d}{d\xi}N_i(\xi_j) = 0, \quad 0 \le \xi \le 1;\ i,j = 1,2;$$

$$\overline{N}_i(\xi_j) = 0, \quad \frac{d}{d\xi}\overline{N}_i(\xi_j) = \delta_{ij}. \qquad (16.39)$$

The potential and its normal derivative are expressed as

$$\phi(x) = \sum_i \left(\phi_i\,N_i(\xi) + \phi'_{i,t}\frac{dx}{d\xi}\overline{N}_i(\xi) \right),$$

$$\phi'_n(x) = \sum_i \left(\phi'_{i,n}\,N_i(\xi) + \phi''_{i,nt}\frac{dx}{d\xi}\overline{N}_i(\xi) \right). \qquad (16.40)$$

Here $\phi'_{i,t}$ corresponds to $\partial\phi(\xi)/\partial t_i$, the tangential derivative along the contour. The normal derivative $\phi'_n(\xi)$ is expressed in terms of its values at the nodes, $\phi'_{i,n}(\xi)$, and its tangential derivatives at the nodes $\phi''_{i,nt}(\xi)$. The extra comma in the subscript here is to indicate that the derivatives of the function ϕ are being evaluated along the normal or the tangential (or both) directions. In solving one problem we seem to have created a larger one in that both ϕ and ϕ'_n are written in terms of nodal values as well as their *tangential* derivatives. The main advantage is that the corner nodes can be treated exactly. The normal and tangential derivatives from the element to the left of the node, at a corner, are simply related to the tangential and normal derivatives at the corner from the element at the right. Also, at corners with an interior angle of $p\pi/2$, where $p = 0, 1, 2$, or 3, the cross-derivative $\phi''_{i,nt}$ for these angles can be related on adjacent elements. Returning to the right-angled corner, we have four unique variables. This is true for all nodes along the contour.

The number of equations generated by the BEM is still $n = nelem$. Hence it is clear that a large number of extra points (as many as $3n$) may be needed, and procedure 4 above needs to be invoked a number of times to obtain the right number of equations. Usually, additional relations between the derivatives at the boundary can be invoked for some of the derivative variables; this can be done based on any relevant symmetries in the problem.

A somewhat different approach is employed by Durodola and Fenner.[11] In order to avoid additional complications at the corner nodes they define cubic one-sided Hermite interpolation polynomials in a three-noded element. The corner node and a second at the center of the element have only one degree of freedom (the nodal value of the potential) while the node at the other extremity of the element also has a derivative degree of freedom. Such interpolation polynomials can be generated by the methods discussed in Chapter 3. They are also available in the literature.[11]

16.5.4 Setting up the matrix equation

The number of equations is usually less than the number of variables in the problem before any BCs are imposed. Let us consider the case of a simple boundary value problem, treated in the linear interpolation approximation, where there are n equations and n BCs.

The integrals appearing in equation (16.30) are evaluated and are assigned to appropriate matrix elements in equation (16.33). Depending on the BCs, these matrix elements are then rearranged so that the unknown variables are on the left side of the equation, with all known terms taken to the right side. This leads to a matrix equation

$$\mathbf{M} \cdot \mathbf{v} = \mathbf{b}, \qquad (16.41)$$

where the vector of unknown variables $\mathbf{v}$ in general includes both the unknown nodal parameters for the potential and the normal derivative, and the vector array $\mathbf{b}$ on the right side contains all the known terms.

Redirecting the matrix elements into their correct locations in matrix $\mathbf{M}$ and vector $\mathbf{b}$ and obtaining the final form of the system of equations is one of the error-prone steps for the beginner. If the problem generates small matrices, i.e., *nelem* is of order 20–100, it is useful to start with a $2n \times 2n$ matrix obtained by inserting the row entries for the ith equation into two *consecutive* rows. Thus, rows 1 and 2 have the same entries to start with, and so do rows $2i - 1$ and $2i$. Each pair of such entries should be thought of as being associated with ϕ_i and ϕ_i'. We now apply the BCs.

Suppose that ϕ_i is known to be $\bar{\phi}_{i0}$ for $i = i_0$. We then focus our attention on the pair of equations in rows $2i_0 - 1$ and $2i_0$. We take the entries in column $2i_0 - 1$, multiply them by $\bar{\phi}_{i0}$, and move the product into the corresponding elements of the column vector on the right side with a change of sign. Having moved the known terms arising from $\bar{\phi}_{i0}$ in the equations to the right side, we now enter zeros in column $2i_0 - 1$ of the $2n \times 2n$ matrix. Next, the row entries in the row $2i_0 - 1$ are also equated to zero. Since we want to keep the size of the matrix the same (to avoid invoking complicated programming tricks), we now set the diagonal element to unity in the $(2i_0 - 1)$th row of the matrix. We want to keep the location of the variable ϕ_i in the arrays intact and have it take on its given value of $\bar{\phi}_{i0}$. So we now set the $(2i_0 - 1)$th element of the vector on the right side to the known value ϕ_{i0}. This procedure has been designated in Chapter 5 by us as applying *benediction* to the matrix. The variable ϕ_{i0}' is then the only unknown associated with the node i_0 and it is solved for when the matrix equation (16.41) is numerically inverted for $\mathbf{v}$. Meanwhile, ϕ_i for $i = i_0$ solves to its given value $\bar{\phi}_{i0}$. When the above procedure is carried out for all the pairs of equations, we have a $2n \times 2n$ matrix with n rows and columns of zero entries except for their unit diagonal elements. The programming for the benediction procedure is rather simple and is sufficiently flexible to adapt to each new problem with its corresponding BCs. Clearly, the book-keeping suggested here becomes prohibitively expensive as the matrix sizes increase; however, it has been found to be very useful in the initial stages of developing computer programs in Fortran for the BEM. We note that, with the freedom to manipulate linked-lists

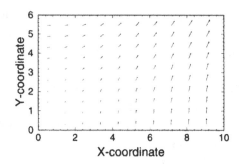

Fig. 16.10 A vector plot of the electric field in the interior region for the Neumann BC.

in C language computer codes, applying BCs and reducing matrix sizes, by eliminating rows or columns appropriately, no longer pose any difficulty. It is also possible to use sparse matrix techniques developed over the past three decades to avoid referring to matrix elements that are zero and simply not allocate memory for storing these matrix elements.

16.5.5   Construction of interior solution

Once the values of the potential and its normal derivative are determined on the boundary, we are free to take the reference point $\mathbf{r}$, where the potential is to be calculated, to make it coincide with any given point in the interior region. Setting up a grid of points in the rectangular region for the examples we have considered, we then calculate the potential $\phi(\mathbf{r})$ using the same boundary integral as in equation (16.30) except that $\mathbf{r}$ is no longer on the boundary. The interior points may be chosen in any convenient manner. For example, a graphical representation of the interior solution may require a uniform grid to be set up in the interior region.

We have

$$
\begin{aligned}
\phi(\mathbf{r}) = \sum_{iel=1}^{nelem} & \left[\left\{ \frac{\ell_{iel}}{2\pi} \int_0^1 d\xi \left(\frac{\cos\theta'}{|\mathbf{r}-\mathbf{r}'|}(1-\xi) \right) \right\} \phi_a^{(iel)} \right. \\
& + \left(\frac{\ell_{iel}}{2\pi} \int_0^1 d\xi \frac{\cos\theta'}{|\mathbf{r}-\mathbf{r}'|} \xi \right) \phi_b^{(iel)} \\
& - \left(\frac{\ell_{iel}}{2\pi} \int_0^1 d\xi \ln(|\mathbf{r}-\mathbf{r}'|)(1-\xi) \right) \phi_a'^{(iel)} \\
& \left. - \left(\frac{\ell_{iel}}{2\pi} \int_0^1 d\xi \ln(|\mathbf{r}-\mathbf{r}'|)\xi \right) \phi_b'^{(iel)} \right].
\end{aligned}
\tag{16.42}
$$

All of the integrals occurring in the above equation can be evaluated in a straightforward manner since the destination point $\mathbf{r}$ does not lie on the contour itself. Only the predetermined values of the potential and its derivative at the boundary occur on the right side of equation (16.42).

For illustrative purposes, we show in Fig. 16.10 the vector plot of the electric field, rather than the potential itself, in the rectangular region for the Neumann problem. Adjacent grid points were used to determine the derivative of the potential along the x- and y-directions and the vector electric field is so determined.

16.6   A worked example

Let us suppose we wish to determine the potential distribution and the electric field between two parallel plates of infinite area. Let there be a potential difference of 100 V between the plates that are held apart at a distance of 1 m.

It is useful to explore the details of the BEM by modeling this simple problem. We consider a square region of unit sides with just one element on each side, as shown in Fig. 16.11. We apply a potential of 100 V on the side with element number four and set the opposite side, element number 2, at zero potential. Even though the potentials along sides 1 and 3 are not known, this is a Dirichlet boundary value problem since the potentials are given at all the nodes. It has four corner nodes where the normal derivatives in

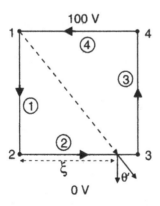

100 V

0 V

Fig. 16.11   A square region with one boundary element per side. The elements are labeled by circled numbers. The angle between $\mathbf{r}_1 - \mathbf{r}'$ and the normal $\mathbf{n}'$ to the path is θ'.

the adjoining elements are partially known: since the potentials are known to be constant along elements 2 and 4, we take the normal derivatives at the corners in elements 1 and 3 to be zero. Thus the corner nodes have nonzero derivatives only in the "vertical" direction in Fig. 16.11, and we solve for these variables. This accounts for all the derivative variables at the corners. We solve for the unknown quantities at the nodes within the framework of the BEM. This problem now physically corresponds to our having a slice of width 1 m out of an infinite parallel-plate capacitor.

With the destination point as node 1, we evaluate the contributions from the path integral, element by element. Using the notation of equation (16.33), we have $\mathcal{H}_{11} = -0.5$, and $\mathcal{H}_{12} = 0$ since the angle θ' between the normal to the path, $\mathbf{n}'$, and the vector $\mathbf{r}_1 - \mathbf{r}'$ is $\pi/2$. We need not evaluate $\mathcal{G}_{11}$ and $\mathcal{G}_{12}$ since their coefficients $\phi'_{1,n}$ and $\phi'_{2,n}$ are zero in this element.

Next, in element 2 we have $\cos\theta' = 1/\sqrt{1 + \xi'^2}$, so that

$$\mathcal{H}_{12}\,\phi_2 = \frac{1}{2\pi} \int_0^1 d\xi' \frac{1 - \xi'}{1 + \xi'^2} \phi_2$$

$$= \left(\frac{1}{8} - \frac{1}{4\pi} \ln 2 \right) \phi_2.$$

The contribution from element 3 is given by

$$\mathcal{H}_{13}\,\phi_3 = \frac{1}{2\pi} \int d\xi' \frac{\xi'}{1 + \xi'^2} \phi_3$$

$$= \frac{\ln 2}{4\pi} \phi_3.$$

The contributions to the G_{1j} terms of equation (16.33) are

$$\mathcal{G}_{12}\,\phi'_{2,n} = -\frac{1}{2\pi} \int_0^1 d\xi' \ln(\sqrt{1 + \xi'^2})\,(1 - \xi')\,\phi'_{2,n}$$

$$= \left(\frac{3}{8\pi} - \frac{1}{8} \right) \phi'_{2,n}.$$

$$\mathcal{G}_{13}\,\phi'_{3,n} = -\frac{1}{2\pi} \int_0^1 d\xi' \ln(\sqrt{1 + \xi'^2})\,\xi'\,\phi'_{3,n}$$

$$= -\left(\frac{\ln 2}{4\pi} - \frac{1}{8\pi} \right) \phi'_{3,n}.$$

The contribution from element 3 to the path integral is given by the expression

$$\left[\frac{\ln 2}{4\pi} \phi_3 + \left(\frac{1}{8} - \frac{\ln 2}{4\pi} \right) \phi_4 - \left(\frac{\ln 2}{4\pi} - \frac{1}{8\pi} \right) \phi'_{3,n} + \left(\frac{3}{8\pi} - \frac{1}{8} \right) \phi'_{4,n} \right]$$

and the contribution from element 4 is given by

$$\frac{1}{8\pi}\phi'_{4,n} + \frac{3}{8\pi}\phi'_{1,n}.$$

The equation associated with the destination point at node 1 is then

$$-0.25\phi_1 + 0.1194\phi'_1 + 0.0698\phi_2 - 0.0056\phi'_2 + 0.1103\phi_3$$
$$- 0.0154\phi'_3 + 0.0698\phi_4 + 0.0398\phi'_4 = 0.$$

Three additional equations can be obtained in a similar manner by taking the nodal points 2, 3, and 4 as the destination points. In these four equations, we insert the known values $\phi_1 = 100$, $\phi_2 = 0$, $\phi_3 = 0$, and $\phi_4 = 100$; we then transfer the known terms to the right side of the equations. After rounding off to display the numbers we have

$$\begin{pmatrix} 0.1194 & -0.0056 & -0.0154 & 0.0398 \\ -0.0056 & 0.1194 & 0.0399 & -0.0154 \\ -0.0154 & 0.0398 & 0.1194 & -0.0056 \\ 0.0398 & -0.0154 & -0.0056 & 0.1194 \end{pmatrix} \cdot \begin{pmatrix} \phi'_1 \\ \phi'_2 \\ \phi'_3 \\ \phi'_4 \end{pmatrix} = \begin{pmatrix} 18.016 \\ -18.016 \\ -18.016 \\ 18.016 \end{pmatrix}.$$

Simple Gaussian elimination then provides the solution

$$\phi'_i = \{100.0, -100.0, -100.0, 100.0\}.$$

We remind the reader that these variables are for the normal derivatives at nodes along the "vertical" direction in Fig. 16.11. Now that we have all the nodal values along the boundary we can generate the solution in the interior at a grid of points as mentioned earlier. For further examples the reader may consult Refs. 12–14.

16.7 Two sum rules

The boundary integral equation (16.3) can be used to derive two sum rules that are of interest. The first sum rule states that the line integral of the normal derivative of the Green's function around the closed path is a constant. The second sum rule is a manifestation of Gauss's law of electrostatics.

Sum rule 1

If a constant λ is added to any given solution to Laplace's equation, it still satisfies the same equation.[¶] This has an interesting consequence

---

[¶] Such a result does not hold if the differential equation has terms with no derivatives in it. For example, in the case of Helmholtz's equation this sum rule will not be satisfied.

in equation (16.33). Substituting $\phi_i \rightarrow \phi_i + \lambda$ into equation (16.33) we obtain

$$\sum_{j=1}^{nelem} [\mathcal{H}_{ij}(\phi_j + \lambda) + \mathcal{G}_{ij}\phi_j'] = 0. \tag{16.43}$$

Since equation (16.33) is still valid, we can equate the coefficients of λ to obtain a sum rule:

$$\sum_{j=1}^{nelem} \mathcal{H}_{ij} = 0. \tag{16.44}$$

This is readily verified, using the worked example, by summing the coefficients of ϕ_i. In equation (16.43), this relation is seen to be valid to the number of digits displayed. As a consequence, the term $\mathcal{H}_{ii}$ (no sum implied) can be expressed as

$$\mathcal{H}_{ii} = -\sum_{j(\neq i)}^{nelem} \mathcal{H}_{ij}. \tag{16.45}$$

This shows that the least convergent integral can be cast as a sum of terms which are more readily evaluated numerically. This relation can be used to improve the accuracy of the boundary element calculations. It also demonstrates the fact that $\mathcal{H}_{ii}$ is convergent.

Sum rule 2

If a constant ρ is added to the Green's function, it still satisfies Poisson's equation with a "unit" source. Let us substitute $g \rightarrow g + \rho$ into equation (16.33). We obtain

$$\sum_{j=1}^{nelem} (\mathcal{H}_{ij}\phi_j + \mathcal{G}_{ij}\phi_j' + \rho\phi_j') = 0, \tag{16.46}$$

leading to

$$\sum_{j=1}^{nelem} \phi_j' = 0. \tag{16.47}$$

This relation is the discretized version of Gauss's law of electrostatics

$$\oint \mathbf{E} \cdot d\mathbf{S} = 0, \tag{16.48}$$

for a charge-free region in vacuum.

16.8   Comparing the BEM with the FEM

The sources of error in the BEM are the evaluation of integrals with integrable end point singularities, and in the solution of the linear equations arising from the discretization. In the post-processing stage in the BEM, again, the interior solution is obtained by integrating over the Green's function and its derivative. These integrals do not have singular integrands; however, maintaining accuracy in the integrations can demand Gauss quadrature of high order when the interior points are close to the boundary. In contrast, in the FEM, the element matrices can usually be calculated with double-precision accuracy using Gauss quadrature. The price to be paid, as mentioned earlier, is the large sparse matrix equation that needs to be solved. The reconstruction of the solution everywhere from the nodal values obtained in the FEM can be accomplished using the same interpolation functions that were employed in the finite element representation of the potential in the individual elements, and this poses no difficulty.

The computational errors in the two approaches are not easily compared since fewer elements are usually needed for the BEM. To quantify such a comparison, the solution to Laplace's equation by both methods is calculated over a rectangular region of dimensions 10×20 with a potential along one of the longer sides given by $V(0, y) = \sin(\pi y/20)$, and with the other three sides held at zero potential. With 10 elements per side and using cubic Hermite interpolation polynomials, the BEM required a filled 76×76 matrix to be solved. The 76 equations arise from the expressions for the potential at the 40 nodal points together with 36 additional points inserted along the sides to generate enough equations to solve for all the unknown variables. The solution obtained by the BEM in the interior region gave an accuracy of 1 part in 10^5. The error here mainly arises in the boundary element discretization, since the element integrals were done *analytically*. In the FEM, on the other hand, the entire region is divided into 10×10 elements. The 2D interpolation polynomials are constructed using products of cubic Hermite interpolation polynomials in x and in y. Now each of the 121 nodal points has four degrees of freedom associated with it, corresponding to the potential, its partial derivatives along x and y, and the second-order cross-derivative. The matrix size, after applying BCs, is 444×444. The results at the interior nodes were accurate to 1 part in 10^6–10^7.

Table **16.1** Comparison of the BEM and the
FEM for the solution of Laplace's equation for a
rectangle with simple BCs. (See the text for details.)

	BEM	FEM
Cubic Hermite elements	4 sides×10	10×10
Number of nodes	40	100
Additional points	36	—
Matrix size	76×76	444×444
Error	10^{-5}	10^{-7}

The numerical errors in the FEM, as mentioned earlier, are associated
again with the discretization and with the matrix inversion, and not
with the Gauss integrations; we used eight-point Gauss quadrature
for evaluating the element matrices. The difference in the convergence with increasing number of elements is due to the BEM being
a self-consistency condition while the FEM is a variational method
with quadratic convergence. This discussion is summarized in Table 16.1. A final comparative comment is that the FEM is more
flexible than the BEM since it can handle both inhomogeneities and
nonlinearities with relative ease.

16.9   Problems

1. Consider a square region of unit dimensions with a potential of
   100 V on the top and bottom sides, as in Fig. 16.12. The normal derivative of the potential is zero on the other two sides,
   and there are no charges in the interior of the region. Let there
   be one element on each side, with a single node at the center of
   each element. We seek the solution for the nodal values of the
   potential at nodes 1 and 3, and the values of the normal derivative at nodes 2 and 4. Assume constant interpolation in each
   element. The use of constant interpolation with the node at the
   center of the element avoids issues mentioned in Section 16.5.3.

   (a) Show that the discretized boundary integral leads to the

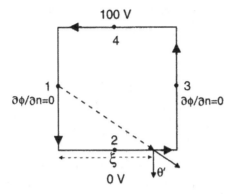

Fig. 16.12 The discretization of the boundary of the square region into four elements. The direction of the line integration is indicated by arrows. Constant interpolation is assumed.

relation

$$-0.5\,\phi_i + \sum_{j=1}^{4}\left[\frac{1}{2\pi}\int_0^1 d\ell'\,\frac{(\mathbf{r}'-\mathbf{r}_i)\cdot\mathbf{n}'}{|\mathbf{r}'-\mathbf{r}_i|^2}\phi_j\right.$$

$$\left.-\frac{1}{2\pi}\int_0^1 d\ell'\,\ln(|\mathbf{r}'-\mathbf{r}_i)|)\phi'_j\right]=0,$$

which can be expressed as

$$\sum_{j=1}^{4}[\mathcal{H}_{ij}\phi_j + \mathcal{G}_{ij}\phi'_j]=0.$$

(b) Verify that the eight nodal variables ϕ_i and ϕ'_i of part (a) satisfy the four equations

$$\sum_{j=1}^{4}\mathcal{H}_{ij}\phi_j + \sum_{j=1}^{4}\mathcal{G}_{ij}\phi'_j = 0,$$

where

$$\mathcal{H}_{11} = \mathcal{H}_{22} = \mathcal{H}_{33} = \mathcal{H}_{44} = -0.5,$$
$$\mathcal{H}_{12} = \mathcal{H}_{23} = \mathcal{H}_{34} = \mathcal{H}_{41},$$
$$\mathcal{H}_{13} = \mathcal{H}_{24} = \mathcal{H}_{31} = \mathcal{H}_{42},$$
$$\mathcal{H}_{14} = \mathcal{H}_{21} = \mathcal{H}_{32} = \mathcal{H}_{43},$$

with the values $\mathcal{H}_{12} = 0.176\,21 = \mathcal{H}_{14}$ and $\mathcal{H}_{13} = 0.147\,58$. The matrix elements of $\mathcal{G}_{ij}$ satisfy the relations

$$\mathcal{G}_{11} = \mathcal{G}_{22} = \mathcal{G}_{33} = \mathcal{G}_{44},$$
$$\mathcal{G}_{12} = \mathcal{G}_{23} = \mathcal{G}_{34} = \mathcal{G}_{41},$$
$$\mathcal{G}_{13} = \mathcal{G}_{24} = \mathcal{G}_{31} = \mathcal{G}_{42},$$
$$\mathcal{G}_{14} = \mathcal{G}_{21} = \mathcal{G}_{32} = \mathcal{G}_{43},$$

with $\mathcal{G}_{11} = 0.269\,47$, $\mathcal{G}_{12} = 0.053\,29$, $\mathcal{G}_{13} = -6.1859 \times 10^{-3}$, and $\mathcal{G}_{14} = 0.053\,29$. These relations arise from the symmetry of the configuration in this problem.

(c) Substitute the boundary values $\phi_2 = 0\,\text{V}$, $\phi_4 = 100\,\text{V}$, $\phi_1' = \phi_3' = 0$, and transfer the constant terms to the right side of the equation. Solve for the other four unknown nodal values, and show that the result is $\phi_1 = \phi_3 = 50\,\text{V}$, and $-\phi_2' = \phi_4' = 117.461\,\text{V/m}$. (The approximation is clearly inadequate for the normal derivatives.) ‖

2. We wish to obtain the potential in the shaded region shown in Fig. 16.13. The potentials on the positive x- and y-axes are V_0 and $-V_0$, respectively. Develop a strategy to account for the discontinuity in the potential at the origin in the BEM. The boundary elements at infinity in the first quadrant are to be ignored. Compare your results with the analytic solution

$$V = V_0 \left\{ 1 - \frac{2}{\pi} \tan^{-1} \left(\frac{2xy}{x^2 - y^2} \right) \right\}.$$

3. Consider the same physical region with the same BCs as in Problem 1, but with a charge distribution in the interior given by $\rho(x, y) = Q_0$, for $-0.25 \le x, y \le 0.25$, and zero elsewhere. Now the potential satisfies Poisson's equation

$$\nabla^2 \phi(x, y) = -4\pi \rho(x, y).$$

Develop the theory for including the source term in the boundary element approach.

‖The text by Gipson[12] may be consulted.

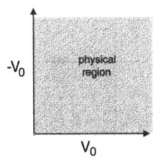

Fig. 16.13 The physical region corresponds to the first quadrant, with the positive x- and y-axes being lines of constant potential.

4. *A 3D problem*: A charge q is placed in front of a plane grounded conductor at a distance d from it, as in Fig. 16.14. It is desired to find the electrostatic potential everywhere in the semi-infinite region $z \geq 0$ to the right of the plane and containing the charge.

(a) Solve Poisson's equation by the boundary integral method and show that in the presence of the charge density $\rho(\mathbf{r}) = q\,\delta(\mathbf{r} - d\hat{\mathbf{z}})$ the potential is given by

$$\phi(\mathbf{r}) = \frac{q}{|\mathbf{r} - d\hat{\mathbf{z}}|} + \frac{1}{4\pi} \int ds' \left(g(\mathbf{r}, \mathbf{r}') \frac{\partial \phi(\mathbf{r}')}{\partial n'} - \phi(\mathbf{r}') \frac{\partial g(\mathbf{r}, \mathbf{r}')}{\partial n'} \right)$$

where ds' is a surface element on the conducting plane, and $\mathbf{r}'$ is the vector distance from the origin, placed on the plane, to any point on the plane. The enclosing surface at infinity does not contribute to the surface integral.

(b) With $\phi = 0$ on the conducting surface, numerically solve for the normal derivative of the potential on the surface using the BEM. Discretize the planar area and evaluate the contributions of the element integrals to the boundary integral. Solve for the normal derivative of the potential at the nodal points on the surface. Truncate the integral beyond a radial distance $\mathbf{r}' = 50$. Compare your answer with the analytical result

$$\frac{\partial}{\partial n'}\phi(x, y, z = 0) = -\frac{qd}{2\,(x^2 + y^2 + d^2)^{3/2}}.$$

(c) Obtain an approximate value for the potential in the physical region using the values given by the BEM for the normal

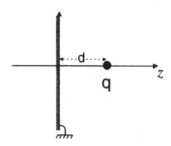

Fig. 16.14 A point charge q is placed at a distance d from the grounded conducting plane $z = 0$. The physical region corresponds to $z \geq 0$.

derivative of the potential on the surface. Compare your result with the analytical result

$$\phi(\mathbf{r}) = \frac{q}{|\mathbf{r} - d\hat{\mathbf{z}}|} - \frac{q}{|\mathbf{r} + d\hat{\mathbf{z}}|}.$$

References

[1] J. D. Jackson, *Classical Electrodynamics*, 2nd ed. (Wiley, New York, 1975).

[2] P. M. Morse and H. Feshbach, *Methods of Theoretical Physics* (McGraw-Hill, New York, 1953).

[3] M. A. Jaswon and G. T. Symm, *Integral Equation Methods in Potential Theory and Electrostatics* (Academic Press, New York, 1977).

[4] P. K. Banerjee and R. Butterfield, *Boundary Element Methods in Engineering Science* (McGraw-Hill, London, 1981).

[5] C. A. Brebbia, J. C. Telles, and L. C. Wrobel, *Boundary Element Techniques* (Springer-Verlag, Berlin, 1984).

[6] K. J. Binns, P. J. Lawrenson, and C. W. Trowbridge, *The Analytical and Numerical Solution of Electric and Magnetic Fields* (Wiley, New York, 1992).

[7] J. H. Kane, *Boundary Element Analysis in Engineering Continuum Mechanics* (Prentice Hall, Englewood Cliffs, NJ, 1994).

[8] A. H. Stroud and D. Secrest, *Gaussian Quadrature Formulas* (Prentice Hall, Englewood Cliffs, NJ, 1966).

[9] The author thanks R. Goloskie for suggesting the insertion of additional destination points to generate more equations. Also see:

K. Tomlinson, C. Bradley, and A. Pullan, *Int. J. Numer. Methods Eng.* **39**, 451 (1996).

[10] R. Goloskie, T. Thio, and L. R. Ram-Mohan, *Comput. Phys.* **10**, Sept/Oct (1996).

[11] J. F. Durodola and R. T. Fenner, *Int. J. Numer. Methods Eng.* **30**, 1051 (1990).

[12] G. S. Gipson, *Boundary Element Fundamentals* (Computational Mechanics, Southampton, UK, 1987).

[13] G. Beer and J. O. Watson, *Introduction to Finite Element and Boundary Element Methods for Engineers* (Wiley, New York, 1992).

[14] F. Hartmann, *Introduction to Boundary Elements: Theory and Applications* (Springer-Verlag, Berlin, 1989).

17

The BEM and surface plasmons

17.1 Introduction

The boundary element method (BEM) can be used to advantage for solving differential equations in a region occupied by more than one material. Here the integral equations are formulated for each subregion, and the matrices obtained on discretization of the contour integrals are put together in a manner analogous to overlaying element matrices in the finite element method (FEM). In this chapter, we begin by solving Laplace's equation for a physical domain having two rectangular regions with dielectric constants ϵ_I and ϵ_{II} in a constant electric field. The example makes use of the calculations of Section 16.6, where the solution at the boundary is given in terms of the linear interpolation between nodal values of the potential, and also the normal derivatives of the potential, at the extremities of a boundary element. We then discuss the issues in using Hermite interpolation polynomials in a multiregion BEM. We note here that when the physical region has a large aspect ratio and is elongated in one direction the boundary integrals will be nearly equal for destination points along the shorter side with source points at the farther end. This leads to ill-conditioning of the final matrices. The problem can be alleviated by breaking up the region into subregions, each with its own boundary integral. Again, a multiregion BEM can be employed.

We show that the BEM is ideally suited for the calculation of the frequencies of surface plasmon modes. Electrons in a metal can sustain density oscillations, called plasma oscillations. The frequency of the quantized plasma oscillation, the plasmon, in bulk metal is $\omega_P = (4\pi n e^2/\epsilon_\infty m^*)^{1/2}$, where n is the number density of electrons, m^* is the electron's effective mass in the metal, and ϵ_∞ is the high frequency dielectric constant associated with the ions.[†] In a confined

---

[†]For metals we may use $\epsilon_\infty \simeq 1$ in the infrared to the ultraviolet region of the spectrum. For doped semiconductors, however, this approximation will not be adequate.

geometry, the electronic plasma has surface modes of vibration whose frequency of oscillation depends on the shape of the surface of the metallic region. For example, the surface plasmon (which may also be called a plasmon–polariton) on the surface of a wire has a natural frequency $\Omega_{SP} = \omega_P/\sqrt{2}$ in the limit of large wavevectors.

In the long wavelength limit, the electrostatic approximation can be invoked and the frequency of the normal mode of the confined plasmon can be calculated by solving Laplace's equation with electrostatic boundary conditions (BCs). A very brief introduction to the theory of plasma oscillations and surface plasmons precedes the numerical calculations. The illustrative calculations include the evaluation of the resonant condition for surface plasmons and the electrostatic field distribution in (i) a single cylindrical wire, (ii) two parallel cylindrical wires (in order to explore the effect of proximity of one wire to the other on the plasmon frequency), and (iii) a metallic wire placed over a semiconducting substrate. These calculations are relevant for the understanding of the remarkable phenomenon of surface-enhanced Raman scattering (SERS). In SERS, molecules adhering to rough metallic surfaces are found to emit Raman radiation with intensities enhanced by factors of 10^6. A reason for the

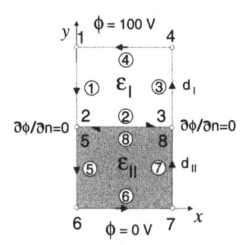

Fig. 17.1 A rectangular physical domain with two regions having dielectric constants ϵ_I and ϵ_{II}. The directions of the line integrations and the location of nodes (open circles with numbers) associated with each region are indicated. The boundary elements are labeled with circled numbers.

dramatic enhancement of the scattered intensity is the resonant enhancement of the local electromagnetic field at the frequency of the surface plasmon.

17.2 Multiregion BEM: two regions

Consider the two rectangular regions of thicknesses d_I and d_{II} with dielectric constants ϵ_I and ϵ_{II}, respectively, as shown in Fig. 17.1. Let there be a total potential drop of $100\,\text{V}$ across the layers. Laplace's equation in each region can be solved using a potential with a linear dependence on the coordinate y. We impose the BCs $\phi_I{=}100\,\text{V}$ at $y = d_I + d_{II}$, and $\phi_{II}{=}0\,\text{V}$ at $y = 0$. At the interface, the continuity condition $\phi_I(d_{II}) = \phi_{II}(d_{II})$ and the continuity condition $\epsilon_I\phi_I'(d_{II}) = \epsilon_{II}\phi_{II}'(d_{II})$, both of which arise from the continuity of the normal component of the electric displacement vector, are used. The Neumann BC $\partial\phi/\partial n = 0$ is assumed to hold along the sides. The analytic solution

$$\phi_I(y) = (100\,\text{V}) \times \frac{d_{II}\,\epsilon_I + (y - d_{II})\,\epsilon_{II}}{d_{II}\,\epsilon_I + d_I\,\epsilon_{II}},$$

$$\phi_{II}(y) = (100\,\text{V}) \times \frac{y\,\epsilon_I}{d_{II}\,\epsilon_I + d_I\,\epsilon_{II}},$$

(17.1)

is readily obtained. For the moment, we observe that in the limit $(d_{II}\,\epsilon_I + d_I\,\epsilon_{II}) \to 0$ the solutions in equations (17.2) diverge. If region I corresponds to free space with $\epsilon_I = 1$, the above condition corresponds to $\epsilon_{II} = -d_{II}/d_I$. We return to this point in the following. We now solve this problem by employing the BEM.

17.2.1 Linear interpolation

We wish to reuse our calculations of the boundary element integrals appearing in the BEM for a single region in Section 16.6. For this reason, we suppose that both regions are again of unit dimensions. Let there be one element per side, with linear interpolation being used for the potential ϕ and its normal derivative $\partial\phi/\partial n$. The node numbers and element numbers (circled) are shown in Fig. 17.1. The normal derivatives at nodes associated with elements 1 and 3 are taken to be zero, and we solve for the derivative variables of elements 2 and 4. Similarly, the normal derivatives at the nodes in elements 5 and 7 are zero, and we solve for the normal derivatives at the nodes in elements 6 and 8. In the absence of free charges in either region the integral equations for the two regions are identical, and we have four equations for the eight variables in each region. Thus, for region I we have

$$-0.25\,\phi_1 + 0.1194\,\phi_1' + 0.0698\,\phi_2 - 0.0056\,\phi_2' + 0.1103\,\phi_3$$
$$- 0.0154\,\phi_3' + 0.0698\,\phi_4 + 0.0398\,\phi_4' = 0,$$
$$0.0698\,\phi_1 - 0.0056\,\phi_1' - 0.25\,\phi_2 + 0.1194\,\phi_2' + 0.0698\,\phi_3$$
$$+ 0.0398\,\phi_3' + 0.1103\,\phi_4 - 0.0154\,\phi_4' = 0,$$
$$0.1103\,\phi_1 - 0.0154\,\phi_1' + 0.0698\,\phi_2 + 0.0398\,\phi_2' - 0.25\,\phi_3$$
$$+ 0.1194\,\phi_3' + 0.0698\,\phi_4 - 0.0056\,\phi_4' = 0, \qquad (17.2)$$
$$0.0698\,\phi_1 + 0.0398\,\phi_1' + 0.1103\,\phi_2 - 0.0154\,\phi_2' + 0.0698\,\phi_3$$
$$- 0.0056\,\phi_3' 0.25 - \phi_4 + 0.1194\,\phi_4' = 0.$$

We have a common boundary for regions I and II and the nodal variables ϕ_2, ϕ_3 are the same as ϕ_5, ϕ_8, respectively. This follows from the continuity of the potential across the dielectric interface. The BC for the normal component of the electric field at the interface between the two regions, $\epsilon_I E_{In} = \epsilon_{II} E_{IIn}$, leads to the relations

$$\epsilon_I\,\phi_2' = -\epsilon_{II}\,\phi_5'; \qquad \epsilon_I\,\phi_3' = -\epsilon_{II}\,\phi_8'. \qquad (17.3)$$

The additional negative sign on the right sides of equations (17.3) arises from the opposing directions of the outward-drawn normals for the two regions. On using these relations the four equations from the BEM for region II can be written as

$$-0.25\,\phi_2 - \frac{\epsilon_I}{\epsilon_{II}}0.1194\,\phi_2' + 0.0698\,\phi_6 - 0.0056\,\phi_6' + 0.1103\,\phi_7$$
$$- 0.0154\,\phi_7' + 0.0698\,\phi_3 - \frac{\epsilon_I}{\epsilon_{II}}0.0398\,\phi_3' = 0,$$

$$0.0698\,\phi_2 - 0.0056\,\phi_2' - 0.25\,\phi_6 + 0.1194\,\phi_6' + 0.0698\,\phi_7$$
$$+ 0.0398\,\phi_7' + 0.1103\,\phi_3 + \frac{\epsilon_I}{\epsilon_{II}}0.0154\,\phi_3' = 0,$$

$$(17.4)$$

$$0.1103\,\phi_2 + \frac{\epsilon_I}{\epsilon_{II}}0.0154\,\phi_2' + 0.0698\,\phi_6 + 0.0398\,\phi_6' - 0.25\,\phi_7$$
$$+ 0.1194\,\phi_7' + 0.0698\,\phi_3 + \frac{\epsilon_I}{\epsilon_{II}}0.0056\,\phi_3' = 0,$$

$$0.0698\,\phi_2 - \frac{\epsilon_I}{\epsilon_{II}}0.0398\,\phi_2' + 0.1103\,\phi_6 - 0.0154\,\phi_6' + 0.0698\,\phi_7$$
$$- 0.0056\,\phi_7' - 0.25\,\phi_3 - \frac{\epsilon_I}{\epsilon_{II}}0.1194\,\phi_3' = 0.$$

The eight equations (17.3) and (17.5) together have 12 variables before the Dirichlet BCs for the problem are imposed. We substitute the boundary values $\phi_1 = \phi_4 = 100$ V, and $\phi_6 = \phi_7 = 0$, and transfer the corresponding constant terms to the right sides of the equations. We are then left with eight unknown variables to be determined from these equations. For numerical computations, let us assume that $\epsilon_I = 1$ and $\epsilon_{II} = 10$. The solution of these eight equations leads to the result

$$\begin{aligned}
\phi_2 = \phi_3 &= 9.0909\,\text{V}, \\
\phi_1' = \phi_4' &= 90.909\,\text{V/m}, \\
\phi_2' = \phi_3' &= -90.909\,\text{V/m}, \\
\phi_6' = \phi_7' &= -9.0909\,\text{V/m}.
\end{aligned} \tag{17.5}$$

The same results are obtained from equations (17.2) for $d_I = d_{II} = 1$ m. The fact that the approximation of linear interpolation reproduces the analytical solution is not surprising since the potential has only a linear dependence on y.

17.2.2 Hermite interpolation

We have alluded to the use of Hermite interpolation polynomials in Section 16.5.3 as a means of overcoming the problem of defining unique nodal variables for normal derivatives at corner nodes in the BEM. Here we discuss issues that arise when implementing a multiregion BEM with Hermite interpolation. In Hermite interpolation, recall that the potential function $\phi(x,y)$ is expressed on the boundary element with two nodes at its extremities in terms of the nodal values of the potential ϕ_i and the tangential derivatives $\phi_{i,t}$. The normal derivative of the potential $\phi_n'(x,y)$ on the boundary is again expressed in terms of Hermite interpolation polynomials using the nodal variables $\phi_{i,n}'$ and $\phi_{i,nt}''$. The tangential derivatives are directed along the contour in the same sense (counterclockwise) as the path integral. Referring to Fig. 17.1, we have to consider anew the BCs at the overlapping nodes, such as nodes 2 and 5 from the two regions. Relations between the four nodal variables at nodes 2 and 5 of elements 2 and 8 are provided by

$$\phi_2 = \phi_5, \qquad \epsilon_I \phi'_{2,n} = -\epsilon_{II} \phi'_{5,n},$$
$$\phi'_{2,t} = -\phi'_{5,t}, \qquad \epsilon_I \phi'_{2,nt} = +\epsilon_{II} \phi'_{5,nt}. \qquad (17.6)$$

The outward-drawn normals for the two regions are antiparallel to each other, so that the normal derivatives of the potentials multiplied by the dielectric constants are related as shown above. The line integrals are directed in opposite directions at nodes 2 and 5, and hence the relation between the tangential derivatives has a negative sign. The cross-derivatives scaled by the dielectric function on each side of the interface are equal. These continuity conditions at the interface together with the ones for nodal variables from nodes 3 and 8 leave us with 24 unknown parameters.

As with linear interpolation, we obtain eight equations from the discretized boundary integrals for the two regions. We now apply the BCs in this case. In element 4, at nodes 1 and 4, the potentials are known, and the tangential derivatives are zero. This leaves two nodal variables ($\phi_{i,n}$ and $\phi_{i,nt}$) at each node to be determined. The same consideration holds for nodes 6 and 7 in element 6. At the overlapping pairs of nodes {2,5} and {3,8}, only the normal derivatives of elements 1 and 3 are known ($\partial \phi / \partial n = 0$), and there are three unknown nodal variables as yet undetermined at these nodes. We then have at hand 14 unknown nodal variables with just eight relations among them. Six additional equations are needed in this example, and we can assign three arbitrary destination points along each of the boundary elements 4 and 6 in order to generate these equations.

In problems with more complex potential distributions, the use of Hermite interpolation polynomials leads to proper treatment of the corner nodes, and generally leads to solutions that are more accurate. All the numerical results shown in the following sections are based on the BEM with cubic Hermite interpolation.

17.3 Bulk and surface plasmons

Electrical conduction in a metal arises from the presence of nearly free electrons occupying the volume of the metal. The ions in the metal provide a uniform positively charged background which neutralizes the charge of the electrons. Under equilibrium conditions the density of the electrons, n_0, is uniform throughout the volume. The

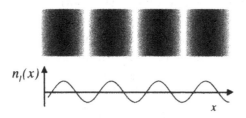

Fig. 17.2 Longitudinal density oscillations in the electron plasma. Here $n_1(x)$ is the deviation from the equilibrium density.

electron gas, also called a plasma, can sustain longitudinal oscillations in the number density.[1] This is shown schematically in Fig. 17.2. In such oscillations in metals the deviation from the equilibrium value, $n_1(\mathbf{r}, t)$, is usually much smaller than n_0. As mentioned in the introduction, a quantum of such an excitation of the electron gas is called a plasmon.

17.3.1 Bulk plasma oscillations

Once a local deviation, $n_1(\mathbf{r}, t)$, of the number density is generated, a strong restoring Coulomb force from the positively charged ions is experienced by the displaced electrons. This gives rise to charge oscillations due to an overshoot beyond the equilibrium configuration.

Let us suppose that a periodic oscillation, $n_1(\mathbf{r})e^{-i\omega t}$, of the electron density is present in the metal. The restoring force from the ions is an oscillating electric field $\mathbf{E}(\mathbf{r}, t) = \mathbf{E}_0(\mathbf{r})e^{-i\omega t}$. The field satisfies Poisson's equation

$$\nabla \cdot \mathbf{E}(\mathbf{r}, t) = -\frac{4\pi|e|}{\epsilon_\infty} n_1(\mathbf{r}, t), \qquad (17.7)$$

where the electron's charge is $-|e|$. Given an effective mass m^* for the electron in the metal, its equation of motion is

$$m^* \left(\ddot{\mathbf{r}} + \frac{\dot{\mathbf{r}}}{\tau} \right) = -|e|\mathbf{E}_0\, e^{-i\omega t}. \qquad (17.8)$$

The oscillation will be naturally damped due to electronic collisions and scattering from ions and defects in the crystalline structure, and this effect has been characterized by the damping time τ. With a trial solution for $\mathbf{r}$ having the same time dependence we see that

$$\mathbf{r} = \frac{|e|}{m^*\omega(\omega + i/\tau)}\, \mathbf{E}_0\, e^{-i\omega t}.$$

We are interested in determining the contribution of this internal degree of freedom to the dielectric function. The contribution from the ions to the dielectric function of the metal is denoted by ϵ_∞. In the following, we will be concerned with metals for which we may set $\epsilon_\infty = 1$. The dipole moment per unit volume is

$$\mathbf{P} = -\frac{n_0 e^2}{m^* \omega(\omega + i/\tau)} \mathbf{E},$$

and we have

$$\epsilon(\omega) = 1 + 4\pi \frac{|\mathbf{P}|}{|\mathbf{E}|}$$

$$= 1 - \frac{4\pi n_0 e^2}{m^* \omega(\omega + i/\tau)}. \tag{17.9}$$

This is known as the Drude form of the dielectric function for a metal. The zeros of the real part of the dielectric function give the frequencies of the normal modes of the electromagnetic system, and assuming that the damping constant is small, the frequency of the plasma oscillation is obtained as

$$\omega_P^2 = \frac{4\pi n_0 e^2}{m^*}. \tag{17.10}$$

The deviation of the number density from equilibrium is small, so that $n_1 \ll n_0$. We can therefore write the current density as $\mathbf{j}(\mathbf{r}, t) \sim -n_0 |e| \dot{\mathbf{r}}$, and the equation of motion, equation (17.8), can be recast as an equation for the current density. This relation together with the equation of continuity of charge

$$\nabla \cdot \mathbf{j} - |e| \frac{\partial n_1}{\partial t} = 0,$$

and Poisson's equation (17.7), allows us to show[2] that the density oscillations satisfy the differential equation

$$\frac{\partial^2 n_1}{\partial t^2} + \frac{1}{\tau} \frac{\partial n_1}{\partial t} + \omega_P^2 \, n_1 = 0. \tag{17.11}$$

This again clearly displays the fact that the plasma oscillations occur at frequency ω_P. We will not consider here effects of diffusion that lead to a wavevector dependence for the plasmon mode.[3]

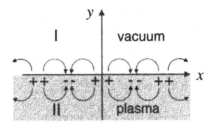

Fig. 17.3 Charge distributions in the surface plasmon, a surface electro-magnetic wave propagating in the x-direction.

Plasma oscillations in bulk metals are longitudinal oscillations and do not couple to radiation. They are observed through electron scattering or energy loss spectroscopy. For example, in silver (Ag) the plasmon energy, $\hbar\omega_P$, is 3.75 eV, while for aluminum (Al) it is 15.3 eV.

17.3.2   Surface plasmons at a single planar interface

At the surface of a metal it is possible to have confined or guided electromagnetic waves that couple to electron density oscillations in the metal. These mixed modes are called plasmon–polaritons, or simply surface plasmons. Such excitations also do not couple to free radiation. However, plasma oscillations between two surfaces such as on thin metallic films can couple to radiation fields. This is discussed in the following.

Let us consider a semi-infinite region (region II) $y < 0$ occupied by a metal, bounded by the $y = 0$ planar surface (Fig. 17.3). Let its dielectric function be of the Drude form,

$$\epsilon(\omega) = 1 - \omega_P^2/\omega^2. \qquad (17.12)$$

Let the region $y \geq 0$ be the vacuum (region I) with dielectric constant $\epsilon = 1$. We solve for the dispersion relation[†] for the guided electro-magnetic waves propagating along the x-direction with a wavevector k, and falling off exponentially away from the surface.

In the absence of external fields there are two possible surface modes of which only the transverse magnetic (TM) mode is of inter-

---

[†]The relation between the frequency and the wavevector of a normal mode is called its dispersion relation.

est, the other (TE) vanishing when the media have the same magnetic permeabilities.[4] The TM mode is defined by

$$\mathbf{E} = (E_x(y),\ E_y(y),\ 0)\ e^{ikx-i\omega t}$$
$$\mathbf{B} = (0,\ 0,\ B_z(y))\ e^{ikx-i\omega t}. \tag{17.13}$$

In the two regions, we let the dependence on y be $e^{-\alpha_I y}$ for $y > 0$, and $e^{+\alpha_{II} y}$ for $y < 0$. The absence of external charges in either region, and Gauss's theorem $\nabla \cdot \mathbf{E} = 0$, lead to the relations

$$E_{Iy} = ik\,E_{Ix}/\alpha_I,$$
$$E_{IIy} = -ik\,E_{IIx}/\alpha_{II}. \tag{17.14}$$

The continuity of the tangential component of the electric field, and the normal component of the displacement field, at the interface can be used to obtain

$$E_{Ix} = E_{IIx},$$
$$E_{Iy} = \epsilon(\omega)\,E_{IIy}. \tag{17.15}$$

The divergence conditions of equation (17.14) are then substituted into (17.15) to provide the dispersion relation

$$\epsilon(\omega) = 1 - \frac{\omega_P^2}{\omega^2} = -\alpha_{II}/\alpha_I. \tag{17.16}$$

The electric fields in the two regions satisfy Maxwell's wave equation, so that

$$k^2 - \alpha_I^2 - \omega^2/c^2 = 0,$$
$$k^2 - \alpha_{II}^2 - \epsilon(\omega)\,\omega^2/c^2 = 0. \tag{17.17}$$

We eliminate the α in equation (17.16) using equation (17.17) to obtain the surface plasmon dispersion relation

$$\Omega_{SP}^2 = \frac{1}{2\omega_P^2} \left[2k^2 c^2 \omega_P^2 + \omega_P^4 \right.$$
$$\left. \pm \sqrt{\left(2k^2 c^2 \omega_P^2 + \omega_P^4\right)^2 - 4k^2 c^2 \omega_P^6} \right]. \tag{17.18}$$

The dispersion curves for the surface plasmon frequency Ω_{SP} are shown in Fig. 17.4. The upper branch, called the Brewster mode,

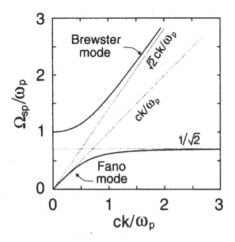

Fig. 17.4 Dispersion curves for the surface plasmon at the surface of a metal.

is not of interest here.[5,6] The lower branch is known as the Fano mode. Neither mode couples to external electromagnetic fields because the dispersion relations (17.18) do not cross the photon's dispersion relation in free space. In the limit $(ck/\omega_P) \gg 1$, the short wavelength limit, the lower branch of the dispersion relation asymptotically reaches the value first derived by Ritchie[7]

$$\Omega_{SP} = \omega_P/\sqrt{2}. \tag{17.19}$$

This result is valid only in the *short* wavelength limit. In this limit, a concern is the validity of the Drude approximation for the dielectric function of the metal; a more appropriate form to use would be the Lindhard function based on linear response theory.[8-10] Ritchie's result, equation (17.19), is usually obtained in the literature directly from the solution of Laplace's equation with electrostatic BCs in the *long* wavelength limit. This does not correspond to the long wavelength limit of the nonretarded calculation. The usual (incorrect) derivation is as follows. In the absence of any charges, except at the surface, the potentials satisfy Laplace's equation in each region. In the absence of external fields the potentials must die out at $y \to \pm\infty$, and the method of separation of variables permits electrostatic potentials to be of the form $\phi_{I,II}(x,y) = (\phi_0)_{I,II} \, e^{ikx}e^{-|k|y}$. The conditions for the continuity of the potential and the normal component of the displacement vector correspond to

$$(\phi_0)_I = (\phi_0)_{II},$$
$$k \, (\phi_0)_I = -k \, \epsilon(\omega) \, (\phi_0)_{II},$$

and the amplitudes $(\phi_0)_{I,II}$ are nonzero if

$$\begin{vmatrix} 1 & -1 \\ 1 & \epsilon(\omega) \end{vmatrix} = 0, \qquad (17.20)$$

again leading to equation (17.19). We note that this result is independent of k, and for small values of k it is clearly incorrect. The field penetrates more and more on either side of the interface as k is made smaller. This electrostatic calculation is wrong and cannot be justified for the single interface.

Note that in the long wavelength limit a single interface does not provide a scale of length that is smaller than the wavelength of radiation. However, in more complex geometries where the physical system provides a length scale that is of finite extent (such as a cylindrical wire), this long wavelength limit provides a proper procedure that leads to meaningful results in this limit. When there is more than one interface present separated by a distance smaller than the wavelength of the incident radiation (as in a thin metallic slab), we again have the electrostatic limit yielding a reasonably accurate dispersion relation for surface electromagnetic excitations in the system. In the following examples we will continue to use the electrostatic approach for calculating the surface plasmon dispersion for a thin metal slab, and for arrangements of cylindrical metallic wires. In these cases we do have a high frequency mode that couples to radiation (thin slab) or has a characteristic scale of length that permits our legitimately taking the long wavelength limit (thin circular wire).

If the system is excited at frequency Ω_{SP} the local potential and the field at the surface should be much larger than the input values. Recall that the condition for the solutions in equations (17.2) to diverge with $d_I = d_{II}$ corresponds precisely to $\epsilon(\omega) = -1$. We anticipate therefore that the BEM, with simple BCs, can be used to solve for the electrostatic potentials everywhere for metallic regions with complex surfaces. With a sweep along negative values for the dielectric function for the metallic regions, we can obtain the resonant conditions for field enhancement. While the calculations are done using electrostatic methods, the dielectric function is varied to simulate a variation of the frequency. This procedure provides us with a powerful numerical method for determining the surface plasmon frequency within the electrostatic approximation.

We note that continuity of the normal component of the displacement field and the Drude form of the dielectric function for the metal gives the following condition at the surface:

$$(E_I)_n - (E_{II})_n = -(\omega_P^2/\omega^2)(E_{II})_n = 4\pi\sigma. \qquad (17.21)$$

This relation implies that in the electrostatic limit there exists an oscillating surface charge density σ. Thus, by choosing appropriate charge distributions oscillating with the coordinate x one can include a wavevector dependence to surface plasmons in our boundary element calculations. This is one way of including the wavevector dependence in the calculations. A more complete calculation for surface plasmons can also be done by fully modeling electromagnetic scattering in the boundary element framework. The latter approach is beyond the scope of this discussion.

17.3.3 Surface plasmons for slab geometry

Let us consider a metal slab of thickness $2d$, region II of Fig. 17.6 below, and having a dielectric function $\epsilon(\omega) = 1 - \omega_P^2/\omega^2$. We obtain the surface plasmon dispersion relations for this structure by solving Maxwell's equations together with the usual BCs at the surfaces of the slab. There are two possible sets of solutions depending on whether the modes are confined modes or radiating modes coupled to external radiation.

We choose traveling wave solutions along the x-direction to write the electromagnetic fields in the TM mode in the form

$$\mathbf{E}(\mathbf{r}) = (E_x(y), E_y(y), 0)\, e^{i(k_x x - \omega t)},$$
$$\mathbf{B}(\mathbf{r}) = (0, 0, B_z(y))\, e^{i(k_x x - \omega t)}. \qquad (17.22)$$

Maxwell's equations and the wave equation are

$$\frac{d}{dy}E_y(y) = -ik_x E_x(y),$$
$$B_z(y) = -\frac{\omega\,\epsilon(\omega)}{k_x\, c}E_x(y), \qquad (17.23)$$

and

$$\frac{d^2}{dy^2} E_x(y) + k_y^2 E_x(y) = 0. \tag{17.24}$$

Let us assume that we are considering radiation modes for which the wavevector k_{yI}, in region I outside the slab, is real. By symmetry, the y-dependence for the fields may be expressed as

$$E_x(y) = \begin{cases} C_{II} \left(e^{ik_y y} \pm e^{-ik_y y} \right), & |y| < d; \\ C_I\, e^{ik_{yI} y}, & |y| > d; \end{cases} \tag{17.25}$$

and

$$E_y(y) = \begin{cases} -\dfrac{k_x}{k_y} C_{II} \left(e^{ik_y y} \mp e^{-ik_y y} \right), & |y| < d; \\ -\dfrac{k_x}{k_{yI}} C_I\, e^{ik_{yI} y}, & |y| > d. \end{cases} \tag{17.26}$$

Here $k_y^2 = (\omega^2 \epsilon/c^2 - k_x^2)$, while the wavevector in vacuum is given by $k_{yI}^2 = (\omega^2/c^2 - k_x^2)$. The $\pm$ signs refer to even and odd modes, respectively. We now employ the continuity of (i) the tangential component of the electric field and (ii) the normal components of the electric displacement vector

$$E_{xI} = E_{xII},$$
$$E_{yI} = \epsilon(\omega)\, E_{yII},$$

and derive the dispersion relations[11]

$$\epsilon(\omega) = \frac{k_y}{k_{yI}} \left(\frac{e^{ik_y d} \pm e^{-ik_y d}}{e^{ik_y d} \mp e^{-ik_y d}} \right). \tag{17.27}$$

The odd radiating mode is of interest, since for small k_x and in the limit of a thin film ($k_P d \ll 1$, with $k_P = \omega_P/c$), the dispersion relation reduces to

$$\Omega_- \simeq \omega_P \left(1 + \frac{1}{2} d^2\, k_x^2 \right). \tag{17.28}$$

The energy and momentum of this surface plasmon mode, called the Ferrell mode,[12] matches the photon dispersion cone when light having the wavevector k_- is at a particular angle θ to the y-axis.

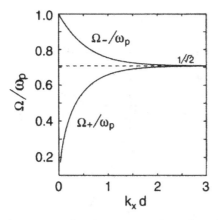

Fig. 17.5 Dispersion curves for the symmetric and antisymmetric plasma modes in a thin metallic slab in the electrostatic approximation.

Under these conditions the external radiation and a particular internal plasmon mode can couple to each other while conserving energy and momentum. In this case, an internal plasmon excitation, having a frequency Ω_- and wavevector k_-, can emit radiation out of the film on both sides along a cone at the angle θ to the y-axis.[12,13] The odd modes have charge distributions with local dipole moments directed essentially perpendicular to the surfaces, and these dipoles can be thought of as being the sources of the external radiation. Later experiments[14] have verified this effect in thin films. The even mode does not couple to radiation, with its dispersion curve being the lower branch in Fig. 17.5.

A similar radiative plasmon mode should exist in thin films of heavily doped narrow-bandgap semiconductors. In direct bandgap materials, excess electrons excited into the conduction band from the valence band, by means of optical laser excitation, can relax back into the valence band via the emission of radiation with energy corresponding to the bandgap. However, in very narrow-gap materials such as $Pb_{1-x}Sn_xTe$ and $Hg_{1-x}Cd_xTe$, instead of photon emission, other internal modes such as plasmon modes get excited. It was therefore suggested by Wolff that a thin film of such a heavily doped semiconductor be used to generate stimulated emission of radiation at the plasmon frequency, and this could provide a source of coherent radiation in the very far-infrared region of the spectrum.[15] That such radiation can be excited in HgCdTe has been confirmed experimentally by McManus.[16] A new situation arises when the bandgap

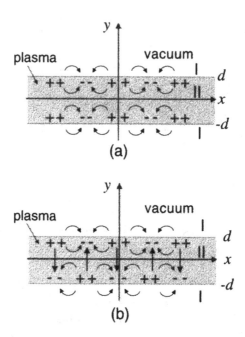

Fig. 17.6 Charge distributions in a thin metallic slab for (a) symmetric and (b) antisymmetric surface plasmon modes.

matches the longitudinal-optic phonon. Now, the electron–hole recombination process may lead to the emission of mixed modes of plasmons and longitudinal-optic phonon modes, instead of radiation. The theory of such excitations and their use in generating stimulated emission of plasmon–LO phonon modes is described in Ref. 17. This phenomenon has been experimentally verified in HgCdTe.[18]

The nonradiating modes supported by the slab plasma have imaginary wavevectors, and the fields fall off exponentially as one progresses away from the surfaces of the slab in the y-direction. A recalculation with k_y, k_{yI} replaced by $i\kappa_y, i\kappa_{yI}$ leads to the dispersion formula

$$\epsilon(\omega) = \frac{\kappa_y}{\kappa_{yI}} \begin{cases} \coth \kappa_y d \\ \tanh \kappa_y d \end{cases} \tag{17.29}$$

for symmetric and antisymmetric modes, respectively. In the nonretarded approximation, κ_y and κ_{yI} tend to k_x, and in this limit the dispersion formula simplifies considerably. We obtain

$$\epsilon(\omega) = \begin{cases} \coth \kappa_y d \\ \tanh \kappa_y d \end{cases} \qquad (17.30)$$

for the even and odd modes in the slab. The surface plasmon modes have the frequencies

$$\Omega_{\pm}^2 = \omega_p^2 \Big/ \left[1 + \begin{cases} \coth k_x d \\ \tanh k_x d \end{cases} \right]. \qquad (17.31)$$

The dispersion curves are shown in Fig. 17.5, where the surface plasmon frequencies are graphed as a function of $k_x d$.

The corresponding charge distributions in the thin slab are shown in Figs. 17.6a and b. The two modes correspond to the surface plasmons from each planar surface. For thin films the two modes considerably influence one another and their eigenfrequencies are different, whereas in the limit of large thickness the two modes act independently and asymptotically approach the value of $\omega_P/\sqrt{2}$. We note that the electrostatic approximation for radiating modes leads to the same dispersion relation. In this instance, it correctly predicts that the upper branch Ω_- crosses the photon's dispersion relation close to the plasmon frequency ω_P, in agreement with equation (17.28).

Calculations for radiating modes in configurations other than the slab geometry have not, in general, been performed as yet. The procedure would be to consider multiply-connected metallic regions in proximity to each other, and surrounded by a "physical region" in the form of a very large circle. The BCs at this circle correspond to outward-propagating electromagnetic waves of a given wavelength. As the dielectric function of the metal is varied the local fields would go through resonances associated with radiating plasmon modes.

We have discussed the conditions for the validity of the electrostatic method at length in order to put the BEM calculations in a proper perspective. We note that there are similar issues in connection with surface phonons and interface phonons.[19]

17.3.4 Surface plasmons in a cylindrical wire

Electromagnetic radiation incident normal to the axis of a metallic wire can excite surface electromagnetic waves, or surface plasmons, on the cylindrical surface of the wire. We wish to determine the frequency of this excitation. In the long wavelength limit, the electrostatic approximation will again be used to obtain the surface plasmon frequency for wires with circular cross-section. We begin by considering a single wire, of radius a, in a constant external electric field $E_0\hat{x}$ perpendicular to its axis. Let the electrostatic potentials inside and outside the wire, ϕ_i and ϕ_o, take the forms

$$\phi_i(r,\theta) = c_1 r \cos\theta,$$
$$\phi_o(r,\theta) = -E_0 r \cos\theta + c_2 \frac{1}{r}\cos\theta. \qquad (17.32)$$

We are, in effect, allowing for local polarization of the wire in the external field. The BCs

$$\phi_i(a) = \phi_o(a),$$
$$\epsilon(\omega)\frac{\partial}{\partial r}\phi_i(a) = \frac{\partial}{\partial r}\phi_o(a), \qquad (17.33)$$

allow us to determine the coefficients

$$c_1 = -\frac{2E_0}{\epsilon(\omega)+1},$$
$$\qquad (17.34)$$
$$c_2 = \frac{\epsilon(\omega)-1}{\epsilon(\omega)+1}E_0 a^2.$$

The resonant frequency is then given by the vanishing of the denominator, $\epsilon(\omega)+1$, which occurs at $\Omega_{SP} = \omega_P/\sqrt{2}$.

BEM calculation for a single wire

The equivalent calculation was performed using the BEM.[20] The physical region is shown in the inset in Fig. 17.7. The potential on the right side of the square is taken to be $+100\,\text{V}$ and the left side is held at $-100\,\text{V}$. The calculation proceeds within the BEM by defining three regions. The circular region at the center is occupied by the metallic wire, while the regions above and below it are the vacuum, with a dielectric constant of unity. The circle is connected to the left and right sides of the square region by two lines that are used to define closed contour paths for the three regions. The right and left edges of the region are each divided into four elements of equal length, while two elements are used along the top and bottom edges of the square. The circular path is divided into eight curved elements with the angle as the parametric variable for the boundary integration. The potential and its normal derivative

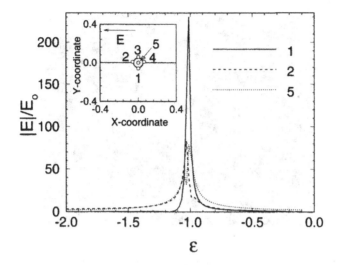

Fig. 17.7 The physical region containing the cross-section of a circular metallic wire is shown in the inset, with the points 1–5 where the electric field is calculated as a function of the dielectric constant. The vacuum regions outside the wire have $\epsilon = 1$. The magnitude of the electric field as a function of the dielectric constant ϵ of the wire at the points 1, 2, and 5 is shown. The graphs are equivalent to the variation of the field with the frequency of the incident field in the "electrostatic" approximation.

are represented in all elements by the use of Hermite interpolation polynomials with nodal variables including normal, tangential, and second-order cross-derivatives. The discretized boundary integrals lead to fewer equations than there are variables, and extra points are chosen along the sides as discussed earlier.

The dielectric function of the wire is varied and the local electric fields, at the points 1–5 shown in the inset in Fig. 17.7, are calculated. The variation of the electric fields with ϵ at points 1, 2, and 5 is also shown. We see that the resonant condition for the field corresponds to $\epsilon = -1.05$. There is a dramatic increase in the magnitude of the local field as the resonance is approached. The local field varies in magnitude and direction fairly rapidly as the dielectric function is swept over negative values of ϵ. This variation of the dielectric function is equivalent to a sweep in the frequency ω of an incident oscillating field, since $\epsilon = 1 - \omega_P^2/\omega^2$. A contour plot of the actual field amplitude at $0.698\,\omega_P$, which corresponds to the resonance frequency, is shown in Fig. 17.8. We note that the physical

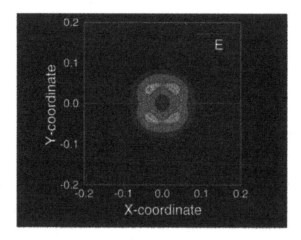

Fig. 17.8 A contour map of the magnitude of the electrostatic field in a circular wire for $\epsilon = -1.05$, at the surface plasmon resonance. The four-fold symmetry of the field distribution is due to the square shape of the exterior region. See the discussion in the text.

region was assumed to be limited to the finite rectangular region; if the boundary of the finite domain is close to the wire, it could have a distorting effect on the calculated field. The field distribution in Fig. 17.8 shows this effect of the proximity of the physical boundary, and the resonance frequency is off by 1% from the value expected, mainly due to this effect. Figure 17.8 also emphasizes the fact that the enhanced field is indeed essentially confined to the surface. Note that the square boundary is not necessary in the BEM. It has been used in these calculations in order to provide a simple illustration of the computational method with simple concepts. In practice, we can put the external square boundary at infinity, where the potential is set to zero. This allows one to obtain the solution with "open boundaries," an advantage that parallels the use of "infinite" elements in the FEM (see Chapter 3 where a 1D version of infinite elements is discussed).

It is evident that we are determining an eigenfrequency for the plasmon by searching for the value of the dielectric function where the field is resonantly enhanced. Hence care has to be taken to ensure that the matrices being inverted in the solution process are not ill-conditioned. Poor conditioning of the coefficient matrices can be

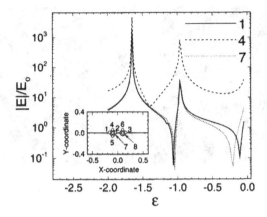

Fig. 17.9 The physical region with two parallel circular metallic wires placed in a constant electric field acting perpendicular to their axes is shown in the inset. The vacuum regions external to the wires have $\epsilon = 1$. The variation of the local electric field with the dielectric constant is evaluated at points 1–8. The variation of the magnitude of the local electric field at points 1, 4, and 7 versus the dielectric constant of the wires for an external field acting vertically downward is shown.

ameliorated by using complex arithmetic in the programming, by invoking the damping term in the dielectric function, as in equation (17.9), and by using higher precision (double precision) in the programming. Other alternatives are suggested in Ref. 21.

17.3.5   Two metallic wires

In the case of two cylindrical metallic wires, it is possible to solve Laplace's equation exactly in bipolar coordinates using the theory of complex variables.[22] The dispersion formulas for the symmetric and antisymmetric modes are equivalent to the form given in equation (17.31), with the variable d replaced by the value of a bipolar coordinate. Here, we prefer to develop the boundary element approach.

We define a rectangular physical region in which the wires with circular cross-sections are placed, as in the inset in Fig. 17.9. (The introduction of a global rectangular region is just for a simpler appreciation of the BEM. It is also possible to let the outer boundary go to infinity.) It is convenient to break up the physical region into four subregions; the top and the bottom regions are the vacuum, and the circular regions are occupied by the cylindrical wires. Eight

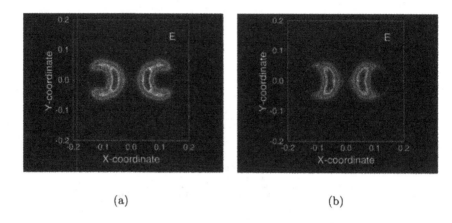

(a)   (b)

Fig. 17.10 Contour plots of the magnitude of the local electric field in two adjacent parallel wires for a frequency corresponding to (a) $\epsilon = -0.966$, and (b) $\epsilon = -1.66$, in an external field pointing vertically downwards. The horizontal line along the middle is part of the path for integration in the BEM.

boundary elements are placed along each of the circular edges of the two wires, and four elements are used along the sides of the square defining the physical region. The results of calculations presented here were performed using cubic Hermite interpolation polynomials.

We have investigated two possible configurations, the first with the external field directed vertically down, and the second with an electric field acting horizontally towards the left. With the field in the vertical direction, the top and bottom sides of the square are held at $\pm 100\,\mathrm{V}$. Even after imposing the BCs, there are more undetermined parameters than there are equations obtained by placing the destination point at each of the nodes. (This is also true of the calculation for a single wire on employing Hermite interpolation polynomials.) The additional equations necessary for determining all the nodal parameters are obtained by placing destination points along the top and bottom of the physical region where the potential is known.

In Fig. 17.9, we show the variation of the magnitude of local electric fields, evaluated at points 1, 4, and 7 of the inset in Fig. 17.9, as a function of the dielectric constant. There are two values of the dielectric constant for which the local field displays a resonant enhancement. These values correspond to two frequencies of an applied oscillating field for which internal plasma modes are excited in this configuration. The frequencies correspond to the symmetric and the antisymmetric modes of the surface plasmon oscillations in the two wires. Using the Drude dielectric function we find $\Omega_+ = 0.613\,\omega_P$ for the lower frequency mode and $\Omega_- = 0.713\,\omega_P$ for the higher frequency mode. The fact that these resonances are not at $\omega_P/\sqrt{2}$ shows that the surface plasmon modes associated with the individual cylindrical wires are interacting with one another, leading to mixed modes. Of these, one is at a frequency higher than $\omega_P/\sqrt{2}$, and the other is pushed to a lower value. This is analogous to the interacting normal modes in a thin slab. Contour plots of the field enhancement $|E|/E_0$ at the higher frequency, Fig. 17.10a, and at the lower frequency, Fig. 17.10b, clearly show the effect of the interaction between the plasma modes of each of the wires. The plots again illustrate the surface nature of the field enhancement.

For a horizontal external field, the sides of the square physical region are held at potentials $\pm100\,$V. The calculation is repeated for this field configuration, and we evaluate the electric field magnitudes at points 1, 5, and 7 of Fig. 17.9.

The results are shown in Fig. 17.11. The largest field enhancement, which now occurs at a frequency $\Omega_+ = 0.648\,\omega_P$, is at point 5. In Fig. 17.11b, the contour plot of the electric field enhancement is shown for the horizontal field.

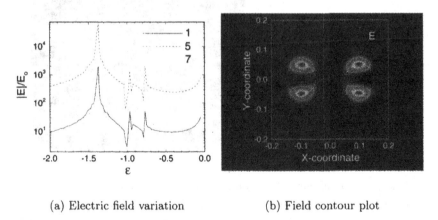

(a) Electric field variation   (b) Field contour plot

Fig. 17.11 (a) The electric field at points 1, 5, and 7, of Fig. 17.9, for two parallel wires in an external horizontal electric field. (b) A contour plot of the magnitude of the local electric field for $\epsilon = -1.38$ for an electric field directed horizontally to the left.

When Fig. 17.11b is compared with Fig. 17.8b, it is evident that the symmetry of the magnitude of the local field around the horizontal and vertical diameters in the single wire in an external field is now distorted. The proximity of the two wires leads to local fields of one wire distorting the local field of the other. The mode mixing between the plasmons from the two wires lowers the frequency to the value Ω_+ given above.

17.3.6   Metal wire on a substrate

Recent developments in submicron technology allow one to lay down metallic structures on semiconducting substrates (see Fig. 17.12a). The local field enhancement for an applied external electromagnetic field has been calculated[20] using the BEM.

The calculations for the field enhancement were performed using the boundary elements shown, and use was made of Hermite interpolation for the potential and its normal derivative. The electrostatic calculations for surface plasmons are relevant so long as the characteristic length scale a for the cross-sections of the wires is much smaller than the wavelength of the incident radiation. The substrate

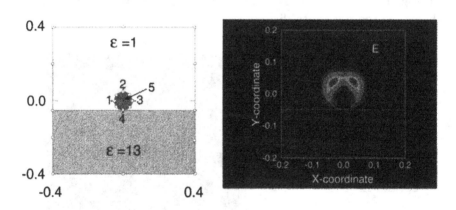

Fig. 17.12 A metal wire placed on a semiconductor substrate (left). The local electric field is calculated at points 1–5 as the dielectric function of the wire is varied in seeking resonant conditions for field enhancements. A contour plot (right) for the magnitude of the electric field enhancement in a metallic wire placed on a semiconductor substrate (dielectric constant $\epsilon = 13$), with an applied horizontal electric field. The calculation is for a value of the dielectric function of the wire of $\epsilon = -1.06$.

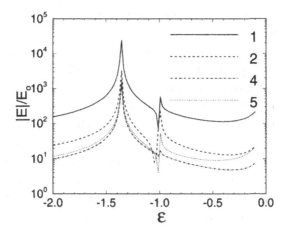

Fig. 17.13 The variation of the magnitude of local electric field, at points 1, 2, 4, and 5 of Fig. 17.12a, with the dielectric function of the metal wire placed on a semiconductor substrate. The external field acts vertically downwards.

is assumed to have a dielectric constant of $\epsilon = 13$, typical of a group IV or III−V semiconductor. The local field near the periphery of the wire is enhanced by a factor of 10^4 for a value of $\epsilon = -1.06$ for the metallic region, for an applied field directed horizontally. There is only one resonance, at $\epsilon = -1$, for this configuration. The contour

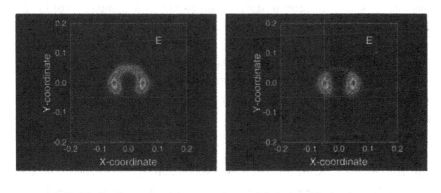

(a) Field plot for $\epsilon = -0.99$ (b) Field plot for $\epsilon = -1.35$

Fig. 17.14 Contour plots for the magnitude of the electric field enhancement for a metallic wire on a semiconductor substrate with a dielectric constant $\epsilon = 13$. The field has been calculated for (a) $\epsilon = -0.99$ and (b) $\epsilon = -1.35$.

plot for the magnitude of the electric field in Fig. 17.12 clearly shows the effect of the substrate.

This may be compared with Fig. 17.8. With a vertical field, the local fields at the points 1, 2, 4, and 5 (see Fig. 17.12a) are calculated as a function of the dielectric constant of the wire. The results are shown in Fig. 17.13, where we find two dielectric values for the wire at $\epsilon = -0.99$ and $\epsilon = -1.35$ at which there is substantial field enhancement. The corresponding field contour plots are shown in Figs. 17.14a and 17.14b. These examples illustrate the fact that the local field enhancement is strongly influenced by the actual configuration of metallic and other dielectric regions. The effect of the depolarization field from the semiconductor substrate can also be noticed.

17.3.7   Plasmons in other confining geometries

It should be evident from the preceding examples that the BEM can be used to determine the normal modes of surface plasmons confined to any configuration of metallic regions in proximity to one another. We have employed the long wavelength approximation in the calculation for the resonant frequency of the surface plasmon. We have been interested in the resonant enhancements of the electromagnetic field ($\lambda \sim 5000$–8000 Å) in metallic wires and dots defined by nanofabrication of typical dimensions on the order of $\sim$500 Å, for which the long wavelength static approximation is more than adequate. Nonlocal effects generated by spatial dispersion in the dynamic response to the externally applied electromagnetic field are usually incorporated using the Lindhard dielectric function obtained using the random-phase approximation.[9] We have ignored the effects of nonlocality since we have employed the long wavelength approximation for the present calculation.

While we have limited our presentation to 2D problems, it is straightforward, but more tedious, to extend our considerations to three dimensions. A number of new issues have to be addressed in 3D calculations. The Greens' function in three dimensions differs from the one in two dimensions, and is given by $G(\mathbf{r}, \mathbf{r}') = 1/|\mathbf{r} - \mathbf{r}'|$. The surface integrals comprising the boundary integral in three dimensions are in general on curved surfaces. The integrals require careful handling to account for the singular nature of the Green's function and its normal derivative when the destination point lies in the same element as the one being evaluated. Further details may be obtained in the text by Brebbia *et al.*[23]

There are a limited number of analytical calculations for the normal modes of the surface plasmon in 3D configurations. The plasma resonances in a metallic sphere are well known and these provide the initial test for any new computer program for 3D calculations. The surface plasmons have the spectrum $\Omega_{SP} = \omega_P/\sqrt{(2 + 1/m)}$ where $m = 1, 2, 3, \ldots$. The lowest energy mode corresponds to $\omega_P/\sqrt{3}$ and the spectrum converges to $\omega_P/\sqrt{2}$ in the large m limit. The m "quantum number" specifies the multipolarity of the normal modes.

Using bispherical coordinates, it is possible to solve Laplace's equation and determine the plasma resonances in two metallic spheres or in a sphere near a planar metallic region.[24] More recent developments in the application of hypercomplex variables,[25] again to solve Laplace's equation, for arrays of spheres show that these systems with symmetry can be used to provide stringent tests for BEM calculations in three dimensions. The BEM computer codes can then be used to investigate the surface plasmon spectrum in more complex configurations of surfaces. We note that a full electromagnetic scattering calculation with an infinite, unbounded exterior region can be performed within the framework of the BEM. The method is described in detail by Kagami and Fukai.[26] The corresponding Green's function for Helmholtz's equation needs to be employed in the calculations.

17.4   Surface-enhanced Raman scattering

Over two decades ago, experimental observations of dramatic enhancements by factors of the order of 10^6 in the intensity of Raman scattering from molecules adsorbed on specially prepared rough surfaces of metal substrates were reported.[27,28] The enhanced scattering was specially pronounced with silver (Ag) globules deposited on a substrate. Theoretical considerations[29] show that this enhanced scattering occurs because the molecule experiences the local field which is amplified by the plasma resonance from the confined carriers in Ag. The Ag globule, enclosing an electronic plasma, acts as an electromagnetic resonator, which amplifies the local field near the surface when the incident radiation has a frequency matching the natural frequency of the plasma oscillations. Other mechanisms contributing to the enhanced scattering have also been considered and have been reviewed in Refs. 30–32. Developments in MBE permit the construction of regular arrays of Ag globules deposited on semi-conductor structures.[33] Recently, the problem has received renewed attention in this context from Pendry and co-workers.[34] It is clear that the BEM will provide the proper framework for the calculation of field enhancements near such arrays of irregularly shaped metallic structures. The problem continues to be of interest in connection with applications, for example, to improved performance of solar cells, and for calculations of depolarization effects in arrangements of packed dielectric spheres at microwave frequencies.[35]

17.5 Problems

1. Consider a region made up of two squares, each of unit dimensions with dielectric constants $\epsilon_I = 1$ and $\epsilon_{II} = 10$, respectively. Let each side of the squares have just one element. Use constant interpolation in the boundary elements and place the nodes at the center of the elements. (See Fig. 17.1; now the nodes are at the center of the elements.)

    (a) Evaluate the discretized boundary integral in each region, taking each node as the destination point.

    (b) Apply the interface BCs, for the continuity of the potential and the normal component of the electric displacement field, to eliminate a pair of nodal variables at the interface.

    (c) Form the set of simultaneous equations for the unknown nodal variables and solve them. Compare your results with the results in the text obtained using linear interpolation.[§]

2. Consider a metallic wire of circular cross-section, with radius 0.08 cm, embedded halfway into a semiconductor substrate of dielectric constant $\epsilon = 13$. The dielectric function of the metal is $\epsilon(\omega) = 1 - \omega_p^2/\omega^2$. Set up a physical region similar to the one in the inset panel in Fig. 17.7. By holding opposite edges of the square at voltages $\pm 100\,\mathrm{V}$ consider the effect of an external field directed in (i) the horizontal, and (ii) the vertical directions.

    (a) Apply the BEM to this system with three regions. Develop computer codes using constant interpolation within the boundary elements.

[§] Also compare with the discussion of a related problem in Gipson,[36] p169.

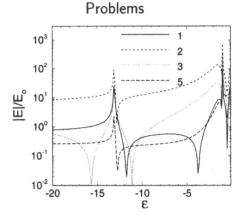

Fig. 17.15 Local field enhancement in a cylindrical wire half-embedded in a semiconductor substrate, of dielectric constant $\epsilon = 13$, for an applied vertical field, at points 1, 2, 3, and 5 of Fig. 17.7.

(b) Obtain the solution for the local potential at nodes placed at the center of the elements with constant interpolation.

(c) Evaluate the potential, on a grid of points in the interior region, by placing the destination point at each grid point.

(d) Vary the dielectric function of the metal over negative values $-15 \leq \epsilon < 0$ so as to determine resonant enhancements for the local potential arising from surface plasmon modes. Evaluate

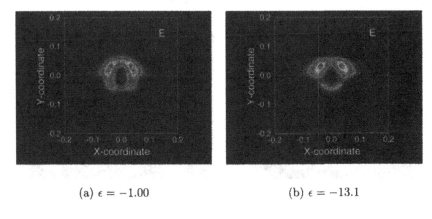

(a) $\epsilon = -1.00$ (b) $\epsilon = -13.1$

Fig. 17.16 Contour plots of the local electric field in a wire half-embedded in a semiconductor substrate for a vertical electric field, for a frequency corresponding to (a) $\epsilon_{wire} = -1.0$, and (b) $\epsilon_{wire} = -13.1$. The substrate has a dielectric constant $\epsilon = 13$.

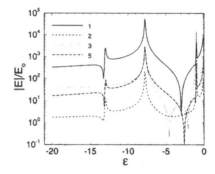

Fig. 17.17 Local field variation with the dielectric function in a cylindrical wire half-embedded in a semiconductor substrate of dielectric constant $\epsilon = 13$, for an applied horizontal field, at points 1, 2, 3, and 5 of Fig. 17.7.

the electric field at the points 1–5 of Fig. 17.7 by divided differences of the potential at adjacent points. Show that there are two resonances for the vertical field direction (see Fig. 17.15) and that there are three resonances for the horizontal electric field (as in Fig. 17.17). Compare your results with the field strengths shown in Fig. 17.15 for the vertical field and Fig. 17.17 for the horizontal field.

(e) Show, by numerical calculations, that the depolarization fields in the dielectric substrate, in general, tend to push the region with field enhancement away from the substrate. Note

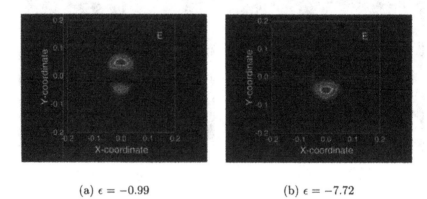

(a) $\epsilon = -0.99$ (b) $\epsilon = -7.72$

Fig. 17.18 Local field enhancement in a cylindrical wire half-embedded in a semiconductor substrate of dielectric constant $\epsilon = 13$, for an applied horizontal field (a) at a frequency corresponding to $\epsilon_{wire} = -0.99$, and (b) a dielectric constant for the wire of $\epsilon_{wire} = -7.72$.

that the low frequency modes for the field polarization parallel to
the surface have field enhancements within the semiconductor.
(See Figs. 17.16 and 17.18.)

Results obtained using Hermite interpolation for the boundary
elements are shown in Figs. 17.16 and 17.18. These figures illustrate the crucial role played by the substrate, and show that a
horizontal field leads to three resonances. The lower frequency
resonances correspond to plasmon configurations which permit
the field to concentrate in the substrate.

3. Laplace's equation is separable in bipolar coordinates which are
defined by

$$w = \xi + i\theta = \ln\left(\frac{z+a}{z-a}\right),$$

where $z = x + iy$ is a complex variable. A number of interesting geometries can be considered for the determination of
surface plasmon modes in the nonretarded limit.[1] The electrostatic problem with two parallel cylindrical wires whose circular
cross-section is defined by the circles of radii $\xi = \xi_1 = -\xi_2$
(Fig. 17.19a), is exactly solvable in bipolar coordinates. Let the
shaded regions have the dielectric constant $\epsilon(\omega) = 1 - \omega_p^2/\omega^2$.

(a) Consider the shaded regions $\xi \leq \xi_2$, $\xi \geq \xi_1$, with $\xi_1 = -\xi_2 \geq$
0. Show that the solutions to Laplace's equation everywhere can
be written in the form

$$\phi(\xi, \theta) = (A_m \sin m\theta + B_m \cos m\theta) \times e^{\pm m\xi}.$$

Here $m = 1, 2, 3, \ldots$ is the constant of separation.

(b) Assume symmetric and antisymmetric solutions for the potential of the form

$$\phi(\xi, \theta) = \phi_i \cosh m\xi \begin{cases} \cos m\theta \\ \sin m\theta \end{cases}.$$

Let the potential tend to zero as $x, y \to \infty$ outside the metallic
wires

$$\phi(\xi, \theta) = \phi_1 e^{-m\xi} \begin{cases} \cos m\theta \\ \sin m\theta \end{cases} \qquad \xi > \xi_1,$$

[1]Problems 3–6 are based on P. A. Wolff and L. R. Ram-Mohan (unpublished
research notes, 1981).

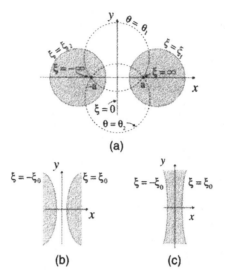

Fig. 17.19 (a) The bipolar coordinate system. Two metal cylindrical wires of very large radii with surfaces nearly touching are shown in (b). In (c), a double-concave metal slab with a large radius of curvature for the sides is shown.

and

$$\phi(\xi, \theta) = \phi_2\, e^{m\xi} \begin{cases} \cos m\theta \\ \sin m\theta \end{cases} \qquad \xi < \xi_2.$$

Apply electrostatic BCs at $\xi = \xi_1$ and $\xi = \xi_2$, and show that the surface plasmon dispersion relation is given by

$$\Omega_{SP}^2 = \omega_P^2 \frac{1}{1 + \begin{cases} \tanh m\xi_1 \\ \coth m\xi_1 \end{cases}},$$

for symmetric and antisymmetric modes.

4. Consider the two metallic cylinders of Problem 3 in the limit of large radii of curvature in comparison with the distance of separation of the surfaces of the cylindrical wires (see Fig. 17.19b). Show that in the limit $\xi_1 \to 0$ (decreasing the gap between the surfaces)

(i) the even modes of the "gap" surface plasmon have the frequency $\Omega_{SP} \simeq \omega_P$;

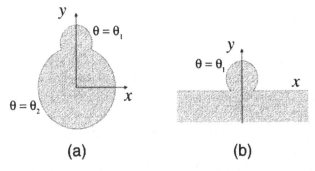

Fig. 17.20 (a) A cylindrical wire with a small boss on it. (b) A metal boss on the surface of a planar metallic semi-infinite region.

(ii) the frequency of the odd modes goes to the limits

$$\Omega_{SP}^2 = \begin{cases} m\,\xi_1\,\omega_P^2, & \xi_1 \to 0, \\ \omega_P^2/2, & m \to \infty. \end{cases} \qquad (17.35)$$

5. In the case of the double-concave metallic slab of Fig. 17.19, using bipolar coordinates, show that the symmetric and antisymmetric surface plasmon modes satisfy the dispersion relations

$$\Omega_{SP}^2 = \omega_P^2 \frac{1}{1 + \left\{ \begin{array}{c} \coth m\,\xi_1 \\ \tanh m\,\xi_1 \end{array} \right\}}.$$

6. Figure 17.20 shows two metal structures in which a small cylindrical boss is present (a) on a cylinder of larger radius, and (b) on a semi-infinite metallic region. Assume that a surface plasmon exists in the metallic region and that the field dies off at infinity in vacuum, away from the surface of the metal. Define an appropriate *finite* physical region in bipolar coordinates to enclose the metallic regions. Choose BCs on the periphery of the region and use the BEM to solve for the potential and the electric field everywhere in the electrostatic approximation. By varying the dielectric constant of the metallic region determine the resonant frequency for the surface plasmons in these configurations. By varying the BCs determine the dispersion of the surface plasmons. (Exact solutions to Laplace's equation for this configuration can be obtained in bipolar coordinates.)

References

[1] C. Kittel, *Introduction to Solid State Physics*, 6th ed. (Wiley, New York, 1986). For a readable elementary treatment of plasmas, see: F. F. Chen, *Introduction to Plasma Physics* (Plenum Press, New York, 1974).

[2] P. A. Wolff, "Notes on plasma wave excitations in semiconductors," unpublished (1980).

[3] P. M. Platzman and P. A. Wolff, *Waves and Interactions in Solid State Plasmas*, Solid State Physics Suppl. 13, ed. H. Ehrenreich, F. Seitz, and D. Turnbull (Academic Press, New York, 1973).

[4] R. E. Collin, *Field Theory of Guided Waves* (McGraw-Hill, New York, 1960).

[5] R. F. Wallis, in *Electromagnetic Surface Excitations*, ed. R. F. Wallis and G. I. Stegeman (Springer-Verlag, Berlin, 1985), pp1–7.

[6] A. D. Boardman, in *Electromagnetic Surface Modes*, ed. A. D. Boardman (Wiley, New York, 1982), p1.

[7] R. H. Ritchie, *Phys. Rev.* **106**, 874 (1957).

[8] J. Lindhard, *Kgl. Danske Videnskab. Selskab, Mat.-Fys. Medd.* **28**, No. 8 (1954).

[9] A. L. Fetter and J. D. Walecka, *Quantum Theory of Many-Paricle Systems* (McGraw-Hill, New York, 1971).

[10] D. Pines, *Elementary Excitations in Solids* (W. A. Benjamin, New York, 1964).

[11] K. L. Kliewer and R. Fuchs, *Phys. Rev.* **144**, 495 (1966); *ibid.*, **150**, 573 (1966), **153** 498 (1966).

[12] R. A. Ferrell, *Phys. Rev.* **111**, 1214 (1958).

[13] H. Raether, in *Physics of Thin Films*, ed. G. Hass, M. H. Francombe, and R. W. Hoffman, Vol. 9 (Academic Press, New York, 1977), pp145–261.

[14] W. Steinmann, *Phys. Rev. Lett.* **5**, 470 (1960). E. T. Arakawa, R. J. Herickhoff, and R. D. Birkhoff, *Phys. Rev. Lett.* **12**, 319 (1964).

[15] P. A. Wolff, in *Proceedings of the Conference on the Physics of Semimetals and Narrow Gap Semiconductors*, ed. D. L. Carter and R. T. Bate (Pergamon, New York, 1971).

[16] J. B. McManus, *Appl. Phys. Lett.* **41**, 692 (1982).

[17] R. B. Sohn, L. R. Ram-Mohan, H. Xie, and P. A. Wolff, *Phys. Rev.* B **42**, 3608 (1990). H. Xie, L. R. Ram-Mohan, and P. A. Wolff, *Phys. Rev.* B **42**, 3620 (1990).

[18] F. Fuchs, H. Schneider, P. Koidl, K. Schwarz, H. Walcher, and R.

Triboulet, *Phys. Rev. Lett.* **67**, 1310 (1991).

[19] P. A. Knipp and T. L. Reinecke, *Phys. Rev.* B **48** 12338 (1993); *Phys. Rev.* B **46** 10310 (1992).

[20] R. Goloskie, T. Thio, and L. R. Ram-Mohan, *Comput. Phys.* **10**, Sept/Oct (1996).

[21] W. D. Murphy, V. Rokhlin, and M. S. Vassiliou, *J. Appl. Phys.* **67**, 6061 (1990).

[22] P. M. Morse and H. Feshbach, *Methods of Mathematical Physics* (McGraw-Hill, New York, 1953), p1210.

[23] C. A. Brebbia, J. C. Telles, and L. C. Wrobel, *Boundary Element Techniques* (Springer-Verlag, Berlin, 1984).

[24] P. K. Aravind and H. Metiu, *J. Phys. Chem.* **86**, 5076 (1982).

[25] A. V. Vagov, A. Radchik, and G. B. Smith, *Phys. Rev. Lett.* **73**, 1035 (1994).

[26] S. Kagami and I. Fukai, *IEEE Trans. Microwave Theory Tech.* **32**, 455 (1984).

[27] M. Fleischman, P. J. Hendra, and A. J. McQuillan, *Chem. Phys. Lett.* **26**, 163 (1974).

[28] R. P. van Duyne, in *Chemical and Biochemical Applications of Lasers*, ed. C. B. Moore, Vol. 4 (Academic Press, New York, 1978).

[29] S. L. McCall, P. M. Platzman, and P. A. Wolff, *Phys. Lett.* **77 A**, 381 (1980).

[30] M. Moskovits, *Rev. Mod. Phys.* **57**, 783 (1985).

[31] A. Otto, in *Light Scattering in Solids IV*, ed. M. Cardona and G. Günterodt (Springer-Verlag, Berlin 1983).

[32] H. Metiu, in *Surface Enhanced Raman Scattering*, ed. R. K. Chang and T. E. Furtak (Plenum Press, New York, 1982).

[33] P. F. Liao, J. G. Bergman, D. S. Chemla, A. Wokaun, J. Melngailis, A. M. Hawryluk, and N. P. Economou, *Chem. Phys. Lett.* **82**, 355 (1981).

[34] J. B. Pendry, A. J. Holden, W. J. Stewart, and I. Youngs, *Phys. Rev. Lett.* **76**, 4773 (1996); F. J. Garcia-Vidal and J. B. Pendry, *Phys. Rev. Lett.* **77** 1163 (1996).

[35] J. E. Sansonetti and J. K. Furdyna, *Phys. Rev.* **22**, 2866 (1980).

[36] G. S. Gipson, *Boundary Element Fundamentals* (Computational Mechanics, Southampton, UK, 1987).

18

The BEM and quantum applications

18.1 Introduction

In this chapter, we consider quantum applications of the boundary element method (BEM). We apply the BEM to the propagation of electrons through mesoscopic devices of dimensions less than 10^{-6} m, with thin metallic wire leads of widths ≤ 1000 Å that are modeled as 2D waveguides. We also show how the BEM can be used to calculate quantum mechanical scattering in two dimensions. Bound state problems are of great interest in quantum mechanics, and we apply the BEM in this chapter to the eigenvalue problem of determining states of electrons bound in the transverse direction in quantum wires and in other 2D potential wells. Recently, it has been shown that two completely different 2D resonant cavities in which Helmholtz's equation holds for the electromagnetic field or the electron wavefunction can be "isospectral." In other words, the two areas of different shape can have the same spectrum. We report on this very interesting effect, and suggest that the BEM may be the ideal method for computing the spectrum of such isospectral areas.

In these applications, we shall assume that the potential energy function of the electron is piecewise constant. This allows us to represent Schrödinger's equation as Helmholtz's equation. The Green's function for the corresponding inhomogeneous equation is known. The complications that arise when the potential varies as a function of the coordinates demand iterative procedures. In the final section, we present concluding remarks on the BEM and give references to additional resources.

18.2 2D electron waveguides

Consider the scattering of electrons injected into a device that has dimensions less than the mean free path of the electrons. An electron can propagate through such a device without inelastic scattering. The wavefunction of the electron will therefore maintain a definite phase, and undergo quantum interference effects in maneuvering through the device. Hence the conductance of the device will be given by quantum mechanical considerations. Such structures, whose typical dimension is less than 1 μm, are said to be mesoscopic in size.[1]

The electrons in the perfectly conducting leads in such a device are assumed to be in local thermodynamic equilibrium. The current is considered to be driven through the device by employing unequal chemical potentials at reservoirs to which the leads are connected. Landauer[2] has shown that the conductance G of such a mesoscopic device can be given in terms of the quantum transmission and reflection coefficients for electrons incident on the device. Hence Schrödinger's equation can be used to solve for the transmission and reflection of electrons. The transport within the device is elastic, and is determined by the confining potential that is specified by the device geometry. The leads and the device can be modeled as 2D electron waveguides attached to a waveguide junction. For a two-terminal device, the conductance is given by

$$G = \frac{2e^2}{\hbar} \sum_{\mu\nu} |t_{\mu,\nu}(E_F)|^2, \tag{18.1}$$

where the electron charge is denoted by e, $\hbar$ is Planck's constant divided by 2π, and the transmission amplitudes are labeled by $t_{\mu\nu}$. The transverse modes at the input and output ports are labeled ν and μ, and E_F is the Fermi energy of the electron distribution at the leads. The quantum contact resistance corresponding to equation (18.1) has been experimentally verified.[3,4] In the following, the leads are referred to as ports or as probes.

Here our concern will be limited to the calculation of the transmission and reflection coefficients by the BEM.[†] The BEM reduces the problem to the solution of an integral equation over the periphery of the 2D waveguides and the junction. This reduction in dimensionality leads to a smaller matrix representation of the problem when the boundary integral is discretized, compared with the FEM scheme of breaking up the area into elements, and is a more efficient computational scheme as long as we have piecewise constant potentials. In

---

[†]The problem of electrons propagating through 2D channels was considered earlier (Chapter 13) within the framework of the FEM.

this case, Schrödinger's equation then reduces to Helmholtz's equation in two dimensions with constant potentials. The corresponding infinite domain Green's function is a Hankel function.

We begin by applying the BEM to the calculation of electron transmission through the simplest structure, a perfect 2D quantum waveguide of infinite length with parallel sides. The physical region is assumed to be comprised of a finite length of this infinite waveguide. While there is no reflection in this case, with no scattering in the waveguide, we shall allow for its possibility in the development of the algorithm.[‡] Let the width of the waveguide be d along the y-direction, and let the length of the channel be L along the x-axis. Here, Schrödinger's equation obeyed by the electron's wavefunction $\psi(\mathbf{r})$ is

$$-\frac{\hbar^2}{2m}\left(\frac{\partial^2}{\partial x^2} + \frac{\partial^2}{\partial y^2}\right)\psi(x,y) - (E - V_0)\,\psi(x,y) = 0. \qquad (18.2)$$

Here m is the mass of the electron. Since the potential energy is determined only up to a constant we may set $V_0 = 0$ in the confining channel. This assumes that we have only one region with a constant potential. The incoming electron is assumed to have a definite energy

$$E = \hbar^2 k^2/2m = \hbar^2[(k_x^{(\nu)})^2 + (k_y^{(\nu)})^2)]/2m, \qquad (18.3)$$

where $k_x^{(\nu)}$ is the wavenumber along the direction of propagation, and $k_y^{(\nu)} = \nu\pi/d$, $n = 1, 2, 3, \ldots$, is the quantum number representing the quantization due to confinement in the transverse direction. The wavevector k_x is determined by the total energy and the transverse energy, so that it depends on the mode index ν. We define the "transverse energy" as $E_\perp = \hbar^2 k_y^{(\nu)\,2}/2m$.

The wavefunction is zero at the side walls of the waveguide. At the output port, the current density is directed to the right so that the wavefunction is of the form $\psi_t \sim \exp\left(ik_x^{(\mu)}L\right)$, where μ is an index for the transverse mode across the output port. At the input port, on the left side, the incoming current is specified by us. We set the incoming wavefunction to be $\psi_{inc} = \exp\left(ik_x^{(\nu_0)}x\right)\sqrt{2/d}\,\sin\left(\nu_0\pi y/d\right)$, assuming

[‡]If there is scattering, or if there is a physical boundary present in the waveguide between two materials, we would have to include evanescent modes that fall off exponentially from the boundary on both sides in order to satisfy BCs. We address this issue later on.

that only a single mode is incident at the input port. The reflected wavefunction is of the general form $\psi_r \sim \exp\left(-ik_x^\nu x\right)$, with $x = 0$, at the input port and has to be determined. The wavefunctions ψ_r and ψ_t can be scattered into transverse modes other than the specific mode ν_0 of the incident wavefunction. We reiterate that we are assuming that we have an infinite 2D waveguide and that we are studying the wave propagation in a small section of the waveguide of length L. The incoming wave is assumed to comprise a single mode. We wish to determine the reflection and transmission coefficients for each of the energetically possible modes at the input and output ports, respectively.

Schrödinger's equation, equation (18.2), for the electron in a waveguide is equivalent to Helmholtz's equation

$$\left(\nabla^2 + k^2\right)\psi(x, y) = 0. \tag{18.4}$$

In Appendix C, we have shown that the Green's function satisfying the same equation with a source term,

$$(\nabla^2 + k^2)\,g(\mathbf{r}, \mathbf{r}') = -4\pi\,\delta(\mathbf{r} - \mathbf{r}'), \tag{18.5}$$

has the form

$$g(\mathbf{r}, \mathbf{r}') = i\pi H_0^{(1)}(k|\mathbf{r} - \mathbf{r}'|). \tag{18.6}$$

The source term in equation (18.5) has been written with a factor of -4π; other choices simply lead to a trivial redefinition of g. The choice of $H_0^{(1)} \equiv J_0 + iN_0$ is made, instead of the Hankel function of the second kind $H_0^{(2)}$, because in the limit $kr \to \infty$ the function $H_0^{(1)}(kr)$ has the asymptotic form $\sim e^{ikr}/\sqrt{r}$. This corresponds to an outgoing wave, appropriate to a scattering process.[§]

As in Chapter 16, we multiply equation (18.4) by $g(\mathbf{r}, \mathbf{r}')$ and equation (18.5) by $\psi(\mathbf{r})$, and subtract one equation from the other to obtain

$$\psi(\mathbf{r}) = \frac{1}{4\pi}\oint d\ell'\left[g(\mathbf{r}, \mathbf{r}')\frac{\partial}{\partial n'}\psi(\mathbf{r}') - \psi(\mathbf{r}')\frac{\partial}{\partial n'}g(\mathbf{r}, \mathbf{r}')\right]. \tag{18.7}$$

Next, the boundary integral is split up into integrals over *nelem* boundary elements. We shall suppose that in each of the elements

---

[§]This is analogous to the choice of the Green's function for scattering problems in three dimensions of $\sim e^{ikr}/r$ with outgoing waves at infinity, instead of e^{-ikr}/r, that corresponds to incoming waves.

the unknown wavefunction is expressed in terms of the values of the function $\psi(\mathbf{r}')$ at the nodes using a linear interpolation scheme. Let the local interpolation parameter range over $0 \leq \xi \leq 1$. The function in the ith element is then expressible in terms of the nodal values ψ_i and ψ_{i+1} as

$$\psi(\mathbf{r}') = (1 - \xi')\,\psi_i + \xi'\,\psi_{i+1}. \tag{18.8}$$

A similar interpolation is used for the normal derivative of the wavefunction along the boundary.

The limiting forms for the Hankel function and its derivative

$$\lim_{z \to 0} H_0^{(1)}(z) \sim \frac{2i}{\pi} \ln z,$$

$$\lim_{z \to 0} \frac{d}{dz} H_0^{(1)}(z) \sim \frac{2i}{\pi z}, \tag{18.9}$$

show that the singular part of this Green's function is equivalent to that of the Green's function for the Poisson problem in two dimensions, discussed in Chapter 16. Using Green's function and the derivative relation $dH_0^{(1)}(z)/dz = -H_1^{(1)}(z)$ in equation (18.7), we have

$$\psi(\mathbf{r}) = \frac{i}{4} \sum_{iel}^{nelem} \ell_{iel} \int_0^1 d\xi' \left[H_0^{(1)}(k|\mathbf{r} - \mathbf{r}'|) \left\{ (1 - \xi')\,\psi_{iel}' + \xi'\,\psi_{iel+1}' \right\} \right.$$

$$\left. + k\,\cos\theta'\,H_1^{(1)}(k|\mathbf{r} - \mathbf{r}'|) \left\{ (1 - \xi')\,\psi_{iel} + \xi'\,\psi_{iel+1} \right\} \right]. \tag{18.10}$$

Here θ' is the angle between $(\mathbf{r}' - \mathbf{r})$ and the unit vector $\mathbf{n}'$ that is the normal to the boundary at $\mathbf{r}'$. The length of the element has been denoted by ℓ_{iel}, where iel is a running index for the sum over elements. Note that $\psi_{nelem+1}$ is the same as ψ_1, since the path of the integration is closed.

The vector $\mathbf{r}$ to the destination point is now allowed to coincide with each of the nodal points, in turn, to generate $nelem$ equations.

As $|\mathbf{r}_i - \mathbf{r}'| \to 0$, we encounter a singularity at the end point $\mathbf{r}_i$. However, this poses no difficulty in the integrals when the singularity is ameliorated by shifting the path of integration outward in an infinitesimal circular arc around $\mathbf{r}_i$. We are thereby making the destination point an interior point, as discussed in Chapter 16. Let us denote the contributions of the integrals over such distortions in the path about $\mathbf{r}_i$ by C_i.[¶]

The integrals in equation (18.10) over ξ' have a finite range and the integrands include Hankel functions of order 0 or 1. Accurate numerical representations of the Bessel functions J_0, J_1 and N_0, N_1 are available as part of standard mathematical subroutine packages. Hence these integrals are very easily evaluated by making use of adaptive Gauss quadrature. The integrals with Bessel functions of the second kind develop an end point integrable singularity when the destination point lies on the boundary element over which the integration is being performed. Adaptive quadrature is capable of evaluating such integrals as well if the range is terminated a distance $\Delta \sim 10^{-6}$–10^{-9} before the end-point singularity is reached. The accuracy of this approximation essentially depends on the value of Δ. In the standard multiplicative Gauss quadrature, the end point is not one of the Gauss points so that the singularity is avoided anyway.

Writing the integrals in the above equation more compactly, we have

$$(1 - C_i)\,\psi_i = \sum_{iel}^{nelem} \big[\,\mathcal{G}_{i,iel}\,\psi'_{iel} + \mathcal{G}_{i,iel+1}\,\psi'_{iel+1}$$
$$+\ \mathcal{H}_{i,iel}\,\psi_{iel} + \mathcal{H}_{i,iel+1}\,\psi_{iel+1}\,\big].$$

We regroup terms having common coefficients ψ_j and ψ'_j. With

$$\mathcal{H}_{ij} = \mathcal{H}_{i,iel} + \mathcal{H}_{i,iel-1} - (1 - C_i)\delta_{ij},$$
$$\mathcal{G}_{ij} = \mathcal{G}_{i,iel} + \mathcal{G}_{i,iel-1}, \tag{18.11}$$

we have

$$\sum_{j=1}^{nelem} [\,\mathcal{H}_{ij}\,\psi_j + \mathcal{G}_{ij}\,\psi'_j\,] = 0. \tag{18.12}$$

This relation, apart from notational changes, is the same as that obtained by Kagami and Fukai[5] in their treatment of the propagation of electromagnetic waves in a 2D waveguide in the transverse electric (TE) configuration. The wavefunction and the Green's function being complex quantities, we must provide double-precision complex assignments for the matrices and vector arrays appearing in the computer programming.

[¶]Recall from Chapter 16 that $C_i = \Delta\theta_i/2\pi$ where $\Delta\theta_i$ is the angle subtended at the center by the circular arc of the distorted contour.

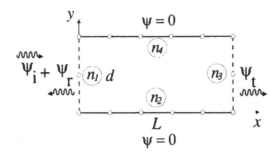

Fig. 18.1 Electron propagation in a 2D waveguide. Open circles denote nodes for boundary elements, and the labels n_i specify the number of elements on the four sides.

18.2.1 Implementing boundary conditions

For concreteness, let us suppose that there are n_1 and n_3 elements along the input and output ports, respectively. Let there be n_2 and n_4 elements on the side walls, as in Fig. 18.1. The number of nodes along each side is, of course, one greater than the number of elements.

At each of the corners we must specify the wavefunction, together with the values of the normal derivatives corresponding to the two elements that meet at a right angle to one another. The wavefunction vanishes at the corners (wall node). The normal derivative at the corner node for the element in a port also vanishes, again since the wavefunctions are zero along the side walls. Thus, of the three parameters at any corner, only the derivative of the wavefunction normal to the side wall is a free parameter, and we are left with four unknown parameters at the four corners.

As we have noted, we have n_2 elements along side 2, and n_4 elements along side 4. This corresponds to $n_2 + 1$ and $n_4 + 1$ nodes along these side walls. Excluding the corner nodes, we have $(n_2 - 1)$ and $(n_4 - 1)$ nodes along the two walls. At these nodes, the wavefunction is set to zero and the normal derivatives have to be determined (Dirichlet BC).

At the output port, for the n_3 elements we have $n_3 + 1$ nodes of which two are corner nodes. We thus have $(n_3 - 1)$ nodes where both the output wavefunction and its normal derivative are unknown. We wish to *enforce* the fact that the wavefunction at this port is an outgoing wave. This can be done, as in our treatment of the same problem using the FEM, by explicitly performing a *modal* (or spectral) analysis of the wavefunction. The outgoing wave is a linear

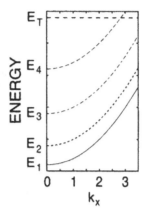

Fig. 18.2 Energy dispersion of electron modes in a 2D waveguide.

superposition of states with the same energy belonging to the different modes (associated with the transverse quantization) allowed for that energy. Each of these modal states can be explicitly chosen to be propagating in the outward direction. We replace the nodal variables ψ and ψ' by the transmission amplitudes $t_{(\mu)}$ in our analysis, and set up the global matrix in terms of these variables. To anticipate the following discussion, we then solve the global matrix equations for the amplitudes $t_{(\mu)}$. Once these transmission amplitudes are known, the full wavefunction can be reconstructed at each node as a superposition of the modal wavefunctions.

The wavefunction and its longitudinal, or normal, derivative at the output port can be written in the form

$$\psi_t(x = L, y) = \sum_{\mu}^{N_{(t)}} t_{(\mu)} \sqrt{\frac{2}{d}} \sin\left(\frac{\mu\pi y}{d}\right) e^{ik_x^{(\mu)} L},$$

(18.13)

$$\frac{\partial}{\partial x}\psi_t(x = L, y) = \sum_{\mu}^{N_{(t)}} t_{(\mu)} \, ik_x^{(\mu)} \sqrt{\frac{2}{d}} \sin\left(\frac{\mu\pi y}{d}\right) e^{ik_x^{(\mu)} L}.$$

The subscript t on ψ refers to it being the transmitted wave. Here $N_{(t)}$ corresponds to the largest allowed value of the mode μ consistent with the given total energy, E_T. For example, in Fig. 18.2, the number of allowed modes $N_{(t)}$ with energy $E_\perp \leq E_T$ is four. The amplitudes $t_{(\mu)}$ are the complex transmission amplitudes in the different modes. We now map the $2n_3 - 2$ nodal variables ψ_i, ψ_i' along the output port into the above unknown parameters $t_{(\mu)}$, thereby ensuring that the wavefunction at the output port corresponds to an outgoing wave only. This suggests that the number of unknowns $2n_3 - 2$ should be at least as large as $N_{(t)}$. If more modes are needed in the analysis, we increase the number of elements along the ports. As has been demonstrated with the FEM formulation, a convenient way of mapping from one set to the other is to define an appropriate transformation matrix $\mathbf{M}_{(t)}$. Let us express the amplitude for the wavefunction and its derivative at the first output node above the corner node at $(x = L, y = 0)$ in terms of the $N_{(t)}$ modes. We begin by defining the matrix elements

$$M_{1,i} = \sqrt{2/d}\,\sin(\mu_i \pi y_1/d)\,e^{ik_x^{(\mu_i)}L}, \qquad i = 1, \ldots, N_{(t)};$$

(18.14)

$$M_{n_3,i} = ik_x^{(\mu_i)}\sqrt{2/d}\,\sin(\mu_i \pi y_1/d)\,e^{ik_x^{(\mu_i)}L}, \quad i = 1, \ldots, N_{(t)}.$$

This definition is repeated for each pair $M_{p,i}$ and $M_{n_3+p-1,i}$, where p ranges over $1, \ldots, n_3 - 1$, at the nodes located along the output port. We then have

$$
\begin{pmatrix}
\psi_1 \\
\psi_2 \\
\psi_3 \\
\vdots \\
\psi'_{n_3-2} \\
\psi'_{n_3-1}
\end{pmatrix}
=
\begin{pmatrix}
M_{11} & M_{12} & M_{13} & \ldots M_{1,N_{(t)}} \\
M_{21} & M_{22} & M_{23} & \ldots M_{2,N_{(t)}} \\
\vdots & \vdots & \vdots & \vdots \\
M_{2n_3-2,1} & M_{2n_3-2,2} & M_{2n_3-2,3} & \ldots M_{2n_3-2,N_{(t)}}
\end{pmatrix}
\begin{pmatrix}
t_1 \\
t_2 \\
t_3 \\
\vdots \\
t_{N_{(t)}}
\end{pmatrix},
$$

which can be put in the more compact form

$$\begin{pmatrix} \mathbf{\Psi}_{(t)} \\ \mathbf{\Psi}'_{(t)} \end{pmatrix} = \mathbf{M}_{(t)} \cdot \mathcal{T}_{N_{(t)}}.$$

(18.15)

The $(2n_3 - 2) \times N_{(t)}$ transformation matrix $\mathbf{M}_{(t)}$ can be used in the construction of the global matrix to replace ψ_i, ψ'_i by $t_{(\mu)}$, as shown below. It is again used at the post-processing stage where the solutions for $t_{(\mu)}$ need to be transformed into the nodal values of the wavefunction and its normal derivative.

At the input port, we know that portion of the wavefunction which corresponds to the incoming wavefunction. This incoming wave is assumed to be in a specific mode, corresponding to our having prepared an initial state. At each of the $(2n_1 - 2)$ input nodes we again express the nodal variables ψ_i, ψ_i' in terms of $\psi_r(x = 0, y)$ and $\psi_{inc}(x = 0, y)$ and their normal derivatives. The terms multiplied by the incident amplitude and its normal derivative are transferred to the right side of equation (18.12).

The reflected wave is again cast in terms of the allowed modes at the input port, with reflection amplitudes $r_{(i)}$ for each mode. The normal derivatives are evaluated with the normal directed along the $-\hat{x}$-direction for the input port. Suppose the number of modes in the reflected wave is $N_{(r)}$. We define the matrix elements for the transformation matrix $\mathbf{M}_{(r)}$ for the reflected waves, at the first node above the corner node at $(x = 0, y = 0)$ in the input port, to be

$$M_{1,i} = \sqrt{2/d}\, \sin(\nu_i \pi y_1/d)\, e^{ik_x^{(\nu_i)} x}|_{x=0}; \qquad i = 1, \dots, N_{(r)};$$

$$M_{n_1,i} = -ik_x^{(\nu_i)}\sqrt{2/d}\, \sin(\nu_i \pi y_1/d)\, e^{ik_x^{(\nu_i)} x}|_{x=0}; \quad i = 1, \dots, N_{(r)}.$$

The negative sign in the second equation arises from the negative direction of the normal. This definition is repeated at each node for the pair $M_{p,i}$ and $M_{n_1+p-1,i}$, where p now ranges over $1, \dots, n_1 - 1$, the nodes along the input port. We then have

$$\begin{pmatrix} \psi_1 \\ \psi_2 \\ \psi_3 \\ \vdots \\ \psi_{n_1-2}' \\ \psi_{n_1-1}' \end{pmatrix} = \begin{pmatrix} M_{11} & M_{12} & M_{13} & \dots M_{1,N_{(r)}} \\ M_{21} & M_{22} & M_{23} & \dots M_{2,N_{(r)}} \\ \vdots & \vdots & \vdots & \vdots \\ M_{2n_3-1,1} & M_{2n_3-1,2} & M_{2n_3-1,3} & \dots M_{2n_3-1,N_{(r)}} \end{pmatrix} \begin{pmatrix} r_1 \\ r_2 \\ r_3 \\ \vdots \\ r_{N_{(r)}} \end{pmatrix}.$$

This relation may be written as

$$\begin{pmatrix} \boldsymbol{\Psi}_{(r)} \\ \boldsymbol{\Psi}_{(r)}' \end{pmatrix} = \mathbf{M}_{(r)} \cdot \mathcal{R}_{N_{(r)}}. \qquad (18.16)$$

In brief, the number of equations generated by the boundary element discretization is $(n_1 + n_2 + n_3 + n_4)$. Referring to Fig. 18.1, the total number of parameters that have to be determined corresponds to the following. We have $(n_2 - 1) + (n_4 - 1)$ unknown normal derivatives of the wavefunction at nodes placed along the side walls. There are four normal derivatives to be determined at the four corner nodes. The number of unknowns along the input and output ports is

$(2n_1 - 2) + (2n_3 - 2)$, corresponding to the reflected and transmitted wavefunctions and their normal derivatives. If we pick the maximum number of transverse modes $N_{(r)} = 2n_1 - 2$ and $N_{(t)} = 2n_3 - 2$ in the reflected and transmitted wavefunctions, we have at hand an additional $(n_1 + n_3 - 2)$ unknown parameters. Thus, over and above the number of equations generated by the BEM, we must construct $(n_1 + n_3 - 2)$ additional equations. As discussed in Chapter 16, we choose a corresponding number of additional destination points in order to generate such extra equations. These points can be conveniently located along the two side walls where the wavefunction is known, and is zero. Naturally, if fewer modes are expected to participate in the scattering process we can correspondingly decrease the number of extra points. This calculation can be automated in computer codes fairly easily.

We rewrite equation (18.12) by identifying the terms arising from the side walls (s), the ports $(r$ or $t)$, the corners (c), and the incident wave (inc). The additional equations from the extra destination points involve the same nodal parameters and these additional equations are implicitly included in the following notation. We have

$$\mathcal{G}^{(s)} \cdot \Psi'_{(s)} + \mathcal{H}^{(r)} \cdot \Psi_{(r)} + \mathcal{G}^{(r)} \cdot \Psi'_{(r)} + \mathcal{H}^{(t)} \cdot \Psi_{(t)} + \mathcal{G}^{(t)} \cdot \Psi'_{(t)}$$
$$+ \mathcal{G}^{(c)} \cdot \Psi'_{(c)} = -\mathcal{H}^{(r)} \cdot \Psi_{(inc)} - \mathcal{G}^{(r)} \cdot \Psi'_{(inc)}.$$

Here we have accounted for the fact that $\Psi_{(inc)}, \Psi'_{(inc)}$ at the input port are known by transferring these constant terms to the right side in equation (18.17). We also have $\Psi_{(s)} = 0$ along the side-walls, and $\Psi_{(c)} = 0$ at the corners. Gathering together the reflection terms and also the transmission terms, we write

$$\mathcal{G}^{(s)} \Psi'_{(s)} + \mathcal{G}^{(c)} \Psi'_{(c)} + \left(\mathcal{H}^{(r)}, \mathcal{G}^{(r)} \right) \cdot \begin{pmatrix} \Psi_{(r)} \\ \Psi'_{(r)} \end{pmatrix}$$
$$+ \left(\mathcal{H}^{(t)}, \mathcal{G}^{(t)} \right) \cdot \begin{pmatrix} \Psi_{(t)} \\ \Psi'_{(t)} \end{pmatrix} = -\mathcal{H}^{(inc)} \Psi_{(inc)} - \mathcal{G}^{(inc)} \Psi'_{(inc)}.$$

Replacing the reflected and transmitted wavefunctions in terms of the modal amplitudes, we have

$$\mathcal{G}^{(s)} \Psi'_{(s)} + \mathcal{G}^{(c)} \Psi'_{(c)} + \left(\mathcal{H}^{(r)}, \mathcal{G}^{(r)} \right) \cdot \mathbf{M}_{(r)} \cdot \mathcal{R} + \left(\mathcal{H}^{(t)}, \mathcal{G}^{(t)} \right) \cdot \mathbf{M}_{(t)} \cdot \mathcal{T}$$
$$= -\mathcal{H}^{(inc)} \Psi_{(inc)} - \mathcal{G}^{(inc)} \Psi'_{(inc)}. \tag{18.17}$$

This set of equations is solved for all the unknown parameters. The wavefunction amplitudes and their normal derivatives at the nodes along the ports can be derived by using equations (18.15) and (18.16) once $\mathcal{T}, \mathcal{R}$ are determined.

The number of nodes along the side walls must be large enough that the wavefunction $\sim \exp(ik_x x)$ can be adequately represented with all its oscillations in the lateral direction using piecewise polynomial representation in the elements. The number of nodes in the input and output ports must be large enough to provide a representation of the transverse modes. Again, highly excited modes in the waveguide will require a correspondingly large number of boundary elements along the edge of the waveguide.

Once the amplitudes $r_{(\nu)}$, $t_{(\mu)}$ are known, we can derive the reflection and transmission coefficients

$$R = \sum_{\nu} \frac{I_\nu^{(r)}}{I^{(inc)}} = \sum_{\nu} \frac{k_x^\nu}{k_x^{\nu_0}} |r_{(\nu)}|^2,$$

$$(18.18)$$

$$T = \sum_{\mu} \frac{I_\mu^{(t)}}{I^{(inc)}} = \sum_{\mu} \frac{k_x^\mu}{k_x^{\mu_0}} |t_{(\mu)}|^2,$$

respectively. Here, the reflected and transmitted currents in each modal channel are denoted by $I_\nu^{(r)}$, $I_\mu^{(t)}$, and the incident current is

$$I^{(inc)} = \frac{\hbar}{2im} \left(\int_0^d dy \, \{\psi_{inc}^*(\mathbf{r})\nabla\psi_{inc}(\mathbf{r}) - \nabla\psi_{inc}^*(\mathbf{r})\psi_{inc}(\mathbf{r})\} \right).$$

If we allow more modes in the analysis than are appropriate for a given energy, the additional higher energy modes will correspond to damped modes. For example, given an input wave of energy E_T of Fig. 18.2, if we allow more than four modes in the calculation, the higher modes will have imaginary propagation constants $k_x = i\kappa = \sqrt{E_T - E_\perp}$. These evanescent modes do not contribute to the asymptotic current far away from the "interaction zone" where any scattering actually takes place. However, they may be necessary to satisfy the BCs. For example, consider a perfect semi-infinite 2D waveguide that is terminated abruptly at some coordinate. The mode(s) propagating in the waveguide now couple to all the modes associated with the exterior region. The modes in the exterior region that have wavenumbers above the energy $E_\perp$ and are evanescent modes will also be excited at the end of the waveguide. However, they will not contribute to the current at large distances from the output port.

All of the above considerations can be automated in a computer program, with input options for the number of elements along each side, and the number of modes to be included at each port. As the computer code is developed, it is also useful to keep in mind further applications. For example, we can distort the side walls to represent junctions, reservoirs, pinches in the waveguide, and so on. Kagami and Fukai[5] display results for the reflection coefficient for the lowest mode in cases where the waveguide is terminated at the output port in various configurations such as flanged, unflanged, and flared sides. With open-ended waveguides the boundary at the output port is distorted outward to infinity, so that in effect the contribution to the boundary integral from the output port vanishes. Koshiba and Suzuki[6] perform similar calculations for waveguide junctions. Their results for the TE electromagnetic modes apply immediately to electron propagation in waveguides with the same geometry. The reader is referred to their articles for further details.

Waveguide analysis for electromagnetic fields has been performed in a variety of theoretical[7-9] and computational ways.[10-12] The classic handbook by Marcuvitz[13] may be consulted for a survey of theoretical and computational approaches of an earlier period of research in this area. All of these techniques can be carried over to the analysis of 2D and 3D electron waveguides.

18.2.2 Multiregion waveguide problems

We find it convenient to break up the physical region into subregions and resort to a multiregion BEM under three conditions. The first case arises when the waveguide has $d \ll L$, leading to a small aspect ratio. Now the boundary integrals over elements at the output port at $x = L$, for example, will be nearly the same for two destination

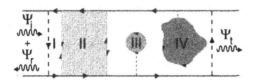

Fig. 18.3 Electron propagation in a 2D waveguide with multiple regions. The wavevector in each region is constant over the region.

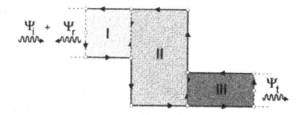

Fig. 18.4 Electron propagation in a 2D double-bend waveguide. Open circles denote nodes for boundary elements. The path of integration for each region is directed counterclockwise. The matrices for the boundary regions are overlaid to obtain a global matrix.

points along the input port at $x = 0$. This makes the global matrix ill-conditioned. The problem can be overcome by breaking up the full waveguide into smaller regions so that the aspect ratio of the subregions is closer to unity. Secondly, the waveguide may contain a number of different materials, as in Fig. 18.3, each requiring a boundary integral. The third situation corresponds to more complex geometry for the waveguide itself, with bends and variations in thickness, as in Fig. 18.4.

In each region, labeled by α, the wavevector k_α can be obtained in terms of the total energy and the local potential V_α. Using $k_\alpha^2 = 2m(E - V_\alpha)/\hbar^2$, and the corresponding Green's function $H_0^{(1)}(k_\alpha r)$, we calculate the boundary integral. If $E < V_\alpha$, the Green's function is $2K_0(\kappa_\alpha r)$,[||] where $\kappa_\alpha^2 = 2m(V_\alpha - E)/\hbar^2$ and K_0 is the modified Bessel function of order 0. The final global matrix is obtained by overlaying the individual matrices for the regions in a manner consistent with the BCs at inter-region boundaries. The set of simultaneous equations so obtained is solved by standard methods for the unknown nodal parameters.

18.2.3 Multiple ports and transmission

The Landauer formula, equation (18.1), has been extended to junctions having multiple ports.[2] An example of a multiport junction is shown in Fig. 18.5. The 2D waveguides labeled 1–5 represent the leads, and the mesoscopic device to which the leads are attached is the waveguide junction. In Fig. 18.5, as an example, we show

[||]Since $i\pi H_0^{(1)}(i\kappa r) = 2K_0(\kappa r)$ and $dK_0(x)/dx = -K_1(x)$, the details and the notational aspects of the calculation remain nearly unchanged by this change of the Green's function.

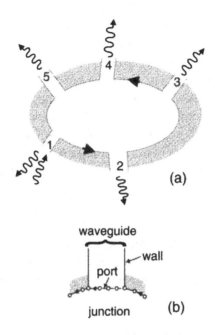

Fig. 18.5 (a) A mesoscopic device having a junction with five ports for electron waveguides. Wavy arrows indicate that electrons are incident at port 1 and are scattered into all ports. The short arrows indicate the direction of line integration in the BEM. (b) Detail of a port between a waveguide and the junction. The arrow-heads indicate the direction of integration. Open circles denote the nodes of boundary elements.

electrons incident through the waveguide labeled 1, and being transmitted through all the waveguides.

It is straightforward to extend the boundary element approach to the calculation of the transmission coefficients $t_{\{i,\mu\};\{j,\nu\}}$, where the indices j, i refer to the output and input ports, and μ, ν refer to the mode indices associated with the transverse components of wavefunctions in the corresponding ports. The path for the boundary integral is along the periphery of the junction region and across each port. Ports refer to the openings between the junction region and the waveguides. At each port, a local coordinate system is defined in order to specify the form of the incoming and outgoing waves and to describe the effect of the quantization in the transverse direction on the wavefunctions at each port. For simplicity, let us assume that the side walls are perpendicular to the boundary of the junction and that the walls at the ports are parallel to each other. It then suffices to employ boundary elements directly at the ports.

The boundary integral over the entire periphery of the junction, including the ports, is discretized into boundary elements. Each element has two nodes at its extremities, and variables are assigned at these nodes for the wavefunction and its normal derivative. These nodal variables are employed in linear or Hermite interpolation in order to represent the wavefunction over the element. Recall that we do not have BCs directly for the wavefunction or its derivative at any of the ports; however, we do have to specify whether the wave is propagating into the junction or is being scattered outward. As in Section 18.2.1, we re-express the nodal variables in the elements at each of the ports in terms of the amplitudes of the corresponding modes associated with the transverse quantization. The wavefunction is assumed to be zero at the wall nodes of the junction. The unknown variables in the problem are the normal derivative of the wavefunction along the walls and the modal amplitudes at the nodes in the ports. The known (unit) amplitude of the mode at the input port generates the known vector on the right side of the set of simultaneous equations.

Recently, the boundary integral method was implemented for electron waveguides with multiple ports by using wavefunctions for the junction region, instead of the Green's function.[14] In contrast to the method presented here, the use of wavefunctions in the junction requires a prior knowledge of a complete set of wavefunctions for that area; the size of the coefficient matrix in the final set of equations depends on the number of modes included in the complete set of wavefunctions for the junction region. This use of wavefunctions for the entire junction is equivalent to using "global" wavefunctions in a manner paralleling the usage of global wavefunctions in the Rayleigh–Ritz type of variational procedure. In our approach we have the added advantage that the junction region could be made up of a number of subregions (see Section 18.2.2), in each of which the electron's potential energy is a constant. The matrix representing the coefficients in the final set of simultaneous equations in our derivation is approximately $2n \times 2n$, where n is the number of boundary elements at the periphery, including the port openings. The actual matrix size depends on the number of transverse modes we wish to consider at each *port*. A finite element approach to the same problem has been espoused by Lent and Kirkner.[15] Their method leads to simultaneous equations with a coefficient matrix that is large and sparse; the size of their matrix depends on the level of discretization of the entire physical region.

18.3   The BEM and 2D scattering

We have focused so far on boundary element calculations with finite domains: a physical structure has been enclosed typically in a rectangular box of finite size in order to define the external fields or the BCs at the periphery. While this pedagogical device is convenient, it is not necessary. In fact, the BEM provides a very natural approach to the calculation of scattering into unbounded regions. We follow the work of Kagami and Fukai[5] which represents an implementation of the Kirchhoff approximation to scattering and diffraction.[16] In investigating 2D scattering, we are concerned with a plane wave incident on a local potential of finite extent from which the scattered waves emanate outwards, as in Fig. 18.6. The physical region is split up into two regions, the interior region I with the local potential enclosed by a boundary, and the exterior region II defined by the same boundary together with a boundary at infinity. The boundary integral contains no contribution from the boundary at infinity where the wavefunction is zero, and hence our focus will be on the finite boundary around the local potential. The infinite domain Green's functions for each of the regions will be different depending on the potential energy function in that region; this requires a calculation of the infinite domain Green's function for each domain, This may not be straightforward for arbitrary potentials; however, for piecewise constant potentials we again have the Hankel function $H_0^{(1)}(k_\alpha r)$ as

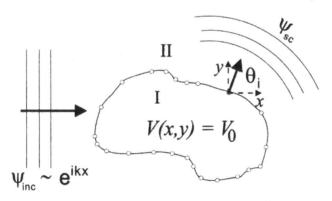

Fig. 18.6 Scattering from a 2D potential $V(x,y) = V_0$ over an arbitrarily shaped region.

the corresponding Green's function in each region, each function being characterized by a corresponding wavenumber k_α. If $k_\alpha = i\kappa_\alpha$ is pure imaginary, the Green's function changes over from $i\pi H_0^{(1)}(k_\alpha r)$ to $2K_0(\kappa_\alpha r)$.

Let us consider electrons propagating along the x-direction with a wavefunction of unit amplitude, and incident on a region of arbitrary shape with a potential $V(x, y) = V_0$ (see Fig. 18.6).

The wavefunction in region I is a superposition of the incident and the scattered wavefunctions, and the source points along the boundary are considered to give rise to both the waves. In region II, we account for only the scattered wavefunction generated by source points on the boundary. The boundary element calculation proceeds by the evaluation of the discretized boundary integrals for the two regions. The boundary integral for the interior region is evaluated in the counterclockwise direction, as usual. For the exterior region, we use the same boundary and evaluate the line integral in the clockwise direction. The Green's functions differ in the two integrals since k_I and k_{II} are given by

$$k_I^2 = 2m(E - V_0)/\hbar^2,$$
$$k_{II}^2 = 2mE/\hbar^2. \tag{18.19}$$

Following the procedures described in Chapters 16 and 17, we obtain the relations

$$\mathcal{H}_{ij}^{(I)} \psi_j^{(I)} + \mathcal{G}_{ij}^{(I)} \psi_j^{'(I)} = 0,$$
$$\mathcal{H}_{ij}^{(II)} \psi_j^{(II)} + \mathcal{G}_{ij}^{(II)} \psi_j^{'(II)} = 0. \tag{18.20}$$

Let us consider the nodal variables for the interior region. These can be written as

$$\psi_i^{(I)} = \psi_{i,inc}^{(I)} + \psi_{i,sc}^{(I)},$$
$$\psi_i^{'(I)} = \psi_{i,inc}^{'(I)} + \psi_{i,sc}^{'(I)}. \tag{18.21}$$

For the ith node, the normal derivative of the incident wavefunction $\exp(ikx_i)$ will be given by

$$\hat{n} \cdot \nabla e^{ikx_i} = ik\, e^{ikx_i} \cos\theta_i, \tag{18.22}$$

where θ_i is the angle at the ith node between the normal to the boundary and the x-direction, as shown in Fig. 18.6. We then have

$$\psi_i^{(I)} = e^{ikx_i} + \psi_{i,sc}^{(I)},$$
$$\psi_i^{'(I)} = ik\, e^{ikx_i} \cos\theta_i + \psi_{i,sc}^{'(I)}. \qquad (18.23)$$

Let us define $\hat{\Psi}_{i,inc} \equiv \exp(ikx_i)$ and $\hat{\Psi}'_{i,inc} \equiv ik \exp(ikx_i) \cos\theta_i$.

The nodal values for the exterior region are

$$\psi_i^{(II)} = \psi_{i,sc}^{(II)} \equiv \psi_{i,sc}^{(I)},$$
$$\psi_i^{'(II)} = \psi_{i,sc}^{'(II)} \equiv -\psi_{i,sc}^{'(I)}. \qquad (18.24)$$

Now equations (18.20) can be written in terms of the nodal values of the scattered wavefunctions in a block-matrix form as

$$\begin{pmatrix} \mathcal{H}^{(I)} & -\mathcal{G}^{(I)} \\ \mathcal{H}^{(II)} & \mathcal{G}^{(II)} \end{pmatrix} \begin{pmatrix} \psi_{sc}^{(II)} \\ \psi_{sc}^{'(II)} \end{pmatrix} = -\begin{pmatrix} \mathcal{H}^{(I)} \cdot \hat{\Psi}_{inc} + \mathcal{G}^{(I)} \cdot \hat{\Psi}'_{inc} \\ 0 \end{pmatrix}.$$

The amplitude for the scattered wave and its normal derivative are obtained by solving the above set of simultaneous equations. We have chosen to work with the nodal amplitudes of the scattered wave in the exterior region; it is, of course, possible to directly solve for the scattered wave nodal amplitudes for the interior region first.

We see that in the present scattering problem the nodes for both the exterior and the interior regions are the same. At each node the interior region has two parts to the wavefunction and its normal derivative, arising from the incident wave and the scattered wave. The known amplitude for the incident wave generates the "driving terms" for the simultaneous equations satisfied by the scattered wavefunction and its normal derivative at the boundary. As usual, once the nodal variables have been determined, the wavefunction can be reconstructed in the interior region as well as in the exterior region.

The traditional quantum mechanical calculation for the asymptotic scattering amplitude can be reproduced in the framework of the BEM by taking a second circular boundary of large radius R_0. On this boundary, the full wavefunction is given by the sum of the incident plane wave and the asymptotic form of the outgoing scattered wave modified by a scattering amplitude $f(\phi)$, where ϕ is the polar angle measured from the x-axis with the origin located at the center of the large circular boundary. We have

$$\psi(\mathbf{r}) = e^{ikx} + \frac{e^{ikr}}{\sqrt{r}}\, f(\phi). \qquad (18.25)$$

Thus, at every node on both the inner surface and the surface at R_0 the incoming part of the wavefunction is known, and the incident wave is not neglected on the finite boundary when the boundary integral is set up for the exterior region. The discretization of the two integral equations for the two regions leads to two matrix equations. These are coupled when we impose the continuity conditions at the interface of regions I and II. The terms associated with the known incident wave are transferred to the right side of the equation before solving for the unknown amplitudes. The details of the derivation of the final equations are similar to the above treatment of Kirchhoff's approximation. In this numerical approach, we obtain the local scattering amplitudes near the scattering center as well as the asymptotic amplitude $f(\phi)$ on the boundary at R_0. Again, a complete reconstruction of the scattered wave at all points in regions I and II can be performed using the integral equations for destination points located away from the boundaries and inside the corresponding regions.

Potentials varying with the coordinate

When the potential is not piecewise constant, an iterative approach will have to be used. We write

$$\left(\nabla^2 + \frac{2mE}{\hbar^2}\right)\psi(\mathbf{r}) = +\frac{2m}{\hbar^2}\,V(\mathbf{r})\,\psi(\mathbf{r}),$$

$$\left(\nabla^2 + \frac{2mE}{\hbar^2}\right)g(\mathbf{r},\mathbf{r}') = -4\pi\delta(\mathbf{r}-\mathbf{r}'), \qquad (18.26)$$

and multiply the first equation in (18.26) by g and the second by ψ, and subtract one from the other. On integration over the entire physical domain we obtain

$$\psi(\mathbf{r}) = \frac{1}{4\pi}\oint d\ell'\left(g(\mathbf{r},\mathbf{r}')\frac{\partial}{\partial n'}\psi(\mathbf{r}') - \psi(\mathbf{r}')\frac{\partial}{\partial n'}g(\mathbf{r},\mathbf{r}')\right)$$

$$-\frac{1}{4\pi}\left(\frac{2m}{\hbar^2}\right)\iint d^2r'\,g(\mathbf{r},\mathbf{r}')\,V(\mathbf{r}')\,\psi(\mathbf{r}'). \qquad (18.27)$$

Obviously, the double integral on the right side cannot be evaluated unless $\psi(\mathbf{r})$ is already known. We begin the iteration process by setting the lowest order iterate $\psi^{(0)}(\mathbf{r}) = 0$ in the double integral. Then the boundary integral is discretized and evaluated as before without any potential energy term. The new solution $\psi^{(1)}$ is used in the next iteration in the double integral. So the iterated relation is

$$\mathcal{H}_{ij}\psi_j^{(p)} + \mathcal{G}_{ij}\psi_j'^{(p)} = \frac{1}{4\pi}\left(\frac{2m}{\hbar^2}\right)\iint d^2r'\, g(\mathbf{r},\mathbf{r}')V(\mathbf{r}')\psi^{(p-1)}(\mathbf{r}'),$$

in which p is the order of iteration. The iterative process must be repeated till convergence is achieved. This approach is, in fact, a numerical implementation for the expansion for potential scattering.

18.4   Eigenvalue problems and the BEM

Consider an electron confined in a 2D potential energy well of finite extent, with the potential being a constant outside this region. If the energy of the electron is below the constant potential outside the well, its wavefunction will decay exponentially in the barrier region and it will have a discrete energy spectrum. In such a bound state problem, we are interested in obtaining the energy eigenvalues and wavefunctions of the physical system. We will limit our consideration to only piecewise constant potentials. Given a multiregion problem, we make use of the infinite domain Green's function for Schrödinger's equation with the appropriate constant potential in each subregion in the BEM.

18.4.1   Quantum wires

An interesting practical problem of this type is the calculation of the energy levels of an electron in the conduction band of a semiconductor quantum wire consisting of a GaAs wire of arbitrary cross-section embedded in AlGaAs. The straight wire is assumed to have a uniform cross-section so that we have at hand a 2D problem. The conduction bandedge in AlGaAs is higher in energy than the bandedge in GaAs, so that an electron in the conduction band of GaAs is confined in the GaAs region, with the AlGaAs region acting as a barrier of height V_0. The barrier height V_0 depends on the concentration of Al in AlGaAs.

We treat this as a problem with an infinite physical domain that consists of two subregions. The interior region is enclosed by a curve along the edge of a cross-section perpendicular to the axis of the quantum wire and has a potential $V = 0$. The exterior region is defined by the same contour together with a curve at infinity, and has a potential $V = V_0$. The wavenumbers for the electron in the two regions are defined by $k_I^2 = 2m_I^* E/\hbar^2$ and $\kappa_{II}^2 = 2m_{II}^*(V_0 - E)/\hbar^2$, where the m^* are the effective masses of the electron in the two regions. The effective mass of the electron in semiconductors varies considerably from the free-electron mass m_0. In GaAs $m^*/m_0 = 0.0665$ and in AlGaAs (with 30% Al) it is $m^*/m_0 \simeq 0.084$. The effective mass arises from a many-body interaction of the electron with all other electrons in the crystal. For 30% Al in AlGaAs the barrier height in the AlGaAs region relative to the GaAs region is $V_0 \simeq 0.276\,\text{eV}$. The Green's function for the interior region is $i\pi H_0^{(1)}(kr)$, and for the exterior region it is $2K_0(\kappa r)$.

For a wire oriented in the z-direction, the wavefunction of the electron is $\Psi(x, y, z) = \psi(x, y)\exp(ik_z z)$. The transverse part of the wavefunction, $\psi(x, y)$, cannot be separated into factors dependent on one coordinate only since the potential is nonseparable in the xy plane. Schrödinger's equation obeyed by $\psi(x, y)$ must now be solved using the BEM. (A finite element formulation of the same problem has been considered in Chapter 12.) The calculation follows essentially the same steps as described earlier. We break up the boundary integral into integrations over a number of boundary elements. The boundary element calculations for the interior region lead to the set of equations

$$\mathcal{H}_{ij}^{(I)}\psi_j^{(I)} + \mathcal{G}_{ij}^{(I)}\psi_j'^{(I)} = 0. \tag{18.28}$$

For the exterior region, the integral along the boundary at infinity vanishes because both ψ and ψ' vanish there. The finite boundary about the GaAs region is traversed in the clockwise direction to evaluate the boundary elements, and we have

$$\mathcal{H}_{ij}^{(II)}\psi_j^{(II)} + \mathcal{G}_{ij}^{(II)}\psi_j'^{(II)} = 0. \tag{18.29}$$

In each region the appropriate Green's function is used in the calculation. The interface BCs are (i) the continuity of the wavefunction, and (ii) the continuity of the probability current across the boundary. If the effective masses of the electron in the two media were the same, the latter condition would correspond to the continuity of the normal derivative of the wavefunction. In the present case the difference in the masses in the two regions leads to the so-called "mass derivative" continuity condition in which $(1/m^*)\psi'$ is continuous across the interface. These conditions, together with the fact that ψ' represents the derivative along the outward-drawn normal, provide us with the following relations for the nodal parameters:

$$\psi_j^{(I)} = \psi_j^{(II)},$$
$$\frac{1}{m_I^*}\psi_j'^{(I)} = -\frac{1}{m_{II}^*}\psi_j'^{(II)}. \tag{18.30}$$

The equations for the two regions can be put together in the block-matrix form

$$\begin{pmatrix} \mathcal{H}^{(I)} & \mathcal{G}^{(I)} \\ \mathcal{H}^{(II)} & -\frac{m_{II}^*}{m_I^*}\mathcal{G}^{(II)} \end{pmatrix} \cdot \begin{pmatrix} \psi^{(I)} \\ \psi'^{(I)} \end{pmatrix} = 0. \tag{18.31}$$

Note that for *nelem* elements along the finite boundary, we have at hand $2 \times nelem$ unknown nodal variables ψ_j, ψ_j'. This leads to a total of $2 \times nelem$ equations, as we take each nodal point to be the destination point, in turn, for both the interior *and* the exterior regions. The two sets of *nelem* equations are not the same since the Green's functions in the two regions are different.

The matrix elements in equation (18.31) are implicit functions of the unknown energy of the bound state. This set of equations has a nontrivial solution if and only if the determinant $\Delta(E)$ of the coefficient matrix is zero. We therefore vary the energy over the range $0 \le E \le V_0$ and numerically locate the roots of $\Delta(E) = 0$. Each of the roots is an allowed energy eigenvalue of a bound state of the system. Multiple roots imply the presence of degenerate bound states.

The wavefunctions are obtained by setting the energy to one of the eigenvalues as determined by root finding, and setting one of the nodal values, for example ψ_1, to unity. Equation (18.31) is now used to determine the other nodal variables in terms of this constant. Once all the nodal variables are determined, we evaluate the boundary integrals using these values as input, separately for the interior and the exterior regions, in order to obtain the wavefunction everywhere. This is done by allowing the destination point in the boundary integral to range over the interior and exterior domains, respectively. This yields all stationary states whose wavefunctions do not contain features finer than those capable of being generated by the boundary elements. A large number of elements are necessary to properly represent highly excited states. Once the full wavefunction is constructed, it can be normalized to unity over the entire physical

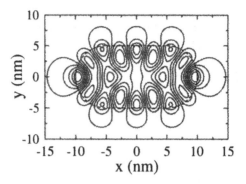

Fig. 18.7 A contour plot for the probability density of an electron in a highly excited state in a quantum wire of oval cross-section. The dashed curve represents the periphery of the wire region. (Reproduced from Knipp and Reinecke.[17])

region.

Examples of such calculations, using the BEM, have been presented for the electronic states in quantum semiconductor heterostructures by Knipp and Reinecke.[17] They considered a quantum wire of GaAs, of oval cross-section, embedded in GaAlAs which forms the barrier region. Figure 18.7 shows the contour plot of the probability density of a highly excited bound state. It is interesting that the probability distribution is largest along the edge of the oval region for the state depicted. Accidental degeneracy for electronic levels,

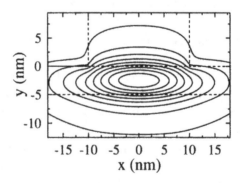

Fig. 18.8 A contour plot for the probability density of an electron in the ground state of a barrier modulated wire whose shape is indicated by the dashed line. GaAs comprises the quantum well, AlGaAs forms the substrate and the cap regions, and AlAs the rest of the region. (Reproduced from Knipp and Reinecke.[17])

level crossings versus anticrossing, and other similar issues can also be addressed, as in Chapter 12. In Fig. 18.8, a contour plot for the probability density for the electronic ground state in a complex quantum well heterostructure is shown. In the heterostructure, an AlGaAs substrate provides the barrier on the lower side. On the opposite side, a central cap layer of AlGaAs, abutted by AlAs regions, provides the barrier. Knipp and Reinecke[17] also obtain solutions for the confined acoustic phonons in heterostructures. These phonons obey the wave equations for acoustic waves that are amenable to solution by the BEM.

As an exercise for further study, it would be interesting to extend the work of Kojima et al.,[18] who have considered bound states of electrons in quantum wires with an L-shaped cross-section using the FEM. While they employed an infinite potential well, the BEM can be applied to the same problem for the investigation of barrier penetration effects for electrons in such structures.

We note that earlier work on the determination of eigenvalues by the BEM is represented by articles on elastodynamics[19] and wave propagation.[20,21]

18.4.2 Hearing the shape of a drum

Given the eigenvalue spectrum of a 2D region, is it possible to determine the shape of the region? This question was asked by Kac[22] in a more dramatic manner: "Can one hear the shape of a drum?" In 1992, Gordon et al.[23] were able to answer the question in the negative, and gave examples of two regions of equal area differing in shape and having exactly the same spectrum. In the spectral analysis of (2D) cavity resonators it is well known that the topology, and especially the shape of the boundary, plays an important role in determining the features of the spectrum. The area and the perimeter of the region are known to determine the smoothness of the spectral density as a function of the energy. However, the isospectrality displayed by two areas suggests that one can have an additional symmetry based on a topological transformation relating one isospectral area to another. The two polygonal regions considered by Gordon et al. are shown in Fig 18.9.

They use the concept of transplantation[24] to transform one region into the other to connect eigenfunction pairs for each eigenvalue. This transplantation, for the areas shown in Fig. 18.9, involves rotation and translation of sections of the triangular areas from one geom-

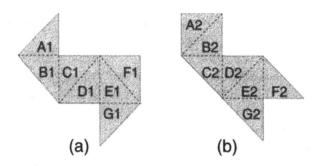

Fig. 18.9 Two isospectral microwave cavities used in the experimental work of Sridhar and Kudrolli.[26] Regions A1–G1 get mapped onto A2–G2 to obtain the wavefunctions of the isospectral partners.

etry to the other. An additional symmetry operation is the inversion about the center line through a 90° vertex. Given one eigenfunction for region 1 it is possible to construct its partner in every triangular subregion in region 2, by overlaying combinations of wavefunctions from various triangular segments of region 1. The combinations are required to satisfy BCs and match values and derivative values at interfaces between the pieces. Hence the new solution satisfies the differential equation in region 2 and has the same eigenvalue. The procedure has to be invertible, thereby completing the proof. Further details of this very intriguing "equivalence" of the two areas are given by Driscoll.[25]

In remarkable experiments using microwave cavities shaped as in Fig. 18.9, Sridhar and Kudrolli[26] were able to demonstrate the exact equivalence of the spectra for the two microwave cavities for up to the first 54 eigenvalues to a few parts in 10^4. The electromagnetic fields were in the transverse magnetic (TM) mode inside cavities made of copper, with the electric field being zero at the boundary (a Dirichlet problem). The resonant frequency was determined from a maximum in the transmission spectra, and the wavefunction was mapped out by using a cavity perturbation technique.[27] A numerical verification of this isospectrality using the BEM has been performed by Knipp.[28] The modulus squared of the wavefunctions for the first, third, and the sixth states in two regions are shown in Fig. 18.10. A large number of isospectral pairs of regions have been found, demonstrating the fact that the discovery of transplantation symmetry allows one to progress beyond the example presented here.

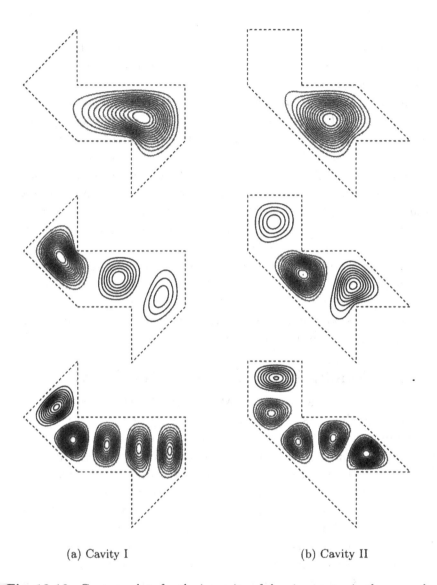

(a) Cavity I   (b) Cavity II

Fig. 18.10 Contour plots for the intensity of the eigenstates in the ground states, the third excited states, and the sixth (!) excited states in two isospectral regions labeled Cavity I and II. These results, based on BEM calculations performed by Knipp,[28] agree with the experimental measurements of Sridhar and Kudrolli.[26]

Isospectrality also occurs in organic molecules with different structures or "topology."[29] In conjugated molecules the carbon bonds correspond to a connectivity graph. The π-orbital electrons in such molecules have wavefunctions and energies determined by a tight-binding model for the overlap integrals between orbitals belonging to nearest neighbors.

In such molecules the energy levels of the electrons are determined by a tight-binding Hamiltonian that is essentially a connectivity matrix for the "connectedness" of the carbon atoms. The problem of determining the energy spectrum of electrons in conjugated molecules is of interest to organic chemists as well as to graph theorists. Isospectrality in molecules with completely different structures is another example, besides the isospectrality of Helmholtz's equation mentioned above, and of the fascinating role played by geometry and topology in quantum mechanics.

We note here that the boundary integral method is an effective means of calculating energy eigenvalues of Helmholtz's equation in 2D resonant cavities. Essentially identical issues arise in the solution of Helmholtz's equation in the context of sound waves, electromagnetic waves, or quantum mechanics.

18.5 Concluding remarks on the BEM

The great advantage of the BEM over the FEM is that only the boundary of the physical domain has to be discretized. In problems in two dimensions, this has a clear advantage in its simplicity of programming, in the size of matrices generated in the method, and also in the post-processing stage where the final output can be generated and manipulated more easily. Once the solution on the boundary is generated, the interior solution can be constructed fairly quickly by taking a grid of points as the destination points in the BEM. The BEM can be used in the case of exterior problems with unbounded domains that have a finite interior boundary. Only the finite boundary is needed in solving the problem. This is clearly an advantage over the FEM.

It is interesting to note that we can take derivatives of the boundary integral equation to directly evaluate the derivatives of the unknown function. For example, we have

$$\nabla_{\mathbf{r}}\, \phi(\mathbf{r}) = \frac{1}{4\pi} \oint d\ell' \left[\frac{\partial}{\partial \mathbf{n}'} \phi(\mathbf{r}') \, \nabla_{\mathbf{r}} \left(g(\mathbf{r}, \mathbf{r}') \right) \right.$$

$$- \nabla_{\mathbf{r}} \left(\frac{\partial}{\partial \mathbf{n}'} g(\mathbf{r}, \mathbf{r}') \right) \phi(\mathbf{r}') \Bigg]. \tag{18.32}$$

While this is a valid equation, the derivatives of the Green's function become more and more singular with each derivative, making the numerical evaluation less reliable for points close to the boundary. However, at interior points away from the periphery, we can discretize this equation and calculate the gradient of the potential using the values of the function and its normal derivative at the nodes, as determined by the standard boundary integral. The convergence of the integrals away from the boundary is fairly rapid. For the Laplace equation, Liu and Rudolphi have shown that the integral equations can be reformulated so that the integrals for both the potential and its derivative (the above equation) contain only weakly singular terms.[30]

The boundary integral method crucially depends on knowledge of the Green's function in the open domain. These Green's functions can only be obtained for linear differential equations with constant coefficients or with some very specific variable coefficients. We have seen, in equation (18.27), that this limitation of the method leads to an iterative procedure for potential scattering when the potential varies with the coordinate. It could turn out, however, that the iterative procedure is computationally less intensive than a straight FEM calculation for specific problems. Again, iterative methods will be needed for nonlinear problems. The general approach to such iteration has already been discussed in the framework of the FEM.

The boundary integral equation for 2D scattering from a thin screen has been investigated in the literature. The problem reduces to a "hypersingular" boundary integral equation. A similar problem arises in elastomechanics in the presence of cracks in the material.[31,32]

We have not treated time-dependent propagations and interactions. In a few instances, the Green's functions are known in closed form. For example, the free-particle Green's function obeys the inhomogeneous Schrödinger equation

$$\left(-i\hbar \frac{\partial}{\partial t} - \frac{\hbar^2}{2m} \nabla^2 + V(\mathbf{r}) \right) g(\mathbf{r} - \mathbf{r}', t - t')$$
$$= -i\hbar \delta(\mathbf{r} - \mathbf{r}') \delta(t - t'), \tag{18.33}$$

with $V(\mathbf{r}) = 0$, and satisfies the BC that $g(\mathbf{r} - \mathbf{r}', t - t') = 0$ for $t < t'$.

It is given by

$$g(\mathbf{r} - \mathbf{r}', t - t') = \left(\sqrt{\frac{m}{2\pi i \hbar (t - t')}} \right)^{d} \exp \left(\frac{im(\mathbf{r} - \mathbf{r}')^2}{2\hbar(t - t')} \right). \quad (18.34)$$

Here d is the number of spatial dimensions in the problem. This raises the possibility of exploring time-dependent phenomena using boundary elements using spatial-temporal boundary elements. This has been proposed previously for problems in heat conduction.[33] Other strategies for time-dependent problems are presented in the literature.[34]

In this presentation of three chapters devoted to the BEM, we have not discussed the convergence properties of the BEM.[35] An error estimation procedure can be used to define a set of criteria for refining the boundary element discretization in order to improve the accuracy. As in the finite element approach, we have three methods for improving the accuracy of the BEM. The h (refinement) adaptive method corresponds to subdividing elements where necessary; the p (enrichment) adaptive method increases the degree of the polynomial interpolation scheme employed in a given element; and the r (relocation) adaptive method redefines the location of existing nodes in order to improve accuracy.[36] A posteriori error estimates and adaptive refinement strategies for the BEM have been considered earlier.[37,38]

We have also laid aside the topic of linking the FEM and BEM[39] in multiregion problems. This linkage, while conceptually straightforward, requires care in ensuring inter-region continuity. It is useful to treat certain multiregion problems by this mixed method, especially if physical properties (e.g., the potential energy of the particle) in certain subregions are coordinate dependent. Then the FEM can be invoked for the inhomogeneous regions, and the local FEM matrices can be overlaid with the matrices generated in the BEM in order to ensure inter-region continuity. Such linkages will naturally lead to nonsymmetric global matrices. However, it was demonstrated that it is possible to develop algorithms to obtain symmetric global matrices.[40,41] We have also not included in our presentation the promising area of 3D applications of the BEM to quantum mechanics.

Research in the BEM and its applications to various fields in science and engineering is an ongoing endeavor and is being pursued very vigorously. While boundary element technology and computer programming methods are still evolving, it is evident from the number of published papers and reviews that the BEM is coming into its own, and is becoming an important adjunct to the finite element approach to solving physical problems. The reader is directed to recent developments in the BEM which are reported regularly in review articles[42,43] and in conference proceedings.[44]

18.6   Problems

1. Consider an electron in a 2D potential

$$V(r, \phi) = \begin{cases} 0, & r \le a, \\ \infty, & r > a, \end{cases}$$

with $a = 100\,\text{Å}$. The wavefunction of the electron is zero at the barrier.

(a) Formulate a solution for Schrödinger's equation

$$\nabla^2 \psi(r, \phi) + \frac{2mE}{\hbar^2}\, \psi(r, \phi) = 0, \qquad r \le a,$$

using the BEM with a contour along the circular boundary at $r = a$. Discretize the boundary integral into $nelem = 20\text{--}100$ elements and employ linear interpolation in the angular variable ϕ to express the integral equation in terms of a sum over boundary element integrals.

(b) Show that, prior to imposing the BCs, the boundary integral can be written in the form

$$\psi_i = -\frac{i}{4} \sum_{iel=1}^{nelem} \ell_{iel} \int_0^1 d\xi' \left[H_0^{(1)}(k|\mathbf{r}_i - \mathbf{r}'|) \right.$$
$$\times \{\psi'_{iel}(1 - \xi') + \psi'_{iel+1}\xi'\} + k\,\cos\theta'\, H_1^{(1)}(k|\mathbf{r}_i - \mathbf{r}'|)$$
$$\left. \times \{\psi_{iel}(1 - \xi') + \psi_{iel+1}\xi'\}\right].$$

Here the length of the element is $\ell_{iel} = 2\pi a/nelem$ and the integration variable ξ' is measured along the circular path. The index i ranges over the destination nodes to yield $nelem$ equations. The parameter k appearing in the Hankel functions is given by $k^2 = 2mE/\hbar^2$. Distort the contour outward in a semicircular

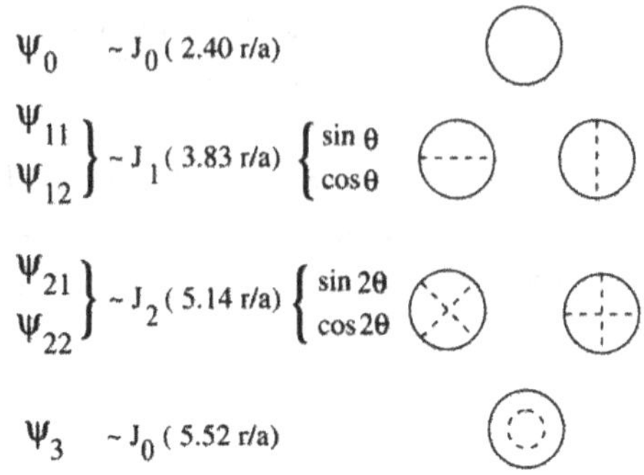

$\Psi_0 \quad \sim J_0(2.40\ r/a)$

$\left.\begin{matrix}\Psi_{11}\\ \Psi_{12}\end{matrix}\right\} \sim J_1(3.83\ r/a) \begin{cases} \sin\theta \\ \cos\theta \end{cases}$

$\left.\begin{matrix}\Psi_{21}\\ \Psi_{22}\end{matrix}\right\} \sim J_2(5.14\ r/a) \begin{cases} \sin 2\theta \\ \cos 2\theta \end{cases}$

$\Psi_3 \quad \sim J_0(5.52\ r/a)$

Fig. 18.11 The analytical expressions for wavefunctions and the corresponding "nodal" diagrams for the lowest four energy levels of an electron in a circular box. The dashed lines across the circle represent the location of the zero of the wavefunction. Here "node" refers to a zero of the wavefunction.

path around the destination point if it lies on the element being evaluated. Evaluate the additional contribution over this small semicircular path about the destination point. Express the above equation in the compact form

$$\sum_{j}^{nelem} [\mathcal{H}_{ij}\,\psi_j + \mathcal{G}_{ij}\,\psi'_j] = 0.$$

(c) With the wavefunction vanishing at the boundary, the boundary integral reduces to

$$\sum_{j}^{nelem} \mathcal{G}_{ij}\,\psi'_j = 0.$$

Obtain the eigenvalues of the electron in the circular box by locating the zeros of the determinant of the $nelem \times nelem$ matrix $\mathcal{G}$ as the energy E in Schrödinger's equation is varied.

(d) For the ground state, we can anticipate circular symmetry, and hence all the ψ'_i are the same. Determine the ground state energy by locating the zero of the sum over row elements of $\mathcal{G}$. Does each row give the same ground state energy?

(e) Determine the next two eigenvalues by the procedure of part (c). (These higher excited states do not have circular symmetry and are doubly degenerate in energy.) Obtain the corresponding eigenfunctions throughout the circular region.

(f) Compare your results for the eigenvalues with the analytical result $E_n = \hbar^2 k_{n\alpha}^2 / 2m$, where $k_{n\alpha}$ are determined by the roots of $J_n(ka) = 0$. (The problem is analogous to the vibrational modes of a drumhead.)** The lowest few roots of the Bessel functions of order 0, 1, and 2 are

$$J_0(x) = 0: \quad x_\alpha \simeq 2.40, \; 5.52, \; 8.65, \; \ldots$$
$$J_1(x) = 0: \quad x_\alpha \simeq 3.83, \; 7.02, \; 10.17, \; \ldots$$
$$J_2(x) = 0: \quad x_\alpha \simeq 5.14, \; 8.42, \; 11.62, \; \ldots.$$

Analytical expressions for the wavefunctions of the lowest few states are shown in Fig. 18.11. A double degeneracy occurs in the second and third energy levels. The dotted lines in the circles represent zeros ("nodes") of the wavefunction.

2. Electrons of energy $E = 0.75\,\text{eV}$ are incident from the left with a wavefunction of unit amplitude, $\psi(\mathbf{r}) = \exp(ik_x x)$, on a 2D scattering potential

$$V(r, \phi) = \begin{cases} 0, & r \leq a, \\ V_0, & r > a, \end{cases}$$

where $a = 100\,\text{Å}$, and $V_0 = 0.5\,\text{eV}$.

(a) Solve the 2D Schrödinger equation for the scattered wavefunction using the BEM. Take a contour along the edge of the potential at $r = a$ and break up the contour into 100 boundary elements. Employ linear interpolation in the boundary elements. Evaluate the scattered wavefunction and its normal derivative at nodal points on the contour, as discussed in the text.

(b) Construct the wavefunction in the interior region $r \leq a$, as a function of the polar angle ϕ.

(c) Determine the scattered probability current numerically at $r = 1000\,\text{Å}$.

(d) Construct a table for the differential scattering cross-section as a function of the polar angle ϕ.

**De May[45] has evaluated the lowest eigenvalue using constant interpolation in the boundary elements.

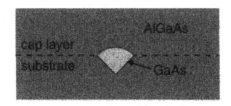

Fig. 18.12 A GaAs quantum wire with a pie-shaped area of cross-section, embedded in AlGaAs.

(e) Are there resonances in the scattering cross-section as the energy is varied over the range $0 \le E \le 1\,\text{eV}$? Remember to switch the form of the Green's function for $E < V_0$ to the one for $E > V_0$, as you vary the energy.

(f) Numerically verify the 2D version of the optical theorem.[46,47]

3. A GaAs quantum wire has been grown by depositing the GaAs on a V-groove etched into the surface of an AlGaAs substrate. The wire has a pie-shaped cross-section with an interior angle of $90°$, as shown in Fig. 18.12. The straight edges are $100\,\text{Å}$ in length. A cap layer of AlGaAs has been grown to completely enclose the wire. The AlGaAs medium has 30% Al and is the barrier region for electrons in the conduction band of the GaAs wire. The barrier height is $V_0 \simeq 0.276\,\text{eV}$. The electron's effective mass in GaAs is $m^*/m_0 = 0.0665$, and it is $m^*/m_0 \simeq 0.084$ in the AlGaAs barrier region.

(a) Solve the 2D Schrödinger equation for the transversely bound states of the electron in the 2D well in the GaAs region. Use the BEM, with linear interpolation in the elements, to obtain the set of homogeneous simultaneous equations obeyed by the nodal values of the wavefunction and its normal derivative. Determine the energy eigenvalues of bound states by locating the zeros of the coefficient matrix.

(b) Obtain wavefunctions of the first two energy levels. Determine the probability of the electron being in the barrier region for the two states.

4. Schrödinger's equation for a particle in an infinitely deep 2D equilateral triangular well is a nonseparable differential equation. Let the triangular region be defined by $y = 0$, $y = \sqrt{3}x$,

and $y = \sqrt{3}(a - x)$, where a $= 100$ Å. The wavefunction vanishes at these boundaries.

Convert the problem to the solution of an integral equation. Discretize the boundary integral into integrations over linear segments on each side of the triangle. Use constant interpolation over each element (in order to avoid any problems in programming the coefficients of the variables at corner nodes), and set up the matrix equation for the normal derivatives of the wavefunction at the nodes.

(b) Obtain the first seven distinct eigenvalues in terms of the natural unit of energy for the problem: $E_0 = 2\hbar^2/(3ma^2)$. Solve for the eigenfunctions and compare the shapes of the degenerate eigenfunctions.[††]

References

[1] N. G. van Kampen, in *Statistical Physics*, Proceedings of the IU-PAP International Conference, ed. L. Pel and Szepflanszy (North-Holland, Amsterdam, 1976). For a recent review see S. Datta, *Electronic Transport in Mesoscopic Systems* (Cambridge University Press, Cambridge, 1995).

[2] R. Landauer, *Z. Phys.* B **68**, 217 (1987); M. Büttiker, Y. Imry, R. Landauer, and S. Pinhas, *Phys. Rev.* B **31**, 6207 (1985).

[3] B. J. van Wees, H. van Houten, C. W. J. Beenakker, J. G. Williamson, L. P. Kouwenhoven, D. van der Marel, and C. T. Foxon, *Phys. Rev. Lett.* **60**, 848 (1988).

[4] D. A. Wharam, T. J. Thornton, R. Newbury, M. Pepper, H. Ahmed, J. E. F. Frost, D. G. Hasko, D. C. Peacock, D. A. Ritchie, and G. A. C. Jones, *J. Phys.* C **21**, L209 (1988).

[5] S. Kagami and I. Fukai, *IEEE Trans. Microwave Theory Tech.* **32**, 455 (1984). (Reprinted in Ref. 10.)

[6] M. Koshiba and M. Suzuki, *IEEE Trans. Microwave Theory Tech.* **MTT-34**, 301 (1986). (Reprinted in Ref. 10.)

[7] J. Schwinger and D. S. Saxon, *Discontinuities in Waveguides* (Gordon and Breach, New York, 1968).

[8] D. S. Jones, *The Theory of Electromagnetism*, (Pergamon, Oxford, 1964).

[††]This problem can be solved analytically using group theory and Fourier series.[48]

[9] R. E. Collin, *Field Theory of Guided Waves*, (McGraw-Hill, New York, 1960).

[10] *Numerical Methods for Passive Microwave and Millimeter Wave Structures*, a reprint collection ed. R. Sorrentino (IEEE Press, New York, 1989).

[11] R. F. Harrington, *Field Computation by Moment Methods* (Krieger, Melbourne, Florida, 1982).

[12] *Numerical Techniques for Microwave and Millimeter-wave Passive Structures*, ed. T. Itoh (Wiley, New York, 1989).

[13] N. Marcuvitz, *Waveguide Handbook*, M.I.T. Radiation Laboratory Series, Vol. 10 (McGraw-Hill, New York, 1951).

[14] H. R. Frohne, M. J. McLennan, and S. Datta, *J. Appl. Phys.* **66**, 2699 (1989).

[15] C. S. Lent and D. J. Kirkner, *J. Appl. Phys.* **67**, 6353 (1990).

[16] J. D. Jackson, *Classical Electrodynamics*, 2nd ed. (Wiley, New York, 1975), Chapter 9, p427.

[17] P. A. Knipp and T. L. Reinecke, *Superlattices Microstruct.* **16**, 201 (1994); *Phys. Rev.* B **54**, 1880 (1996). Figures 18.7 and 18.8 were generously provided in electronic form by the authors.

[18] K. Kojima, K. Mitsunaga, and K. Kyuma, *Appl. Phys. Lett.* **55**, 882 (1989).

[19] Y. Niwa, S. Kobayashi, and M. Kitahara, in *Developments in Boundary Element Methods*, Vol. 2, ed. P. K. Banerjee and R. P. Shaw (Applied Science Publishers, London, 1979).

[20] R. P. Shaw, in *Developments in Boundary Element Methods*, Vol. 1, ed. P. K. Banerjee and R. Butterfield (Applied Science Publishers, London, 1979).

[21] G. R. C. Tai and R. P. Shaw, *J. Acoust. Soc. Am.* **56**, 796 (1974).

[22] M. Kac, *Am. Math. Mon.* **73**, 1 (1966).

[23] C. Gordon, D. Webb, and S. Wolpert, *Bull. Am. Math. Soc.* **27**, 134 (1992).

[24] P. Berard, *Math. Ann.* **292**, 547 (1992).

[25] T. A. Driscoll, "Eigenmodes of isospectral drums," *SIAM Rev.* **39**, pp1–17 (1997).

[26] S. Sridhar and A. Kudrolli, *Phys. Rev. Lett.* **72**, 2175 (1994).

[27] S. Sridhar, *Phys. Rev. Lett.* **67**, 785 (1991).

[28] The author wishes to thank Professor P. Knipp for generously providing him with Fig. 18.10, which is a beautiful illustration of the remarkable power of the BEM.

[29] D. M. Cvetkovic, M. Doob, and H. Sachs, *Spectra of Graphs* (Aca-

demic Press, New York, 1980).

[30] Y. Liu and T. J. Rudolphi, *Eng. Anal. Boundary Elem.* **8**, 301 (1991).

[31] P. A. Martin and F. J. Rizzo, *Proc. R. Soc. London* A **421**, 341 (1989).

[32] W. L. Wendland and E. P Stephan, *Arch. Ration. Mech. Anal.* **112**, 363 (1990).

[33] Y. P. Chang, C. S. Kang, and D. J. Chen, *Int. J. Heat Mass Transfer* **16**, 1905 (1973).

[34] F. J. Rizzo and D. J. Shippy, *AIAA J.* **8**, 2004 (1970); T. A. Cruse and F. J. Rizzo, *J. Math. Anal. Appl.* **22**, 244 (1968); R. P. Shaw, *ASME J. Appl. Mech.* **42** 147 (1975); D. Nardini and C. A. Brebbia, in *Boundary Elements VII*, Proceedings of the 7th International Conference, Lake Como, Vol. 1, ed. C. A. Brebbia and G. Maier (Computational Mechanics, Springer-Verlag, Berlin, 1985); C. A. Brebbia, in *Advanced Topics in Boundary Element Analysis*, ed. T. A. Cruse, A. B. Pifko, and H. Arme (ASME, New York, 1985).

[35] W. L. Wendland, in *Mathematics of Finite Elements and Applications*, ed. J. Whiteman, Vol. 5 (Academic Press, London, 1985).

[36] S. I. Umetani, *Adaptive Boundary Elements in Elastomechanics*, in *Topics in Engineering*, Vol. 5, ed. C. A. Brebbia and J. J. Conner (Computational Mechanics, Southampton, UK, 1988).

[37] E. Rank, in *Proceedings of the International Conference on Adaptive Refinements in Finite Element Computations*, ed. I. Babuska (Lisbon, 1984); in *Accuracy Estimates and Adaptive Refinements in Finite Element Computations*, ed. I. Babuska, O. C. Zienkiewicz, J. Gago, and E. R. de A. Oliveira (Wiley, New York, 1986); in *Adaptive Finite and Boundary Element Methods*, ed. C. A. Brebbia and M. H. Aliabadi (Computational Mechanics, Elsevier Applied Science, Southampton, UK, 1993).

[38] J. J. Rencis and R. L. Mullen, *Comput. Mech.* **3**, 309 (1988); J. J. Rencis and K. Y. Jong, *Comput. Methods Appl. Mech. Eng.* **73**, 295 (1989); J. J. Rencis, D. A. Hopkins, and C. C. Chamis, *Finite Elem. Anal. Des.* **9**, 229 (1991).

[39] O. C. Zienkiewicz, D. W. Kelly, and P. Bettess, in *International Symposium on Innovative Numerical Analysis in Applied Engineering Science*, Versailles, France (1977); *Int. J. Numer. Methods Eng.* **11**, 355 (1977).

[40] D. W. Kelly, G. G. W. Mustoe, and O. C. Zienkiewicz, in *Devel-*

opments in Boundary Element Methods, Vol. 1, ed. P. K. Banerjee and R. Butterfield (Applied Science Publishers, London, 1979); G. G. W. Mustoe, F. Volait, and O. C. Zienkiewicz, *Res Mech.* **4**, 57 (1982).

[41] J. Ma and M. Le, in *First Conference of BEM in Engineering*, Chongqing, China (1985).

[42] *Topics in Boundary Element Research*, ed. C. A. Brebbia, Vol. 1 (Springer-Verlag, Berlin, 1984), and other volumes in this series.

[43] *Developments in Boundary Element Methods*, Vol. 1, ed. P. K. Banerjee and R. Butterfield (Applied Science Publishers, London, 1979); Vol. 2, ed. P. K. Banerjee and R. P. Shaw (Applied Science Publishers, London, 1982); Vol. 4, ed. P. K. Banerjee and J. O. Watson (Elsevier Applied Science, London, 1986); and other volumes in this series.

[44] *Boundary Element Technology Conference: BETECH86*, proceedings ed. J. J. Connor and C. A. Brebbia (Computational Mechanics, Southampton, UK, 1986), and more recent conference proceedings.

[45] G. De Mey, *Int. J. Numer. Methods Eng.* **10**, 59 (1976).

[46] I. R. Lapidus, *Am. J. Phys.* **50**, 45, (1982); *ibid.*, **54**, 459 (1986); P. A. Maurone and T. K. Lim, *Am. J. Phys.* **51**, 856 (1983); I. Galbraith, Y. S. Ching, and E. Abraham, *Am. J. Phys.* **52**, 60 (1984).

[47] S. K. Adhikari, *Am. J. Phys.* **54**, 362 (1986).

[48] W. K. Li and S. M. Blinder, *J. Chem. Educ.* **64**, 131 (1987).

Part VI

Appendices

A

Gauss quadrature

A.1 Introduction

Gauss quadrature is an efficient method for obtaining numerical estimates for definite integrals. The integral reduces to the sum of the integrand evaluated at special points x_i, called Gauss points, multiplied by weight factors w_i. What is remarkable is that in n-point Gauss quadrature, the integration would be exact if the integrand were a polynomial of degree $2n - 1$. We describe below the theory of Gauss quadrature using Legendre orthogonal polynomials, giving details of how the special points and weight factors are determined. Next, an adaptive quadrature procedure is described that permits obtaining numerical estimates for integrals with very high accuracy. The details of implementing Gauss quadrature for the Cauchy principal value integrals are given. This appendix includes exercises that illustrate further the applicability of the method.

A.2 Gauss–Legendre quadrature

Let us suppose that we wish to obtain a numerical value for the integral

$$I = \int_{-1}^{1} f(x)\, dx. \tag{A.1}$$

The general integral

$$J = \int_{a}^{b} g(z)\, dz \tag{A.2}$$

can always be reduced to the standard form in equation (A.1) by mapping the range $[a, b]$ into $[-1, 1]$. We assume that the range of integration $[a, b]$ is finite. We can use the substitution

$$z = \left[\frac{b+a}{2} + \frac{b-a}{2}x\right],$$

$$x = \frac{2z - (a + b)}{b - a}, \tag{A.3}$$

to convert equation (A.2) into our standard form equation (A.1). In the following, we focus our attention on equation (A.1).

In Gauss integration, we express equation (A.1) in the approximate form

$$I = \int_{-1}^{1} f(x)\,dx \cong \sum_{i=1}^{n} w_i f(x_i), \qquad (A.4)$$

where w_i and x_i are chosen in a special way as discussed in the next section. If $f(x)$ were a polynomial of degree $2n - 1$, then the approximation (A.4) becomes an exact relation. We show this in the following.[1,2]

First we approximate $f(x)$ in terms of an interpolating polynomial and a remainder. The motivation for this is that interpolating polynomials would be easier to integrate than the original function. The interpolating polynomial is taken to be of the Lagrange type. We write the nth-degree polynomial as

$$L_i(x) = \prod_{\substack{j=1 \\ j \neq i}}^{n} \frac{(x - x_j)}{(x_i - x_j)}, \qquad (A.5)$$

and

$$f(x) = \sum_{i=1}^{n} f(x_i) L_i(x) + R_n(x). \qquad (A.6)$$

The polynomials $L_i(x)$ are the same as the shape functions $N_i(x)$ discussed in Chapter 3, and are such that

$$L_i(x_j) = \delta_{ij}. \qquad (A.7)$$

Here δ_{ij} is the Kronecker delta which equals zero when $i \neq j$, and is unity for $i = j$. Given equation (A.7), we see that the first term on the right side of equation (A.6) exactly equals the left side at the n points x_i. The remainder term R_n has to be chosen such that it has factors which ensure that $R_n(x) = 0$ at x_i. It can be shown that

$$R_n(x) = \left[\prod_{i=1}^{n} (x - x_i) \right] \frac{f^{(n)}(\xi)}{n!}, \qquad -1 < \xi < 1. \qquad (A.8)$$

Here, ξ is a given number in the interval $[-1, 1]$. In the following, we choose $\xi \equiv x$ *merely to study how well the relation* (A.4) *holds,* and also so that we may re-express the n functions $\prod_{i=1}^{n}(x - x_i)$, and $L_i(x)$, in terms of Legendre polynomials.

A.3   Gauss–Legendre base points and weights

Suppose first that $f(x)$ is a polynomial of degree $2n$. Then the nth derivative $f^{(n)}(x)$ is a polynomial of degree n. Let $f^{(n)}(x)/n! = h_n(x)$, so that equation (A.6) may be written as

$$f(x) = \sum_{i=1}^{n} f(x_i) L_i(x) + \left[\prod_{i=1}^{n} (x - x_i) \right] h_n(x). \tag{A.9}$$

On integrating $f(x)$ over $[-1, 1]$, we have

$$I = \int_{-1}^{1} f(x)\, dx = \sum_{i=1}^{n} f(x_i) \int_{-1}^{1} L_i(x)\, dx$$

$$+ \int_{-1}^{1} dx\, h_n(x) \left[\prod_{i=1}^{n} (x - x_i) \right],$$

$$\equiv \bar{I} + \tilde{I}. \tag{A.10}$$

Let

$$\int_{-1}^{1} L_i(x)\, dx = w_i, \tag{A.11}$$

and expand $h_n(x)$ in terms of the Legendre polynomials[†]

$$h_n(x) = \sum_{j=0}^{n} \beta_j P_j(x). \tag{A.12}$$

Similarly, let us write

$$\prod_{i=1}^{n} (x - x_i) = \sum_{j=0}^{n} \alpha_j P_j(x). \tag{A.13}$$

We can use the orthogonality of $P_n(x)$ to write

$$I = \bar{I} + \tilde{I} = \sum_{i=1}^{n} f(x_i) w_i + \sum_{i=0}^{n} \alpha_i \beta_i \int_{-1}^{1} P_i^2(x)\, dx.$$

$$= \sum_{i=1}^{n} f(x_i) w_i + \sum_{i=0}^{n} \alpha_i \beta_i \frac{2}{2i + 1}. \tag{A.14}$$

[†] The properties of Legendre polynomials used in this section are collected together in Section A.7, at the end of this appendix.

Now suppose we choose x_i to be the roots of $P_n(x)$. We are then identifying the product $\prod_{i=1}^{n}(x - x_i)$ as being proportional to $P_n(x)$:

$$\prod_{i=1}^{n}(x - x_i) = \frac{2^n(n!)^2}{(2n)!} P_n(x). \tag{A.15}$$

Hence in equation (A.13), all the α_i for $0 \leq i \leq n - 1$ are automatically zero; furthermore, for $i = n$ we have

$$\alpha_n = \frac{2^n(n!)^2}{(2n!)}.$$

This choice of x_i reduces $\tilde{I}$ to just one term, the nth term. We then have

$$\tilde{I} = \int_{-1}^{1} \alpha_n \beta_n P_n^2(x)\, dx = \alpha_n \beta_n \frac{2}{2n + 1}.$$

If $f(x)$ were a polynomial of degree $2n - 1$ instead of $2n$, we would have $\beta_n = 0$, since the nth derivative of $f(x)$ will be of degree $n - 1$. Thus, by using the roots x_i, we have shown that I can be evaluated exactly in this case; it is now given by

$$I = \sum_i f(x_i)\, w_i, \tag{A.16}$$

for $f(x)$ being a polynomial of degree $2n - 1$.

We now calculate the weights w_i. We have

$$w_i = \int_{-1}^{1} L_i(x)\, dx = \int_{-1}^{1} dx \prod_{\substack{j=1 \\ j \neq i}}^{n} \frac{(x - x_j)}{(x_i - x_j)}. \tag{A.17}$$

Here x_i, x_j are the roots of $P_n(x)$.

Let us define $S(x) \equiv \prod_{j=1}^{n}(x - x_j)$. It is straightforward to show that

$$L_i(x) = \frac{S(x)}{S'(x_i)(x - x_i)} = \frac{P_n(x)}{P_n'(x_i)(x - x_i)}, \tag{A.18}$$

so that

$$w_i = \int_{-1}^{1} dx \frac{P_n(x)}{P_n'(x_i)(x - x_i)}. \tag{A.19}$$

Now we use the Christoffel identity (equation (A.31)), with $t = x_i$, where x_i is a zero of $P_n(x)$, to obtain

$$(x_i - x) \sum_{j=0}^{n} (2j + 1) P_j(x) P_j(x_i) = (n + 1)[P_{n+1}(x_i)P_n(x)$$

$$- P_n(x_i)P_{n+1}(x)].$$

The second term on the right vanishes since x_i are the roots of $P_n(x)$. We thus have

$$\sum_{j=0}^{n}(2j+1)P_j(x)P_j(x_i) = -\frac{(n+1)P_{n+1}(x_i)P_n(x)}{(x-x_i)}.$$

We use this relation in equation (A.18) to obtain

$$\int_{-1}^{1}\frac{P_n(x)}{(x-x_i)}\,dx = -\frac{1}{(n+1)P_{n+1}(x_i)}\sum_{j=0}^{n}(2j+1)\int_{-1}^{1}P_j(x)P_j(x_i)\,dx$$

$$= \frac{-2}{(n+1)P_{n+1}(x_i)},$$

as only the $j = 0$ term survives. The recursion relation, equation (A.29), with $x = x_i$, then allows us to write

$$(n+1)P_{n+1}(x_i) = -nP_{n-1}(x_i),$$

so that

$$w_i = \left(\frac{2}{nP_{n-1}(x_i)}\right)\cdot\frac{1}{P_n'(x_i)}.$$

This can be cast in a more convenient form using the recursion relation equation (A.30). With $x = x_i$, a root of $P_n(x)$, we have

$$(1-x_i^2)P_n'(x_i) + nx_i\,P_n(x_i) = nP_{n-1}(x_i).$$

Here the second term on the left side, containing the factor $P_n(x_i)$, is zero. Hence

$$w_i = \frac{2(1-x_i^2)}{n^2\,[P_{n-1}(x_i)]^2}. \tag{A.20}$$

In summary, we have shown that we can obtain an exact expression for $\int_{-1}^{1}f(x)\,dx$ if $f(x)$ is a polynomial of degree $2n-1$, with evaluation of the function at just n points. The use of side conditions that determine x_i and w_i lends high accuracy to this method. This is the basic advantage of the Gauss integration method. The error in this procedure for evaluating the integral may be estimated by noting that the approximation in representing the function $f(x)$ in terms of the interpolation polynomials is of the order $f^{(2n)}(\xi)/(2n)!$. Hence the actual error can be estimated; the result is[3]

$$\varepsilon = \frac{2^{2n+1}(n!)^4}{(2n+1)[(2n)!]^3}f^{(2n)}(\xi), \qquad -1 < \xi < 1.$$

The weights w_i and base points x_i are best evaluated once and for all, and kept ready in tabulated form for various values of n. An

Table A.1 Table of Gauss base points and weights of order 4, 6, 8, 10, and 12. Only positive base points are given. The negative base points have the same weights as the corresponding given positive points. (From Ref. 4.)

Order	Base points	Weights
4	0.861136311594052	0.347854845137453
	0.339981043584856	0.652145154862546
6	0.932469514203152	0.171324492379170
	0.661209386466264	0.360761573048138
	0.238619186083196	0.467913934572691
8	0.960289856497536	0.101228536290376
	0.796666477413626	0.222381034453374
	0.525532409916328	0.313706645877887
	0.183434642495649	0.362683783378361
10	0.973906528517171	0.066671344308688
	0.865063366688984	0.149451349150580
	0.679409568299024	0.219086362515982
	0.433395394129247	0.269266719309996
	0.148874338981631	0.295524224714752
12	0.981560634246719	0.047175336386511
	0.904117256370474	0.106939325995318
	0.769902674194304	0.160078328543346
	0.587317954286617	0.203167426723065
	0.367831498998180	0.233492536538354
	0.125233408511468	0.249147045813402

extensive tabulation is given by Stroud and Secrest.[4] The Gauss–Legendre base points and weights for a select few orders are reproduced in Tables A.1 and A.2.

Table A.2 Table of Gauss base points and weights of order 14 and 16. Only positive base points are given. The negative base points have the same weights as the corresponding given positive points. (From Ref. 4.)

Order	Base points	Weights
14	0.986283808696812	0.035119460331751
	0.928434883663573	0.080158087159760
	0.827201315069764	0.121518570687903
	0.687292904811685	0.157203167158193
	0.515248636358154	0.185538397477937
	0.319112368927889	0.205198463721295
	0.108054948707343	0.215263853463157
16	0.989400934991649	0.027152459411754
	0.944575023073232	0.062253523938647
	0.865631202387831	0.095158511682492
	0.755404408355003	0.124628971255533
	0.617876244402643	0.149595988816576
	0.458016777657227	0.169156519395002
	0.281603550779258	0.182603415044923
	0.095012509837637	0.189450610455068

A.4   An algorithm for adaptive quadrature

The procedure described earlier shows that if $f(x)$ is a polynomial of degree $2n - 1$, or less, we have an exact result:

$$\int_{-1}^{1} f(x)\, dx = \sum_{i=1}^{n} w_i\, f(x_i). \qquad (A.21)$$

If $f(x)$ is a polynomial of degree larger than $2n - 1$, or if it is a transcendental function, the above equation holds only approximately.

To improve the accuracy, one has to go to a larger and larger number of base points. An alternate way is to breakup the region $[-1, 1]$ into smaller parts, and use the n-point quadrature in each, since the function would be better approximated over small ranges by a polynomial of degree $(2n - 1)$. We can do this breakup of the range of integration more effectively by employing an adaptive procedure

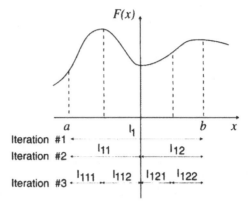

Fig. A.1 Adaptive Gauss iteration implemented by bisection of the range of the integral.

in which the breakup is done in a recursive manner automatically, with a check at each level to see if the desired accuracy is achieved. This is tested by evaluating the difference in the value of the integral between two successive recursions and comparing it with an input value for the "error tolerance."

Consider an integration $I = \int_{-1}^{1} f(x)\,dx$. We obtain an initial guess for I on approximating it by $I_0 = \sum w_i f(x_i)$. Next, we divide $[-1, 1]$ into two parts and map each of the regions $[-1, 0]$ and $[0, 1]$ separately into two new regions each with a standard range $[-1, 1]$, and reuse the Gauss procedure on each of them. Suppose the sum is different from I. We evaluate $I_0 - (I_{11} + I_{12})$ and see if it is less than the required error tolerance. If it is, then we accept the value of I as given by $I = I_{11} + I_{12}$, and exit the procedure. Otherwise we subdivide each region further and compare the integral I_{12} with $I_{121} + I_{122}$, and so on. This is shown schematically in Fig. A.1. The error tolerance over the original range can be scaled by a factor which is the ratio of the new "subrange" to the original range. This allows us to test whether further division of the subinterval is necessary. When done in this manner, the breakup of the regions occurs only where it is demanded by the required error tolerance for that region. The values of all the subintegrals are summed from a storage array to obtain the final answer. The error estimate is a very rough one. It adds up the difference between a given level of "divisioning" with the ones from the next level. Each term in the sum over such differences is of course less than the scaled value of the error tolerance for that subdivision.

Eventually, the desired accuracy is achieved, with the function being adequately represented by polynomials over small regions. We then sum up the integrals over the subregions and obtain our final result. The procedure is called "adaptive" because the region is subdivided and Gauss iterations are implemented where they are required. We can keep the "order of the Gauss quadrature" fairly small at, say, $n = 8$.

A.5   Other Gauss formulas

We have seen the derivation of one Gauss formula, traditionally called the Gauss–Legendre formula

$$\int_{-1}^{1} F(z)\,dz \;\cong\; \sum_{i=1}^{n} F(z_i)\,w_i.$$

The procedure can be extended to integrations in which it is desirable to represent the integrand using other sets of complete functions in specific domains:

$$\int_{-1}^{1} \frac{1}{\sqrt{1-x^2}} f(z)\,dz \;\cong\; \sum_{i=1}^{n} w_i f(z_i) \;:\; \text{Gauss–Chebychev formula}$$

$$\int_{0}^{\infty} e^{-z} f(z)\,dz \;\cong\; \sum_{i=1}^{n} w_i f(z_i) \;\;:\; \text{Gauss–Laguerre formula}$$

$$\int_{-\infty}^{\infty} e^{-z^2} f(z)\,dz \cong \sum_{i=1}^{n} w_i f(z_i) \;:\; \text{Gauss–Hermite formula}.$$

The z_i and w_i are, of course, unique for each procedure. They are given in tabular form by Stroud and Secrest,[4] and by Davis and Rabinowitz.[5]

In Table A.3, we provide a few of the lowest order Gauss weights and points for the integral $\int_0^1 \ln(1/x) f(x)\,dx$ that frequently appears in boundary element calculations.

Table A.3 The base points and weights for the Gauss integration $\int_0^1 \ln(1/x)f(x)dx \simeq \sum_i^n f(x_i)w_i$.

Order (n)	Base points	Weights
2	0.112008806166976	0.718539319030384
	0.602276908118738	0.281460680969615
4	0.0414484801993832	0.383464068145135
	0.245274914320602	0.386875317774762
	0.556165453560275	0.190435126950142
	0.848982394532985	0.0392254871299598
8	0.0133202441608924	0.164416604728002
	0.0797504290138949	0.237525610023306
	0.197871029326188	0.226841984431919
	0.354153994351909	0.175754079006070
	0.529458575234917	0.112924030246759
	0.701814529939099	0.0578722107177820
	0.849379320441106	0.0209790737421329
	0.953326450056359	0.00368640710402761

A.6 The Cauchy principal value of an integral

The algorithm for adaptive Gauss quadrature is very effective in estimating integrals involving integrable singularities. This is of interest in the boundary element method in which integrals over functions with integrable end point singularities occur. Here we consider an integral which exists in the sense of a Cauchy principal value. For example, suppose we wish to evaluate the Cauchy principal value of an integral

$$I = \mathcal{P}\int_a^b \frac{f(x)}{x - x_0}\,dx; \qquad a < x_0 < b, \qquad \text{(A.22)}$$

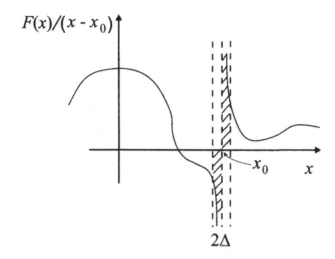

Fig. A.2 An arbitrary function $f(x)/(x - x_0)$, where $f(x)$ is a smooth function, showing a cancellation of the areas under the curve in the neighborhood of x_0.

with $f(x)$ being a differentiable function. Here $\mathcal{P}$ denotes the Cauchy principal value of the integral. We write I as

$$
I = \int_a^{x_0-\Delta} \frac{f(x)}{x - x_0} \, dx + \int_{x_0+\Delta}^b \frac{f(x)}{x - x} \, dx + \mathcal{P} \int_{x_0-\Delta}^{x_0+\Delta} \frac{f(x)}{x - x_0} \, dx
$$
$$
= I_1 + I_2 + I_3.
$$

The Cauchy value is obtained when we take $\Delta \to 0$ at the same rate simultaneously on both sides of the singularity. Then the integral can exist and be finite because the areas under the curve on either side of x_0 cancel out. This is shown schematically in Fig. A.2. To evaluate this integral numerically, we stay away from x_0 and evaluate I_1 and I_2 using the adaptive Gauss quadrature procedure. Next, we consider I_3 with a finite value for Δ, say, $\sim 10^{-4}$.

We change the integration variable in order to put the limits symmetrically about x_0. Let $x - x_0 = \xi$ so that $x = \xi + x_0$ and $-\Delta \le \xi \le \Delta$. We have

$$
I_3 = \mathcal{P} \int_{x_0-\Delta}^{x_0+\Delta} \frac{f(x)}{x - x_0} \, dx = \mathcal{P} \int_{-\Delta}^{\Delta} d\xi \, \frac{f(\xi + x_0)}{\xi}.
$$

Expanding $f(\xi + x_0)$ in a Taylor series about x_0, we obtain

$$I_3 = P \int_{-\Delta}^{\Delta} \frac{d\xi}{\xi} \left(f(x_0) + f'(x_0)\,\xi \;+\; \frac{f''(x_0)}{2!}\,\xi^2 \;+\; \frac{f'''(x_0)}{3!}\,\xi^3 + \ldots \right)$$

$$= f(x_0)\, P \int_{-\Delta}^{\Delta} \frac{d\xi}{\xi} \;+\; [f'(x_0)\,\xi]\Big|_{-\Delta}^{\Delta} \;+\; \left[\frac{f''(x_0)}{2!}\, \frac{1}{2}\, \xi^2 \right]\Big|_{-\Delta}^{\Delta}$$

$$+ \left[\frac{f'''(x_0)}{3!}\, \frac{\xi^3}{3} \right]\Big|_{-\Delta}^{\Delta} \;+\; \mathcal{O}(\xi^4)$$

$$= 0 \;+\; 2\Delta\, f'(x_0) \;+\; \frac{f'''(x_0)}{3!}\, \frac{2\Delta^3}{3} \;+\; \ldots \ . \tag{A.23}$$

We have used the result that the Cauchy principal value of the first term vanishes. With $\Delta \sim 10^{-4}$ and $f(x)$ a "smooth function," we do not usually have to go any further with the series development. For further developments and more examples, see Zwillinger.[6]

A.7   Properties of Legendre functions

Legendre polynomials are defined as

$$P_n(x) = \frac{1}{2^n n!}\, \frac{d^n}{dx^n}(x^2 - 1)^n, \tag{A.24}$$

with

$$P_0(x) = 1.$$

These polynomials satisfy the relation

$$\int_{-1}^{1} x^k P_n(x)\, dx = 0, \qquad \text{for } k = 0,\, 1,\, \ldots,\, n - 1. \tag{A.25}$$

Proof. On integrating by parts, we obtain

$$\int_{-1}^{1} x^k \frac{d^n}{dx^n}(x^2 - 1)^n\, dx = \left[x^k \frac{d^{n-1}}{dx^{n-1}}(x^2 - 1)^n \right]\Bigg|_{x=-1}^{x=+1}$$

$$- \int_{-1}^{1} k x^{k-1} \frac{d^{n-1}}{dx^{n-1}}(x^2 - 1)^n\, dx.$$

The first term on the right is zero. On continuing the integration by parts we get the result equation (A.25).

We can show that[7,8]

$$\int_{-1}^{1} x^n P_n(x)\, dx = \frac{2^{n+1}(n!)^2}{(2n+1)!},$$

(A.26)

and

$$\int_{-1}^{1} [P_n(x)]^2\, dx = \frac{2}{2n+1}.$$

(A.27)

Equations (A.25)–(A.27) show that

$$\int_{-1}^{1} P_m(x)P_n(x)\, dx = \frac{2}{2n+1}\,\delta_{mn}.$$

(A.28)

The Legendre functions satisfy the recursion relations

$$(n+1)P_{n+1}(x) = (2n+1)\, x\, P_n(x) - n\, P_{n-1}(x),$$

(A.29)

$$(1-x^2)P_n'(x) + nx\, P_n(x) = nP_{n-1}(x).$$

(A.30)

Christoffel's identity

Consider equation (A.29) multiplied by $P_n(t)$:

$$[(2n+1)xP_n(x)]\, P_n(t) = [(n+1)P_{n+1}(x) + nP_{n-1}(x)]\, P_n(t).$$

Now write this after interchanging x and t, and subtract one equation from the other. We then have

$$(2n+1)(t-x)P_n(x)P_n(t) = (n+1)\left[P_{n+1}(t)P_n(x) - P_n(t)P_{n+1}(x)\right]$$

$$- n\left[P_n(t)P_{n-1}(x) - P_{n-1}(t)P_n(x)\right].$$

On summing over the index n from 0 to n, we obtain

$$(t-x)\sum_{i=0}^{n}(2i+1)P_i(x)P_i(t) = (n+1)\left[P_{n+1}(t)P_n(x)\right.$$

$$\left. - P_n(t)P_{n+1}(x)\right].$$

(A.31)

A.8 Problems

1. Show that
$$\int_1^3 \frac{(\sin x)^2}{x}\, dx \simeq 0.794\,825\,18,$$
by Gauss–Legendre quadrature.

2. (a) Show numerically that
$$-\int_0^1 \frac{\ln(1-x)}{x}\, dx = \frac{\pi^2}{6}.$$
Note the singularity at $x = 1$, and the need for L'Hospital's rule for the limit $x \to 0$.

Compare your result from Gauss–Legendre quadrature with that obtained using a product rule for logarithmic integrands over the range $[0,1]$ for which the coordinates and the weights are given in Table A.3.

(b) Evaluate numerically the left side of
$$\int_0^1 \ln x \ln(1-x) = 2 - \pi^2/6.$$

3. Compare (a) the adaptive Gauss method with (b) the usual trapezoidal rule and (c) Simpson's rule to verify that
$$\frac{2}{\sqrt{\pi}} \int_0^2 e^{-x^2}\, dx \simeq 0.995\,322\,265\,0.$$

Compare the number of arithmetic operations needed to achieve the same accuracy by these methods. Also, compare adaptive Gauss–Legendre quadrature with a straight application of Gauss-Hermite quadrature.

4. Verify numerically that
$$\int_0^\infty e^{-2x} \ln x\, dx = -0.635\,181.$$

Again, compare results from adaptive Gauss–Legendre quadrature with Gauss–Laguerre quadrature. For adaptive quadrature, the integral over the range $[0, \infty]$ can be recast into two integrals, one extending up to a number R_0, and the other in which a mapping $x \to 1/y$ is used to convert the infinite range into one over $[0, 1/R_0]$.

5. Extend the adaptive Gauss quadrature procedure to evaluate the integral of a *complex* function along any *straight line* on the complex plane.

6. Extend the adaptive integration procedure to evaluate integrals over the 2D range $[-1 \le x \le 1]$ and $[-1 \le y \le 1]$. Notice that this method becomes computationally intensive fairly quickly as error tolerance is lowered.[‡]

7. Give derivations[9] paralleling the one presented in the text for Gauss–Legendre evaluation of Gauss points and weights, for the determination of Gauss points and weights for (a) Gauss–Hermite, (b) Gauss–Laguerre, (c) Gauss–Jacobi, and (d) Gauss–Chebychev integration procedures.

8. Write a brief report on the method of product integration for integrals involving spherical Bessel functions.[10]

9. Obtain the volume of a unit sphere by three-dimensional Gauss–Legendre quadature.

References

[1] S. D. Conte and C. de Boor, *Elementary Numerical Analysis* (McGraw-Hill, New York, 1980).

[2] W. H. Press, B. P. Flannery, S. A. Teukolsky, and W. T. Vetterling, *Numerical Recipes* (Cambridge University Press, Cambridge, 1988).

[3] M. Abramowitz and I. A. Stegun, *Handbook of Mathematical Functions* (Dover, New York, 1965).

[4] A. H. Stroud and D. Secrest, *Gaussian Quadrature Formulas* (Prentice Hall, Englewood Cliffs, NJ, 1966).

[5] P. S. Davis and P. Rabinowitz, *Numerical Integration* (Blaisdell, Waltham, MA, 1967).

[6] D. Zwillinger, *Handbook of Integration* (Jones and Bartlett, Boston, 1992).

[‡]The adaptive Gauss quadrature procedures do not reuse the previous evaluations of the integrand at Gauss points in the next level of "iteration." This seems to make the adaptive programs, based on the strategy discussed in the text, less efficient. However, the actual number of function evaluations is usually less than that for Romberg integration for the same error tolerance.

[7] R. V. Churchill, *Fourier Series and Boundary Value Problems*, 2nd ed. (McGraw-Hill, New York, 1963).

[8] P. M. Morse and H. Feshbach, *Methods of Mathematical Physics* (McGraw-Hill, New York, 1953).

[9] F. B. Hildebrand, *Introduction to Numerical Analysis*, 2nd ed. (Dover, New York, 1987).

[10] D. R. Lehman, W. C. Parke, and L. C. Maximon, *J. Math. Phys.* **22**, 1399 (1981).

B

Generalized functions

B.1 The Dirac δ-function

The Dirac delta-function, which is denoted by $\delta(x)$, is one of a wide class of "generalized functions," also called "distributions," or "functionals."[1-3]

The Dirac δ-function in one dimension is defined as

$$\delta(x) = 0, \qquad \text{for } x \neq 0,$$
$$= \infty, \qquad \text{for } x = 0, \qquad \text{(B.1)}$$

and

$$\int_{-\infty}^{\infty} \delta(x)\, dx = \int_a^b \delta(x)\, dx = 1, \quad \text{for } a < 0 < b. \qquad \text{(B.2)}$$

The properties (B.1) and (B.2) lead to the result that

$$\int f(x)\,\delta(x - x_0)\, dx = f(x_0), \qquad -\infty < x_0 < \infty \qquad \text{(B.3)}$$

and

$$\int f(x)\,\delta'(x - x_0)\, dx = -f'(x_0), \qquad \text{(B.4)}$$

for any "smooth" function $f(x)$.

The usual functions defined within the framework of classical function theory do not have these properties. The δ-function is a singular function specified at only one point. It is zero for all $x \neq 0$, while its value and its derivatives are not zero at the point $x = 0$.

Such functions fortunately occur only in the intermediate stages in solving problems in mathematical physics. Invariably, the final results involve only integrals of such functions multiplied by some well-behaved function. Thus, it is actually unnecessary to ask what the δ-function itself is; it suffices to note that its properties are reflected onto the functions with which it is multiplied and integrated. A colorful way of saying this is that the δ-functions are specified by the company they keep![4]

From equations (B.1)–(B.4), using the change of variables $y = ax$, we have

$$\int \delta(ax)\, f(x)\, dx = \int \delta(y)\, f\left(\frac{y}{a}\right) \frac{dy}{a},$$

so that

$$\int \delta(ax)\, f(x)\, dx = \frac{1}{a}\, f(0). \tag{B.5}$$

This relation is usually written in an abstract or operator form as

$$\delta(ax) = \frac{1}{a}\, \delta(x).$$

We note that the δ-function is an even function of its argument. Hence, if $a \le 0$, we can write $ax = -|a|x$, and $\delta(ax) = \delta(|a|x)$, so that

$$\delta(ax) = \frac{1}{|a|}\, \delta(x). \tag{B.6}$$

We can generalize the above to $\delta(f(x))$, where $f(x)$ has simple zeros at $x = x_i$ in the region over which integration is carried out. Near any one of its zeros, x_i, we can expand $f(x)$ in a Taylor series as

$$f(x) = f(x_i) + f'(x_i)\, (x - x_i) + \ldots,$$

with the first term being zero. Let us now evaluate

$$\int \delta(f(x))\, g(x)\, dx,$$

where $g(x)$ is any "smooth" function. For simplicity, assume that $f(x)$ has only one zero, at $x = x_i$, within the range of integration. We then have

$$\int \delta(f(x))\, g(x)\, dx = \int \delta(f'(x_i)\, (x - x_i))\, g(x)\, dx.$$

Using equation (B.6) we obtain

$$\int \delta(f(x))\, g(x)\, dx = \frac{1}{|f'(x_i)|} \int \delta(x - x_i)\, g(x)\, dx$$

$$= \frac{1}{|f'(x_i)|}\, g(x_i). \tag{B.7}$$

If we have many zeros for the function $f(x)$, then $\delta(f(x))$ will be nonzero only at those values of $x = x_i$ which are within the range of integration. We can write

$$\int \delta(f(x))\, g(x)\, dx = \sum_i \frac{1}{|df/dx|_{x=x_i}}\, g(x_i), \qquad \text{(B.8)}$$

or, in an operational sense,

$$\delta(f(x)) = \sum_i \left(\frac{1}{|f'(x_i)|}\, \delta(x - x_i) \right). \qquad \text{(B.9)}$$

The result, equation (B.9), can be obtained even more directly by a transformation of variables as we show below.

We evaluate $\int \delta(f(x))\, g(x)\, dx$ by substituting $f(x) = y$. Let the inverse relation $x = \mathcal{F}^{-1}y$ be the solution of x in terms of y. Then the change of variables leads to (i) $\delta(f(x)) \rightarrow \delta(y)$, (ii) $g(x) \rightarrow g(\mathcal{F}^{-1}y)$, and (iii) $dx \rightarrow dy/|f'(\mathcal{F}^{-1}y)|$. Here the modulus sign on the f' factor is present because the δ-function is even in its argument. These substitutions lead to

$$\int dx\, \delta(f(x))\, g(x) = \int dy\, \frac{1}{|f'(\mathcal{F}^{-1}y)|}\, \delta(y)\, g(\mathcal{F}^{-1}y)$$

$$= \left[\frac{g(\mathcal{F}^{-1}y)}{|f'(\mathcal{F}^{-1}y)|} \right]_{y=0}.$$

Here $y = 0$ corresponds to $f(x) = 0$. Denoting x_i to be the zeros of $f(x)$, we see that the integral reduces to

$$\sum_i g(x_i)/|f'(x_i)|,$$

so that we have the relation

$$\delta(f(x)) = \sum_i \frac{\delta(x - x_i)}{|f'(x_i)|}. \qquad \text{(B.10)}$$

Examples

1. $\delta(x^2 - a^2)$:

   Here $f(x) = x^2 - a^2$ has zeros at $x = \pm a$, with $|f'(x)|_{x=x_i} = |2x|_{x=\pm a} = |2a|$.

Hence we have

$$\delta(x^2 - a^2) = \frac{1}{|2a|}[\delta(x+a) + \delta(x-a)]. \qquad \text{(B.11)}$$

2. $\delta(\cos\theta - \cos\theta_0)$

Here $f(\theta) = \cos\theta - \cos\theta_0$, with $0 \leq \theta \leq 2\pi$, has a zero at $\theta = \theta_0$. Hence we have $f'(\theta) = -\sin\theta$, and $|f'(\theta)|_{\theta=\theta_0} = \sin\theta_0$. We then obtain

$$\delta(\cos\theta - \cos\theta_0) = \frac{1}{\sin\theta_0}\delta(\theta - \theta_0). \qquad \text{(B.12)}$$

B.2   The δ-function as the limit of a "normal" function

The δ-function can be obtained from a sequence of regular functions which converge to it. Such a sequence is called a δ-convergent sequence. We first illustrate the issue with examples.

1. Consider the function

$$f_\epsilon(x) = \frac{1}{\pi}\frac{\epsilon}{x^2 + \epsilon^2}.$$

We have shown in Fig. B.1 the function $f_\epsilon(x)$ for various values of ϵ.

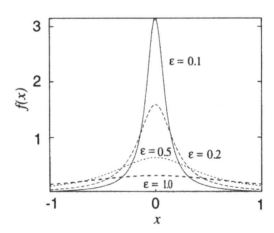

Fig. B.1 The sequence of functions of the form $\epsilon/(x^2 + \epsilon^2)$ approaches the δ-function in the limit $\epsilon \to 0$.

Since

$$\int_a^b f_\epsilon(x)\, dx = \frac{1}{\pi}\left(\arctan\frac{b}{\epsilon} - \arctan\frac{a}{\epsilon}\right),$$

we see that the $f_\epsilon(x)$ is clearly a δ-convergent sequence for $a \neq b$ and $\epsilon \to 0$. Thus

$$\lim_{\epsilon \to 0} \frac{1}{\pi}\frac{\epsilon}{(x^2 + \epsilon^2)} \to \delta(x). \qquad (\text{B.13})$$

2. The δ-function can be defined as the limit of a Gaussian function. Consider the Gaussian function

$$f_\sigma(x) = \frac{1}{2\sqrt{\pi\sigma}}e^{-x^2/4\sigma}, \quad \sigma > 0,$$

in the limit $\sigma \to 0$. First, we have $f_\sigma(x) > 0$, and secondly, for any a, b we have

$$\int_a^b f_\sigma(x)\, dx \leq \int_{-\infty}^{\infty} \frac{1}{2\sqrt{\pi\sigma}}\, e^{-x^2/4\sigma}\, dx = 1.$$

Hence, for any $a < 0 < b$,

$$\lim_{\sigma \to 0}\int_a^b f_\sigma(x)\, dx = \lim_{\sigma \to 0}\int_{a/\sqrt{\sigma}}^{b/\sqrt{\sigma}} \frac{1}{2\sqrt{\pi}}e^{-y^2/4}\, dy$$
$$= 1.$$

Again, for any $b > 0$, we have

$$\int_b^\infty \frac{1}{2\sqrt{\pi\sigma}}e^{-x^2/4\sigma}\, dx < \frac{1}{2\sqrt{\pi\sigma}}\int_b^\infty e^{-x^2/4\sigma}\left(\frac{x}{2\sigma}\right)\left(\frac{2\sigma}{b}\right)\, dx$$
$$= \frac{\sqrt{\sigma}}{b\sqrt{\pi}}e^{-b^2/4\sigma}.$$

This expression converges to zero as $\sigma \to 0$. This result is true also for any segment $(-\infty, a)$ for $a < 0$. We see that our sequence of functions $f_\sigma(x)$ is indeed a δ-convergent sequence with

$$\lim_{\sigma \to 0}\frac{1}{2\sqrt{\pi\sigma}}e^{-x^2/4\sigma} \to \delta(x). \qquad (\text{B.14})$$

3. The function $f_\nu(x) = (1/\pi)\sin(\nu x)/x$, with $0 < \nu < \infty$, is a δ-convergent sequence in ν. Here we observe that

$$\int_{-\infty}^{\infty} f_\nu(x)\,dx = \int_{-\infty}^{\infty} \frac{1}{\pi}\frac{\sin \nu x}{x}\,dx = 1. \qquad (B.15)$$

Also, for any $b > a > 0$, the integrals

$$\int_a^b f_\nu(x)\,dx = \frac{1}{\pi}\int_a^b \frac{\sin \nu x}{x}\,dx = \frac{1}{\pi}\int_{a\nu}^{b\nu} \frac{\sin y}{y}\,dy$$

tend to zero as $\nu \to \infty$ (since the region of integration tends to zero and the integrand is oscillatory and tending to zero). This is also true for any integration region from a to b where $a < b < 0$. Furthermore,

$$\left| \frac{1}{\pi}\int_a^b \frac{\sin \nu x}{x}\,dx \right| = \left| \frac{1}{\pi}\int_{a\nu}^{b\nu} \frac{\sin y}{y}\,dy \right|$$

is bounded uniformly in (a,b) for all ν. Hence we have

$$\lim_{\nu\to\infty} f_\nu(x) = \lim_{\nu\to\infty} \frac{1}{\pi}\frac{\sin \nu x}{x} = \delta(x). \qquad (B.16)$$

Since

$$\frac{1}{\pi}\frac{\sin \nu x}{x} = \frac{1}{2\pi}\int_{-\nu}^{\nu} e^{i\xi x}\,d\xi,$$

we deduce that

$$\lim_{\nu\to\infty}\int_{-\nu}^{\nu} e^{i\xi x}\,d\xi = 2\pi\delta(x) = \int_{-\infty}^{\infty} e^{i\xi x}\,d\xi. \qquad (B.17)$$

The above relation states that the Fourier transform of unity is the δ-function.

B.3 δ-functions in three dimensions

The δ-function in three dimensions is defined as a product of three one-dimensional δ-functions. Thus

$$\delta^{(3)}(\mathbf{r} - \mathbf{R}) = \delta(x - X)\,\delta(y - Y)\,\delta(z - Z), \qquad (B.18)$$

and

$$\int \delta^{(3)}(\mathbf{r} - \mathbf{R})\,dx\,dy\,dz = 1. \qquad (B.19)$$

When we change from one orthogonal coordinate system to another we have to account for the Jacobian of transformations. For example, let $\mathbf{r} = (x, y, z) \rightarrow (r, \theta, \phi)$ and $\mathbf{R} = (X, Y, Z) \rightarrow (R, \Theta, \Phi)$. We assume that both the points located at $\mathbf{r}$ and $\mathbf{R}$ are in the volume of integration. Then in rectangular coordinates

$$\int dx\, dy\, dz\, [\delta(x - X)\, \delta(y - Y)\, \delta(z - Z)] = 1,$$

and in spherical polar coordinates we need

$$\int r^2 dr\, \sin\theta\, d\theta\, d\phi\, \delta^{(3)}(\mathbf{r} - \mathbf{R}) = 1.$$

This requires that

$$\delta^{(3)}(\mathbf{r} - \mathbf{R}) \equiv \frac{\delta(r - R)\, \delta(\theta - \Theta)\, \delta(\phi - \Phi)}{r^2 \sin\theta}. \qquad (\text{B.20})$$

This relation is readily appreciated if we observe that the one-dimensional δ-function has the units of inverse length in order that its integral be dimensionless. The normalization with respect to integration then allows us to pick the correct scale factors. Again, for the cylindrical coordinate system with $\mathbf{r} \rightarrow (\rho, \phi, z)$ and $\mathbf{R} \rightarrow (\rho', \phi', z')$, we require that

$$\delta^{(3)}(\mathbf{r} - \mathbf{R}) = \frac{1}{\rho} \delta(\rho - \rho')\delta(\phi - \phi')\, \delta(z - z'). \qquad (\text{B.21})$$

B.4   Other generalized functions

B.4.1   The step-function $\theta(x)$

The step-function (also called the Heaviside function) $\theta(x)$, shown in Fig. B.2, is defined as

$$\theta(x) = \begin{cases} 0, & x < 0 \\ 1, & x > 0. \end{cases} \qquad (\text{B.22})$$

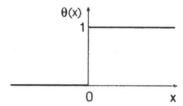

Fig. B.2 The step-function $\theta(x)$.

It is obvious that

$$\int_{-\infty}^{\infty} \theta(x)\, f(x)\, dx \; = \; \int_{0}^{\infty} f(x)\, dx. \qquad (B.23)$$

We also have the result, for any well-defined function $f(x)$ (which helpfully dies out at infinity), that

$$\int_{-\infty}^{\infty} \theta'(x)\, f(x)\, dx \; = \; - \int_{-\infty}^{\infty} \theta(x)\, f'(x)\, dx$$

$$= \; - \int_{0}^{\infty} f'(x)\, dx = \; f(0). \qquad (B.24)$$

We therefore deduce that

$$\frac{d}{dx}\, \theta(x) \; = \; \theta'(x) = \delta(x). \qquad (B.25)$$

B.4.2   The sign-function $\varepsilon(x)$

The sign-function $\varepsilon(x)$ is defined as

$$\varepsilon(x) \; = \; \begin{cases} -1, & x < 0 \\ +1, & x > 0. \end{cases} \qquad (B.26)$$

From this it follows that

$$\frac{d}{dx}\varepsilon(x) = \frac{d}{dx}\left(\theta(x) - \theta(-x) \right) = 2\,\delta(x). \qquad (B.27)$$

The function $\varepsilon(x)$ is shown in Fig. B.3.

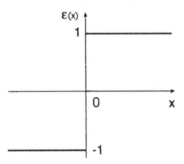

Fig. B.3 The sign-function $\varepsilon(x)$.

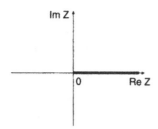

Fig. B.4 The branch cut for $\ln z$ on the complex z-plane.

B.4.3   The Plemelj formula

For completeness, we include a derivation of the Plemelj formula[5] which is used frequently in working with dispersion relations (Hilbert transforms):

$$\lim_{\eta \to 0} \frac{1}{x \pm i\eta} = \mathcal{P}\frac{1}{x} \mp i\pi\,\delta(x). \tag{B.28}$$

We prove this by considering the function defined by

$$\ln(x + i0^+) = \begin{cases} \ln|x| + i\pi, & \text{for } x < 0, \\ \ln|x|, & \text{for } x > 0. \end{cases} \tag{B.29}$$

This is just the limit of the complex function $\ln(z) = \ln(x + iy)$, as $y \to 0$. This function is analytic in the upper half plane (or rather on the plane with the exclusion of the branch cut shown in Fig. B.4), and we have specified the branch cut to be along the ray $(x > 0, \theta = 0)$. (In other words, we have excluded all the Riemann sheets except the principal one.)

Here

$$\ln(x + iy) = \ln\sqrt{x^2 + y^2} + i\arctan\left(\frac{y}{x}\right) \tag{B.30}$$

converges to the function $\ln(x + i0^+)$ as $y \to 0$. The first term on the right side of equation (B.30) converges to $\ln|x|$, and the second term has its modulus bounded by π and converges to the function $h(x)$ given by

$$h(x) = \begin{cases} i\pi, & \text{for } x < 0, \\ 0, & \text{for } x > 0, \end{cases} \tag{B.31}$$

so that

$$h(x) = i\pi\,\theta(-x).$$

The derivative of $\ln|x|$ is $1/x$, and the derivative of $h(x)$ is $-\pi\,\delta(x)$.

Thus

$$\frac{d}{dx} \ln(x + i0^+) = \lim_{y \to 0^+} \frac{d}{dx} \ln(x + iy) = \lim_{y \to 0^+} \frac{1}{x + iy},$$

giving the result

$$\lim_{\eta \to 0^+} \frac{1}{x + i\eta} = \frac{1}{x} - i\pi\,\delta(x). \tag{B.32}$$

These distributions have meaning only in the generalized sense, and are to be understood as appearing inside integrals. It is for this reason that the first term on the right side of equation (B.32) is written as $\mathcal{P}(1/x)$, where $\mathcal{P}$ denotes the Cauchy principal value of the integral. In an operational sense, we then have relation (B.28) to be valid.

B.4.4 An integral representation for $\theta(z)$

The integral representation for $\theta(x)$ can be shown to be

$$\theta(\pm x) = \lim_{\eta \to 0} \left\{ \pm \frac{i}{2\pi} \int_{-\infty}^{\infty} d\omega\, \frac{e^{-i\omega x}}{\omega \pm i\eta} \right\}. \tag{B.33}$$

The integral is evaluated by closing the contour on the complex ω-plane. For $x > 0$, the contour is closed on the lower half complex ω-plane, and for $x < 0$ the contour is closed in the upper half plane. The validity of the integral representation is readily checked by using the method of residues. For $x < 0$, there are no poles in the upper half plane in the integrand, and so $\theta(x) = 0$ in this case. For $x > 0$, using a residue evaluation we verify that $\theta(x) = 1$.

B.5 Problems

1. (a) A mass M is distributed uniformly as a thin spherical shell of radius R. Show that the mass density is

$$\rho(\mathbf{r}) = \frac{M}{4\pi r^2}\,\delta(r - R).$$

(b) A mass M of a material is formed into a uniform disk of radius R located at the origin. Show that its mass density in cylindrical coordinates (ρ, ϕ, z) is

$$\rho(\mathbf{r}) = \frac{M}{\pi R^2}\,\delta(z)\,\theta(R - \rho),$$

and that in spherical polar coordinates it is given by

$$\rho(\mathbf{r}) = \frac{3M}{2\pi R^3}\,\theta(R-r)\,\delta(\cos\theta).$$

2. Two metallic hemispherical shells of radius R are insulated from each other while placed coaxially. The two hemispheres are at potential $\pm V_0$. Verify that the potential on the surface of the structure in spherical polar coordinates is given by

$$V(\mathbf{r}) = V_0\,\delta(r-R)\,\varepsilon(\theta-\pi/2),$$

where ε is the sign function.

3. A charge Q is distributed on a spherical shell of radius R on which a cap defined by the cone $\theta=\alpha$ at the north pole has been removed.[†] We require an expression for the charge density such that in the limits $\alpha\to 0$ and $\alpha\to\pi$ the charge density integrates to a total charge Q. Show that

$$\rho(\mathbf{r}) = \frac{Q}{2\pi R^2}\,\frac{\delta(r-R)\,\theta(\theta-\alpha)\,\theta(\pi-\theta)}{(1+\cos\alpha)}$$

satisfies these requirements.

4. The one-particle Green's function for an electron gas at temperature T is given by

$$i\,G^0(\mathbf{x},\mathbf{x}';t-t') = i\int\frac{d\omega}{2\pi}\,e^{-i\omega(t-t')}\,G^0(\mathbf{x},\mathbf{x}';\omega)$$

where

$$i\,G^0(\mathbf{x},\mathbf{x}';\omega) = i\hbar\sum_\nu \phi_\nu(\mathbf{x})\phi_\nu(\mathbf{x}')\left[\frac{1-f(\mu_F,E_\nu,T)}{\hbar\omega-E_\nu+i\eta}\right.$$
$$\left.+\frac{f(\mu_F,E_\nu,T)}{\hbar\omega-E_\nu-i\eta}\right],$$

and μ_F is the Fermi energy. Here $f(\mu_F,E_\nu,T)$ is the Fermi–Dirac distribution function

$$f(\mu_F,E_\nu,T) = \frac{1}{e^{(E_\nu-\mu_F)/k_BT}+1},$$

[†]This is the starting point of Problem 3.2 in Jackson;[6] other examples in the same chapter provide further illustrations in the use of generalized functions to describe charge distributions.

with k_B being the Boltzmann constant. Using the Plemelj relation show that the number density $n(\mathbf{x})$ of electrons is given by

$$n(\mathbf{x}) = -i \lim_{t'-t \to 0^+} \left\{ \lim_{\mathbf{x} \to \mathbf{x}'} G^0(\mathbf{x}, \mathbf{x}'; t - t') \right\}$$

$$= \sum_\nu |\phi_\nu(\mathbf{x})|^2 f(\mu_F - E_\nu).$$

5. The Lindhard dielectric function for an electron gas at zero temperature is defined by

$$\Pi^0(\mathbf{q}, q_0) = \frac{2}{(2\pi)^3 \hbar} \int d^3p \, \theta(|\mathbf{q} + \mathbf{p}| - p_F) \, \theta(p_F - p)$$

$$\times \left(\frac{1}{\hbar q_0 + E_\mathbf{p} - E_{\mathbf{p}+\mathbf{q}} + i\eta} - \frac{1}{\hbar q_0 + E_{\mathbf{p}+\mathbf{q}} - E_\mathbf{p} - i\eta} \right).$$

Here $E_\mathbf{p} = \hbar^2 p^2 / 2m$ and p_F is the Fermi wavevector of the electron. Use the Plemelj relation to evaluate $Im\,\Pi$ and show that[7]

$$Im\,\Pi(\mathbf{q}, q_0) = -\frac{p_F^2 m}{4\pi \hbar^3 q}(1 - y_0^2),$$

where $0 \le y_0 \le 1$, and

$$y_0 = \max \left\{ \begin{array}{l} \sqrt{1 - \frac{\hbar q_0}{E_F}} \\ \left(\frac{q}{2p_F} - \frac{\hbar q_0}{2E_F(q/p_F)} \right) \\ \left(\frac{\hbar q_0}{2E_F(q/p_F)} - \frac{q}{2p_F} \right). \end{array} \right.$$

References

[1] I. M. Gel'fand and G. E. Shilov, *Generalized Functions*, Vols. I & II (Academic Press, New York, 1964); M. J. Lighthill, *Fourier Analysis and Generalized Functions* (Cambridge University Press, Cambridge, 1968); H. Bremermann, *Distributions, Complex Variables, and Fourier Transforms* (Addison-Wesley, Reading, MA, 1965).

[2] A. Papoulis, *The Fourier Integral and its Applications* (McGraw-Hill, New York, 1962).

[3] B. Friedman, *Principles and Techniques of Applied Mathematics* (Wiley, New York, 1956).

[4] J. R. Klauder and E. C. G. Sudarshan, *Fundamentals of Quantum Optics* (Benjamin, New York, 1968).

[5] J. Plemelj, *Monatsh. Math. Phys.* **19**, 205 (1908). T. Carleman, *Ark. Mat. Astron. Fys.* **16**, 26 (1922). See also, A. I. Markushevich, *Theory of Functions of a Complex Variable*, ed. R. A. Silverman (Prentice Hall, Englewood Cliffs, NJ, 1965).

[6] J. D. Jackson, *Classical Electrodynamics*, 2nd ed. (Wiley, New York, 1975).

[7] A. L. Fetter and J. D. Walecka, *Quantum Theory of Many-Particle Systems* (McGraw-Hill, New York, 1971).

C

Green's functions

C.1 Introduction

It can be shown that many physical quantities obey differential equations governing their local variation. For example, the electrostatic potential $\Phi(\mathbf{r})$ satisfies Laplace's equation, $\nabla^2\Phi(\mathbf{r}) = 0$, in regions free of charge distributions. The differential equation is integrated to obtain a solution at each point in the physical domain for the physical quantity. We provide here an introduction to the Green's function method for the solution of inhomógeneous differential equations.[†] We begin by considering one-dimensional examples to illustrate this method. The properties of Green's functions and the standard methods used to construct them are described. Next, examples from electrostatics are used to illustrate different types of boundary conditions (BCs) and the role of boundaries in defining the form of the solution. The textbook by Jackson[1] provides a detailed discussion of Green's functions for electrostatics, and our considerations here are meant to supplement that presentation. The Green's functions appearing in electrodynamics and the Green's solutions for the wave equation in one, two, and three dimensions are derived in order to provide the Green's functions for the scattering of waves. This presentation ends with a brief discussion on the application of Green's functions to convert differential equations into integral equations. We draw attention to a very readable, and brief, historical note on George Green.[2]

Let us begin with a one-dimensional example of a differential equation

$$\frac{d^2}{dx^2}y(x) - 3\frac{d}{dx}y(x) + 2y(x) = x\,e^x, \qquad \text{(C.1)}$$

[†]The Green's function is also referred to as the "fundamental solution" of the inhomogeneous differential equation.

obeyed by a function $y(x)$. This differential equation with constant coefficients has an inhomogeneous term on the right side, called the source term. Denoting d/dx as $\mathcal{D}$ and the right side as $f(x)$, we have

$$\mathcal{L}(\mathcal{D})\,y(x) = f(x), \qquad\qquad (C.2)$$

where $\mathcal{L}(\mathcal{D})$ is the function of the differential operator appearing on the left side of the equation. The general solution of this equation is the sum of the particular integral $y_{PI}(x)$, which takes account of the inhomogeneous term, and the *complementary function* $y_{CF}(x)$, which is the solution of the homogeneous equation $\mathcal{L}(\mathcal{D})y(x) = 0$. The operator $\mathcal{L}(\mathcal{D})$ factorizes into $(\mathcal{D} - 2)(\mathcal{D} - 1)$. Using standard methods described in books on differential equations[3,4] we obtain

$$y_{PI}(x) = [\mathcal{L}(\mathcal{D})]^{-1} f(x)$$

$$= \left[\frac{1}{(\mathcal{D} - 2)} - \frac{1}{(\mathcal{D} - 1)} \right] x\, e^x$$

$$= \int \left(e^{2x - 2x'} - e^{x - x'} \right) x' e^{x'} dx'$$

$$= -(1 + x)e^x - x^2\, e^x /2.$$

The complementary function is obtained by inspecting the factors of $\mathcal{L}(\mathcal{D})$

$$y_{CF}(x) = C_1 e^{2x} + C_2 e^x. \qquad\qquad (C.3)$$

The quantities C_1, C_2 are to be determined by the BCs appropriate to the problem.

The function in the integrand $\mathcal{G}(x - x') \equiv (e^{2x - 2x'} - e^{x - x'})$ has the same form whatever the source function is on the right side of equation (C.1). The significance of this can be understood by replacing the source term by a "unit" function, the δ-function. We can formally write[5]

$$\mathcal{L}(\mathcal{D})\,\mathcal{G} = 1, \qquad\qquad (C.4)$$

with a solution

$$\mathcal{G} = [\mathcal{L}(\mathcal{D})]^{-1} 1, \qquad\qquad (C.5)$$

so that $\mathcal{G}$ represents the "inverse" of the differential operator. The solution to equation (C.1) is given by

$$y = [\mathcal{L}(\mathcal{D})]^{-1} \mathbf{F},$$

where $\mathbf{F}$ represents the inhomogeneous term. In the coordinate-space representation, we write the above equation as

$$y(x) = \int \mathcal{G}(x - x')x'e^{x'}\,dx. \tag{C.6}$$

This solution can always be complemented by the solutions, Y_{CF}, of the homogeneous differential equation. Thus, once the Green's function is determined for the differential equation it is applicable to the same differential equation with any function as the inhomogeneous term on the right side of the equation. This general procedure of first solving for $\mathcal{G}(x - x')$, and then integrating over this function multiplied by the actual inhomogeneity, is known as the Green's function method of solving inhomogeneous differential equations. The function $\mathcal{G}$ is specific to the given inhomogeneous differential equation. A problem may be inhomogeneous because the differential equation has an inhomogeneous term, as in the above example, or because the BCs are inhomogeneous.[6] A transformation from one form of inhomogeneity to the other is possible.

C.2 Properties of Green's functions

Let us investigate the properties of Green's functions using a specific example. Consider an elastic string of unit length which is held fixed at its ends, and is subjected to a unit force at a point $x = x'$, where $0 \le x, x' \le 1$. The displacement $y(x)$ satisfies the differential equation

$$\frac{d^2}{dx^2}y(x) = -\delta(x - x'). \tag{C.7}$$

Let the BCs for $y(x)$ be

$$y(x) = 0, \quad \text{for} \quad \begin{cases} x = 0 \\ x = 1. \end{cases} \tag{C.8}$$

Equation (C.7) can be integrated from $(x' - \epsilon)$ to $(x' + \epsilon)$ across the discontinuity, to obtain

$$\lim_{\epsilon \to 0} \left\{ \frac{d}{dx} y(x)\bigg|_{x=x'+\epsilon} - \frac{d}{dx} y(x)\bigg|_{x=x'-\epsilon} \right\} = -1. \tag{C.9}$$

Observe that in Fig. C.1 we have a discontinuity in $y'(x)$ but none in $y(x)$. The general solution to equation (C.7) is of the form

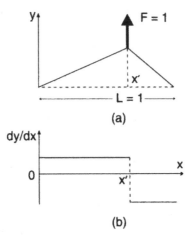

Fig. C.1 (a) A string of unit length with a unit force acting at $x = x'$. (b) The slope of the string as a function of the coordinate x.

$y = mx + c$. However, there are two separate regions $0 \le x < x'$ and $x' < x \le 1$ where this general form has to be separately employed. We can satisfy the BCs (C.8) by having

$$y(x) = \begin{cases} A\,x, & x < x', \\ B\,(1-x), & x > x'. \end{cases} \quad \text{(C.10)}$$

The continuity in y together with the discontinuity in its derivative at $x = x'$ given by equation (C.9) require that

$$A\,x' = B\,(1 - x'),$$
$$-A - B = -1. \quad \text{(C.11)}$$

We can therefore identify the constants as $A = (1 - x')$ and $B = x'$, and hence

$$y(x) \equiv g(x, x') = \begin{cases} x\,(1-x'), & x \le x', \\ x'\,(1-x), & x \ge x'. \end{cases} \quad \text{(C.12)}$$

Here $g(x, x')$ is the Green's function for the differential equation $d^2 y/dx^2 = -\delta(x - x')$. It gives the response at x for a unit "source" placed at x' and it satisfies the imposed BCs.

We now enumerate the properties of Green's functions using the above example as an illustration.[6]

1. The Green's functions are symmetrical in x, x' so that the relation $g(x, x') = g(x', x)$ holds. This is called the reciprocity relation. It states that the response at x due to a unit point disturbance at x' is the same as the response at x' due to a unit disturbance at x. In other words, the exchange of the source and observer does not change $g(x, x')$. This property is explicitly satisfied in our example, in equation (C.12).

2. Consider a new problem with different BCs as in Fig. C.2. A solution of the type $Px + Q$ which satisfies the conditions

$$y = a, \qquad \text{for } x = 0,$$
$$y = b, \qquad \text{for } x = 1, \qquad\qquad (C.13)$$

is given by

$$y(x) = a \left[\frac{\partial}{\partial x'} g(x, x') \right]_{x'=0} - b \left[\frac{\partial}{\partial x'} g(x, x') \right]_{x'=1}$$
$$= a\,(1 - x) + b\,x. \qquad\qquad (C.14)$$

We see that the standard Green's function can be used to generate other solutions to the differential equation satisfying different BCs; this has been achieved by shifting the discontinuity in our standard solution $g(x, x')$ to the boundaries.

3. Consider a solution of $d^2y/dx^2 = -\sum_j f_j \delta(x - x_j)$ which is a linear combination of a number of such Green's functions corresponding to a set of sources at x_j:

$$y(x) = \sum_{j=1}^{n} f_j\, g(x, x'_j); \qquad\qquad (C.15)$$

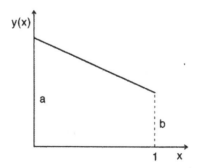

Fig. C.2 The function $Y(x)$ with nonzero displacements at $x = 0, 1$.

here f_j are arbitrary constants. Now the derivative drops an amount f_1 at x_1', and f_2 at x_2', and so on. If we replace the sum by an integral in order to represent a very large number of "source" points x_j' we obtain

$$d^2y(x)/dx^2 = -f(x). \tag{C.16}$$

The solution of the above equation is given by

$$y(x) = \int_0^1 g(x, x') f(x') \, dx'$$
$$= \int_0^x (1-x) x' f(x') \, dx' + \int_x^1 x (1-x') f(x') \, dx'. \tag{C.17}$$

Let us verify this result by differentiating $y(x)$ using the general formula (Leibnitz's theorem)[‡]

$$\frac{d}{dt} \int_{\alpha(t)}^{\beta(t)} f(x,t) \, dx = f(\beta) \frac{\partial \beta(t)}{\partial t} - f(\alpha) \frac{\partial \alpha(t)}{\partial t}$$
$$+ \int_{\alpha(t)}^{\beta(t)} \frac{\partial f(x,t)}{\partial t} \, dx. \tag{C.18}$$

We obtain

$$\frac{dy(x)}{dx} = -\int_0^x x' f(x') \, dx' + \int_x^1 (1-x') f(x') \, dx', \tag{C.19}$$

and

$$\frac{d^2}{dx^2} y(x) = -x f(x) - (1-x) f(x) = -f(x), \tag{C.20}$$

[‡]We can readily derive the rule for differentiation when the integration limits depend on the variable by using properties of the step-function. The step-function

$$\theta(x - \alpha(t)) = \begin{cases} 0, & \text{for } x < \alpha(t), \\ 1, & \text{for } x > \alpha(t), \end{cases}$$

has the derivative

$$\frac{d}{dt}\theta(x - \alpha(t)) = -\frac{\partial \alpha(t)}{\partial t} \delta(x - \alpha(t)).$$

We begin by writing

$$\frac{d}{dt} \int_{\alpha(t)}^{\beta(t)} f(x,t) \, dx = \frac{d}{dt} \int dx \, f(x,t) \, [\theta(x - \alpha(t)) - \theta(x - \beta(t))].$$

On now freely pulling the derivative inside the integral, we obtain the stated result of Leibnitz.

assuming that $f(x)$ is continuous in the interval $[0,1]$. Thus, our so-lution equation (C.17) is a solution of the inhomogeneous differential equation (C.16).

Conversely, we can state that the solution of the differential equation (C.16), with BCs $y = 0$ at $x = 0, 1$, and with an inhomogeneity $-f(x)$ which is continuous in $[0,1]$, is given by

$$y(x) = \int_0^1 g(x, x')\, f(x')\, dx'. \tag{C.21}$$

Such a solution is continuous in $[0,1]$ and has continuous first and second derivatives. Thus the Green's functions allow us to solve inhomogeneous differential equations with the same BCs, in which the inhomogeneity is any given smooth function.

C.3 Sturm–Liouville differential operators

Let us now restate the considerations of the previous section in more general terms. We wish to find the solution to the differential equation of the Sturm–Liouville type with an inhomogeneous term $f(x)$,

$$\mathcal{H}(\mathcal{D})\,\Phi(x) \; - \; \lambda\,\Phi(x) \; = \; f(x), \tag{C.22}$$

over a domain Ω where $\mathcal{H}$ is a Hermitian differential operator, with $\Phi(x)$ subject to the usual type of homogeneous BCs. Here, λ is a given constant.

To accomplish this, we first consider the homogeneous problem with the same BCs, and suppose that we have eigensolutions $u_i(x)$ with eigenvalues λ_i (which are not equal to λ):

$$\mathcal{H}\,u_i(x) \; = \; \lambda_i\,u_i(x). \tag{C.23}$$

We say that $\mathcal{H}$ is Hermitian if

$$\int_\Omega u^*(x)\,\mathcal{H}\,v(x)\, dx = \left[\int_\Omega v^*(x)\,\mathcal{H}\,u(x)\, dx \right]^*,$$

where u, v are any two solutions of the differential equation

$$\mathcal{H}(\mathcal{D})\Phi(x) \; - \; \lambda\,\Phi(x) \; = \; 0.$$

We can show that if $\mathcal{L}$ is Hermitian then the eigenvalues λ_i are real and the eigenfunctions form an orthogonal set. For any two eigenfunctions satisfying

$$\mathcal{H}\,u_i(x) \; = \; \lambda_i\,u_i(x),$$
$$\mathcal{H}\,u_j(x) \; = \; \lambda_j\,u_j(x),$$

let us construct the quantities

$$\int_\Omega u_j^*(x)\,\mathcal{H}\,u_i(x)\,dx \;=\; \lambda_i \int_\Omega u_j^*(x)\,u_i(x)\,dx$$

$$\int_\Omega u_i^*(x)\,\mathcal{H}\,u_j(x)\,dx \;=\; \lambda_j \int_\Omega u_i^*(x)\,u_j(x)\,dx.$$

Since $\mathcal{H}$ is Hermitian, the left hand sides of the above two equations are complex conjugates of each other, and hence

$$(\lambda_i - \lambda_j^*) \int_\Omega u_j^*(x)\,u_i(x)\,dx = 0.$$

$$(C.24)$$

Thus, either $i = j$, in which case $\lambda_i = \lambda_j^*$, or $i \neq j$ leading to

$$\int_\Omega u_j^*(x)\,u_i(x)\,dx \;=\; 0,$$

and the eigenfunctions are orthogonal. We can also show that they form a complete set of orthogonal functions, so that any function can be expanded in terms of this set of functions.[4,7]

The existence of the orthogonal basis allows us to represent the most general solution to equation (C.23) as a linear superposition of the eigenfunctions u_i

$$u(x) \;=\; \sum_i c_i u_i(x).$$

Now, in order to solve the inhomogeneous equation (C.22), we again expand $\Phi(x)$ and $f(x)$ in terms of the eigenfunctions of $\mathcal{H}$:

$$\Phi(x) \;=\; \sum_n a_n u_n(x);$$

$$f(x) \;=\; \sum_m b_m u_m(x).$$

Substituting these expansions into equation (C.22), we have

$$\sum_n a_n (\lambda_n - \lambda)\, u_n(x) \;=\; \sum_m b_m u_m(x). \qquad (C.25)$$

Since the u_n form a complete, linearly independent, orthogonal set of functions, we can equate the coefficients to obtain

$$a_n \;=\; \frac{b_n}{\lambda_n - \lambda}.$$

Thus

$$\Phi(x) = \sum_n \frac{u_n(x)}{\lambda_n - \lambda} \int u_n^*(x')\, f(x')\, dx'$$

$$= \int \left\{ \sum_n \frac{u_n(x)\, u_n^*(x')}{\lambda_n - \lambda} \right\} f(x')\, dx'. \qquad (C.26)$$

If we define the Green's function to be

$$G(x, x') = \sum_n \frac{u_n(x) u_n^*(x')}{\lambda_n - \lambda}, \qquad (C.27)$$

we see that our solution for $\Phi(x)$ has the same integral representation as in the examples considered in the earlier sections. The Green's function is determined by (i) a differential Hermitian operator, (ii) a physical domain defining the range of the variables, and (iii) BCs appropriate to the problem.

The above Green's function $G(x, x')$ satisfies the differential equation

$$(\mathcal{H}(\mathcal{D}) - \lambda)\, G(x, x') = \delta(x - x'), \qquad (C.28)$$

as is readily seen by using equation (C.23). The left side can be rewritten as $\sum_n u_n(x)u_n^*(x')$, which is just the δ-function represented in the function space of the eigenfunctions satisfying the specified BCs. We verify this as follows. We note that for any function $h(x)$ expanded in terms of $u_n(x)$, and we have

$$h(x) = \sum_n u_n(x) \int dx'\, u_n^*(x')\, h(x')$$

$$= \int dx'\, h(x') \sum_n u_n(x)\, u_n^*(x')$$

$$= \int dx'\, h(x')\, \delta(x - x').$$

The relation

$$\delta(x - x') = \sum_n u_n(x)u_n^*(x') \qquad (C.29)$$

is called the completeness relation, or the closure property of the eigenfunctions.

We have thus shown that we can obtain the Green's function for the problem with unit inhomogeneity, and, hence, the solution to the same differential equation with any inhomogeneity.

Examples

1. Suppose we are given a vibrating string of length L, with its ends fixed, subject to a time-dependent driving force at x' proportional to $\delta(x - x')\, e^{-i\omega t}$. The homogeneous equation

$$\left(\frac{\partial^2}{\partial x^2} - \frac{1}{c^2}\frac{\partial^2}{\partial t^2}\right) u(x,t) = 0 \qquad (C.30)$$

is used to obtain the eigenfunctions. The time-independent part is obtained by substituting $u(x,t) = u(x)e^{-i\omega t}$. We then have

$$\frac{d^2}{dx^2}\, u(x) + k^2\, u(x) = 0, \qquad (C.31)$$

where $k^2 \equiv \omega^2/c^2 = -\lambda$. The BCs are $u(L) = u(0) = 0$.

The eigenfunctions are $u_n(x) = \sqrt{(2/L)}\sin(n\pi x/L)$ with eigenvalues given by

$$-\lambda_n = k_n^2 = (n/\pi L)^2\,; \quad n = 1,2,3,\ldots.$$

We can at once write the Green's function solution to

$$\left(\frac{d^2}{dx^2} + k^2\right) G(x,x') = +\delta(x - x') \qquad (C.32)$$

as

$$
\begin{aligned}
G(x,x') &= \sum \frac{u_n(x)u_n(x')}{\lambda_n - \lambda} \\
&= \frac{2}{L}\sum \frac{\sin(n\pi x/L)\,\sin(n\pi x'/L)}{k^2 - (n\pi/L)^2}.
\end{aligned} \qquad (C.33)
$$

The above Green's function equation (C.33) clearly exhibits the reciprocity property. The infinities of G as $\lambda \to \lambda_n$ are just the infinities which occur when a non-dissipative vibrating system is driven at one of its resonant frequencies! In nature, such singularities will be "smoothed out" by sources actually being extended sources (i.e., they are generally not point sources) and by effects of damping or by other dissipative effects.

2. We could also have obtained $G(x,x')$ by the method of matching the solutions at $x = x'$. Let us follow this procedure here, along the lines of our very first example (Section C.2). For $x \neq x'$, the

inhomogeneity of equation (C.32) is absent (since $\delta(x - x') = 0$, for $x \neq x'$). We solve

$$\frac{d^2}{dx^2} G(x, x') + k^2 G(x, x') = 0 \qquad (\text{C.34})$$

in the two regions $x > x', x < x'$ to obtain solutions consistent with the boundary conditions

$$G(x, x') = \begin{cases} a \sin kx, & x < x', \\ b \sin k(x - L), & x > x'. \end{cases} \qquad (\text{C.35})$$

To determine a and b, we use the information that G is continuous at $x = x'$. We integrate the differential equation from $(x' - \epsilon)$ to $(x' + \epsilon)$ over the variable x

$$\int_{x'-\epsilon}^{x'+\epsilon} dx \, \frac{d^2}{dx^2} G(x, x') + \int_{x'-\epsilon}^{x'+\epsilon} dx \, k^2 G(x, x')$$

$$= \int_{x'-\epsilon}^{x'+\epsilon} \delta(x - x') \, dx = 1.$$

Therefore

$$\left[\frac{dG(x, x')}{dx} \right]_{x=x'-\epsilon}^{x=x'+\epsilon} + 2\epsilon \left[k^2 G(x, x') \right]_{x=x'} = 1. \qquad (\text{C.36})$$

Since G is continuous, the second term tends to zero as $\epsilon \to 0$, and we have

$$\frac{dG}{dx} \bigg]_{x=x'-\epsilon}^{x=x'+\epsilon} = 1. \qquad (\text{C.37})$$

Thus, the first derivative has a unit discontinuity at $x = x'$. This relation together with the continuity of G at $x = x'$ can be used to obtain a and b:

$$a \sin kx' = b \sin k(x' - L) ,$$
$$kb \cos(k(x' - L)) - ka \cos kx' = +1,$$

giving

$$a = \frac{\sin k(x' - L)}{k \sin kL},$$

$$b = \frac{\sin kx'}{k \sin kL}.$$

We thus have

$$G(x, x') = \frac{1}{k \sin kL} \begin{cases} \sin kx \sin k(x' - L), & 0 \le x \le x', \\[2mm] \sin k(x - L) \sin kx', & x' \le x \le L, \end{cases}$$

$$\equiv \left(\frac{1}{k \sin kL} \right) \sin kx_< \sin k(L - x)_<. \qquad (C.38)$$

Here we have denoted $x_<$ as the minimum of (x, x') and the minimum of $\{L - x, L - x'\}$ is represented by $(L - x)_<$.

3. An alternate way of obtaining the Green's function is to use the method of variation of coefficients. We express the Green's function in terms of solutions of the corresponding homogeneous equation as

$$g(x, x') = \alpha(x, x') \cos kx + \beta(x, x') \sin kx. \qquad (C.39)$$

We differentiate $g(x, x')$ once to obtain the relation

$$\frac{d}{dx} g(x, x') = -k(\alpha \sin kx - \beta \cos kx)$$
$$+ \left\{ \frac{d\alpha}{dx} \cos kx + \frac{d\beta}{dx} \sin kx \right\}. \qquad (C.40)$$

We provide a constraint on the form of α and β by setting the second term in braces in equation (C.40) to zero

$$\frac{d\alpha}{dx} \cos kx + \frac{d\beta}{dx} \sin kx = 0. \qquad (C.41)$$

The second derivative of $g(x, x')$ is now obtained from equation (C.40) and compared with the differential equation satisfied by $g(x, x')$ to obtain

$$-k \left\{ \frac{d\alpha}{dx} \sin kx - \frac{d\beta}{dx} \cos kx \right\} = \delta(x - x'). \qquad (C.42)$$

Equations (C.41,C.42) are used to solve for the derivatives of α and β. We have

$$\frac{d\alpha}{dx} = -\frac{1}{k} \sin kx \, \delta(x - x'),$$
$$\frac{d\beta}{dx} = \frac{1}{k} \cos kx \, \delta(x - x'). \qquad (C.43)$$

These relations can be integrated to obtain

$$
\alpha = C_1 - \begin{cases} \dfrac{1}{k}\sin kx', & x > x', \\[2mm] 0, & x < x', \end{cases}
$$

$$
\beta = C_2 + \begin{cases} \dfrac{1}{k}\cos kx', & x > x', \\[2mm] 0, & x < x'. \end{cases}
$$

The Green's function is expressed as a sum of this particular solution and the complementary function which is a linear combination of the two solutions of the homogeneous equation

$$
g(x, x') = C_1 \cos kx + C_2 \sin kx + \frac{1}{k} \begin{cases} \sin k(x - x'), & x > x', \\[2mm] 0, & x < x'. \end{cases}
$$

The BCs $g(0, x') = 0 = g(L, x')$ require that $C_1 = 0$ and $C_2 = -(1/k)\sin k(L - x')/\sin kL$, leading to the Green's function of equation (C.38).

4. Finally, we have yet another way of solving for G. In obtaining G, we have to face two problems: (i) the singular nature of G has to be reproduced, and (ii) the given BCs have to be satisfied. We can separate the two issues by setting

$$
G(x, x') = u(x, x') + v(x, x') \tag{C.44}
$$

where u includes the singular behavior at $x = x'$ without necessarily satisfying the BCs, while v is a "smooth" function at $x = x'$ which is adjusted so as to make $G(x, x')$ satisfy the BC. Here v must be a solution of the homogeneous differential equation. This method is particularly useful in numerical approaches to the solution of inhomogeneous differential equations, and has been extensively employed in boundary element analysis.[8–10] Further theoretical developments may be perused in Ref. 11. This completes our brief presentation of the theory of Green's functions. The following sections provide further examples in typical applications.

C.4 Green's functions in electrostatics

Coulomb's law of force between two charges leads to a definition for the electric field at any point $\mathbf{r}$ due to a charge q located at $\mathbf{r}'$. We have

$$\mathbf{E}(\mathbf{r}) = q\frac{(\mathbf{r} - \mathbf{r}')}{|\mathbf{r} - \mathbf{r}'|^3}. \tag{C.45}$$

The longitudinal nature of the static electric field is guaranteed by the fact that $\nabla \times \mathbf{E}(\mathbf{r}) = 0$. This allows us to define the vector field as the negative gradient of a scalar function, the potential function $\Phi(\mathbf{r})$, and we have

$$\mathbf{E}(\mathbf{r}) = -\nabla\Phi(\mathbf{r}).$$

In the following, we shall set the dielectric constant ϵ of the region of interest to be unity. The differential form of Gauss's law of electrostatics,

$$\nabla \cdot \mathbf{E}(\mathbf{r}) = 4\pi\,\rho(\mathbf{r})$$

can then be written in terms of the potential function, and we find that $\Phi(\mathbf{r})$ satisfies Poisson's equation

$$\nabla^2\Phi(\mathbf{r}) = -4\pi\,\rho(\mathbf{r}). \tag{C.46}$$

In typical problems, we are given a distribution of charge $\rho(\mathbf{r})$ and are required to obtain the electric field or the electrostatic potential generated by the charge distribution. The solution of Poisson's equation for appropriate BCs is thus the central problem of electrostatics.[1] Being an inhomogeneous differential equation, we anticipate that the Green's function method can be used to solve for $\Phi(\mathbf{r})$ if we are given a charge distribution $\rho(\mathbf{r})$.

The Green's function for the open, infinite domain is readily obtained as follows. The electric field, equation (C.45), can be written as

$$\mathbf{E}(\mathbf{r}) = -\nabla\frac{q}{|\mathbf{r} - \mathbf{r}'|}$$
$$= -\nabla\Phi(\mathbf{r}).$$

We identify the potential function as

$$\Phi(\mathbf{r}) = \frac{q}{|\mathbf{r} - \mathbf{r}'|}, \tag{C.47}$$

and we have immediately solved Poisson's equation for a point charge $\rho(\mathbf{r}) = q\,\delta(\mathbf{r} - \mathbf{r}')$. With $q = 1$, we identify the Green's function as

$$G(\mathbf{r} - \mathbf{r}') \equiv \Phi(\mathbf{r})|_{q=1} = \frac{1}{|\mathbf{r} - \mathbf{r}'|}. \qquad (\text{C.48})$$

This is the Green's function for the open domain with the boundaries at infinity, where far away from the point charge at $\mathbf{r}'$, the potential is seen to be zero. Then, any general charge distribution $\rho(\mathbf{r})$ generates a potential which is given by

$$\Phi(\mathbf{r}) = \int d^3r'\rho(\mathbf{r}')\,G(\mathbf{r} - \mathbf{r}'). \qquad (\text{C.49})$$

The integral form of the solution is seen to arise from the Green's function technique; on the other hand, it is also a manifestation of the linear superposition principle for the potential, a fact which can be appreciated when the above integral is discretized to a sum over the contributions of the charge elements $(\rho(\mathbf{r}')\,d^3r')$ to the potential, using equation (C.48).

Let us verify that the Green's function $G(\mathbf{r}, \mathbf{r}') = 1/|\mathbf{r} - \mathbf{r}'|$ is indeed a solution of

$$\nabla^2 G(\mathbf{r}, \mathbf{r}') = -4\pi\delta(\mathbf{r} - \mathbf{r}'). \qquad (\text{C.50})$$

We do this in two steps. First, we consider the case when $\mathbf{r} \neq \mathbf{r}'$, so that the right side of equation (C.50) is zero, and $G(\mathbf{r}, \mathbf{r}')$ satisfies Laplace's equation. With $\mathbf{r} \neq \mathbf{r}'$, we shift the origin to $\mathbf{r}'$ and directly evaluate $\nabla^2 G(\mathbf{r})$. We obtain

$$\begin{aligned}
\nabla^2 G(r) &= \nabla^2 \frac{1}{r} \\
&= \frac{1}{r}\frac{d^2}{dr^2}\left(r \cdot \frac{1}{r}\right) \\
&\equiv \frac{1}{r^2}\frac{d}{dr}\left(r^2\frac{d}{dr}\left(\frac{1}{r}\right)\right) = 0.
\end{aligned}$$

Next, we must show that for $\mathbf{r} = \mathbf{r}'$, the Laplacian of $G(\mathbf{r}, \mathbf{r}')$ reproduces the δ-function. Here G is singular at $\mathbf{r} = \mathbf{r}'$; we can, however, show the equality of the two sides of the relation equation (C.50) by integrating both sides over a small spherical volume of radius ϵ centered at $\mathbf{r}'$. The volume integration over the right side gives -4π.

Gauss's theorem from vector analysis is used to convert the volume integral on the left side into a surface integral, and we obtain

$$\int_{V(\epsilon)} \nabla^2 \frac{1}{r} d^3r = \int_{V(\epsilon)} \nabla \cdot \left(\nabla \frac{1}{r} \right) d^3r$$

$$= \int_{S(\epsilon)} \nabla \frac{1}{r} \cdot \hat{n} \, dS$$

$$= \int_{S(\epsilon)} \frac{d}{dr} \left(\frac{1}{r} \right) r^2 \, d\Omega$$

$$= -4\pi.$$

Here, $S(\epsilon)$ is the surface enclosing the volume $V(\epsilon)$, and $d\Omega$ is the element of solid angle subtended by an element of area $dS = r^2 \, d\Omega$. We have thus verified that the Green's function of equation (C.48) is the correct solution to equation C.50).

Finite domains

If we have a *finite region* we will need to re-establish the corresponding form of the Green's function which will satisfy the new BCs. Let us suppose that we have a physical region of volume V bounded by a surface S.

We first show our solution for the potential function that is unique, up to an additive constant, if we specify only $\Phi(\mathbf{r}_S)$ or $\partial \Phi(\mathbf{r}_S)/\partial n|_S$ at the surface. Here either the potential function or its derivative along the outward-drawn normal at the surface S may be given on various sections of the surface.

We prove the uniqueness of the solution by first supposing that the contrary is true, and that we have two solutions, $\Phi_1(\mathbf{r})$ and $\Phi_2(\mathbf{r})$, to Poisson's equation satisfying the same BCs. If this is so, then the difference $\phi(\mathbf{r}) = \Phi_1(\mathbf{r}) - \Phi_2(\mathbf{r})$ clearly satisfies Laplace's equation $\nabla^2 \phi = 0$; furthermore, on the boundary, either $\phi_S = 0$ or $\partial \phi/\partial n|_S = 0$. Now consider the quantity

$$\int_V d^3r \, \nabla \cdot (\phi \nabla \phi) = \int_V d^3r \, [\phi \nabla^2 \phi + \nabla \phi \cdot \nabla \phi]. \tag{C.51}$$

Since $\nabla^2 \phi = 0$, the first term on the right side is zero. Also, $\nabla \phi \cdot \nabla \phi = |\nabla \phi|^2$. Furthermore, we can convert the left side to a surface integral, again using Gauss's theorem. Equation (C.51) then reduces to

$$\oint_S (\phi \nabla \phi).\hat{n}\, ds \equiv \oint_S \phi \frac{\partial \phi}{\partial n}\, ds = \int_V |\nabla \phi|^2\, d^3r. \qquad \text{(C.52)}$$

Our BCs specify either $\phi = 0$ or $\partial \phi/\partial n = 0$ on the various sections of the surface, so that

$$\int_V d^3r\, |\nabla \phi|^2 = 0.$$

We thus find that $\nabla \phi = 0$ inside V. In other words, we have $\phi = \Phi_1 - \Phi_2$ is a constant everywhere. If the BC is $\Phi_1 = \Phi_2$ on S, the constant is zero. If $\partial \Phi_1/\partial n = \partial \Phi_2/\partial n$ on S, we find that the solution Φ is unique up to a constant. When Φ_S is specified we have the Dirichlet BC, and when $(\partial \Phi/\partial n)_S$ is specified it is called a Neumann BC.

Once $G(\mathbf{r}, \mathbf{r}')$ which fulfills the boundary requirements is known, we can construct the solution to Poisson's equation for any arbitrary charge distribution $\rho(\mathbf{r})$ present in the volume V. The determination of G is not always straightforward; the examples and problems contained in Jackson[1] attest to the effort needed. We will show that there is an alternative, numerical, approach to solving Poisson's equation, known as the boundary integral method.[8–10]

C.5   Boundary integral solutions: a comment

Let us suppose that we have a finite volume V bounded by a surface S on which the BCs are: (i) $\Phi(\mathbf{r}_S)$ is specified on S_1, and (ii) $\partial \Phi(\mathbf{r})/\partial n|_{r_S}$ is given on S_2, where $S = S_1 + S_2$. Let the charge distribution inside V be $\rho(\mathbf{r})$. The potential on S_1 and the normal component of the electric field on S_2 have contributions from the charges inside the surface as well as from other charges *exterior* to V. Both of these together give rise to the specified values of Φ_S and $\partial \Phi/\partial n|_S = -\mathbf{E}_{S_2} \cdot \hat{n}$.

We wish to avoid a full development of the actual Green's function for this problem. We have seen earlier (equation (C.44)) that the Green's function can be expressed as the sum of a singular solution representing the solution of the differential equation with a unit inhomogeneity placed in an open domain and a function which satisfies the homogeneous differential equation

$$
\begin{aligned}
G(\mathbf{r}, \mathbf{r}') &= \frac{1}{|\mathbf{r} - \mathbf{r}'|} + F(\mathbf{r}, \mathbf{r}') \\
&\equiv G_\infty(\mathbf{r}') + F(\mathbf{r}, \mathbf{r}').
\end{aligned} \tag{C.53}
$$

Generally, the Green's function for the infinite domain is straightforward to obtain. The function $F(\mathbf{r}, \mathbf{r}')$ provides the ability for G to satisfy the given BCs on the surface defining the finite domain, and G_∞ takes care of the singular nature of the Green's function as $\mathbf{r} \to \mathbf{r}'$. Since this singular nature of G is the crucial aspect of solving inhomogeneous differential equations, we would expect that it is possible to use G_∞ alone, thereby bypassing the difficult issue of having to solve for F. In fact, Green's theorem of vector analysis allows us to write the solution in terms of G_∞ and Φ, $\partial\Phi/\partial n$ on the boundary. Using this theorem we show below that a *self-consistency relation* is obeyed by Φ. This self-consistency condition can be used to numerically solve for the potential in the finite physical domain.

Given two scalar functions $\Phi(\mathbf{r})$ and $\Psi(\mathbf{r})$ defined in V, we have

$$
\int_V d^3 r' \, \nabla \cdot (\Phi \nabla \Psi \cdot \hat{n}) = \int_S ds' \, \Phi \frac{\partial \Psi}{\partial n},
$$

or

$$
\int_V d^3 r' \, (\Phi \nabla^2 \Psi + \nabla \Phi \cdot \nabla \Psi) = \int_S ds' \, \Phi \frac{\partial \Psi}{\partial n}. \tag{C.54}
$$

In a similar manner, on interchanging Ψ and Φ, we obtain

$$
\int_V d^3 r' \, (\Psi \nabla^2 \Phi + \nabla \Psi \cdot \nabla \Phi) = \int_S ds' \, \Psi \frac{\partial \Phi}{\partial n}. \tag{C.55}
$$

Subtracting (C.55) from (C.54), we obtain Green's theorem:

$$
\int_V d^3 r' \, (\Phi \nabla^2 \Psi - \Psi \nabla^2 \Phi) = \int_S ds' \left(\Phi \frac{\partial \Psi}{\partial n} - \Psi \frac{\partial \Phi}{\partial n} \right). \tag{C.56}
$$

Now let us suppose that Φ satisfies

$$
\nabla^2 \Phi(\mathbf{r}) = -4\pi \rho(\mathbf{r}) \tag{C.57}
$$

inside V, and let

$$
\Psi \equiv G_\infty(\mathbf{r}, \mathbf{r}') = 1/|\mathbf{r} - \mathbf{r}'|, \tag{C.58}
$$

where $\mathbf{r}$ and $\mathbf{r}'$ are interior to V. Substituting these relations into equation (C.56) and using $\nabla^2 G_\infty = -4\pi\delta(\mathbf{r} - \mathbf{r}')$, we find

$$\int_V d^3r' \delta(\mathbf{r} - \mathbf{r}')\Phi(\mathbf{r}') = \int_V d^3r' \frac{\rho(\mathbf{r}')}{|\mathbf{r} - \mathbf{r}'|} + \frac{1}{4\pi}\int_S ds' \frac{1}{|\mathbf{r} - \mathbf{r}'|}\left(\frac{\partial\Phi}{\partial n}\right)_{\mathbf{r}'=\mathbf{r}_S}$$

$$- \frac{1}{4\pi}\int_S ds' \, \Phi(\mathbf{r}') \frac{\partial}{\partial n}\left(\frac{1}{|\mathbf{r} - \mathbf{r}'|}\right),$$

or

$$\Phi(\mathbf{r}) = \int_V d^3r' \frac{\rho(\mathbf{r}')}{|\mathbf{r} - \mathbf{r}'|} + \frac{1}{4\pi}\int_S ds' \frac{\partial\Phi/\partial n}{|\mathbf{r} - \mathbf{r}'|}$$

$$+ \frac{1}{4\pi}\int_S ds' \, \Phi(\mathbf{r}') \frac{(\mathbf{r} - \mathbf{r}').\hat{n}}{|\mathbf{r} - \mathbf{r}'|^3}. \tag{C.59}$$

We can use equation (C.59) to obtain a numerical solution for the potential everywhere using the boundary element method, as discussed in Chapter 16. For further details, the reader may consult Refs. 8–10.

Equation (C.59) requires further interpretation. First, the left hand side is zero if $\mathbf{r}$ is outside the volume, because of the properties of the δ-function. In the interior, Φ is expressed as the sum of (i) the Coulomb potential due to a charge distribution within V, and (ii) two terms which give the effect of all the charges outside V.

There is an equivalent interpretation which requires our focusing our attention only on the region V, and replacing the exterior charges by a charge density σ on the surface S,

$$\sigma(\mathbf{r}_s) = (1/4\pi)\partial\Phi/\partial n,$$

and a surface distribution of electric dipoles with density

$$\mathbf{D}(\mathbf{r}_s) = (1/4\pi)\Phi(\mathbf{r}_s)\hat{n}.$$

The relevant geometry is shown in Fig. C.3. Note that the potential due to these surface charge distributions in the exterior region, $\tilde{V}$, is not the same potential as that due to the true (original) charge distribution. In fact, we show below that the new surface charge distributions enforce the remarkable condition that Φ and $\mathbf{E}$ are zero outside V. The charges σ, and the dipoles $\mathbf{D}$, are to be used only for the interior solution. Using well-known Gaussian pill-box arguments[1]

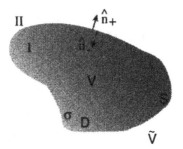

Fig. C.3 The domain V of interest, the normal vectors defining the direction of **E**. The surface charge σ and the surface dipole density **D** are also shown.

used in conjunction with Gauss's theorem of electrostatics, we can show that the surface charge layer leads to

$$E_{In_+} - E_{In_-} = 4\pi\sigma(\mathbf{r}_s),$$

or

$$\frac{\partial\Phi_I}{\partial n}\bigg|_- - \frac{\partial\Phi_{II}}{\partial n}\bigg|_+ = 4\pi\sigma(\mathbf{r}_s).$$

On the other hand,

$$\frac{\partial\Phi_I}{\partial n} = 4\pi\sigma(\mathbf{r}_s),$$

so that

$$\frac{\partial\Phi_{II}}{\partial n_+} = 0$$

on the outside. Similarly, the discontinuity in the potential due to the dipole layer D is[1]

$$\Phi_I - \Phi_{II} = 4\pi D(\mathbf{r}_s).$$

Again, by construction we have

$$\mathbf{D}(\mathbf{r}_s) = \frac{1}{4\pi}\Phi_I(\mathbf{r}_s)\,\hat{n}.$$

Thus $\Phi_{II}(x) = 0$ in the exterior region. Once the charge layers are constructed there are no charges on the outside, so that $\nabla^2\Phi_{II} = 0$, $\Phi_{IIS} = 0$, and $\partial\Phi_{II}/\partial n|_S = 0$. Hence, everywhere on the outside, we have $\Phi_{II} = 0$, and also $E_{IIn} = 0$.

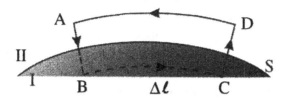

Fig. C.4 The path used to relate E_{It} to E_{IIt}.

Suppose $\Phi(\mathbf{r}_s)$ is given on the surface. We fix our attention on the interior region only. Now we require a distribution of dipoles to obtain the correct potential in V. We shall have (see Fig. C.4)

$$\Phi_A - \Phi_B = 4\pi D(x) \tag{C.60}$$

$$\Phi_D - \Phi_C = 4\pi[D(x) + \nabla(D(x)) \cdot d\mathbf{l}]. \tag{C.61}$$

Now, $\oint \mathbf{E} \cdot d\mathbf{l} = 0$ along the path shown in figure. This leads to the relation

$$-4\pi D(x) + \Phi_C - \Phi_B + 4\pi[D(x) + \nabla D(x) \cdot d\mathbf{l}] + \Phi_A - \Phi_D = 0.$$

On dividing through by $(-dl)$ we obtain

$$E_{IIt} - E_{It} = -4\pi\nabla D(x) \cdot \frac{d\mathbf{l}}{dl}$$
$$\equiv -4\pi\nabla D(x) \cdot \hat{t}. \tag{C.62}$$

Since $E_{It} = 4\pi\nabla D(x) \cdot \hat{t}$ in our case (by construction), we find $E_{IIt} = 0$. We again see that the field in the exterior region is zero.

We have, in effect, "smeared out" the external charges, replaced them by surface charges, and obtained the correct potential inside, given Φ_S and $\partial\Phi/\partial n|_S$. Meanwhile, the potential and the electrostatic field outside have been set to zero.

We note that if S were an equipotential surface then $\hat{n} \times \mathbf{E}_I \equiv 0$, i.e., $E_{It} = 0$, and no dipole distributions are needed to obtain the potential in V. It turns out that if the boundaries are formed by conductors (which are equipotential surfaces) it is sufficient to give the total charge, rather than the surface charge density on the conductors, in order to obtain the potential in V.

The surface S enclosing the volume V was considered to be a real surface defining the shape and volume of the physical region.

We can use arguments similar to the above to show that one may close off any portion of an electrostatic field in a region by a surface, reducing the field and potential outside to zero, and taking account of charges outside by single and double layers of charges placed on the introduced surface. This is what is done in the "method of images." The method of images in electrostatics provides examples which allow us to "visualize" the relationship between the two interpretations discussed here.

The method of images

Consider the simple example of a charge q located a distance a from an infinite grounded ($\Phi = 0$) conducting plane coinciding with the yz plane. With q placed at $(a, 0, 0)$, an induced negative surface charge develops on the conducting plane shown in Fig. C.5. Since the plane is grounded, the corresponding positive charge on the plane (the plane conductor was presumably neutral before the charge q was brought within its vicinity) is "drained" away by the ground connection which keeps the plane at zero potential. The surface charge density σ on the plane conductor, together with the charge q, defines the potential in the physical region $x \geq 0$.

In the method of images, we recognize the fact that we can take away the conducting plane, remove its induced charge, and place a charge q' in the "unphysical" exterior region $z < 0$. The charge q' is chosen such that it, together with the original charge q, reproduces the potential on the plane surface. The choice of $q' = -q$ placed at $(-a, 0, 0)$ ensures that the plane $x = 0$ continues to be a surface with zero potential. The potential from these two charges is

$$\Phi(\mathbf{r}) = \frac{q}{\sqrt{(x-a)^2 + y^2 + z^2}} + \frac{-q}{\sqrt{(x+a)^2 + y^2 + z^2}}. \qquad \text{(C.63)}$$

For $x \geq 0$, this satisfies Laplace's equation (except at $x = a$) and it satisfies the imposed BCs on the conducting plane. By the uniqueness theorem, then, this is the correct solution.

The induced charge on the surface has a surface density

$$\sigma = \frac{1}{4\pi} \frac{\partial \Phi}{\partial n} = -\frac{1}{4\pi} \frac{\partial \Phi}{\partial x}\Big|_{x=0}$$

$$= -\frac{2qa}{4\pi} \left[\frac{1}{(a^2 + y^2 + z^2)^{3/2}} \right]. \qquad \text{(C.64)}$$

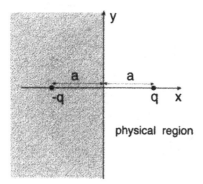

Fig. C.5 The method of images used for obtaining the Green's function for a semi-infinite region with a grounded plane.

It is straightforward to show that the total induced charge on the surface is $-q$.

For a unit charge $q = 1$ placed at a, we can obtain the Green's function from the expression for the potential satisfying the present BCs

$$G(\mathbf{r}, \mathbf{a}) = \frac{1}{|\mathbf{r} - a\hat{x}|} - \frac{1}{|\mathbf{r} + a\hat{x}|}. \tag{C.65}$$

With the Green's function in hand, we can generate the solution of Poisson's equation, in the region $x \geq 0$, for any arbitrary charge distribution present in the physical region.

Poisson's equation with axial symmetry
Consider a charge distribution with axial symmetry, i.e., rotational symmetry about the z-axis. The Green's function depends only on the radial distance $|\rho - \rho'|$ in the xy plane between the "unit" line charge placed at ρ' and the "point of observation" at ρ; in this case, the Green's function does not depend on the coordinate z. For the purpose of normalization, we can suppose that the dimension of the system is L_z along the z-direction. If we shift the origin to the location of the charge, ρ', the Green's function obeys the differential equation

$$\frac{1}{\rho} \frac{\partial}{\partial \rho} \left(\rho \frac{\partial G}{\partial \rho} \right) = -\frac{4\pi}{\rho} \delta(\rho) \, \delta(\phi) \cdot \frac{1}{L_z}. \tag{C.66}$$

Rather than directly solve equation (C.66), we can employ the integral form of Gauss's law to obtain the electric field E, and from it the electrostatic potential Φ, for a line charge with linear charge

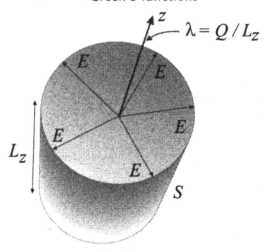

Fig. C.6 The Gaussian surface for obtaining the field due to a line charge.

density $\lambda \equiv Q/L_z$ located at the origin. Construct a cylindrical "Gaussian surface" coaxial with the charge, and capped off at both ends as in Fig. C.6. The electric field is directed radially outwards from the axis of the charge distribution. We evaluate the total flux of the electric field out of the cylindrical box of radius ρ and relate it to the total charge enclosed, as per Gauss's law of electrostatics.[1] Using cgs units, we obtain

$$\oint \mathbf{E} \cdot d\mathbf{S} = 4\pi Q,$$
$$E\, 2\pi\rho\, L_z = 4\pi\lambda\, L_z,$$
$$E = 2\lambda/\rho \;=\; 2\frac{Q}{L_z\rho}. \tag{C.67}$$

We then have

$$\Phi(\rho) \;=\; -\frac{Q}{L_z}\int d\rho\, \frac{2}{\rho}. \tag{C.68}$$

We set $Q = 1$ in the expression for the potential and scale the integration variable by a length R_0. This leads to the Green's function

$$G(\rho) \;=\; -\frac{2}{L_z}\, \ln(\rho/R_0). \tag{C.69}$$

On shifting the origin back to ρ' we have

$$G(\rho - \rho') = -\frac{2}{L_z}\, \ln\left(\frac{|\rho - \rho'|}{R_0}\right). \tag{C.70}$$

The Green's function in three dimensions has the units of $[L]^{-1}$, as can be seen from the differential equation. This is accounted for, in the case of axial symmetry, by the factor L_z appearing in the above equation. We surmise that the Green's function for Poisson's equation in two dimensions is given by the above equation without the factor of L_z.

C.6   Green's functions in electrodynamics

Maxwell's equations in free space are

$$
\begin{aligned}
\nabla \cdot \mathbf{E}(\mathbf{r}, t) &= 4\pi\, \rho(\mathbf{r}, t), \\
\nabla \times \mathbf{E}(\mathbf{r}, t) &= -\frac{1}{c}\frac{\partial}{\partial t}\mathbf{B}(\mathbf{r}, t), \\
\nabla \cdot \mathbf{B}(\mathbf{r}, t) &= 0, \\
\nabla \times \mathbf{B}(\mathbf{r}, t) &= \frac{4\pi}{c}\mathbf{j}(\mathbf{r}, t) + \frac{1}{c}\frac{\partial}{\partial t}\mathbf{E}(\mathbf{r}, t).
\end{aligned}
\tag{C.71}
$$

We wish to find the time-dependent solutions for these equations. The electric and magnetic fields are first expressed in terms of potentials. Here, $\nabla \cdot \mathbf{B} = 0$ implies that a vector field $\mathbf{A}$ can be defined such that

$$
\nabla \times \mathbf{A}(\mathbf{r}, t) = \mathbf{B}(\mathbf{r}, t).
$$

This alone does not determine $\mathbf{A}$ uniquely. From Faraday's law of induction we have

$$
\nabla \times \mathbf{E} + \frac{1}{c}\frac{\partial \mathbf{B}}{\partial t} = 0 = \nabla \times \left(\mathbf{E} + \frac{1}{c}\frac{\partial \mathbf{A}}{\partial t} \right),
$$

so that $(\mathbf{E} + (1/c)\,\partial \mathbf{A}/\partial t)$ can always be written as the gradient of a scalar field. Thus

$$
\mathbf{E}(\mathbf{r}, t) = -\nabla \phi(\mathbf{r}, t) - \frac{1}{c}\frac{\partial}{\partial t}\mathbf{A}(\mathbf{r}, t).
$$

Here $\mathbf{A}(\mathbf{r}, t)$ is the magnetic vector potential and $\phi(\mathbf{r}, t)$ is the electric scalar potential.

Suppose $\lambda(r, t)$ is an arbitrary scalar field. Then the potentials

$$
\begin{aligned}
\mathbf{A}' &= \mathbf{A} + \nabla \lambda, \\
\phi' &= \phi - \frac{1}{c}\frac{\partial \lambda}{\partial t},
\end{aligned}
\tag{C.72}
$$

yield the same fields as $\mathbf{A}$ and ϕ, as can be verified directly. Furthermore,

$$\nabla \cdot \mathbf{B} = 0,$$

$$\nabla \times \mathbf{E} = -\frac{1}{c}\frac{\partial \mathbf{B}}{\partial t},$$

are automatically satisfied. Here λ is called the gauge function. We are free to select the gauge function in a convenient manner, as in the following.

Maxwell's equations can be used to obtain the wave equations for the potentials

$$\nabla^2 \mathbf{A}(\mathbf{r}, t) - \frac{1}{c^2}\frac{\partial^2}{\partial t^2}\mathbf{A}(\mathbf{r}, t) = -\frac{4\pi}{c}\mathbf{j}(\mathbf{r}, t) + \nabla\left(\nabla \cdot \mathbf{A} + \frac{1}{c}\frac{\partial \phi}{\partial t}\right),$$

$$\text{(C.73)}$$

$$\nabla^2 \phi(\mathbf{r}, t) - \frac{1}{c^2}\frac{\partial^2}{\partial t^2}\phi(\mathbf{r}, t) = -4\pi\rho(\mathbf{r}, t) - \frac{1}{c}\frac{\partial}{\partial t}\left(\nabla \cdot \mathbf{A} + \frac{1}{c}\frac{\partial \phi}{\partial t}\right).$$

Identical equations hold for $\mathbf{A}'$ and ϕ'. If we start with $\mathbf{A}'$ and ϕ' we can always find a function $\lambda(\mathbf{r}, t)$ so that

$$\nabla \cdot \mathbf{A} + \frac{1}{c}\frac{\partial \phi}{\partial t} = 0. \qquad \text{(C.74)}$$

For this purpose, it is enough to choose λ such that

$$\nabla^2 \lambda(\mathbf{r}, t) - \frac{1}{c^2}\frac{\partial^2 \lambda(\mathbf{r}, t)}{\partial t^2} = \nabla \cdot \mathbf{A}' + \frac{1}{c}\frac{\partial \phi'}{\partial t}. \qquad \text{(C.75)}$$

If λ satisfies the wave equation so that the left side of equation (C.75) is zero, we have a gauge in which equation (C.74) holds, thereby leading to a decoupling of equation (C.74). We have

$$\left(\nabla^2 - \frac{1}{c^2}\frac{\partial^2}{\partial t^2}\right)\mathbf{A}(\mathbf{r}, t) = -\frac{4\pi}{c}\mathbf{j}(\mathbf{r}, t)$$

$$\left(\nabla^2 - \frac{1}{c^2}\frac{\partial^2}{\partial t^2}\right)\phi(\mathbf{r}, t) = -4\pi\rho(\mathbf{r}, t). \qquad \text{(C.76)}$$

From the solution of equations (C.76) we can construct the electric and magnetic fields obeying Maxwell's equations (C.72).

All the four equations (C.76) are of the same general form. So it suffices to focus our attention on the wave equation satisfied by the scalar potential ϕ.

Let us consider the case of boundaries at infinity. The presence of the source terms on the right side of equations (C.76) suggests that we invoke the Green's function technique. Let the charge density be replaced by a unit source charge

$$\rho(\mathbf{r}, t) = \delta(\mathbf{r} - \mathbf{r}') \, \delta(t - t'), \qquad (C.77)$$

i.e., we have a unit charge at $\mathbf{r} = \mathbf{r}'$ which is switched on and off instantaneously at time t'. (We replace the current source by a vector unit source in a similar manner and have a vector Green's function. In Cartesian coordinates each component may be treated separately.) Clearly, the field produced by such a charge depends only on $\mathbf{r} - \mathbf{r}'$ and $t - t'$. Equations (C.76) in this case lead to

$$\left(\nabla^2 - \frac{1}{c^2} \frac{\partial^2}{\partial t^2}\right) G(\mathbf{r} - \mathbf{r}', t - t') = -4\pi\delta(\mathbf{r} - \mathbf{r}') \, \delta(t - t'), \quad (C.78)$$

and the solution for the potential will be given by

$$\phi(r, t) = \int d^3 r' \, dt' \, G(\mathbf{r} - \mathbf{r}', t - t') \, \rho(\mathbf{r}', t').$$

We have once again reduced the problem to finding a suitable Green's function.

Let us set $\{\mathbf{r}', t'\} = 0$ by shifting the origin for the coordinate and time. The function $G(\mathbf{r}, t)$ is represented as a Fourier integral[§]

$$G(\mathbf{r}, t) = \frac{1}{(2\pi)^4} \int d^3 k \, d\omega \, e^{i(\mathbf{k}\cdot\mathbf{r} - \omega t)} \, G(\mathbf{k}, \omega). \qquad (C.79)$$

The inverse Fourier transform is given by

$$G(\mathbf{k}, \omega) = \int d^3 r \, dt \, e^{-i(\mathbf{k}\cdot\mathbf{r} - \omega t)} \, G(\mathbf{r}, t).$$

The Fourier transform of $\delta(\mathbf{r}) \, \delta(t)$ is unity, so that

$$\delta(\mathbf{r}) \, \delta(t) = \frac{1}{(2\pi)^4} \int d\mathbf{k} \, d\omega \, e^{i(\mathbf{k}\cdot\mathbf{r} - \omega t)}. \qquad (C.80)$$

[§]Observe that we have taken a Fourier transform over four variables $\{\mathbf{r}, t\}$ and hence put in a factor of $(2\pi)^{-4}$. Also, this factor occurs only in one of the transforms – this is a common, if unsymmetrical, choice.

Substituting equations (C.79) and (C.80) into equation (C.78) we have

$$\int dk\, d\omega\, e^{i(\mathbf{k}\cdot\mathbf{r}-\omega t)}\left\{\left(\frac{\omega^2}{c^2}-k^2\right)G(\mathbf{k},\omega)+4\pi\right\} = 0. \quad (C.81)$$

Since $e^{i(\mathbf{k}\cdot\mathbf{r}-\omega t)}$ form a complete set of orthogonal functions, the quantity in parentheses in the integrand must be identically zero; this leads to

$$G(\mathbf{k},\omega) = 4\pi\frac{c^2}{k^2c^2-\omega^2}. \quad (C.82)$$

We have thus reduced the partial differential equation into an algebraic equation by going to the Fourier-transformed space. We now have to transform back to configuration space to determine the spatial–temporal dependence of the Green's function.

We evaluate $G(\mathbf{r},t)$ by Fourier transforming $G(\mathbf{k},\omega)$

$$G(\mathbf{r},t) = -\frac{4\pi c^2}{(2\pi)^4}\int_0^\infty k^2 dk\int_0^\pi \sin\theta\, d\theta\int_0^{2\pi}d\phi\int_{-\infty}^\infty d\omega$$
$$\times\frac{e^{i(kr\cos\theta-\omega t)}}{(\omega+ck)(\omega-ck)}$$
$$= -\frac{c^2}{2\pi^2 ir}\int_{-\infty}^\infty k\, dk\, e^{ikr}\int_{-\infty}^\infty d\omega\frac{e^{-i\omega t}}{(\omega+ck)(\omega-ck)}. \quad (C.83)$$

We cannot obtain $G(\mathbf{r},t)$ from equation (C.83) without additional specification for the evaluation of the ω-integral by properly treating the singularities at $\omega=\pm ck$. The integration procedure is established by the following physical considerations. $G(\mathbf{r},t)$ represents the field arising from switching on a unit charge at $\mathbf{r}=0$ for an instant at $t=0$ and immediately switching it off. So we require $G\equiv 0$ if $t<0$ and $G\neq 0$ if $t>0$.

We first allow ω to be a complex variable and shift the zeros of the denominators from the real ω line into the lower half of the complex ω-plane, by adding a small quantity $i\epsilon$ to each factor in the denominator. The integral is evaluated first and then the limit $\epsilon\to 0$ is taken. We now consider

$$I = \int_{-\infty}^\infty\frac{e^{-i\omega t}d\omega}{(\omega+ck+i\epsilon)(\omega-ck+i\epsilon)}.$$

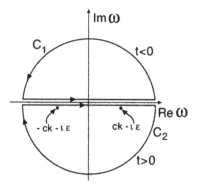

Fig. C.7 The contours to be used for imposing BCs on the Green's function.

If $t < 0$, we close the contour on the upper half ω-plane using the contour C_1 of Fig. C.7. The contour encloses no singularities, and by Cauchy's integral theorem the integral I is zero. For $t > 0$, we close the contour on the lower half plane using the contour C_2 of Fig. C.7. The integral is given by $-2\pi i \times$ (sum of residues at poles located at $\pm ck - i\epsilon$). This leads to

$$I = -2\pi i \left(\frac{e^{-ikct}}{2kc} - \frac{e^{ikct}}{2kc} \right) = -\frac{i\pi}{ck} \left(e^{-ikct} - e^{ikct} \right).$$

Thus, for $t > 0$,

$$G(\mathbf{r}, t) = \frac{c}{2\pi r} \int_{-\infty}^{\infty} dk \left(e^{ik(r-ct)} - e^{ik(r+ct)} \right)$$
$$= \frac{c}{r} \delta(r - ct).$$

Only one of the δ-functions survives here for $t > 0$, since we have $r + ct \neq 0$.

We have

$$G(\mathbf{r} - \mathbf{r}', t - t') = \begin{cases} c\, \delta(|\mathbf{r} - \mathbf{r}'| - c(t - t'))/|\mathbf{r} - \mathbf{r}'|, & t > t', \\ 0 & \text{for } t < t'. \end{cases}$$
$$(\text{C.84})$$

In equation (C.84), we have the so-called retarded Green's function which is nonzero only for $t > t'$. It is important to realize that

BCs have to be included by actually choosing the way in which we close the contour.

We now obtain

$$\phi(\mathbf{r}, t) = \int d^3 r' \frac{\rho(\mathbf{r}', t - \frac{|\mathbf{r} - \mathbf{r}'|}{c})}{|\mathbf{r} - \mathbf{r}'|}, \tag{C.85}$$

and

$$\mathbf{A}(\mathbf{r}, t) = \frac{1}{c} \int d^3 r' \frac{\mathbf{j}(\mathbf{r}', t - \frac{|\mathbf{r} - \mathbf{r}'|}{c})}{|\mathbf{r} - \mathbf{r}'|}. \tag{C.86}$$

These are the retarded solutions for the potentials, known as the Liénard–Wiechert potentials. The $\mathbf{E}$ and $\mathbf{B}$ fields are obtained by taking appropriate derivatives of these potentials.

Green's function for Helmholtz's equation
If we limit ourselves to waves of a given frequency ω, generated by a source term which has the same frequency dependence, the scalar wave equation for $\phi(\mathbf{r}, t) = \phi(\mathbf{r}) e^{-i\omega t}$ reduces to

$$\nabla^2 \phi(\mathbf{r}) + k^2 \phi(\mathbf{r}) = -4\pi \rho(\mathbf{r}), \tag{C.87}$$

where $k = \omega/c$. This is Helmholtz's equation. The solution of

$$(\nabla^2 + k^2) G(\mathbf{r} - \mathbf{r}') = -4\pi \delta(\mathbf{r} - \mathbf{r}') \tag{C.88}$$

is the Green's function

$$G(\mathbf{r} - \mathbf{r}') = e^{ik|\mathbf{r} - \mathbf{r}'|}/|\mathbf{r} - \mathbf{r}'|. \tag{C.89}$$

It is easy to show that this Green's function satisfies the differential equation (C.88) when $\mathbf{r} \neq \mathbf{r}'$. In the limit $\mathbf{r} \to \mathbf{r}'$, we can show that the Green's function has the correct singular behavior by integrating both sides of the differential equation over a small spherical volume, and showing that the integrated forms are equal to each other.

C.7 The wave equation in one dimension

Consider the wave equation

$$\frac{\partial^2}{\partial x^2} G(x, t) - \frac{1}{c^2} \frac{\partial^2}{\partial t^2} G(x, t) = -4\pi \delta(x - x') \delta(t - t'). \tag{C.90}$$

We now wish to obtain the Green's function in the open domain, rather than over $[0, L]$ as we had done in Section C.3. Also, here we wish to include the time dependence explicitly in G.

Let the Fourier integral representation of G be given by

$$G(x,t) = \frac{1}{2\pi} \int dk\, e^{ikx}\, g(k,t).$$

Then

$$\left(-k^2 - \frac{1}{c^2}\frac{\partial^2}{\partial t^2}\right) g(k,t) = -4\pi\,\delta(t-t'). \qquad (C.91)$$

Let the BC be $G(x,t) = 0$, for $t \le 0$. We then have $g(k,t) = 0$ for $t \le 0$.

The homogeneous equation has the solution

$$g(k,t) = A\sin kct, \qquad (C.92)$$

since $g(k,t) = 0$ at $t = 0$. This is shown in Fig. C.8.

Substitute equation (C.92) into equation (C.91) and integrate over time to obtain

$$-\frac{1}{c^2}\frac{\partial}{\partial t} g(k,t)\Big|_{0-}^{0+} = -4\pi.$$

The lower limit gives no contribution since $g(k,t) = 0$ for $t \le 0$. Hence $(Ak/c) = 4\pi$, or $A = (4\pi c/k)$. Thus

$$g(k,t) = \frac{4\pi c}{k}\sin kct,$$

and its Fourier transform is

$$G(x,t) = \frac{1}{2\pi}\int_{-\infty}^{\infty} g(k,t)\,e^{ikx}\,dk = 2c\int_{-\infty}^{\infty}\frac{\sin kct}{k}\,e^{ikx}\,dk$$

$$= \frac{c}{i}\int_{-\infty}^{\infty}\frac{dk}{k}\left(e^{ik(x+ct)} - e^{ik(x-ct)}\right). \qquad (C.93)$$

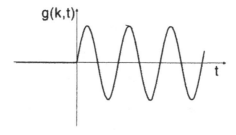

Fig. C.8 The Green's function $g(k,t)$ for the one-dimensional wave equation.

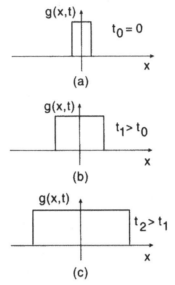

Fig. C.9 The Green's function $g(x,t)$ for the one-dimensional wave equation.

We use Euler's formula to express the exponential functions in terms of sines and cosines. The integrand with the cosine terms is odd under inversion in k and is zero, so that

$$G(x,t) = c \int_{-\infty}^{\infty} \frac{dk}{k} [\sin k(x+ct) - \sin k(x-ct)]. \qquad (C.94)$$

This is the "signal" function and can be written as

$$G(x,t) = \pi c [\epsilon(x+ct) - \epsilon(x-ct)],$$

where the signal function is

$$\epsilon(x) = 2\theta(x) - 1.$$

Expressed in terms of the step-functions, the Green's function is

$$G(x,t) = 2\pi c [\theta(x+ct) - \theta(x-ct)]$$
$$= 2\pi c\, \theta(c^2 t^2 - x^2). \qquad (C.95)$$

Thus, in one dimension, the disturbance propagates in a manner such that the amplitude remains nonzero behind the wavefront, as shown in Fig. C.9. This is a peculiarity of the one-dimensional wave. In three dimensions, the disturbance is nonzero only on an ever-expanding sphere, which constitutes the wavefront.

C.8   The wave equation in two dimensions

We solve

$$\left(\nabla^2 - \frac{1}{c^2}\frac{\partial^2}{\partial t^2}\right) G(\mathbf{r}, t) = -4\pi\,\delta(\mathbf{r})\,\delta(t) \qquad \text{(C.96)}$$

as before, by using the Fourier transform

$$G(\mathbf{r}, t) = \frac{1}{(2\pi)^3} \int d\mathbf{k}\, d\omega\, e^{i\mathbf{k}\cdot\mathbf{r}\,-\,i\omega t}\, G(k, \omega). \qquad \text{(C.97)}$$

From equations (C.96) and (C.97) we obtain

$$G(k, \omega) = -\frac{4\pi c^2}{\omega^2 - c^2 k^2}.$$

On transforming back to configuration space we have

$$G(\mathbf{r}, t) = -\frac{4\pi c^2}{(2\pi)^3} \int_0^\infty k\, dk \int_0^{2\pi} d\phi \int d\omega\, \frac{e^{ikr\cos\phi\,-\,i\omega t}}{(\omega + ck)(\omega - ck)}.$$

The ω integration for the retarded case is precisely as was noted earlier in the three-dimensional case. The integration over ϕ leads to

$$\frac{1}{2\pi}\int_0^{2\pi} d\phi\, e^{ikr\cos\phi} = J_0(kr); \qquad r = \sqrt{x^2 + y^2}.$$

We then have for the retarded Green's function, G_R,

$$G_R(\mathbf{r}, t) = -\frac{c^2}{\pi}\int_0^\infty k\, dk\, J_0(kr)\left(\frac{-i\pi}{ck}\right)\left(e^{-ikct} - e^{+ikct}\right)$$

$$= 2\int_0^\infty dq\, J_0\left(\frac{qr}{c}\right)\sin qt; \qquad (q = ck)$$

$$= 2 \times \begin{cases} 0; & t < r/c, \\[2mm] 1/\sqrt{t^2 - r^2/c^2}; & 0 < r/c < t. \end{cases} \qquad \text{(C.98)}$$

Thus G_R is zero when $t < r/c$, and as is required by causality, it is nonzero for $t > r/c$. Note that G_R falls off with increasing r as $(t^2 - r^2/c^2)^{-1/2}$. Hence, behind the wavefront the disturbance falls off and is nonzero indicating that there is a wake behind the wavefront. This is a feature of the two-dimensional problem.

The Green's function for Helmholtz's equation in two dimensions
Consider Helmholtz's equation in two dimensions

$$(\nabla^2 + k_0^2)\, G(\mathbf{r}, \mathbf{r}') = -4\pi\, \delta(\mathbf{r} - \mathbf{r}')$$

$$\frac{1}{r}\frac{\partial}{\partial r}\left(r\frac{\partial G}{\partial r}\right) + \frac{1}{r^2}\frac{\partial^2 G}{\partial \phi^2} + k_0^2 G = -4\pi\,\delta(\mathbf{r} - \mathbf{r}'). \quad \text{(C.99)}$$

Here $k_0 = \omega/c$ is a given parameter. On using the Fourier integral representation for G, after shifting the origin to $\mathbf{r}' = \{x', y'\}$, we obtain

$$G(\mathbf{r}, \mathbf{r}') = \frac{4\pi}{(2\pi)^2} \int_0^\infty k\, dk \int_0^{2\pi} d\phi\, e^{ikr\cos\phi} \frac{1}{(k^2 - k_0^2)}. \quad \text{(C.100)}$$

The integration of $e^{ikr\cos\phi}$ over ϕ yields¶ $2\pi\, J_0(kr)$. The substitution $\alpha \equiv -ik_0$ gives

$$G(\mathbf{r}, \mathbf{r}') = 2\int_0^\infty \frac{k\,dk\, J_0(kr)}{k^2 + \alpha^2}$$
$$= 2\, K_0(\alpha r). \quad \text{(C.101)}$$

The Bessel function K_0 has the variable α in its argument. Using the relation‖ $K_0(z) = i(\pi/2)H_0^{(1)}(iz)$ we have

$$G(\mathbf{r}, \mathbf{r}') = i\pi\, H_0^{(1)}(k_0|\mathbf{r} - \mathbf{r}'|). \quad \text{(C.102)}$$

This function is employed in two-dimensional waveguide problems within the framework of the boundary element method to model wave propagation in confined regions.

C.9 Green's functions and integral equations

Suppose we wish to solve the homogeneous differential equation

$$\nabla^2\, \Psi(\mathbf{r}) = f(\mathbf{r})\, \Psi(\mathbf{r}) \quad \text{(C.103)}$$

in a given volume V, with suitable BCs. Let $G(\mathbf{r}, \mathbf{r}')$ satisfy the differential equation

$$\nabla^2\, G(\mathbf{r}, \mathbf{r}') = -4\pi\, \delta(\mathbf{r}, \mathbf{r}'),$$

with the same BCs.

¶I. S. Gradshteyn and I. M. Ryzik, *Table of Integrals, Series, Products*, trans. and ed. A. Jeffrey (Academic Press, New York, 1980): 6.532.4, p678.

‖I. S. Gradshteyn and I. M. Ryzik, *Table of Integrals, Series, Products*, trans. and ed. A. Jeffrey (Academic Press, New York, 1980): 8.407.1, p952.

Recall that if the right side of the differential equation were a function $h(\mathbf{r})$ independent of $\Psi(\mathbf{r})$, i.e.,

$$\nabla^2 \Psi(\mathbf{r}) = h(\mathbf{r}),$$

we would have a solution in terms of $G(\mathbf{r}, \mathbf{r}')$ in the form

$$\Psi(\mathbf{r}) = -\frac{1}{4\pi} \int d^3 r'\, G(\mathbf{r}, \mathbf{r}')\, h(\mathbf{r}').$$

Now we let

$$h(\mathbf{r}) = f(\mathbf{r})\, \Psi(\mathbf{r}) \tag{C.104}$$

to obtain the integral equation

$$\Psi(\mathbf{r}) = -\frac{1}{4\pi} \int d^3 r'\, G(\mathbf{r}, \mathbf{r}')\, f(\mathbf{r}')\, \Psi(\mathbf{r}'). \tag{C.105}$$

This is an integral representation for Ψ with the BCs built into it via $G(\mathbf{r}, \mathbf{r}')$. We are assuming that $f(\mathbf{r})$ and $h(\mathbf{r})$ are well-behaved functions. We can develop an iterative solution for Ψ by starting with a lowest order solution Ψ_0 which satisfies the differential equation without $f(\mathbf{r})$. Then the next iterate Ψ_1 is obtained by inserting the previous solution Ψ_0 into the right side of equation (C.105).

This series development can be used for potential scattering. Consider Schrödinger's equation

$$-\frac{\hbar^2}{2m} \nabla^2 \Psi(\mathbf{r}) + V(\mathbf{r})\Psi(\mathbf{r}) = E\,\Psi(\mathbf{r}).$$

With the substitution $k^2 = (2mE/\hbar^2)$, we can cast the above equation in the form

$$\nabla^2 \Psi(\mathbf{r}) + k^2 \Psi(\mathbf{r}) = \frac{2m}{\hbar^2} V(\mathbf{r})\, \Psi(\mathbf{r}). \tag{C.106}$$

Apart from the second term proportional to k^2, this is just of the form discussed above. We shall not pursue this line of discussion any further. The reader is referred to Refs. 12 and 13 for further developments.

C.10   Problems

1. Consider the pair of differential equations

$$\frac{d^2}{dx^2} u(x) + k^2 u(x) = f(x),$$

$$\frac{d^2}{dx^2} g(x, x') + k^2 g(x, x') = \delta(x - x'),$$

over the range $0 \le x, x' \le L$, and satisfying the BCs that the solutions for both equations vanish at $x = 0, L$. Multiply the first equation by $g(x, x')$ and the second by $u(x)$, and subtract one equation from the other to show that

$$u(x) = \int_0^L f(x') \, g(x, x') \, dx'.$$

2. Determine the Green's function for the equation

$$\frac{d^2}{dx^2} g(x, x') - k^2 g(x, x') = \delta(x - x')$$

with the BCs $g(0, x') = 0 = g(L, x')$.

3. Show that the Green's function satisfying the equation

$$\frac{d^2}{dx^2} g(x, x') + 2\gamma \frac{d}{dx} g(x, x') + (k^2 + \gamma^2) g(x, x') = -\delta(x - x')$$

together with the BCs $g(0, x') = 0 = g(L, x')$ is given by

$$g(x, x') = \begin{cases} e^{-\gamma(x-x')} \dfrac{\sin kx \, \sin k(L - x')}{k \sin kL}, & x < x', \\[3mm] e^{-\gamma(x-x')} \dfrac{\sin kx' \, \sin k(L - x)}{k \sin kL}, & x > x'. \end{cases}$$

4. Show that the Green's function solution for the equation

$$\frac{d^2}{dx^2} g(x, x') + \frac{1}{x} \frac{d}{dx} g(x, x') + \left(1 - \frac{m^2}{x^2}\right) g(x, x')$$

$$= -\frac{1}{x} \delta(x - x'),$$

for $0 \leq x, x' \leq R$, with the BCs (i) $g(0, x')$ finite, and (ii) $g(R, x') = 0$, is expressible in terms of J_m and N_m, the Bessel functions of the first and second kind, as

$$g(x, x') = \frac{\pi}{2J_m(R)} \begin{cases} J_m(x) \left[N_m(R) J_m(x') - J_m(R) N_m(x') \right], \\ x < x'; \\ J_m(x') \left[N_m(R) J_m(x) - J_m(R) N_m(x) \right], \\ x > x'. \end{cases}$$

5. Consider a general, linear, second-order differential equation of the form

$$\frac{d^2}{dx^2} y(x) + p(x) \frac{d}{dx} y(x) + q(x) y(x) = r(x).$$

Here $p(x)$, $q(x)$, and $r(x)$ are continuous throughout the interval $[a, b]$. Let the homogeneous differential equation have the linearly independent solutions $y_1(x)$ and $y_2(x)$, so that

$$y(x) = C_1 \, y_1(x) + C_2 \, y_2(x) + y_P(x), \qquad \text{(C.107)}$$

where C_1, C_2 are arbitrary constants to be determined by BCs, and y_P is a particular integral, i.e., any solution of the given inhomogeneous differential equation. Use the method of variation of coefficients to determine the coefficients α, β in

$$y_P = \alpha(x) y_1(x) + \beta(x) y_2(x), \qquad \text{(C.108)}$$

as in Section C.3, and show that

$$\alpha(x) = - \int \frac{r(x) y_2(x)}{W(x)} dx,$$

$$\beta(x) = \int \frac{r(x) y_1(x)}{W(x)} dx, \qquad \text{(C.109)}$$

where $W(x) = y_1(x) y_2'(x) - y_1'(x) y_2(x)$ is the Wronskian.

(a) Show that the single-point BCs $y(x_0) = y_0$ and $y'(x_0) = y_0'$, where $a \leq x_0 \leq b$ can be satisfied by the general solution, equation (C.107), with y_P fulfilling the homogeneous BCs $y_P(x_0) = 0$ and $y_P'(x_0) = 0$, and that this can be achieved by writing

$$y_P(x) = \int_{x_0}^{x} g_s(x, x') r(x') dx',$$

where the single-point Green's function is

$$g_s(x, x') = \frac{y_1(x')\, y_2(x) - y_1(x)\, y_2(x')}{W(x')}.$$

Verify that (i) $g_s(x, x')$ satisfies the homogeneous equation, that (ii) $g_s(x', x') = 0$, and that (iii) $dg(x', x')/dx = 1$. Hence y_p satisfies the above-mentioned homogeneous conditions.

(b) Show that the two-point BCs

$$Y_a = A_1\, y(a) + A_2\, y'(a),$$
$$Y_b = B_1\, y(b) + B_2\, y'(b),$$

can be satisfied by the general solution, equation (C.107), with y_P obeying the homogeneous BCs

$$A_1\, y_P(a) + A_2\, y_P'(a) = 0,$$

and

$$B_1\, y_P(b) + B_2\, y_P'(b) = 0.$$

Show that this may be accomplished by requiring

$$y_P(x) = \int_a^b g_T(x, x')\, r(x') dx',$$

with

$$g_T(x, x') = \begin{cases} y_1(x')y_2(x)/W(x'), & a \le x' \le x, \\ y_1(x)y_2(x')/W(x'), & x \le x' \le b, \end{cases}$$

and by requiring $y_1(x)$ and $y_2(x)$ to satisfy

$$A_1\, y_1(a) + A_2\, y_1'(a) = 0,$$
$$B_1\, y_2(b) + B_2\, y_2'(b) = 0.$$

Note: When the particular integral satisfies the homogeneous BCs it is straightforward to use the complementary function (linear combination of solutions of the homogeneous differential equation) to satisfy the inhomogeneous BCs.

References

[1] J. D. Jackson, *Classical Electrodynamics*, 2nd ed. (Wiley, New York, 1975). The reader will also find the following book useful: W. R. Smythe, *Static and Dynamic Electricity*, 3rd ed. (McGraw-Hill, New York, 1969).

[2] S. Doniach and E. H. Sondheimer, *Green's Functions for Solid State Physicists* (Benjamin, New York, 1974).

[3] L. A. Pipes and L. R. Harvill, *Applied Mathematics for Engineers and Physicists* (McGraw-Hill, New York, 1985); also see: J. Mathews and R. L. Walker, *Mathematical Methods of Physics* (Benjamin, New York, 1965).

[4] R. Courant and D. Hilbert, *Methods of Mathematical Physics* (Wiley-Interscience, New York, 1962).

[5] B. Friedman, *Principles and Techniques of Applied Mathematics* (Wiley, New York, 1956).

[6] H. Bateman, *Partial Differential Equations of Mathematical Physics* (Cambridge University Press, New York, 1969).

[7] P. M. Morse and H. Feshbach, *Methods of Theoretical Physics* (McGraw-Hill, New York, 1953).

[8] M. A. Jaswon and G. T. Symm, *Integral Equation Methods in Potential Theory and Electrostatics* (Academic Press, New York, 1977).

[9] P. K. Banerjee and R. Butterfield, *Boundary Element Methods in Engineering Science* (McGraw-Hill, London, 1981).

[10] C. A. Brebbia, J. C. F. Telles, and L. C. Wrobel, *Boundary Element Techniques* (Springer-Verlag, Berlin, 1983).

[11] G. Barton, *Elements of Green's Functions and Propagation* (Clarendon Press, Oxford, 1989).

[12] E. N. Economou, *Green's Functions in Quantum Physics*, 2nd ed. (Springer, New York, 1983).

[13] A. L. Fetter and J. D. Walecka, *Quantum Theory of Many-Particle Systems* (McGraw-Hill, New York, 1971).

D

Physical constants

Table D.1 Physical constants.

Quantity	Symbol	Value
Speed of light	c	$2.997\,924\,58 \times 10^{8}\,\mathrm{m\,s^{-1}}$
Planck constant (reduced)	$\hbar = h/2\pi$	$1.054\,572\,66 \times 10^{-34}\,\mathrm{J\,s}$
		$= 6.582\,122\,0 \times 10^{-16}\,\mathrm{eV\,s}$
Electron charge	$|e|$	$1.602\,177\,33 \times 10^{-19}\,\mathrm{C}$
		$= 4.803\,206\,8 \times 10^{-10}\,\mathrm{esu}$
Electron mass	m_e	$9.109\,389\,7 \times 10^{-31}\,\mathrm{kg}$
	$m_e\,c^2$	$0.510\,999\,06 \times 10^{6}\,\mathrm{eV}$
Permittivity of free space	ϵ_0	$8.854\,187\,817 \times 10^{-12}\,\mathrm{F\,m^{-1}}$
Permeability of free space	μ_0	$4\pi \times 10^{-7}\,\mathrm{N\,A^{-2}}$
Avogadro's number	N_A	$6.022\,136\,7 \times 10^{23}\,\mathrm{mol^{-1}}$
Boltzmann constant	k_B	$1.380\,658 \times 10^{-23}\,\mathrm{J\,K^{-1}}$
		$= (1/11\,604.444\,8)\,\mathrm{eV\,K^{-1}}$

Table D.2 Derived physical constants.

Quantity	Symbol	Value
Conversion constant	$\hbar c$	$1.973\,270\,53 \times 10^{-5}\,\text{eV cm}$
Fine-structure constant	$\alpha = \frac{e^2}{\{4\pi\epsilon_0\}\hbar c}$	$1/137.035\,989\,5$
Bohr radius $(m_{nucleus} = \infty)$	$a_\infty = \frac{\{4\pi\epsilon_0\}\hbar^2}{m_e e^2}$	$0.529\,177\,249 \times 10^{-10}\,\text{m}$
Rydberg energy	$R_\infty = \frac{m_e e^4}{2(\{4\pi\epsilon_0\})^2\hbar^2}$ $= \frac{e^2}{2\times\{4\pi\epsilon_0\}a_\infty}$	$13.605\,698\,1\,\text{eV}$
Bohr magneton	$\mu_B = e\hbar/2m_e$	$5.788\,382\,63 \times 10^{-4}\,\text{eV/}$
Electron cyclotron (frequency/field)	$\omega^e_{cycl}/B = e/m_e$	$1.758\,819\,62 \times 10^{11}\,\text{rad/}$
Landau radius	$\sqrt{\hbar/m_e\omega^e_{cycl}}$	$256.56\dfrac{\overset{\circ}{\text{A}}}{\sqrt{[B_0]}}\ \{B_0[\text{T}]\}$

Table D.3 Conversion factors.

$$
\begin{aligned}
1\ \text{eV} &= 1.602\,177\,33 \times 10^{-19}\,\text{J}\\
1\ \text{erg} &= 10^{-7}\,\text{J}\\
1\ \text{meV} &= 11.604\,4448\,\text{K}\\
300\ \text{K} &= 25.852\,163\,\text{meV}\\
0^\circ\,\text{C} &= 273.15\,\text{K}\\
1\ \text{gauss (G)} &= 10^{-4}\,(\text{T})\\
1\ \text{wavenumber}\ (1/\lambda):\ \text{cm}^{-1} &= 0.123\,984\,24\,\text{meV}\\
&= \{1/8.065\,541\}\ \text{meV}
\end{aligned}
$$

Author index

Subject index

About the author

L. Ramdas Ram-Mohan received his B.Sc. Honors degree in physics from the University of Delhi in 1964, and his Ph.D. at Purdue University in 1971. He is presently Professor of Physics and Electrical & Computer Engineering at Worcester Polytechnic Institute. Ram-Mohan's interests have ranged over elementary particle theory, nuclear matter and phase transitions, theory of metals, and the linear and nonlinear optical properties of quantum semiconductor heterostructures, with over 150 publications. He has worked as a consultant on problems in theoretical physics at MIT, Purdue, Notre Dame, University of Missouri, NEC Research Institute, and at the Wright–Patterson Air Force Base. He has had a continuing interest in computational techniques since 1977 as an adjunct to his theoretical work. He received the WPI Trustees' Awards for creative scholarship (1990) and for outstanding teaching (1994). In 1996 he was awarded the Distinguished Alumni Award by the School of Science at Purdue University. He is a Fellow of the American Physical Society, the Australian Institute of Physics, and the Optical Society of America. He is the President of Quantum Semiconductor Algorithms, a software and consultancy company started by him in 1993.

Printed in the United States
By Bookmasters